Vascular Plants of Texas

Vascular Plants of Texas

A Comprehensive Checklist including Synonymy, Bibliography, and Index

Stanley D. Jones
Joseph K. Wipff
Paul M. Montgomery

University of Texas Press Austin

Printed in the United States of America
First edition, 1997

Requests for permission to reproduce material from this work should be sent to Permissions, University of Texas Press, P.O. Box 7819, Austin, TX 78713-7819.

∞ The paper used in this publication meets the minimum requirements of American National Standard for Information Sciences—Permanence of Paper for Printed Library Materials, ANSI Z39.48-1984.

Library of Congress Cataloging-in-Publication Data

Jones, Stanley D., 1948–
Vascular plants of Texas : a comprehensive checklist including synonymy, bibliography, and index / Stanley D. Jones, Joseph K. Wipff, and Paul M. Montgomery. — 1st ed.
p. cm.
Includes bibliographical references (p.) and index.
ISBN 0-292-74044-1 (cl.: alk. paper).
1. Botany—Texas—Nomenclature. I. Wipff, J. K., 1962– . II. Montgomery, Paul M., 1946– . III. Title.
QK188.J62 1997
581.964'014—dc20 96-28571

This checklist is dedicated to every individual who has ever collected a Texas plant and placed it into an herbarium. The plant collector and the herbarium are essential for all systematic research in botany and serve as the foundation for all botanical research.

Contents

Preface

This is the first synonymized checklist of vascular plants for the State of Texas in which all taxa are indexed. In addition to providing an updated checklist of the native, naturalized, and adventive vascular plants, we are the first to include all cultivated crops and forages, and all of the introduced ornamental perennial vascular plants which have the potential of persisting in Texas.

This synonymized checklist is intended to provide the most recent taxonomic classification or disposition as shown in literature, and, whenever possible, it also reflects information provided by the individuals listed in the Acknowledgments. Except for the families Cyperaceae and Poaceae, this checklist does not necessarily reflect the authors' opinions on the taxonomic classification presented. Since taxonomic opinions vary on the classification of certain groups, and since botanical names are changed as a result of systematic and nomenclatural research, we hope that by providing and indexing the relevant synonyms, this checklist enables individuals to take a familiar scientific name and equate it with the classification that is generally accepted today and find the correct name within that classification.

Since taxonomy is dynamic, the accumulation and interpretation of data never cease. As relationships between organisms are better understood, changes in classification may be required to reflect this new and better understanding. This change in classification may necessitate a change in the nomenclature (names).

Nomenclature is the naming of groups of organisms and the rules governing the application of these names. Nomenclature is sometimes confused with classification and systematics (Stuessy 1990). After groups of organisms have been classified, names must be given to these groups so that communication about particular units will be facilitated and so that continual progress in classification can be made (Hitchcock 1916).

The object of the rules found in the *International Code of Botanical Nomenclature* (1994) "is to put the nomenclature of the past into order and to provide for that of the future; names contrary to a rule cannot be maintained." The *International Code of Botanical Nomenclature* also states that "the only proper reasons for changing a name are either a more profound knowledge of the facts resulting from adequate taxonomic study or the necessity of giving up a nomenclature that is contrary to the rules." Thus, nomenclature, as taxonomy, is dynamic and evolving. Even taxonomists sometimes have difficulty keeping up with nomenclature changes, and without being actively involved in nomenclature, it is difficult to realize that these changes actually promote nomenclatural stability.

Our goals are: to provide an in-depth list of the vascular plants that are known to occur in Texas; to provide the current classification and nomenclature of these plants; and, to make current classification and correct nomenclature readily accessible through the indexing of all names. This way, any person who deals with Texas plants will find this book useful. Plant taxonomists should derive the most use of this checklist; however, archeologists, botanical enthusiasts, ecologists, entomologists, farmers, foresters, herpetologists, horticulturists, land managers, ornithologists, mammologists, naturalists, palynologists, and ranchers will also find this checklist helpful. If you deal with the plants of Texas you will find this book useful. If you deal with other living organisms, you will also find this book useful, because those organisms must deal with plants on a daily basis.

Acknowledgments

We are indebted to the multitude of plant collectors who have collected specimens throughout the history of Texas and who have deposited these specimens in herbaria (public or private). We are also indebted to the host of curators of these herbaria, past and present, who house and care for these specimens. If it were not for their tireless efforts, neither checklists, manuals, nor botanical studies would be possible. We are also deeply indebted to the scores of researchers throughout the world who have published their botanical research and thus made their knowledge available for others to use and added to the overall understanding of the botanical world. Every individual cited in the bibliography has made a contribution to this checklist in one way or another and their contribution is appreciated.

We would like to thank and acknowledge the following individuals who have reviewed portions of our checklist or have answered significant questions concerning nomenclature and/or taxonomic classification. Their time and assistance is greatly appreciated. These individuals and the group they reviewed are: Kinder and Mary L. Chambers [*Bambuseae* (cultivated)]; Karen H. Clary (*Yucca*); Larry Fowler and Polly Lehtonen (federal noxious weeds); Paul A. Fryxell (Malvaceae); Almut G. Jones (*Aster, sensu lato*); Guy L. Nesom (Asteraceae); A. Michael Powell (Cactaceae); Richard K. Rabeler (Caryophyllaceae); S. Galen Smith; (contributed; *Bolboshoenus, Isolepis, Schoenoplectus*, and *Scirpus*); and Connie S. Taylor (*Euthamia* and *Solidago*).

In certain instances we requested reviews from individuals and at a later date a major revision of the same group of plants was published by another author; although we might have followed the most recent published data, it did not diminish the contributions made by the initial reviewer and both reviewers are cited.

We would also like to thank and acknowledge our indebtedness to the following individuals: Anthony A. Reznicek, Charles T. Bryson, J. Richard Carter, and Robert Kral, who answered specific questions about individual *Carex, Cyperus*, and *Rhynchospora* species (Cyperaceae). James S. Alderson provided information on forage grasses (Poaceae) in Texas; Mary E. Barkworth answered specific questions about species and genera of Poaceae; Rick Darke assisted with ornamental grasses; Kevin B. Jensen provided information on the classification of the wheat grasses (*Triticeae* tribe); Michael G. Lelong provided information on *Panicum capillare* (Poaceae); and Bryan K. Simon assisted with *Bromus japonicus* (Poaceae). T. D. Pennington provided insights to the genus *Sideroxylon* (Sapotaceae), while Billie L. Turner answered specific questions concerning *Monarda* (Lamiaceae). William E. Fox III provided a list of specimens at the S. M. Tracy Herbarium (**TAES**) of Asteraceae that were not previously listed as occurring in Texas. These specimens were then identified or verified by Guy L. Nesom and Billie L. Turner. Larry E. Brown (**SBSC**) agreeably allowed us to use several new state records prior to his official publication of these names. Larry also contributed information in the Lemnaceae, Lentibulariaceae, and Sapotaceae families.

We greatly appreciate Shannon Davies of the University of Texas Press for believing in the necessity of this checklist and for her patience, friendliness, and professionalism. Also at the University of Texas Press, we graciously thank Mandy Woods for her patience and her expertise in copyediting, as she greatly improved this book. We are indebted to Sonya M. Skelly and Cecile M. Vallastro for editing our bibliography. Last, but foremost, we acknowledge and express our sincere gratitude to Gretchen D. Jones, who tirelessly edited and re-edited the checklist and provided ideas, encouragement, and hope.

We also believe that a checklist or manual, "in so far as possible, should epitomize and extol the results of the labor of all those who have gone before. If it were otherwise it would be for them a case of 'love's labor lost.' To evolve knowledge of any sort, we should build on the results that have accrued to us through the work of others. We like to think of the present work as an evolvement of this nature, but the authors cannot relinquish the responsibility of its contents. They are solely responsible for any shortcomings that may be found within these covers" (Correll and Johnston 1970).

As with any in-depth publication dealing with complex subject matters, errors and oversights are inevitable. We would greatly appreciate receiving any corrections and additions which would improve future editions of this checklist. Please send this information to the Botanical Research Center, P. O. Box 6717, Bryan, TX 77805-6717 (e-mail addresses: sdjones@bihs.net or wipff@bihs.net).

Introduction

This book is dedicated to the plant collectors of this state, because our knowledge of Texas plants is built upon the efforts of collectors since the early 1800's. Early explorers have described the hazards of collecting in Texas: the swift "northers" which can drop the temperature by 30°F in minutes; the burning heat; the torrential rains; the torment by droves of gadflies by day and incredible swarms of mosquitoes by night; the bitter saline water; the short rations; and, above all, the sense of abandonment (Geiser 1948). "Let the botanist, with unflagging diligence, gather hundreds of specimens by a day's hard work: at night there might come a rain storm that despite every precaution would completely wet not only the specimens in the driers, but all the botanical drying paper as well. Or floods might carry away the drying paper and leave the naturalist stranded a thousand miles from any source of supply, as once happened to August Fendler on an expedition from St. Louis to Santa Fe" (Geiser 1948). Charles Wright describes the difficulties encountered by early botanical collectors in the following letter to Asa Gray written in Quihi, Texas, on 2 June 1849 (published in Geiser 1948).

My Dear Dr.

I wrote you so recently that if I were not full I would keep silence[.] But steam is so high that if I do not blow off fearful consequences may follow[.]

Yesterday morning [June 1, on the banks of the Medina] we had a violent norther cold and accompanied with rain after which and when ready to start my baggage, paper &c was distributed about into three or four waggons[.] It was so packed that it was not much injured[.] This morning about daylight we had another more severe accompanied with hail[.] My collections were nearly all wet and I have had no time to dry them so they will be much damaged[.] My paper is nearly all wet[.] I should not wonder if we have another storm tonight[.]

Now these misfortunes attendant on my dependent situation and I can not prevent them[.] The officers care nothing about my affairs and the waggoners have a little curiosity to gratify by looking on while I change my plants and care no more about it or rather would be pleased if they were sunk in the river and their load would be lightened[.]

You will recollect I suppose a suggestion made to you that I should be equipped with a waggon and horse from which you dissented instancing the labors of other botanists who had made large collections[.] But I venture to say that Drummond did not attempt to save 12–25 specimens of each species or if he did he had an art which I do not possess[.]

The outfit which I proposed seemed to you perhaps large but I am sincerely of the opinion that the entire cost of the outfit might have been clear saved the present year[.] I would rather have a horse and carriage and ten dollars in my pocket than have five hundred as I am so far as it facilitates my operations[.] I have money in my pocket but it does me no good[.] I can buy nothing with it[.] I sit uninvited and see others eating and it is a severe trial to my feelings to thrust myself among them[.] The men have their rations and often none to spare and how am I to get along to El Paso I know not[.] If I had consulted my own feelings alone I should have stopped at San Antonio and turned back[.] But you and Mr. Lowell had expectations which would not have been realized and I felt reluctant to disappoint you[.] You wrote to me [Jan. 17, 1848] of working like a dog[.] I know how you live—then call your situation dog-paradise and mine hog- and ass-paradise combined and you may realize my situation—sleep all night if you can in the rain and walk 12–15 miles next day in the mud and then overhaul [*sic*] a huge packet of soaked plants and dry them by the heat of the clouds[.]

I have been now three or four days in such a state of uncertainty about the possibility of going on that

I have no enjoyment and today I have not saved a specimen—have merely collected some seeds as I walked along the way[.] As for studying the plants I have not attempted it so long that I have almost forgotten[.] I have been vexed enough to cry or swear when thinking that I have the pleasing prospect of being dependent for six months on a parcel of men who call me a fool and wish me at the bottom of the sea[.]

There is a man who is bound for California in our company—provided with a carriage and mules provisions and cooking utensils—independent as a wood sawyer and dependent on others only for safety against enemies[.] If I had such a one my expenses would be very trifling[.]I could collect twice as many specimens of twice as many species and twice as well preserved[.] I could attend to them at any time I pleased in wet or dry weather and have the assurance that the rain at least could be prevented from coming to them[.] I could also take them to Houston or other seaport and put them on shipboard myself and then I would know they would depend for their forwarding on no careless agent[.]

I am fully resolved that this season will close my botanical travels on horseback or on foot if I can not operate to better advantage[.] I'll give it up and turn my attention to something else[.]

I can now only hope that when Capt. French arrives in camp by [*sic*] situation will be improved by an appointment or in some other way . . . You now know my sentiment [*sic*] on the mode of botanizing in this country & if you wish to continue it on my plan I am ready to do all I can . . .

Affectionately yours
Charles Wright

"Asa Gray, professor at Harvard and the foremost American botanist of the time, depended upon numerous collectors, whose work he supported by soliciting subscriptions for the sale of their specimens. Although Charles Wright had written Gray a strong remonstrance in 1849 [above letter], Gray still did not appreciate the grueling trials of the pioneer collectors" (Muller 1980).

The scientists within the herbaria had charge of processing and caring for the large number of specimens being sent, and recording of the botanical information learned from these collections. They had their own set of difficulties, albeit neither life threatening nor grueling, to overcome, most of which, botanists today can still relate to. The following are excerpts from a letter sent to Charles Wright by Asa Gray, written 9 May 1848 (published in Geiser 1948).

Dear Friend:

That I ought to have replied to your letter of the 19th Nov.—to say nothing of that of Sept. 21 & June 18, there is no doubt. The later I have carried in my pocket a good while, hoping to catch a moment somewhere and sometime to write to you. . . . But I have not had an hours leisure not demanded by letters of immediate pressing consequence, or in which I was not too tired to write.

There are many correspondents whom I neglected almost as much as I have you. I have worked like a dog, but my work laid out to be finished last July is not done yet.

[Here Gray discusses the illness and death of his brother from typhoid fever.] This has somewhat interrupted the printing of the last sheets of my *Manual of N. Botany*,—which, with all my efforts at condensation—has extended to almost 800 pages!! (12mo) including the Introduction. It will be difficult to get the vol. within covers. A year's hard labor is bestowed upon it. I hope it will be useful & supply a desideratum. As a consolation for my honest faithfulness in making it tolerably thorough—and so much larger than I expected it would prove, it is now clear that I shall get nothing or next to it for my year's labor,—that, at the price to which it must be kept to get it into our schools & c. there is so little to be made by it that I cannot induce a publisher to pay the heavy bills, except upon terms which swallow up the proceeds—or at the very least I may get $200 if it all sells—a year or two hence.

Meanwhile I have paid the expenses principally incurred on the 1st Vol. of *Illustrated Genera*—which I cant [*sic*] print and finish till the Manual is out,—have run heavily into debt—in respect to these works, which were merely a labor of love for the good of the science & an honorable ambition—and how I am going to get through I cannot well see . . .

I wish I could write to you as you wish, all about Botany &c—I wish I could aid you as I desire, but I fear it is impossible. I must have rest and less anxiety. Two more years like the last would probably destroy me. If I had an assistant or two, to take details off my hands I might stand it: as it is, I cannot. Carey spent 3 months with me last season, and was to study and ticket your Texan coll. in my hands—take a set for his trouble, and Mr. Lowell & Mr. S. T. Carey would take what they needed and pay for them, so that I could pay your book-bill at Fowle's. The utmost Carey found time to do was to throw the coll. into orders—*There they still lie,* in the corner! There perhaps they had best, now, till the coll. of the past season

> reaches me when I will try to study them altogether [*sic*] along with Lindheimer's collections—a set of which still wait for me to study them. Will you wonder that I am a little disheartened, when, in spite of every effort, I made so little progress.—And in 6 weeks, I begin to lecture in college again,—and in April the garden will require more time than I could give it. Such are merely some of the things on my hands, some of my cares! Still I am interested in you, and in your collections, & will do what I can . . . [An extended discussion of botanical matters here follows.]
>
> Forgive my long neglect: accept my apologies . . .
>
> Yours sincerely
> A. Gray

Gray's ambitious goal to produce a complete *Synoptical Flora of North America* placed him in a difficult situation and at a disadvantage, because this required large numbers of specimens for study and lacking the money to buy specimens, Gray had to earn them by naming collections and acting as the agent in their sale to private and public herbaria everywhere. So he had to send out and encourage collectors whose expenses were usually covered by the sale of the plants they collected. This is where the problems would arise, because the greater his success in processing and distributing the collections, and publishing the new taxa to science, the greater the encouragement to collectors. This meant that he was able to send out more collectors, but as the number of collectors increased so did the probability of something going wrong. This made it more difficult to provide prompt and suitable services, which increased the probability that European or American botanists would be able to intervene and isolate him from the specimens he needed. Karl A Geyer's important collection in the Northwest failed to return to the United States: he took it to England where William J. Hooker published the new taxa and then Gray had to buy specimens (Muller 1980).

Muller (1980) also commented that Gray's youthful experiences with the publications of Thomas Nuttall (which he intensely felt owed him credit) and those of Constantine S. Rafinesque (which he regarded as liabilities) conditioned him for strong objection when American botanists began in the early 1850's to publish their own discoveries. This situation culminated, in the 1870's and 1880's, in charges of censorship against Gray, in private publications west of the Mississippi River. "Having long since, and with more justification than good judgment, arrogated to himself the premiership of American botany, and recognizing only his old mentor, John Torrey, and a safely cooperative group of collaborators such as George Engelmann and later Sereno Watson, he did not accept kindly a 'foolishly ambitious, restless, vain, and independent character,' that is to say, a collector who published his own novelties" (Muller 1980).

"Usually the collections of the field botanist were sent back to closet [herbarium] botanists in Europe and the eastern United States instead of being described and classified by the field workers themselves" (Geiser 1948). As a result, few early botanists attained immediate recognition or fame in their profession. Though commemorated in the names of some plants they collected, the early field botanists could expect little reward beyond the joy of the day's work and knowing they had contributed to the advancement of our knowledge. "Too often the scientific explorer has borne the burden of the heat of the day, while the closet [herbarium] naturalist has reaped that whereon he bestowed relatively little labor" (Geiser 1948).

Although Alvar Núñez Cabeza de Vaca explored Texas around 1530, there is little information in his book of travels that can definitely be taken as referring to Texas. None of the later Spanish expeditions into the region that would become Texas brought back any notes of interest to the naturalist. Thus scientific, botanical, exploration in Texas did not begin until the early 1800's.

The work of Jean Louis Berlandier (1805–1851) with the Boundary Commission (November 1827–November 1829) was the first extensive collecting done in Texas and antedates, by five or six years, Thomas Drummond's extensive explorations of the region around San Felipe de Austin and Gonzales. This commission was sent to survey and establish the boundary between the Mexican Republic and the United States in the Mexican Province of Texas. Berlandier intensively explored areas surrounding Laredo, San Antonio de Béxar, Goliad, San Felipe on the Brazos (capital of Austin's First Colony), and Matamoros. The expedition also traveled through Aransas Bay and Gonzales. Berlandier traveled through, or near, the present-day towns of Knippa, Kerrville, Comfort, Boerne, Schulenburg, Columbus, Hempstead, Courtney, and Anderson. After his contract with de Candolle ended, Berlandier settled in Matamoros and between 1830 and 1851 made frequent botanical explorations into various parts of México and Texas.

Berlandier collected heavily around Matamoros and, in 1834 in Texas, along Arroyo Colorado (near the present-day Willacy-Cameron County line), around Santa Gertrudis (near present-day Kingsville), and along the route from San Patricio to Goliad (through present-day San Patricio, Bee, and Refugio counties). He also made extensive collections around Cópano on Cópano Bay, which he did not distinguish from Aransas Bay. Berlandier made important collections on his excursions, in 1834, from Béxar to Guerrero (Presidio del Río Grande) and the adjoining mission San Juan Bautista (Muller 1980).

It is our opinion that Berlandier and not Ferdinand J. Lindheimer should be referred to as the "Father of Texas Botany." The work of Berlandier was "lost" and underestimated, until it was revealed by the work of Geiser (1948) and Muller (1980) and the translation of Berlandier's journal in 1980. The assessment of Berlandier's contribution to Texas botany given by Winkler (1915) represents the common misconception about Berlandier's work and why he has not been considered, until recent history, to have made major contributions to Texas botany. Winkler (1915) reported that Berlandier made only a small collection of about 150 species of plants from 1826 to 1834, and that these specimens ended up in European herbaria. As we now know, Berlandier continued to collect plants from 1834 until his death in 1851.

Muller (1980) presents evidence that while Berlandier was working for de Candolle, he collected and sent back 52,000 botanical specimens (which represented 2,351 accessions) from México and present-day Texas. In addition, Berlandier's collections have been deposited in sufficient enough quantities to have been especially noted in the herbaria of 17 European and 10 United States institutions (Muller 1980).

Berlandier's collections are the first to represent the plants from the arid and subtropical southwestern region. Though ignored for a century, Berlandier still has not lost his botanical significance, because his collections hold priority throughout hundreds of thousands of square miles of Texas and México. "That Berlandier ranks high among plant collectors can scarcely be denied. Not only did he discover much of interest to botanists, but he inspired the efforts of field men for a century" (Muller 1980).

There is much history intertwined with taxonomy and nomenclature. The classification of plants depends upon dried (preserved) specimens, and every name is based upon a plant specimen, which was collected by a person. The scientific name used in classification is more than just a Latin or latinized name with the name of an author (or authors) attached. Unfortunately, authors' names are frequently abbreviated to the point that they are nothing more than meaningless attachments to the name. The scientific name represents not only a group of plants, but also real people and history—that of the collector, the original describing author, and possibly an author who reclassified it; this is our botanical heritage. The collectors faced great hardships and constant dangers to collect the plants that became the specimens that the names in this checklist are based on and the specimens that document which plants are present in Texas. In fact, collectors today may still face hardships and dangers, depending on the part of world that they are exploring.

As McVaugh (1947) stated, we who follow botanical paths today must pause for a moment to salute the earlier botanical explorers and collectors for their hard and dangerous pursuits of their avocations in botany on the frontiers of America. These travelers, who must have loved the lives they lived, provided the materials that the Torreys and Grays needed to write the history and science of botany. Without them and their devotion to botany we should have faired poorly before travel became the easy thing it is now. "Enduring many privations and hardships, and with no other hope of reward than the satisfaction of having contributed to the enrichment of the botanical knowledge concerning a region practically unknown, they toiled faithfully and persistently on" (Winkler 1915).

Format

The basic format of this checklist follows a format similar to that of Kartesz and Kartesz (1980) and Kartesz (1994). Kartesz's basic checklist format is the most user-friendly of the checklists that we examined. We felt that alphabetizing families, genera, species, and infraspecific taxa and using "Sy =" to indicate the synonyms is a superior method of presenting information in a checklist. Therefore, these aspects of their format were used as a guideline in developing the format used in this checklist.

The taxa in this checklist are first arranged by classes according to Cronquist's system of classification (1968, 1981) and are then alphabetized by family, genus, species, and infraspecific rank. Although out of date, Cronquist's classification system is still the most recent and up-to-date system available. This system of classification is taught in most taxonomy classes and is followed by most new manuals (e.g. *Flora North America*, II). The utility of the Cronquistian system has been diminished at the family level by the overwhelming amount of phylogenetic research over the past 15 years. This, coupled with our desire to create a more user-friendly checklist, leads us to list families, genera, and species alphabetically under the classes.

Classes are denoted by the word **CLASS** followed by a colon and then the class name. Classes are in all caps, boldface, and are in the largest font size. Families are in all caps, and are the next largest font size. Genera are in small caps and boldface, and are in the third largest font size. The author(s) of the genera are not in all caps or boldface.

Scientific Names: Species and Infraspecific Rank

The scientific names reflecting the current taxonomic disposition are in boldface italics. The authors of scientific names are neither italicized nor in boldface. After the genus name is spelled out, it is then abbreviated using the first letter of the genus, as is standard practice. Infraspecific ranks are abbreviated as follows: "subsp." = subspecies, "var." = variety, and "f." = *forma* (form).

Authors of Taxa

The following Article (Art.), Article 46.1, from the 1994 *International Code of Botanical Nomenclature* is one of the Code's provisions regarding the citation of authors' names for the purpose of precision: "For the indication of the name of a taxon to be accurate and complete, and in order that the date may be readily verified, it is necessary to cite the name of the author(s) who validly published the name concerned unless the provisions for autonyms apply (Art. 22.1 and 26.1)." An autonym is when the infraspecific name is the same as the specific name (e.g., *Panicum acuminatum* O. Swartz var. *acuminatum*). In this case the infraspecific name is not followed by an author's name (Art. 26.1). This also applies to subdivisions of a genus (Art. 22.1). So, with the goal of precise author citations in mind, we decided not to abbreviate an author's last name.

Authors' last names are spelled out and, at least, the initial of their first name is given, wherever possible. Two or more additional initials are given in order to distinguish between authors with similar names. If two or more authors have the same last name and initials, their first names are spelled out. Although this makes the scientific names longer, the added information is more in line with the spirit of listing authors' names. We also use the original spelling of authors' names, including diacritical signs. Article 60.6 of the 1994 *International Code of Botanical Nomenclature* states that diacritical signs are not used in Latin plant names; however, this Article is not applicable to authors' names. Unfortunately, there are floras and manuals that have applied Article 60.6 to plant authors' names. In certain instances, not only have diacritical signs of plant authors' names been suppressed, the spelling has also been erroneously altered to conform to Latin. In certain countries, removal of the diacritical signs from a person's name is in essence misspelling their name.

Authors of Plant Names (Brummitt and Powell, 1992) is an impressive book that attempts to standardize the abbreviations of author names. It is a definite improvement over the traditional method of citing authors. For example, Brummitt and Powell list 15 authors with the last name Nair. According to their scheme, no author with the name Nair should be cited solely as Nair, as is common practice. For three Nairs, the first initial is *N,* for three it is *V,* for two each it is *K* and *G,* and for one each it is *L, M,* and *R,* while one is spelled out as Nath. The latter could easily be lost under one of the three N. Nairs. Following their scheme, only three Nairs are being differentiated by their first initial. The others have to be differentiated either by using two to three initials or by having their first name spelled out. If there are fifteen Nairs, how many Browns, Johnsons, Smiths, and Joneses are there going to be in twenty-five years? How many of us know the most common names in Ethiopia, Japan, South Korea, South Vietnam, Russia, Portugal, México, France, New Zealand, or lesser-known countries? Although Brummitt and Powell's work is definitely a step in the right direction, there are several problems with their system.

This lacks numerous pre-1992 Old and New World authors. As the population increases, the name game will increase exponentially. For abbreviations of author names to work, all of the following will be necessary: all authors will have to follow Brummitt and Powell's abbreviation scheme; every author worldwide who has published a new name must be listed; and their scheme must be updated frequently to include new authors. The probability that every botanist will purchase a copy of Brummitt and Powell and will follow their abbreviation scheme is very low. Also, it is difficult, and sometimes impossible, to take a traditionally abbreviated author's name and determine the correct author, especially when more than one author has the same last name and first initials. Unless an author's name is precisely cited, one begins a frustrating search to determine the correct publishing author and date, and the first provision of Article 46.1 of the Code, regarding the citation of authors' names for purposes of precision, is not fulfilled.

Spelling out an author's name not only fulfills the need for scientific accuracy and precision, it also makes it possible to readily verify the publication date and greatly assists in finding the original publication, if needed. The need for complete publication citations has already been realized by some scientific journals (e.g., *Systematic Botany* and *Palynology*), which require that journal names be written in full and not abbreviated. We also feel this is necessary for authors' names, especially in reference works. The spelling out of author names does not diminish in any way the need for books, such as Brummitt and Powell (1992), that provide lists of authors of plant names, because these lists are important in determining the number of initials needed to distinguish one author from another. It is for these reasons that we have spelled out authors' names and added the initials needed to separate them.

*The Use of Parentheses and "*ex*" in the Citation of Authors' Names*

When the name of a taxon is transferred to another genus or species, or changes rank, the author(s) who originally published that name is cited in parentheses, followed by the author(s) making the transfer. Article 49 of the 1994 Code states: "When a genus or a taxon of lower rank is altered in rank but retains its name or the final epithet in its name, the author of the earlier, name- or epithet-bringing legitimate name (the author of the basionym) must be cited in parentheses, followed by the name of the author who effected the alteration (the author of the new name). The same holds when a taxon of lower rank than genus is transferred to another genus or species, with or without alteration of rank." For example, when *Schoenus coloratus* C. Linnaeus is transferred to *Dichromena* it is cited as *Dichromena coloratus* (C. Linnaeus) A. Hitchcock. When it is transferred to *Rhynchospora* it is cited as *Rhynchospora* (C. Linnaeus) H. Pfeiffer. Linnaeus is the author of the basionym, and A. Hitchcock and H. Pfeiffer are the authors who transferred the name to different genera.

The word *ex* is mainly used in circumstances in which there is an author (or authors) who only contributes the name of a taxon, while it is a different author (or authors) who actually validly publishes that name. In these cases, the author(s) who ascribed the name is separated from the publishing author(s) by *ex.* Article 46.4 of the 1994 Code states: "A name of a new taxon must be attributed to the author or authors of the publication in which it appears when only the name but not the validating description or diagnosis was ascribed to a different author or different authors. A new combination or a *nomen novum* [an avowed substitute] must be attributed to the author or authors of the publication in which it appears, although it was ascribed to a different author or to different authors, when no separate statement was made that they contributed in some way to the publication. However, in both cases authorship as ascribed, followed by '*ex*', may be inserted before the name(s) of the publishing author(s)." For example, the name of the taxon *Carex cephalophora* G. H. Muhlenberg *ex* C. von Willdenow was contributed by G. H. Muhlenberg, but the validating description was written by C. von Willdenow.

Common Names

We present only scientific names in our checklist. This is not to isolate the general public or individuals not involved in taxonomy. Our reasoning is simple: vernacular (common) names are regional at best. Within any given region a taxon is likely to have several vernacular names. In addition, a given vernacular name is often applicable to several taxa. For example, in Texas, *Sesbania drummondii* (P. Rydberg) V. Cory is known as Drummond's sesbania, rattlebush, poison bean, or coffee bean. However, coffee bean refers to both *Sesbania drummondii* and *S. macrocarpa* G. H. Muhlenberg. Once outside the given region, these already ambiguous names become more unclear. For example, in Texas (and probably most of the United States), *Cynodon dactylon* (C. Linnaeus) C. Persoon is known as "Bermuda grass." But depending upon what region of the world you live in, some of the other common names for this grass are Bahamas grass, dhoub, kiri-hiri, devil's grass, African couch, star grass, and kweek grass. Even if we provide common names, we are not providing a service until common names become internationally standardized; but standardization is the function of scientific names.

Abbreviations and Special Designations

The following is a legend of the abbreviations and special designations found in the checklist, and lists where to find a definition and discussion of these terms.

auct. = *auctorum. See* Misapplied Names.
[**cultivated**] = cultivated plant. *See* Cultivated Plants.
cv. = *cultivarietas:* cultivar name. *See* Cultivated Plants.
ex. See "The Use of Parentheses and '*ex*' in the Citation of Authors' Names" in preceding section, "Authors of Taxa."
f. (after a personal noun) = *filius:* son. *See* Scientific Names.
f. (before an infraspecific epithet) = *forma:* form. *See* Scientific Names.
[**federal noxious weed**] = plant listed on the federal noxious weed list. *See* Federal Noxious Weeds.

[***ined.***] = *ineditus:* unpublished. *See* Unpublished Names.
Hn = horticultural name. *See* Cultivated Plants.
Ma = misapplied name. *See* Misapplied Names.
[***name*** × ***name***] = hybrid. *See* Hybrid Taxa.
nomen conservandum. See Conserved Names.
orthographic *conservandum* = *See* Orthography (Art. 61.1).
orthographic error = *See* Orthography.
orthographic variant = *See* Orthography.
sensu = of authors. *See* Misapplied Names.
[**state flower**]. *See* State Botanical Symbols.
[**state fruit**]. *See* State Botanical Symbols.
[**state grass**]. *See* State Botanical Symbols.
[**state tree**]. *See* State Botanical Symbols.
subsp. = subspecies. *See* Scientific Names.
Sy = synonym. *See* Synonyms.
[**TOES: (roman numeral)**] = *See* Endangered and Threatened Taxa.
typographic error = *See* Orthography (Typography).
var. = variety. *See* Scientific Names.
* in left margin signifies a TOES listing. *See* Endangered and Threatened Taxa.

Conserved Names

Conserved names occur at the family, genus, and species ranks. All plant names used in our checklist that are conserved by the 1994 *International Code of Botanical Nomenclature* are indicated by the Latin words "*nomen conservandum*" following the name that is conserved. In a few instances the spelling has been conserved. This is indicated by the words "orthographic *conservandum.*" We have suppressed all diacritical signs for Latin names, following Article 60.6, but have maintained diaereses, indicating that a vowel is to be pronounced separately from the preceding vowel (e.g., *Isoëtes*), which is permissible under Article 60.6.

Cultivated Plants

The term "[**cultivated**]" (boldface and in brackets) following a scientific name is subjective and represents plants that do not escape cultivation, or cultivated plants that may escape to various degrees, although the majority still exists under cultivation [e.g., *Cortaderia selloana* (J. A. Schultes & J. H. Schultes) P. Ascherson & K. Graebner, *Nerium oleander* C. Linnaeus, and *Zea mays* C. Linnaeus]. We do not distinguish between crops or horticulture plants in our indication of "cultivated." Certain taxa have become naturalized that originated as cultivated plants and continue to be sold as such. However, in our opinion, if these taxa are more prevalent outside cultivation, they should not be listed as cultivated [e.g., *Ligustrum sinense* J. de Louriero and *Sapium sebiferum* (C. Linnaeus) W. Roxburgh].

Cory and Parks (1937) set the precedent of listing cultivated taxa by including *Lagerstroemia indica* C. Linnaeus and *Nerium oleander,* two cultivated plants that do not escape cultivation. Correll and Johnston (1970) listed these two taxa plus several others that fall into the cultivated category, but failed to list those provided by Shinners (1958). Gould (1975) listed *Cortaderia selloana,* another species that does not escape cultivation. Although these authors fail to mention other cultivated taxa, the mentioning of these taxa enriched our knowledge of the Texas flora. Shinners (1958, 1972) listed cultivated plants for the Dallas–Fort Worth area and Silveus (1933) included the cultivated grasses in his *Texas Grasses.* Unfortunately, recent checklists and manuals have failed to include these taxa. They have opted to include some previously cited cultivated plants but have made no attempt to list new ones. We view the exclusion of cultivated plants in Texas as a major blemish. If these plants are not listed in our floras and keyed out in our manuals, then making determinations of these plants is extremely difficult and time consuming.

Horticultural Names

"Hn" is used to designate names used in horticulture and "cv." designates the actual cultivar name used in the horticulture industry. Horticultural names are indicated by "Hn =" and are not italicized or boldfaced. Cultivar names are denoted by "cv." preceding the cultivar name, and the cultivar name is enclosed in quotes (e.g., Hn = *Miscanthus* cv. "Giganteus").

Endangered and Threatened Taxa

Texas Endangered Plant Law

Texas Parks and Wildlife code, Chapter 88, Sections 88.001–88.012 [1981 (amended 1985)]. This law: 1. authorizes the listing of endangered, threatened, and protected plants; 2. prohibits taking for sale from public lands; 3. prohibits taking for sale from private land without a permit; and 4. authorizes adoption of regulations for permitting.

Texas Endangered Plant Regulations

Rules for Protected, Threatened, and Endangered Native Plant Species, Texas Parks and Wildlife Department, Sections 57.401–57.413. Contains an official list of protected species.

In the checklist, taxa listed by the *Texas Organization for Endangered Species* (TOES, 1993) have an asterisk * to the left of the scientific name and a roman numeral in brackets following the name of the author(s). The roman numeral indicates the TOES category. Asterisks are placed by the scientific names as they appear in TOES, regardless of present nomenclature or taxonomic disposition. The definitions of the categories described in the *TOES Endangered, Threatened and Watch Lists of Texas Plants* were revised in Addendum I (November 1994) to read as follows:

Legally Protected Species

Category I: Endangered species. The term "endangered species" is defined as any species that is in danger of extinction throughout all or a significant portion of its range [Section 3(6) of the Federal Endangered Species Act].

Category II: Threatened species. The term "threatened species" is defined as any species that is likely to become an endangered species within the foreseeable future throughout all or a significant portion of its range [Section 3(19) of the Federal Endangered Species Act].

TOES Listed Species

Category III: Texas Endangered. Any species that lacks legal protection and is in danger of extinction or extirpation in Texas.

Category IV: Texas Threatened. Any species that lacks legal protection and that is likely to become endangered within the foreseeable future.

Category V: Watch List. Any species that lacks legal protection and at present has either low population numbers or restricted range in Texas and that is not declining or being restricted in its range but requires attention to ensure that the species does not become endangered or threatened. These species need more research.

Federal Noxious Weeds

Plants are designated and listed as noxious weeds in the Federal Noxious Weed Regulations [7 CFR Part 360.200] because they are within the definition of a "noxious weed" as defined in Section 3(c) of the Federal Noxious Weed Act of 1974 [7 U.S.C. 2802(c)].

Hybrid Taxa

Hybrid taxa (nothotaxa) are indicated by the use of the multiplication sign, ×. Hybrid genera are listed alphabetically within a family. Hybrid species are listed at the end of the alphabetical listing of species under a genus. When the parents are known they are listed in brackets. For example, ***Quercus* × *tharpii*** C. H. Muller [*emoryi* × *graciliformis*]. *Quercus tharpii* is the name applied to this hybrid by C. H. Muller. *Quercus emoryi* and *Q. graciliformis* are its parental stock. All hybrids that have been named and are known to occur in Texas have been listed except for *Quercus*. Given the propensity of *Quercus* to hybridize, we listed any known hybrid if both parents occurred in the state. Many of these *Quercus* hybrids have been reported for Texas, but others have not. In our opinion, it is only a matter of time before they are reported, so we provided their proper name anyway.

Misapplied Names

A misapplied name is when the name of one taxon is erroneously applied to a different taxon. Misapplied names are indicated by "Ma =." Misapplied names are listed alphabetically after the synonyms. The format for presenting misapplied names follows Recommendation 50.D of the *International Code of Botanical Nomenclature* (1994) as closely as possible. For example, Ma = *Woodsia mexicana auct.,* non A. Feé: Texas authors. *Woodsia mexicana* A. Feé is the name of a species that does not occur in Texas. However, this name has been erroneously used by Texas authors for *W. neomexicana* M. Windham, which does occur in Texas. In other words, the species that Texas authors previously called *W. mexicana* is actually *W. neomexicana*. The abbreviation *auct.* is for the Latin word *auctorum,* meaning "of authors."

Orthography

Orthography. Orthography is derived from the Greek words *ortho*—straight—and *graphikos*—capable of painting, drawing, or writing. So, orthography is the spelling of words according to accepted usage. The accepted usage for botanical names is defined by the *International Code of Botanical Nomenclature.*

The following are some of the Articles (Art.) from the 1994 *International Code of Botanical Nomenclature* regarding orthography.

Article 60.1: "The original spelling of a name or epithet is to be retained, except for the correction of typographical or orthographical errors and the standardizations imposed by Art. 60.5 (*u/v* or *i/j* used interchangeably), 60.6 (diacritical signs and ligatures), 60.8 (compounding forms), 60.9 (hyphens), 60.10 (apostrophes), and 60.11 (terminations; see also Art. 32.6), as well as Recommendation [Rec.] 60H."

Article 60.5. "When a name has been published in a

work where the letters *u, v* or *i, j* are used interchangeably or in any other way incompatible with modern practices (e.g., one of those letters is not used or only in capitals), those letters are to be transcribed in conformity with modern botanical usage."

Article 60.6. "Diacritical signs are not used in Latin plant names. In names (either new or old) drawn from words in which such signs appear, the signs are to be suppressed, with the necessary transcription of the letters so modified; for example *ä, ö, ü* become, respectively, *ae, oe, ue; é, è, ê* become *e*, or sometimes *ae; ñ* becomes *n; ø* becomes *oe; å* becomes *ao*. The diaeresis, indicating that a vowel is to be pronounced separately from the preceding vowel (as in *Cephaëlis, Isoëtes*), is permissible; the ligatures *æ* and *œ*, indicating that the letters are to be pronounced together, are to be replaced by the separate letters *ae* and *oe*."

Article 60.8. "The use of a compounding form contrary to Rec. 60G in an adjectival epithet is treated as an error to be corrected."

Article 60.9. "The use of a hyphen in a compound epithet is treated as an error to be corrected by deletion of the hyphen, except if an epithet is formed of words that usually stand independently, or if the letters before and after the hyphen are the same, when a hyphen is permitted (see Art. 23.1 and 23.3)."

Article 60.10. The use of an apostrophe in an epithet is treated as an error to be corrected by deletion of the apostrophe.

Article 60.11. The use of a termination (for example *-i, -ii, -ae, -iae, -anus*, or *-ianus*) contrary to Rec. 60C.1 (but not 60C.2) is treated as an error to be corrected (see also 32.6).

Article 61.1. "Only one orthographical variant of any one name is treated as validly published, the form which appears in the original publication except as provided by in Art. 60 (typographical or orthographical errors and standardizations), Art. 14.11 (conserved spellings), and Art. 32.6 (incorrect Latin terminations)."

Article 61.2. "For the purpose of this *Code*, orthographical variants are the various spelling, compounding, and inflectional forms of a name or its epithet (including typographical errors), only one type being involved."

Article 61.3. "If orthographical variants of a name appear in the original publication, the one that conforms to the rules and best suits the recommendations of Art. 60 is to be retained; otherwise the first author who, in an effectively published text (Art. 29–31), explicitly adopts one of the variants, rejecting the other(s) must be followed."

Article 61.4. "The orthographical variants of a name are to be corrected to the validly published form of that name. Whenever such a variant appears in print, it is to be treated as if it were printed in its corrected form."

Article 61.5. "Confusingly similar names based on the same type are treated as orthographical variants."

Typography

Typography is the art or process of printing with type. So, a typographical error is an error in printed or typewritten matter resulting from striking the improper key of a keyboard, mechanical failure, or the like. See Article 60.1 above.

State Symbols

Bigony (1995) provides a nice overview of the ten state symbols of the natural world and when they were adopted. There are four state botanical symbols: flower, fruit, grass, and tree.

State Flower

In 1901, the 27th Texas Legislature adopted *Lupinus subcarnosus* W. Hooker (bluebonnet) as the state flower. Then in 1971, the legislature extended the state flower status to include all of the *Lupinus* species [*L. concinnus* K. Agardh, *L. havardii* S. Watson, *L. perennis* C. Linnaeus subsp. *gracilis* (T. Nuttall) D. Dunnand, *L. plattensis* S. Watson, *L. subcarnosus, L. texensis* W. Hooker] in Texas.

State Fruit

In 1993, the Texas red grapefruit, *Citrus* × *paradisi* (C. Linnaeus) J. Macfayden [*maxima* × *sinensis*] [C. cv. "Ruby" (redblush)], was designated as the state fruit over *Fragaria* × *ananassa* A. Duchesne (*chiloënsis* × *virginiana*) (cultivated strawberry) and *Prunus persica* (C. Linnaeus) A. Batsch var. *persica* (peach). All three of these are introduced species. Though the Texas red grapefruit was developed in Texas, the parents of this hybrid are introduced.

State Grass

In 1971, the 62nd Texas Legislature adopted *Bouteloua curtipendula* (A. Michaux) J. Torrey (sideoats grama) as the state grass of Texas.

State Tree

In 1919, the State Legislature adopted *Carya illinoinensis* (F. von Wangenheim) K. Koch (pecan) as the state tree of Texas.

Synonyms

A synonym is a name that is not the oldest validly published name for a particular taxon. Synonyms are indicated by "Sy =" and are not italicized or boldfaced. Synonyms are listed alphabetically under the accepted name. In some instances it was necessary to include cultivar names in the synonymy.

Unpublished Names

If "[***ined.***]" follows any scientific name, it indicates that the name combination has not yet been validly published, but is expected to be in the near future. The abbreviation *ined.* is for the Latin word *ineditus,* meaning "unpublished."

Summary of Taxa

Table 1 is a summary of the vascular plants of Texas. This summary indicates that we have recorded 11 classes, 227 families, 1599 genera, 5812 species, 941 infraspecific taxa, and 117 hybrid taxa, of which 46 (38%) are accounted for in one genus, *Quercus* (Fagaceae). The total number of taxa presented in this checklist, including infraspecific taxa and hybrids, is 6871.

The ten largest families, based on the total number of species, are indicated by boldface roman numerals to the right of the family name. Each of the ten largest families is represented by more than 100 species. Families are listed alphabetically under each class. For each family, the number of genera, the number of species, the number of infraspecific taxa (subspecies and varieties), and the number of hybrids are listed. The number of infraspecific taxa does not include the nominate, because the nominate is counted as a species.

The five largest genera, based on the number of species, are: *Carex* (Cyperaceae)—92, *Euphorbia* (Euphorbiaceae)—68, *Cyperus* (Cyperaceae)—61, *Panicum* (Poaceae)—51, and *Quercus* (Fagaceae)—50. *Euphorbia* and *Quercus* are the only genera of these five to be a member of the class Magnoliopsida (dicots). *Quercus* is the only woody genus represented. The other three genera are members of the class Liliopsida (monocots).

The percentage of species having infraspecific taxa (excluding the nominate) was calculated for families having ten or more genera. The families with the greatest number of infraspecific taxa (number of infraspecific taxa divided by number of species) are: Cactaceae (45%), Asteraceae (25%), Solanaceae (25%), Ranunculaceae (23%), and Brassicaceae (21%). The families with the fewest infraspecific taxa are: Arecaceae (0%), Oleaceae (3%), Nyctaginaceae (4%), Orchidaceae (5%), and Amaranthaceae (5%).

The average number of species per genus (total number of species divided by total number of genera) was calculated for families with ten or more genera. The families with the greatest number of species per genus are: Cyperaceae (16.9), Euphorbiaceae (12.8), Amaranthaceae (5.5), Convolvulaceae (5.1), and Cactaceae (4.6). The families with the fewest species per genus are: Araceae (1.3), Bignoneaceae (1.4), Arecaceae (1.7), Apiaceae (1.8), and Rutaceae (2.0).

Summary Table

CLASS	FAMILY	GENERA	SPECIES	INFRASPECIFIC	HYBRIDS
LYCOPODIOPSIDA	Lycopodiaceae	3	5	0	0
ISOËTOPSIDA	Isoëtaceae	1	3	0	0
	Selaginellaceae	1	12	2	1
EQUISETOPSIDA	Equisetaceae	1	4	1	1
PSILOTOPSIDA	Psilotaceae	1	1	0	0
POLYPODIOPSIDA	Anemiaceae	1	1	0	0
	Aspleniaceae	1	5	0	0
	Azollaceae	1	3	0	0
	Blechnaceae	2	3	1	0
	Dennstaedtiaceae	2	2	3	0
	Dryopteridaceae	9	17	2	0
	Lygodiaceae	1	1	0	0
	Marsileaceae	2	4	0	0
	Ophioglossaceae	2	9	1	0
	Osmundaceae	1	2	1	0
	Parkeriaceae	1	1	0	0
	Polypodiaceae	1	3	2	0
	Pteridaceae	7	43	4	1
	Salviniaceae	1	1	0	0
	Thelypteridaceae	3	7	3	0
CYCADOPSIDA	Cycadaceae	1	2	0	0
	Zamiaceae	2	4	0	0
GINKGOÖPSIDA	Ginkgoaceae	1	1	0	0
PINOPSIDA	Araucariaceae	1	2	0	0
	Cupressaceae	5	18	2	1
	Pinaceae	4	19	3	0
	Podocarpaceae	1	2	1	0
	Taxaceae	1	1	0	0
	Taxodiaceae	1	3	0	0
GNETOPSIDA	Ephedraceae	1	6	1	0
MAGNOLIOPSIDA	Acanthaceae	17	56	4	0
	Aceraceae	1	8	4	0

CLASS	FAMILY	GENERA	SPECIES	INFRASPECIFIC	HYBRIDS
	Achatocarpaceae	1	1	0	0
	Aizoaceae	2	6	0	0
	Amaranthaceae	11	60	3	0
	Anacardiaceae	5	18	11	0
	Annonaceae	1	2	0	0
	Apiaceae	46	80	7	1
	Apocynaceae	16	33	6	1
	Aquifoliaceae	1	14	1	1
	Araliaceae	5	8	1	1
	Aristolochiaceae	1	9	0	0
	Asclepiadaceae	5	59	5	0
	Asteraceae (**I**)	196	682	170	2
	Avicenniaceae	1	1	0	0
	Balsaminaceae	1	2	0	0
	Basellaceae	1	4	0	0
	Bataceae	1	1	0	0
	Berberidaceae	3	9	1	1
	Betulaceae	4	6	0	0
	Bignoniaceae	10	14	2	0
	Bixaceae	2	2	0	0
	Bombacaceae	2	2	0	0
	Boraginaceae	20	77	8	0
	Brassicaceae (**VIII**)	48	136	27	0
	Buddlejaceae	3	9	1	0
	Buxaceae	3	4	0	0
	Cabombacae	2	2	0	0
	Cactaceae	20	91	41	3
	Callitrichaceae	1	5	0	0
	Calycanthaceae	1	1	0	0
	Campanulaceae	6	21	11	0
	Cannabaceae	1	1	1	0
	Capparaceae	6	14	3	0
	Caprifoliaceae	7	34	7	1
	Caricaceae	1	1	0	0

CLASS	FAMILY	GENERA	SPECIES	INFRASPECIFIC	HYBRIDS
	Caryophyllaceae	19	59	6	0
	Casuarinaceae	1	3	0	0
	Celastraceae	6	12	2	0
	Ceratophyllaceae	1	2	1	0
	Chenopodiaceae	17	64	11	0
	Cistaceae	2	9	0	0
	Clethraceae	1	1	0	0
	Clusiaceae	2	25	1	0
	Combretaceae	1	1	0	0
	Convolvulaceae	12	61	8	2
	Cornaceae	2	5	0	0
	Crassulaceae	7	19	2	0
	Crossomataceae	1	3	0	0
	Cucurbitaceae	16	32	2	0
	Cuscutaceae	1	18	5	0
	Cyrillaceae	1	1	0	0
	Dilleniaceae	1	1	0	0
	Dipsacaceae	1	1	0	0
	Droseraceae	1	3	0	0
	Ebenaceae	1	3	0	0
	Elaeagnaceae	1	4	0	0
	Elatinaceae	2	3	0	0
	Ericaceae	7	19	3	0
	Euphorbiaceae (**VII**)	19	151	19	0
	Fabaceae (**III**)	86	393	71	2
	Fagaceae	3	53	6	44
	Flacourtiaceae	2	3	0	0
	Fouquieriaceae	1	1	0	0
	Frankeniaceae	1	2	0	0
	Fumariaceae	2	6	4	0
	Garryaceae	1	2	2	0
	Gentianaceae	7	27	1	0
	Geraniaceae	3	10	0	1
	Goodeniaceae	1	1	0	0

CLASS	FAMILY	GENERA	SPECIES	INFRASPECIFIC	HYBRIDS
	Grossulariaceae	3	6	2	0
	Haloragaceae	2	8	3	0
	Hamamelidaceae	3	3	0	0
	Hippocastanaceae	1	3	2	0
	Hydrangeaceae	4	22	3	0
	Hydrophyllaceae	6	32	8	0
	Illiciaceae	1	1	0	0
	Juglandaceae	3	14	2	0
	Krameriaceae	1	4	0	0
	Lamiaceae (**VI**)	40	148	30	2
	Lauraceae	6	9	1	0
	Leitneriaceae	1	1	0	0
	Lentibulariaceae	2	11	0	0
	Linaceae	1	21	3	0
	Loasaceae	3	18	1	0
	Loganiaceae	3	6	0	0
	Lythraceae	10	21	1	1
	Magnoliaceae	3	8	0	1
	Malpighiaceae	7	8	0	0
	Malvaceae	32	97	7	0
	Melastomataceae	2	6	1	0
	Meliaceae	1	1	0	0
	Menispermaceae	3	4	0	0
	Menyanthaceae	1	2	0	0
	Molluginaceae	2	4	0	0
	Monotropaceae	2	3	0	0
	Moraceae	5	9	1	0
	Myricaceae	1	2	0	0
	Myrsinaceae	2	3	0	0
	Myrtaceae	7	15	0	1
	Nelumbonaceae	1	1	0	0
	Nyctaginaceae	13	55	2	1
	Nymphaeaceae	2	5	1	0
	Nyssaceae	1	3	0	0

CLASS	FAMILY	GENERA	SPECIES	INFRASPECIFIC	HYBRIDS
	Oleaceae	10	40	2	3
	Onagraceae	7	70	17	0
	Orobanchaceae	3	7	2	0
	Oxalidaceae	1	14	6	0
	Papaveraceae	5	15	3	0
	Passifloraceae	1	8	1	1
	Pedaliaceae	2	6	1	0
	Phyrmaceae	1	1	0	0
	Phytolaccaceae	4	4	0	0
	Piperaceae	1	1	0	0
	Pittosporaceae	1	1	0	0
	Plantaginaceae	1	13	0	0
	Platanaceae	1	1	0	0
	Plumbaginaceae	3	6	0	0
	Polemoniaceae	7	35	16	0
	Polygalaceae	1	28	7	0
	Polygonaceae	8	66	13	0
	Portulacaceae	4	17	2	0
	Primulaceae	6	13	5	0
	Proteaceae	3	3	0	0
	Punicaceae	1	1	0	0
	Rafflesiaceae	1	1	0	0
	Ranunculaceae	13	55	15	0
	Resedaceae	1	1	0	0
	Rhamnaceae	12	25	2	0
	Rosaceae (**IX**)	32	132	19	7
	Rubiaceae	23	71	11	0
	Rutaceae	13	26	4	2
	Salicaceae	2	20	3	3
	Santalaceae	1	1	0	0
	Sapindaceae	7	13	2	0
	Sapotaceae	2	4	2	0
	Sarraceniaceae	1	1	0	0
	Saururaceae	3	3	0	0

CLASS	FAMILY	GENERA	SPECIES	INFRASPECIFIC	HYBRIDS
	Saxifragaceae	4	7	2	0
	Scrophulariaceae (X)	38	115	18	2
	Simaroubaceae	3	3	1	0
	Solanaceae	21	95	23	0
	Sphenocleaceae	1	1	0	0
	Sterculiaceae	5	10	0	0
	Styracaceae	2	6	1	0
	Symplocaceae	1	1	0	0
	Tamaricaceae	1	5	0	0
	Theaceae	6	8	0	0
	Thymelaeaceae	1	2	0	0
	Tiliaceae	3	4	2	0
	Tropaeolaceae	1	1	0	0
	Turneraceae	1	2	1	0
	Ulmaceae	4	13	1	0
	Urticaceae	4	11	4	0
	Valerianaceae	2	8	0	0
	Verbenaceae	16	64	11	1
	Violaceae	2	15	4	1
	Viscaceae	2	11	3	0
	Vitaceae	5	20	3	2
	Zygophyllaceae	6	13	0	0
LILIOPSIDA	Acoraceae	1	1	0	0
	Agavaceae	9	48	3	0
	Alismataceae	3	13	3	0
	Araceae	11	14	1	0
	Arecaceae	19	32	0	1
	Bromeliaceae	2	5	0	0
	Burmanniaceae	2	3	0	0
	Cannaceae	1	3	0	2
	Commelinaceae	5	25	4	1
	Costaceae	1	2	0	0
	Cymodoceaceae	2	2	0	0

CLASS	FAMILY	GENERA	SPECIES	INFRASPECIFIC	HYBRIDS
	Cyperaceae (**IV**)	18	290	33	2
	Dioscoreaceae	1	4	0	0
	Eriocaulaceae	2	7	0	0
	Hydrocharitaceae	7	7	0	0
	Hypoxidaceae	1	5	0	0
	Iridaceae	16	50	6	2
	Juncaceae	2	33	8	0
	Lemnaceae	4	14	0	0
	Liliaceae (**V**)	60	152	11	6
	Limnocharitaceae	1	1	0	0
	Marantaceae	1	1	0	0
	Mayacaceae	1	1	0	0
	Muscaceae	2	3	0	1
	Najadaceae	1	2	0	0
	Orchidaceae	23	54	3	0
	Poaceae (**II**)	160	655	104	9
	Pontederiaceae	3	6	0	0
	Potamogetonaceae	1	11	1	1
	Ruppiaceae	1	1	0	0
	Smilacaceae	1	10	1	0
	Sparganiaceae	1	2	0	0
	Strelitziaceae	1	2	0	0
	Thyphaceae	1	2	0	0
	Xyridaceae	1	15	2	0
	Zannichelliaceae	1	1	1	0
	Zingiberaceae	4	7	0	0
	TOTAL: 227	1599	5812	939	117

Note: The State of Texas has been commemorated in a total of 139 species names—of which 102 are still recognized and 37 are now in synonymy—and a total of 47 infraspecific (i.e., subspecies or varieties)—of which 32 are still recognized and 15 are now in synonymy.

Checklist

CLASS: LYCOPODIOPSIDA

LYCOPODIACEAE

LYCOPODIELLA J. Holub
L. alopecuroides (C. Linnaeus) R. Cranfill
* Sy = Lycopodium alopecuroides C. Linnaeus [**TOES: IV**]
Sy = L. inundatum C. Linnaeus var. alopecuroides (C. Linnaeus) E. Tuckerman
L. appressa (A. Chapman) R. Cranfill
Sy = Lycopodium alopecuroides C. Linnaeus var. appressum A. Chapman
Sy = L. appressum (A. Chapman) J. Lloyd & L. Underwood
Sy = L. inundatum C. Linnaeus var. appressum A. Chapman
L. prostrata (R. Harper) R. Cranfill
Sy = Lycopodium alopecuroides C. Linnaeus var. pinnatum (A. Chapman) J. Lloyd & L. Underwood *ex* C. A. Brown & D. Correll
Sy = L. inundatum C. Linnaeus var. pinnatum A. Chapman
Sy = L. prostratum R. Harper

PALHINHAEA J. Vasconcellos & J. Franco
P. cernua (C. Linnaeus) J. Vasconcellos & J. Franco
Sy = Lycopodium cernuum C. Linnaeus

PSEUDOLYCOPODIELLA J. Holub
P. caroliniana (C. Linnaeus) J. Holub var. **_caroliniana_**
Sy = Lycopodiella caroliniana (C. Linnaeus) R. Pichi-Sermolli var. caroliniana
Sy = Lycopodium carolinianum C. Linnaeus

CLASS: ISOËTOPSIDA

ISOËTACEAE

ISOËTES C. Linnaeus
I. butleri G. Engelmann
* **_I. lithophila_** N. Pfeiffer [**TOES: V**]
I. melanopoda J. Gay & M. Durieu de Maisonneuve *ex* M. Durieu de Maisonneuve
Sy = I. melanopoda var. pallida G. Engelmann

SELAGINELLACEAE, *nomen conservandum*

SELAGINELLA A. Palisot de Beauvois, *nomen conservandum*
S. apoda (C. Linnaeus) F. Spring
Sy = S. apus F. Spring
S. arenicola L. Underwood subsp. **_riddellii_** (G. Van Eseltine) R. Tryon
Sy = S. riddellii G. Van Eseltine
S. arizonica W. Maxon
S. densa P. Rydberg
Sy = S. rupestris (C. Linnaeus) F. Spring var. densa W. Clute
S. lepidophylla (W. Hooker & R. Greville) F. Spring
S. mutica D. C. Eaton *ex* L. Underwood var. **_limitania_** C. Weatherby
Sy = S. mutica var. texana C. Weatherby
S. mutica D. C. Eaton *ex* L. Underwood var. **_mutica_**
S. peruviana (J. Milde) G. Hieronymus
Sy = S. sheldonii W. Maxon
S. pilifera A. Braun
Sy = S. pilifera var. pringlei (J. G. Baker) C. Morton
S. rupincola L. Underwood
Sy = S. rupestris (C. Linnaeus) F. Spring var. rupincola (L. Underwood) W. Clute

S. underwoodii G. Hieronymus
Sy = S. fendleri (L. Underwood) G. Hieronymus
Sy = S. rupestris (C. Linnaeus) F. Spring var. fendleri L. Underwood
Sy = S. underwoodii var. dolichotricha C. Weatherby
* ***S. viridissima*** C. Weatherby [**TOES: V**]
S. wrightii G. Hieronymus
S. × ***neomexicana*** W. Maxon [***mutica*** × ***rupincola***]

CLASS: EQUISETOPSIDA

EQUISETACEAE

EQUISETUM C. Linnaeus
E. arvense C. Linnaeus
E. hyemale C. Linnaeus subsp. ***affine*** (G. Engelmann) J. Calder & R. Taylor
Sy = E. hyemale C. Linnaeus var. affine (G. Engelmann) A. Eaton
Sy = E. praeltum C. Rafinesque-Schmaltz
Sy = E. robustum A. Brown
E. laevigatum A. Braun
Sy = E. funstonii A. Eaton
Sy = E. kansanum J. Schaffner
Sy = E. laevigatum subsp. funstonii E. Hartman
Sy = E. praeltum C. Rafinesque-Schmaltz var. laevigatum B. Bush
Sy = Hippochaete laevigata O. Farwell
Sy = H. laevigata var. funstonii O. Farwell
E. scirpoides A. Michaux [**cultivated**]
E. × ***ferrissii*** W. Clute [***hyemale*** subsp. ***affine*** × ***laevigatum***]
Sy = E. hyemale var. intermedium A. Eaton

CLASS: PSILOTOPSIDA

PSILOTACEAE

PSILOTUM O. Swartz
P. nudum (C. Linnaeus) A. Palisot de Beauvois
Sy = Lycopodium nudum C. Linnaeus

CLASS: POLYPODIOPSIDA

ANEMIACEAE

ANEMIA O. Swartz, *nomen conservandum*
A. mexicana J. Klotzsch

ASPLENIACEAE

ASPLENIUM C. Linnaeus
A. palmeri W. Maxon
A. platyneuron (C. Linnaeus) N. Britton, E. Sterns, & J. Poggenburg
Sy = Acrostichum platyneuron C. Linnaeus
Sy = Asplenium platyneuron var. bacculum-rubrum (M. Fernald) M. Fernald
Sy = A. platyneuron var. incisum (W. E. Howe *ex* C. H. Peck) B. Robinson
Sy = A. platyneuron var. proliferum D. C. Eaton
A. resiliens G. Kunze
Sy = A. parvulum M. Martens & H. Galeotti, non W. Hooker
Sy = Chamaefilix resiliens O. Farwell
A. septentrionale (C. Linnaeus) K. Hoffmann
Sy = Acropteris septentrionalis J. Link
Sy = Acrostichum septentrionale C. Linnaeus
Sy = Amesium septentrionale E. Newman
Sy = Blechnum septentrionale C. Wallroth
Sy = Scolopendrium septentrionale A. Roth
A. trichomanes C. Linnaeus subsp. ***trichomanes***
Sy = A. melanocaulon C. von Willdenow
Sy = Chamaefilix trichomanes O. Farwell

AZOLLACEAE

AZOLLA J. de Lamarck
A. caroliniana C. von Willdenow
A. filiculoides J. de Lamarck
A. mexicana D. von Schlechtendal & A. von Chamisso *ex* K. Presl

BLECHNACEAE

BLECHNUM C. Linnaeus
B. occidentale C. Linnaeus var. ***minor*** W. Hooker
Sy = B. occidentale var. pubirachis E. Rosenstock

WOODWARDIA J. E. Smith
W. areolata (C. Linnaeus) T. Moore
Sy = Lorinseria areolata (C. Linnaeus) K. Presl
W. virginica (C. Linnaeus) J. E. Smith
Sy = Anchistea virginica (C. Linnaeus) K. Presl

DENNSTAEDTIACEAE

DENNSTAEDTIA J. Bernhardi
D. globulifera (J. Poiret) G. Hieronymus
Sy = Polypodium globuliferum J. Poiret

PTERIDIUM J. Gleditsch *ex* J. Scopoli, *nomen conservandum*
P. ***aquilinum*** (C. Linnaeus) F. Kuhn var. ***latiusculum*** (N. Desvaux) L. Underwood *ex* A. A. Heller
Sy = P. latiusculum (N. Desvaux) G. Hieronymus
P. ***aquilinum*** (C. Linnaeus) F. Kuhn var. ***pseudocaudatum*** (W. Clute) A. A. Heller
Sy = P. latiusculum (N. Desvaux) G. Hieronymus var. pseudocaudatum (W. Clute) A. A. Heller
P. ***aquilinum*** (C. Linnaeus) F. Kuhn var. ***pubescens*** L. Underwood
Sy = P. aquilinum subsp. lanuginosum (A. von Bongard) O. Hultén
Sy = P. aquilinum var. lanuginosum (A. von Bongard) M. Fernald

DRYOPTERIDACEAE, *nomen conservandum*

ATHYRIUM A. Roth
A. filix-femina (C. Linnaeus) A. Roth var. ***asplenioides*** (A. Michaux) O. Farwell
Sy = A. asplenioides (A. Michaux) A. Eaton
Sy = A. filix-femina subsp. asplenioides (A. Michaux) Shiu-ying Hu

CYRTOMIUM K. Presl
C. falcatum (C. von Linné) K. Presl **[cultivated]**
Sy = Polypodium falcatum C. von Linné
Sy = Polystichum falcatum (C. von Linné) F. Diels

CYSTOPTERIS J. Bernhardi, *nomen conservandum*
C. bulbifera (C. Linnaeus) J. Bernhardi
Sy = Aspidium bulbiferum O. Swartz
Sy = Cystea bulbifera D. Watt
Sy = Filicula bulbifera O. Farwell
Sy = Filix bulbifera (C. Linnaeus) L. Underwood
Sy = Nephrodium bulbiferum A. Michaux
Sy = Polypodium bulbiferum C. Linnaeus
C. fragilis (C. Linnaeus) J. Bernhardi
Sy = Aspidium fragile O. Swartz
Sy = Athyrium fragile K. Sprengel
Sy = Cyathea fragilis J. E. Smith
Sy = Cyclopteris fragilis S. Gray
Sy = Cystea fragilis J. E. Smith
Sy = Cystopteris dickieana T. R. Sim
Sy = C. fragilis subsp. dickieana (T. R. Sim) N. Hylander
Sy = C. fragilis var. angustata P. Lawson
Sy = C. fragilis var. woodsioides D. Christ
Sy = Filix fragilis (C. Linnaeus) J.-E. Gilibert
Sy = Polypodium fragilis C. Linnaeus
C. reevesiana D. Lellinger
Sy = C. fragilis (C. Linnaeus) J. Bernhardi subsp. tenuifolia W. Clute
C. utahensis M. Windham & C. Haufler

DRYOPTERIS M. Adanson, *nomen conservandum*
D. cinnamomea (A. Cavanilles) C. Christensen
Sy = Tectaria cinnamomea A. Cavanilles
D. filix-mas (C. Linnaeus) H. Schott
Sy = Filix filix-mas O. Farwell
Sy = Filix-mas filix-mas O. Farwell
Sy = Lastrea filix-mas K. Presl
Sy = Lophodium filix-mas E. Newman
Sy = Nephrodium filix-mas L. C. Richard
Sy = Polypodium filix-mas C. Linnaeus
Sy = Polystichum filix-mas A. Roth
Sy = Tectaria filix-mas A. Cavanilles
Sy = Thelypteris filix-mas J. Nieuwland
* ***D. ludoviciana*** (G. Kunze) J. K. Small **[TOES: IV]**
Sy = Aspidium ludovicianum G. Kunze
Sy = Dryopteris floridana (W. Hooker) G. Kunze

ONOCLEA C. Linnaeus
O. sensibilis C. Linnaeus
Sy = O. sensibilis var. obtusilobata (C. Schkuhr) J. Torrey

PHANEROPHLEBIA K. Presl
P. auriculata L. Underwood
Sy = Cyrtomium auriculatum (L. Underwood) C. Morton
P. umbonata L. Underwood
Sy = Cyrtomium umbonatum (L. Underwood) C. Morton

POLYSTICHUM A. Roth, *nomen conservandum*
P. acrostichoides (A. Michaux) H. Schott
Sy = P. acrostichoides var. lonchitoides R. Brooks
Sy = P. acrostichoides var. schweinitzii (G. Beck) J. K. Small

TECTARIA A. Cavanilles
T. heracleifolia (C. von Willdenow) L. Underwood
Sy = Aspidium heracleifolium C. von Willdenow

WOODSIA R. Brown
W. neomexicana M. Windham
Ma = W. mexicana *auct.*, non A. Feé: Texas authors
W. obtusa (K. Sprengel) J. Torrey subsp. ***obtusa***

W. obtusa (K. Sprengel) J. Torrey subsp. ***occidentalis*** M. Windham
Sy = W. phillipsii M. Windham
W. plummerae J. Lemmon
Sy = W. obtusa (K. Sprengel) J. Torrey var. glandulosa D. C. Eaton & M. Faxon
Sy = W. obtusa var. plummerae (J. Lemmon) W. Maxon
Sy = W. pusilla E. Fournier var. glandulosa (D. C. Eaton & M. Faxon) T. M. Taylor

LYGODIACEAE

LYGODIUM O. Swartz, *nomen conservandum*
L. japonicum (C. Thunberg *ex* J. Murray) O. Swartz

MARSILEACEAE

MARSILEA C. Linnaeus, *nomen conservandum*
M. macropoda G. Engelmann *ex* A. Braun
M. mollis B. Robinson & M. Fernald
Ma = M. mexicana *auct.*, non A. Braun: Texas authors
M. vestita W. Hooker & R. Greville
Sy = M. fournieri C. Christensen
Sy = M. minuata E. Fournier
Sy = M. mucronata A. Braun
Sy = M. oligospora L. Goodding
Sy = M. tenuifolia G. Engelmann *ex* A. Braun
Sy = M. uncinata A. Braun
Sy = M. vestita subsp. tenuifolia (G. Engelmann *ex* A. Braun) D. M. Johnson
Sy = M. vestita var. mucronata (A. Braun) J. G. Baker
Sy = Zaluzanskya vestita K. E. O. Kuntze

PILULARIA C. Linnaeus
P. americana A. Braun

OPHIOGLOSSACEAE

BOTRYCHIUM O. Swartz
B. biternatum (M. de Savigny) L. Underwood
Sy = B. dissectum K. Sprengel var. tenuifolium (L. Underwood) O. Farwell
B. dissectum K. Sprengel
Sy = B. dissectum var. obliquum (G. H. Muhlenberg) W. Clute
Sy = B. dissectum var. oblongifolium (G. Graves) M. Broun
Sy = B. obliquum G. H. Muhlenberg
Sy = B. obliquum var. elongatum B. D. Gilbert & J. W. Haberer
B. lunarioides (C. Linnaeus) O. Swartz
Sy = Botrypus lunarioides A. Michaux
B. virginianum (C. Linnaeus) O. Swartz
Sy = B. virginianum subsp. europaeum (J. Ångström) A. Jávorka
Sy = B. virginianum var. europaeum J. Ångström

OPHIOGLOSSUM C. Linnaeus
O. crotalophoroides T. Walter
Sy = O. crotalophoroides var. nanum C. Osten *ex* J. S. de Lichtenstein
O. engelmannii K. Prantl
O. nudicaule C. von Linné var. ***tenerum*** (G. Mettenius *ex* K. Prantl) R. Clausen
Sy = O. dendroneuron E. St. John
Sy = O. ellipticum W. Hooker & R. Greville
Sy = O. mononeuron E. St. John
Sy = O. nudicaule var. minus R. Clausen
Sy = O. tenerum G. Mettenius *ex* K. Prantl
O. petiolatum W. Hooker
O. vulgatum C. Linnaeus
Sy = O. pycnostichum (M. Fernald) A. Löve & D. Löve
Sy = O. vulgatum var. pycnostichum M. Fernald

OSMUNDACEAE

OSMUNDA C. Linnaeus
O. cinnamomea C. Linnaeus
Sy = O. cinnamomea var. frondosa A. Gray
Sy = O. cinnamomea var. glandulosa C. E. Waters
Sy = O. cinnamomea var. imbricata (G. Kunze) J. Milde
O. regalis C. Linnaeus var. ***spectabilis*** (C. von Willdenow) A. Gray
Sy = O. spectabilis C. von Willdenow

PARKERIACEAE

CERATOPTERIS A. T. de Brongniart
C. thalichtroides (C. Linnaeus) A. T. de Brongniart
Sy = C. deltoidea R. Benedict
Sy = C. siliquosa (C. Linnaeus) E. B. Copeland

POLYPODIACEAE

PLEOPELTIS F. von Humboldt & A. Bonpland *ex* C. von Willdenow

P. polylepis (J. J. Römer *ex* G. Kunze) T. Moore var. ***erythrolepis*** (C. Weatherby) T. Wendt
Sy = P. erythrolepis (C. Weatherby) R. Pichi-Sermolli
Sy = Polypodium erythrolepis C. Weatherby
P. polypodioides (C. Linnaeus) E. Andrews & M. Windham var. ***michauxiana*** (C. Weatherby) E. Andrews & M. Windham
Sy = Polypodium polypodioides (C. Linnaeus) D. Watt subsp. michauxiana (C. Weatherby) E. Andrews & M. Windham
P. riograndensis (T. Wendt) E. Andrews & M. Windham
Sy = Polypodium thyssanolepis A. Braun *ex* J. Klotzsch var. riograndense T. Wendt

PTERIDACEAE

ADIANTUM C. Linnaeus
A. capillus-veneris C. Linnaeus
Sy = A. capillus O. Swartz
Sy = A. capillus-veneris var. modestum (L. Underwood) M. Fernald
Sy = A. capillus-veneris var. protrusum M. Fernald
Sy = A. modestum L. Underwood
Sy = A. rimicola M. Slosson
A. pedatum C. Linnaeus [**cultivated**]
A. tricholepis A. Feé

ARGYROCHOSMA (J. K. Small) M. Windham
A. dealbata (F. Pursh) M. Windham
Sy = Cheilanthes dealbata F. Pursh
Sy = Notholaena dealbata (F. Pursh) G. Kunze
Sy = Pellaea dealbata (F. Pursh) K. Prantl
A. limitanea (W. Maxon) M. Windham subsp. ***mexicana*** (W. Maxon) M. Windham
Sy = Cheilanthes limitanea (W. Maxon) M. Windham var. mexicana (W. Maxon) J. Mickel
Sy = Notholaena limitanea W. Maxon subsp. mexicana W. Maxon
Sy = N. limitanea W. Maxon var. mexicana (W. Maxon) M. Broun
Sy = Pellaea limitanea var. mexicana (W. Maxon) C. Morton
A. microphylla (G. Mettenius *ex* F. Kuhn) M. Windham
Sy = Cheilanthes parvifolia (R. Tryon) J. Mickel
Sy = Notholaena parvifolia R. Tryon
Sy = Pellaea microphylla G. Mettenius *ex* F. Kuhn

ASTROLEPIS D. Benham & M. Windham
A. cochisensis (L. Goodding) D. Benham & M. Windham subsp. ***chihuahuensis*** D. Benham & M. Windham
A. cochisensis (L. Goodding) D. Benham & M. Windham subsp. ***cochisensis***
Sy = Cheilanthes cochisensis (L. Goodding) J. Mickel
Sy = C. integerrima (W. Hooker) J. Mickel
Sy = Notholaena cochisensis L. Goodding
Sy = N. integerrima (W. Hooker) R. Hevly
A. sinuata (M. Lagasca y Segura *ex* O. Swartz) D. Benham & M. Windham subsp. ***mexicana*** D. Windham
A. sinuata (M. Lagasca y Segura *ex* O. Swartz) D. Benham & M. Windham subsp. ***sinuata*** (M. Lagasca y Segura *ex* O. Swartz) D. Benham & M. Windham
Sy = Acrostichum sinuatum M. Lagasca y Segura *ex* O. Swartz
Sy = Cheilanthes sinuata (M. Lagasca y Segura *ex* O. Swartz) K. Domin
Sy = Notholaena sinuata (M. Lagasca y Segura *ex* O. Swartz) G. Kaulfuss
A. windhamii D. Benham

BOMMERIA E. Fournier
B. hispida (G. Mettenius *ex* F. Kuhn) L. Underwood
Sy = B. schaffneri E. Fournier
Sy = Gymnogramma hispida G. Mettenius *ex* F. Kuhn
Sy = Gymnopteris hispida (G. Mettenius) L. Underwood

CHEILANTHES O. Swartz, *nomen conservandum*
C. aemula W. Maxon
C. alabamensis (S. Buckley) G. Kunze
Sy = Pellaea alabamensis (S. Buckley) J. G. Baker *ex* W. Hooker
Sy = Pteris alabamensis S. Buckley
C. bonariensis (C. von Willdenow) G. Proctor
Sy = Acrostichum bonariense C. von Willdenow
Sy = Notholaena aurea (J. Poiret) N. Desvaux
C. eatonii J. G. Baker
Sy = C. castanea W. Maxon
Sy = C. pinkavae T. Reeves
Sy = C. tomentosa J. Link var. eatonii G. Davenport
C. feei T. Moore
Sy = Allosurus vestitus O. Farwell
Sy = Cheilanthes gracilis N. Riehl *ex* G. Mettenius
Sy = Myriopteris gracilis A. Feé

C. fendleri W. Hooker
Sy = Myriopteris fendleri (W. Hooker) E. Fournier
C. horridula W. Maxon
Sy = Pellaea aspera (W. Hooker) J. G. Baker
C. kaulfussii G. Kunze
Sy = C. glandulifera F. Liebmann
Sy = C. viscosa J. Link, non D. Carmichael
C. lanosa (A. Michaux) D. C. Eaton
Sy = C. vestita (K. Sprengel) O. Swartz
Sy = Nephrodium lanosum A. Michaux
C. lendigera (A. Cavanilles) O. Swartz
Sy = Myriopteris lendigera (A. Cavanilles) J. E. Smith
Sy = Pteris lendigera A. Cavanilles
C. leucopoda J. Link
C. lindheimeri W. Hooker
Sy = Allosurus lindheimeri O. Farwell
Sy = Myriopteris lindheimeri (W. Hooker) J. E. Smith
C. tomentosa J. Link
Sy = Myriopteris tomentosa (J. Link) J. E. Smith
C. villosa G. Davenport *ex* W. Maxon
C. wootonii W. Maxon
C. wrightii W. Hooker
Sy = Allosurus wrightii O. Farwell
C. yavapensis T. Reeves *ex* M. Windham

NOTHOLAENA R. Brown
N. aliena W. Maxon
Sy = Cheilanthes aliena (W. Maxon) J. Mickel
N. aschenborniana J. Klotzsch
Sy = Cheilanthes aschenborniana (J. Klotzsch) G. Mettenius
Sy = Notholaena bipinnata F. Liebmann
N. copelandii C. Hall
Sy = N. candida (M. Martens & H. Galeotti) W. Hooker var. copelandii (C. Hall) A. Tryon
N. grayi G. Davenport subsp. ***grayi***
Sy = Cheilanthes grayi (G. Davenport) K. Domin
N. greggii (G. Mettenius *ex* F. Kuhn) W. Maxon
Sy = Cheilanthes greggii (G. Mettenius *ex* F. Kuhn) J. Mickel
N. nealleyi H. Seaton *ex* J. Coulter
Sy = Cheilanthes nealleyi (H. Seaton *ex* J. Coulter) K. Domin
Sy = Notholaena schaffneri (E. Fournier) L. Underwood *ex* G. Davenport var. nealleyi (H. Seaton *ex* J. Coulter) C. Weatherby
N. neglecta W. Maxon
Sy = Cheilanthes neglecta (W. Maxon) J. Mickel
N. standleyi W. Maxon
Sy = Cheilanthes hookeri K. Domin
Sy = C. standleyi (W. Maxon) J. Mickel
Sy = Chrysochosma hookeri (K. Domin) J. Kümmerle
Sy = Notholaena hookeri D. C. Eaton

PELLAEA J. Link, *nomen conservandum*
P. atropurpurea (C. Linnaeus) J. Link
Sy = Pteris atropurpurea C. Linnaeus
P. cordifolia (M. Sessé y Lacasta & J. Mociño) A. R. Smith
Sy = Adiantum cordifolium M. Sessé y Lacasta & J. Mociño
Sy = Pellaea cardiomorpha C. Weatherby
Sy = P. cordata (A. Cavanilles) J. E. Smith
Sy = P. sagittata (A. Cavanilles) J. Link var. cordata (A. Cavanilles) A. Tryon
P. glabella G. Mettenius *ex* F. Kuhn subsp. ***glabella***
Sy = Pellaea atropurpurea (C. Linnaeus) J. Link var. bushii K. Mackenzie
P. intermedia G. Mettenius *ex* F. Kuhn
Sy = Cassebeera intermedia O. Farwell
Sy = Pellaea intermedia var. pubescens G. Mettenius *ex* F. Kuhn
P. ovata (N. Desvaux) C. Weatherby
Sy = P. flexuosa (G. Kaulfuss *ex* D. von Schlechtendal & A. von Chamisso) J. Link
Sy = Pteris ovata N. Desvaux
P. ternifolia (A. Cavanilles) J. Link subsp. ***arizonica*** M. Windham
P. ternifolia (A. Cavanilles) J. Link subsp. ***ternifolia***
Sy = Allosurus ternifolius K. E. O. Kunze *ex* J. Klinge
Sy = Cassebeera ternifolia O. Farwell
Sy = Cheilanthes ternifolia T. Moore
Sy = Notholaena ternifolia A. von Keyserling
Sy = Pteris ternifolia A. Cavanilles
P. truncata L. Goodding
Ma = P. longimucronata *auct.*, non W. Hooker: American authors
P. wrightiana W. Hooker
Sy = P. mucronata D. C. Eaton
Sy = P. ternifolia var. wrightiana (W. Hooker) A. Tryon
P. × wagneri M. Windham

SALVINIACEAE

SALVINIA J. Séguier
S. minima J. G. Baker

THELYPTERIDACEAE

MACROTHELYPTERIS (H. Ito) Ren-Chang Ching
M. torresiana (C. Gaudichaud-Beaupré) Ren-Chang Ching
Sy = Dryopteris uliginosa (G. Kunze) C. Christensen, non (A. Braun *ex* P. Dowell) G. Druce
Sy = Thelypteris torresiana (C. Gaudichaud-Beaupré) A. Alston

PHEGOPTERIS (K. Presl) A. Feé
P. hexagonoptera (A. Michaux) A. Feé
Sy = Dryopteris hexagonoptera (A. Michaux) C. Christensen
Sy = Thelypteris hexagonoptera (A. Michaux) C. Weatherby

THELYPTERIS C. Schmidel, *nomen conservandum*
T. dentata (P. Forsskål) E. St. John
Sy = Cyclosorus dentatus (P. Forsskål) Ren-Chang Ching
Sy = Dryopteris dentata (P. Forsskål) C. Christensen
Sy = Thelypteris reducta J. K. Small *ex* R. St. John
T. hispidula (J. Decaisne) C. F. Reed var. ***versicolor*** (R. St. John) D. Lellinger
Sy = Dryopteris macilenta (E. St. John) D. Correll
Sy = D. versicolor (R. St. John) M. Broun
Sy = Thelypteris macilenta E. St. John
Sy = T. quadrangularis (A. Feé) E. A. Schelpe var. versicolor (R. St. John) A. R. Smith
Sy = T. versicolor R. St. John
T. kunthii (N. Desvaux) C. Morton
Sy = Dryopteris normalis C. Christensen
Sy = D. saxatilis (R. St. John) M. Broun
Sy = Thelypteris macrorhizoma E. St. John
Sy = T. normalis (C. Christensen) G. Moxley
Sy = T. saxatilis R. St. John
Sy = T. unca R. St. John
T. ovata R. St. John var. ***lindheimeri*** (C. Christensen) A. R. Smith
Sy = Dryopteris normalis C. Christensen var. lindheimeri C. Christensen
T. palustris H. Schott var. ***pubescens*** (P. Lawson) M. Fernald
Sy = Dryopteris thelypteris (C. Linnaeus) O. Swartz var. haleana (M. Fernald) M. Broun & C. Weatherby
Sy = D. thelypteris var. pubescens (P. Lawson) A. Prince *ex* C. Weatherby
Sy = Thelypteris confluens (C. Thunberg) C. Morton var. pubescens (P. Lawson) J. Pringle
Sy = T. palustris var. haleana M. Fernald

CLASS: CYCADOPSIDA

CYCADACEAE, *nomen conservandum*

CYCAS C. Linnaeus
C. circinnalis C. Linnaeus [**cultivated**]
C. revoluta C. Thunberg [**cultivated**]

ZAMIACEAE

DIOON J. Lindley, *nomen conservandum*
D. edule J. Lindley [**cultivated**]
D. spinulosum W. T. Dyer [**cultivated**]

ZAMIA C. Linnaeus, *nomen conservandum*
Z. integrifolia C. von Linné [**cultivated**]
Sy = Z. floridana A. P. de Candolle
Sy = Z. silvicola J. K. Small
Sy = Z. umbrosa J. K. Small
Z. pumila C. Linnaeus [**cultivated**]

CLASS: GINKGOÖPSIDA

GINKGOACEAE, *nomen conservandum*

GINKGO C. Linnaeus
G. biloba C. Linnaeus [**cultivated**]

CLASS: PINOPSIDA

ARAUCARIACEAE, *nomen conservandum*

ARAUCARIA A. L. de Jussieu
A. araucana K. Koch [**cultivated**]
Sy = A. imbricata J. Pavón
A. heterophylla (R. A. Salisbury) J. Franco [**cultivated**]
Sy = A. excelsa R. Brown

CUPRESSACEAE, *nomen conservandum*

CHAMICYPARIS E. Spach
C. obtusa (P. von Siebold & J. Zuccarini) S. Endlicher [**cultivated**]

Sy = Thuja obtusa (P. von Siebold & J. Zuccarini) M. T. Masters

× **CUPRESSOCYPARIS** W. Dallimore [***Chamicyparis* × *Cupressus***]

× ***C. leylandii*** (A. B. Jackson & W. Dallimore) W. Dallimore & A. B. Jackson [***Chamicyparis nootkatensis* × *Cupressus macrocarpa***] [**cultivated**]

CUPRESSUS C. Linnaeus

C. arizonica E. Greene

Sy = C. arizonica var. bonita J. Lemmon

C. glabra G. Sudworth [**cultivated**]

Sy = C. arizonica E. Greene var. glabra

C. sempervirens C. Linnaeus [**cultivated**]

JUNIPERUS C. Linnaeus

J. ashei J. Buchholz

Sy = J. occidentalis W. Hooker var. conjungens G. Engelmann

Sy = J. occidentalis var. texana G. Vasey

Ma = J. sabinoides *auct.*, non C. Nees von Esenbeck: *sensu* C. Sargent

J. chinensis C. Linnaeus var. ***chinensis*** [**cultivated**]

Sy = Sabina chinensis (C. Linnaeus) F. Antoine

J. chinensis C. Linnaeus var. ***procumbens*** (P. von Siebold) S. Endlicher [**cultivated**]

Sy = J. procumbens P. von Siebold

Sy = Sabina pacifica T. Nakai

J. conferta F. Parlatore [**cultivated**]

Sy = J. litoralis C. Maximowicz

J. deppeana E. von Steudel var. ***deppeana***

Sy = J. deppeana var. pachyphloea (J. Torrey) M. Martínez

Sy = J. mexicana K. Sprengel

Sy = J. pachyphloea J. Torrey

J. deppeana E. von Steudel var. ***sperryi*** D. Correll

J. erythrocarpa V. Cory

Sy = J. coahuilensis (M. Martínez) H. M. Gaussen *ex* R. Adams

Sy = J. erythrocarpa V. Cory var. coahuilensis M. Martínez

Sy = J. pinchotii G. Sudworth var. erythrocarpa (V. Cory) J. Silba

J. flaccida D. von Schlechtendal

Sy = J. flaccida var. gigantea (B. Roezl) H. M. Gaussen

Sy = J. gracilis S. Endlicher

Sy = Sabina flaccida (D. von Schlechtendal) F. Antoine

J. horizontalis C. Moench [**cultivated**]

Sy = J. horizontalis var. variegata L. Beissner

Sy = J. hudsonica J. Forbes

Sy = J. prostrata C. Persoon

Sy = J. repens T. Nuttall

Sy = J. virginiana C. Linnaeus var. prostrata (C. Persoon) J. Torrey

Sy = Sabina horizontalis (C. Moench) P. Rydberg

Sy = S. prostrata (C. Persoon) F. Antoine

J. monosperma (G. Engelmann) C. Sargent

Sy = J. gymnocarpa (J. Lemmon) V. Cory

Sy = J. mexicana K. Sprengel var. monosperma (G. Engelmann) V. Cory

Sy = J. occidentalis W. Hooker var. gymnocarpa J. Lemmon

Sy = Sabina monosperma (G. Engelmann) P. Rydberg

J. pinchotii G. Sudworth

Sy = J. monosperma (G. Engelmann) C. Sargent var. pinchotii (G. Sudworth) P. van Melle

Sy = J. texensis P. van Melle

J. sabina C. Linnaeus [**cultivated**]

J. scopulorum C. Sargent

Sy = J. scopulorum var. columnaris N. Fassett

Sy = J. virginiana C. Linnaeus subsp. scopulorum (C. Sargent) E. Murray

Sy = J. virginiana var. montana G. Vasey

Sy = J. virginiana var. scopulorum (C. Sargent) J. Lemmon

Sy = Sabina scopulorum (C. Sargent) P. Rydberg

J. virginiana C. Linnaeus var. ***virginiana***

Sy = J. virginiana subsp. crebra (M. Fernald & L. Griscom) E. Murray

Sy = J. virginiana var. crebra M. Fernald & L. Griscom

Sy = Sabina virginiana (C. Linnaeus) F. Antoine

PLATYCLADUS E. Spach

P. orientalis (C. Linnaeus) J. Franco [**cultivated**]

Sy = Biota orientalis (C. Linnaeus) S. Endlicher

Sy = Thuja orientalis C. Linnaeus

THUJA C. Linnaeus

T. occidentalis C. Linnaeus [**cultivated**]

PINACEAE, *nomen conservandum*

CEDRUS C. Trew

C. deodara (W. Roxburgh *ex* D. Don) G. Don [**cultivated**]

Sy = Pinus deodara W. Roxburgh *ex* D. Don

PICEA A. Dietrich
P. pungens G. Engelmann [**cultivated**]
Sy = P. parryana C. Sargent

PINUS C. Linnaeus
P. arizonica G. Engelmann var. *stormiae* M. Martínez
P. cembroides J. Zuccarini
Sy = P. cembroides var. bicolor E. Little
P. echinata P. Miller
P. edulis G. Engelmann
Sy = Caryopitys edulis (G. Engelmann) J. K. Small
Sy = Pinus cembroides var. edulis G. Engelmann
Sy = P. monophylla J. Torrey & J. Frémont var. edulis (G. Engelmann) M. E. Jones
P. eldarica J. Medwedew [**cultivated**]
P. elliottii G. Engelmann var. *elliottii*
Sy = P. heterophylla (S. Elliott) G. Sudworth
Ma = P. caribaea *auct.*, non P. M. Morelet: *sensu* J. K. Small
P. glabra T. Walter [**cultivated**]
P. halepensis P. Miller [**cultivated**]
P. mugo A. Turra [**cultivated**]
P. nigra F. C. Arnold [**cultivated**]
Sy = P. nigra var. austriaca (F. Höss) P. Ascherson & K. Graebner
P. palustris P. Miller
Sy = P. australis F. Michaux
P. ponderosa D. Douglas *ex* P. Lawson & C. Lawson var. *scopulorum* G. Engelmann
Sy = P. brachyptera G. Engelmann
Sy = P. ponderosa subsp. scopulorum (G. Engelmann) E. Murray
Sy = P. scopulorum (G. Engelmann) J. Lemmon
P. remota (E. Little) D. Bailey & F. Hawksworth
Sy = P. cembroides J. Zuccarini var. remota E. Little
P. strobiformis G. Engelmann
Sy = P. ayachuite C. G. Ehrenberg
Sy = P. flexilis var. reflexa G. Engelmann
Sy = P. reflexa (G. Engelmann) G. Engelmann
P. taeda C. Linnaeus
P. thunbergiana J. Franco [**cultivated**]

PSEUDOTSUGA É. Carrière
P. menziesii (C. Mirbel) J. Franco var. *glauca* (H. Mayr) J. Franco
Sy = P. flahaultii F. Flous
Sy = P. globulosa F. Flous
Sy = P. merrillii F. Flous
Sy = P. rehderi F. Flous
Sy = P. taxifolia (A. Lambert) N. Britton var. glauca (L. Beissner) G. Sudworth

PODOCARPACEAE, *nomen conservandum*

PODOCARPUS L'Héritier de Brutelle *ex* C. Persoon
P. macrophyllus (C. Thunberg) R. Sweet var. *maki* S. Endlicher [**cultivated**]
Sy = P. chinensis (W. Roxburgh) R. Sweet
Sy = Taxus chinensis W. Roxburgh
P. sinensis J. Teijsmann & S. Binnendijk [**cultivated**]

TAXACEAE, *nomen conservandum*

TAXUS C. Linnaeus
T. cuspidata P. von Siebold & J. Zuccarini [**cultivated**]

TAXODIACEAE, *nomen conservandum*

TAXODIUM L. C. Richard
T. ascendens A. T. de Brongniart [**cultivated**]
Sy = T. distichum (C. Linnaeus) L. C. Richard var. imbricarium (T. Nuttall) A. Croom
Sy = T. distichum var. nutans (W. Aiton) R. Sweet
T. distichum (C. Linnaeus) L. C. Richard
Sy = Cupressus disticha C. Linnaeus
* *T. mucronatum* M. Tenore [**TOES: III**]
Sy = T. distichum (C. Linnaeus) L. C. Richard var. mexicanum G. Gordon
Sy = T. mexicanum É. Carrière
Sy = T. montezumae J. Decaisne

CLASS: GNETOPSIDA

EPHEDRACEAE, *nomen conservandum*

EPHEDRA C. Linnaeus
E. antisyphilitica J. Berlandier *ex* C. von Meyer
Sy = E. antisyphilitica var. brachycarpa V. Cory
Sy = E. texana E. L. Reed
E. aspera G. Engelmann *ex* S. Watson
Sy = Ephedra nevadensis S. Watson var. aspera (G. Engelmann *ex* S. Watson) L. Benson
Sy = E. peninsularis I. M. Johnston
Sy = E. reedii V. Cory
E. coryi E. L. Reed
E. pedunculata G. Engelmann *ex* S. Watson

E. torreyana S. Watson var. ***powelliorum*** T. Wendt
E. torreyana S. Watson var. ***torreyana***
E. trifurca J. Torrey *ex* S. Watson
Sy = E. trifaria F. Parlatore

CLASS: MAGNOLIOPSIDA

ACANTHACEAE, *nomen conservandum*

ACANTHUS J. de Tournefort *ex* C. Linnaeus
A. mollis S. Graf & F. Noë *ex* C. Nees von Esenbeck [**cultivated**]

ANISACANTHUS C. Nees von Esenbeck
A. linearis (S. H. Hagen) J. Henrickson & E. Lott
Sy = A. insignis A. Gray var. linearis S. H. Hagen
A. puberulus (J. Torrey) J. Henrickson & E. Lott
Sy = A. insignis A. Gray
Sy = Drejera puberulus J. Torrey
A. quadrifidus (M. H. Vahl) C. Nees von Esenbeck var. ***wrightii*** (J. Torrey) J. Henrickson
Sy = A. wrightii (J. Torrey) A. Gray
Sy = Drejera wrightii J. Torrey
Sy = Justicia quadrifida M. H. Vahl

CARLOWRIGHTIA A. Gray, *nomen conservandum*
C. arizonica A. Gray
Sy = C. cordifolia A. Gray
C. linearifolia (J. Torrey) A. Gray
Sy = Schaueria linearifolia J. Torrey
C. mexicana J. Henrickson & T. Daniel
C. parviflora (S. Buckley) D. Wasshausen
Sy = Dianthera parviflora (S. Buckley) A. Gray
Sy = Drejera parviflora S. Buckley
C. parvifolia T. Brandegee
C. serpyllifolia A. Gray
C. texana J. Henrickson & T. Daniel
C. torreyana D. Wasshausen
Sy = C. pubens A. Gray
Sy = Croftia parvifolia (J. Torrey) J. K. Small
Sy = Dianthera parvifolia (J. Torrey) A. Gray
Sy = Ecbolium parvifolium (J. Torrey) K. E. O. Kuntze
Sy = Schaueria parvifolia J. Torrey

CROSSANDRA R. A. Salisbury
C. undulifolia R. A. Salisbury [**cultivated**]
Sy = C. infundibuliformis C. Nees von Esenbeck

DICLIPTERA A. L. de Jussieu, *nomen conservandum*
D. brachiata (F. Pursh) K. Sprengel
Sy = Diapedium brachiatum (F. Pursh) K. E. O. Kuntze
Sy = Dicliptera brachiata var. attentuata A. Gray
Sy = D. brachiata var. glandulosa (G. Scheele) M. Fernald
Sy = D. brachiata var. ruthii M. Fernald
Sy = D. glandulosa G. Scheele
Sy = Justicia brachiata F. Pursh
D. sexangularis (C. Linnaeus) A. L. de Jussieu
Sy = Diapedium assurgens (C. Linnaeus) K. E. O. Kuntze
Sy = Dicliptera assurgens (C. Linnaeus) A. L. de Jussieu
Sy = D. assurgens var. vahliana (C. Nees von Esenbeck) G. Maza
Sy = D. vahliana C. Nees von Esenbeck
Sy = Justicia assurgens C. Linnaeus

DYSCHORISTE C. Nees von Esenbeck
D. crenulata C. Kobuski
D. decumbens (A. Gray) K. E. O. Kuntze
Sy = Calophanes decumbens A. Gray
Sy = C. oblongifolia J. Torrey
D. linearis (J. Torrey & A. Gray) K. E. O. Kuntze
Sy = Calophanes linearis (J. Torrey & A. Gray) A. Gray
Sy = C. oblongifolia J. Torrey var. texensis C. Nees von Esenbeck
Sy = Dipteracanthus linearis J. Torrey & A. Gray

ELYTRARIA A. Michaux
E. bromoides A. Oersted
Sy = E. acuminata (J. K. Small) V. Cory
Sy = Tubiflora acuminata J. K. Small
E. imbricata (M. H. Vahl) C. Persoon
Sy = E. tridentata M. H. Vahl
Sy = Justicia imbricata M. H. Vahl
Sy = Tubiflora squamosa (N. von Jacquin) K. E. O. Kuntze

HYGROPHILA R. Brown
H. lacustris (D. von Schlechtendal & A. von Chamisso) C. Nees von Esenbeck
Sy = Ruellia lacustris D. von Schlechtendal & A. von Chamisso
H. polysperma (W. Roxburgh) T. Anderson [**federal noxious weed**]

JUSTICIA C. Linnaeus
J. americana (C. Linnaeus) M. H. Vahl
Sy = Dianthera americana C. Linnaeus
Sy = D. americana var. subcoriacea (M. Fernald) L. Shinners

Sy = Justicia americana var. subcoriacea M. Fernald
Sy = J. mortuifluminis M. Fernald
Sy = J. umbratilis M. Fernald
J. brandegeana D. Wasshausen & L. B. Smith [**cultivated**]
Sy = Beloperone guttata T. Brandegee
Sy = Calliaspidia guttata (T. Brandegee) N. E. Bremekamp
Sy = Drejerella guttata (T. Brandegee) N. E. Bremekamp
J. longii R. Hilsenbeck
Sy = Adathoda longiflora J. Torrey
Sy = Siphonoglossa longoflora (J. Torrey) A. Gray
J. ovata (T. Walter) G. Lindau var. ***lanceolata*** (A. Chapman) R. Long
Sy = Dianthera lanceolata (A. Chapman) J. K. Small
Sy = D. ovata T. Walter var. lanceolata A. Chapman
Sy = Justicia lanceolata (A. Chapman) J. K. Small
J. pilosella (C. Nees von Esenbeck) R. Hilsenbeck [***ined.***]
Sy = Monechma pilosella C. Nees von Esenbeck
Sy = Siphonoglossa pilosella (C. Nees von Esenbeck) J. Torrey
* ***J. runyonii*** J. K. Small [**TOES: V**]
J. spicigera D. von Schlechtendal [**cultivated**]
J. turneri R. Hilsenbeck [***ined.***]
J. warnockii B. L. Turner
* ***J. wrightii*** A. Gray [**TOES: V**]

NOMAPHILA K. von Blume
N. stricta (M. H. Vahl) C. Nees von Esenbeck
Sy = Hygrophila stricta (C. Nees von Esenbeck) G. Lindau, non J. Hasskarl
Sy = Justicia stricta M. H. Vahl

PSEUDERANTHEMUM L. Radlkofer
P. laxiflorum C. Hubbard *ex* L. Bailey [**cultivated**]

RUELLIA C. Linnaeus
R. brittoniana E. Leonard [**cultivated**]
R. caroliniensis (J. F. Gmelin) E. von Steudel
Sy = R. caroliniensis subsp. caroliniensis
Sy = R. caroliniensis var. caroliniensis
Sy = R. caroliniensis var. salicina M. Fernald
Sy = R. caroliniensis var. semicalva M. Fernald
Sy = R. caroliniensis var. serrulata B. Tharp & F. Barkley
R. corzoi B. Tharp & F. Barkley
Sy = R. drushelii B. Tharp & F. Barkley
R. drummondiana (C. Nees von Esenbeck) A. Gray
Sy = Dipteracanthus drummondianus C. Nees von Esenbeck
Sy = D. lindheimerianus G. Scheele
R. humilis T. Nuttall var. ***depauperata*** B. Tharp & F. Barkley
R. humilis T. Nuttall var. ***humilis***
Sy = R. humilis var. expansa M. Fernald
Sy = R. humilis var. frondosa M. Fernald
Sy = R. humilis var. longiflora (A. Gray) M. Fernald
R. malacosperma J. Greenman [**cultivated**]
R. metzae B. Tharp & F. Barkley
Sy = R. metzae var. marshii B. Tharp & F. Barkley
Sy = R. muelleri B. Tharp & F. Barkley
R. nudiflora (G. Engelmann *ex* A. Gray) I. Urban var. ***nudiflora***
Sy = Dipteracanthus nudiflora G. Engelmann *ex* A. Gray
Sy = Ruellia nudiflora var. hispidula L. Shinners
R. nudiflora (G. Engelmann *ex* A. Gray) I. Urban var. ***runyonii*** (B. Tharp & F. Barkley) B. L. Turner
Sy = R. runyonii B. Tharp & F. Barkley
Sy = R. runyonii var. berlandieri B. Tharp & F. Barkley
R. occidentalis (A. Gray) B. Tharp & F. Barkley
Sy = R. davisiorum B. Tharp & F. Barkley
Sy = R. nudiflora (G. Engelmann & A. Gray) I. Urban var. occidentalis (A. Gray) E. Leonard
Sy = R. occidentalis var. wrightii B. Tharp & F. Barkley
Sy = R. strictopaniculata B. Tharp & F. Barkley
Sy = R. tuberosa (C. Linnaeus) var. occidentalis A. Gray
R. parryi A. Gray
R. pedunculata J. Torrey *ex* A. Gray
Sy = R. pedunculata subsp. pinetorum (M. Fernald) R. Long
Sy = R. pinetorum M. Fernald
R. strepens C. Linnaeus
Sy = R. strepens var. cleistantha A. Gray
R. yucatana (E. Leonard) B. Tharp & F. Barkley
Sy = R. nudiflora (G. Engelmann & A. Gray) I. Urban var. yucatana E. Leonard

SIPHONOGLOSSA A. Oersted
S. greggii J. Greenman & C. Thompson

STENANDRIUM C. Nees von Esenbeck, *nomen conservandum*

S. barbatum J. Torrey & A. Gray
Sy = Gerardia barbata (J. Torrey & A. Gray) S. F. Blake
S. dulce (A. Cavanilles) C. Nees von Esenbeck
Sy = Gerardia dulcis (A. Cavanilles) C. Nees von Esenbeck
Sy = Gerardia dulcis var. floridana (A. Gray) S. F. Blake
Sy = Ruellia dulcis A. Cavanilles
Sy = Stenandrium dulce var. floridanum A. Gray
Sy = S. fascicularis (G. Bentham) D. Wasshausen
Ma = S. floridanum *auct.*, non (A. Gray) J. K. Small

TETRAMERIUM C. Nees von Esenbeck, *nomen conservandum*
T. nervosum C. Nees von Esenbeck
Sy = T. hispidum C. Nees von Esenbeck
Sy = T. nervosum var. hispidum (C. Nees von Esenbeck) J. Torrey

THUNBERGIA A. Retzius, *nomen conservandum*
T. alata W. Bojer *ex* J. Sims [**cultivated**]
T. fragrans W. Roxburgh [**cultivated**]
Sy = T. volubilis C. Persoon
T. grandiflora W. Roxburgh [**cultivated**]

YEATESIA J. K. Small
Y. platystegia (J. Torrey) R. Hilsenbeck
Sy = Tetramerium platystegium J. Torrey
Y. viridiflora (C. Nees von Esenbeck) J. K. Small
Sy = Dicliptera halei J. Riddell
Sy = D. viridiflora (C. Nees von Esenbeck) R. Long
Sy = Gatesia laetevirens (S. Buckley) A. Gray
Sy = Justicia laetevirens S. Buckley
Sy = Rhytiglossa viridiflora C. Nees von Esenbeck
Sy = Yeatesia laetevirens (S. Buckley) J. K. Small

ACERACEAE, *nomen conservandum*

ACER C. Linnaeus
A. barbatum A. Michaux
Sy = A. barbatum var. longii (M. Fernald) M. Fernald
Sy = A. barbatum var. villipes (A. Rehder) W. Ashe
Sy = A. floridanum (A. Chapman) F. Pax
Sy = A. floridanum var. longii M. Fernald
Sy = A. floridanum var. villipes A. Rehder
Sy = A. nigrum F. Michaux. var. floridanum (A. Chapman) F. Fosberg
Sy = A. saccharinum C. Linnaeus var. floridanum A. Chapman
Sy = A. saccharum H. Marshall subsp. floridanum (A. Chapman) Y. Desmarais
Sy = A. saccharum var. floridanum (A. Chapman) J. K. Small & A. A. Heller
Sy = Saccharodendron barbatum (A. Michaux) J. Nieuwland
Sy = S. floridanum (A. Chapman) J. Nieuwland
A. grandidentatum T. Nuttall var. ***sinuosum*** (A. Rehder) E. Little
Sy = A. brachypterum E. Wooton & P. Standley
Sy = A. grandidentatum var. bracypterum (E. Wooton & P. Standley) E. Palmer
Sy = A. saccharum subsp. brachypterum (E. Wooton & P. Standley) E. Murray
Sy = A. saccharum subsp. sinuosum (A. Rehder) C. Sargent
Sy = A. sinuosum A. Rehder
A. leucoderme J. K. Small
Sy = A. nigrum F. Michaux var. leucoderme (J. K. Small) F. Fosberg
Sy = A. saccharum H. Marshall subsp. leucoderme (J. K. Small) Y. Desmarais
Sy = A. saccharum var. leucoderme (J. K. Small) A. Rehder
Sy = Saccharodendron leucoderme (J. K. Small) J. Nieuwland
A. negundo C. Linnaeus var. ***negundo***
Sy = Negundo aceroides (C. Linnaeus) C. Moench
Sy = N. negundo (C. Linnaeus) G. Karsten
Sy = Rulac negundo (C. Linnaeus) A. Hitchcock
A. negundo C. Linnaeus var. ***texanum*** F. Pax
Sy = A. negundo subsp. latifolium (F. Pax) G. Schwerin
A. palmatum C. Thunberg [**cultivated**]
A. rubrum C. Linnaeus var. ***drummondii*** (W. Hooker & G. Arnott *ex* T. Nuttall) C. Sargent
Sy = A. drummondii (W. Hooker & G. Arnott *ex* T. Nuttall) C. Sargent
Sy = A. rubrum subsp. drummondii (W. Hooker & G. Arnott *ex* T. Nuttall) J. K. Small
A. rubrum C. Linnaeus var. ***rubrum***
Sy = A. rubrum var. tomentosum I. Tausch
Sy = A. stenocarpum N. Britton
Sy = Rufacer rubrum (C. Linnaeus) J. K. Small
A. rubrum C. Linnaeus var. ***trilobum*** J. Torrey & A. Gray *ex* K. Koch
Sy = A. carolinianum T. Walter

Sy = A. rubrum var. tridens A. Wood
Sy = Rufacer carolinianum (T. Walter) J. K. Small
A. saccharinum C. Linnaeus [**cultivated**]
Sy = A. dasycarpum F. Ehrhart
Sy = A. saccharinum var. laciniatum F. Pax
Sy = A. saccharinum var. wieri A. Rehder
Sy = Argentacer saccharinum (C. Linnaeus) J. K. Small
A. saccharum H. Marshall var. ***saccharum***
Sy = A. nigrum F. Michaux var. glaucum (F. Schmidt) F. Fosberg
Sy = A. nigrum var. saccharophorum (K. Koch) R. Clausen
Sy = A. saccharum var. glaucum (F. Schmidt) C. Sargent
Sy = Saccharodendron saccharum (H. Marshall) H. Moldenke

ACHATOCARPACEAE, *nomen conservandum*

PHAULOTHAMNUS A. Gray
P. spinescens A. Gray

AIZOACEAE, *nomen conservandum*

SESUVIUM C. Linnaeus
S. maritimum (T. Walter) N. Britton, E. Sterns, & J. Poggenburg
S. portulacastrum (C. Linnaeus) C. Linnaeus
Sy = Portulaca portulacastrum C. Linnaeus
S. sessile C. Persoon
* ***S. trianthemoides*** D. Correll [**TOES: III**]
S. verrucosum C. Rafinesque-Schmaltz
Sy = S. erectum D. Correll
Sy = S. portulacastrum (C. Linnaeus) C. Linnaeus var. subsessile A. Gray
Sy = S. sessile B. Robinson

TRIANTHEMA C. Linnaeus
T. portulacastrum C. Linnaeus
Sy = T. monogynum C. Linnaeus
Sy = T. procumbens P. Miller

AMARANTHACEAE, *nomen conservandum*

ACHYRANTHES C. Linnaeus
* ***A. aspera*** C. Linnaeus var. ***aspera*** [**TOES: III**]
Sy = A. aspera var. indica (C. Linnaeus) P. Miller
Sy = A. indica (C. Linnaeus) P. Miller
Sy = A. obtusifolia J. de Lamarck
Sy = Centrostachys aspera (C. Linnaeus) P. Standley
Sy = C. indica (C. Linnaeus) P. Standley

ALTERNANTHERA P. Forsskål
A. bettzichiana (E. von Regel) P. Standley [**cultivated**]
Sy = Achyranthes bettzichiana (E. von Regel) P. Standley
Sy = A. ficoidea (C. Linnaeus) J. de Lamarck subsp. bettzichiana (E. von Regel) J. G. Baker
Sy = Alternanthera tenella L. Colla var. versicolor (C. Lemaire) J. Veldkamp
Ma = A. tenella subsp. tenella *auct.*, non L. Colla: Hatch et al. 1990
A. caracasana K. Kunth
Sy = Achyranthes peploides (F. von Humboldt & A. Bonpland *ex* J. A. Schultes) N. Britton & Percy Wilson
Sy = Alternanthera peploides (F. von Humboldt & A. Bonpland *ex* J. A. Schultes) I. Urban
A. paranychioides A. de Saint-Hilaire var. ***paranychioides***
Ma = Alternanthera polygonoides *auct.*, non (C. Linnaeus) R. Brown *ex* R. Sweet
A. philoxeroides (K. von Martius) A. Grisebach
Sy = Achyranthes philoxeroides (K. von Martius) P. Standley
A. pungens K. Kunth
Sy = Achyranthes leiantha (M. A. Seubert) P. Standley
Sy = A. repens C. Linnaeus
Sy = Alternanthera achyrantha (C. Linnaeus) R. Brown *ex* R. Sweet
A. sessilis (C. Linnaeus) R. Brown *ex* A. P. de Candolle [**federal noxious weed**]

AMARANTHUS C. Linnaeus
A. acanthochiton J. Sauer
Sy = Acanthochiton wrightii J. Torrey
A. albus C. Linnaeus
Sy = A. albus var. pubescens (E. Uline & W. Bray) M. Fernald
Sy = A. gracecizans C. Linnaeus var. pubescens E. Uline & W. Bray
Sy = A. pubescens (E. Uline & W. Bray) P. Rydberg
Ma = A. gracecizans *auct.*, non C. Linnaeus
A. arenicola I. M. Johnston
A. australis (A. Gray) J. Sauer
Sy = Acinda alabamensis P. Standley
Sy = A. australis A. Gray
Sy = A. cuspidata C. L. Bertero *ex* K. Sprengel
A. bigelovii E. Uline & W. Bray
Sy = A. pringlei S. Watson
Ma = A. torreyi *auct.*, non (A. Gray) S. Watson

A. blitoides S. Watson
A. blitum C. Linnaeus
Sy = A. ascendens C. Linnaeus
Sy = A. lividus C. Linnaeus
Sy = A. lividus subsp. polygonoides (C. Moquin-Tandon) R. Probst
Sy = A. lividus var. polygonoides (C. Moquin-Tandon) A. Thellung
Ma = A. viridis *auct.*, non C. Linnaeus
A. californicus (C. Moquin-Tandon) S. Watson
Sy = A. albomarginatus E. Uline & W. Bray
Sy = A. microphyllus L. Shinners
A. chihuahuensis S. Watson
A. crassipes D. von Schlechtendal
A. cruentus C. Linnaeus
Sy = A. hybridus C. Linnaeus subsp. cruentus (C. Linnaeus) A. Thellung
Sy = A. hybridus var. cruentus (C. Linnaeus) C. Moquin-Tandon
Sy = A. hybridus var. paniculatus (C. Linnaeus) E. Uline & W. Bray
Sy = A. paniculatus C. Linnaeus
A. fimbriatus (J. Torrey) G. Bentham *ex* S. Watson var. ***fimbriatus***
Sy = Amblogyma fimbriata A. Gray
A. greggii S. Watson
Sy = A. annectens S. F. Blake
Sy = A. myrianthus P. Standley
A. hybridus C. Linnaeus
Sy = A. chlorostachys C. von Willdenow
Sy = A. incurvatus J. Grenier & D. Godron
Sy = A. patulus A. Bertolonii
A. hypochondriacus C. Linnaeus
Sy = A. caudatus C. Linnaeus
Sy = A. hybridus C. Linnaeus subsp. hypochondriacus (C. Linnaeus) A. Thellung
Sy = A. leucocarpus S. Watson
A. incurvatus J. Grenier & D. Godron **[cultivated]**
A. obcordatus (A. Gray) P. Standley
A. palmeri S. Watson
Sy = A. palmeri var. glomeratus E. Uline & W. Bray
A. polygonoides C. Linnaeus
Sy = A. berlandieri (C. Moquin-Tandon) E. Uline & W. Bray
Sy = A. polygonoides subsp. berlandieri (C. Moquin-Tandon) A. Thellung
A. powellii S. Watson
Sy = A. bouchonii A. Thellung
Sy = A. bracteosus E. Uline & W. Bray
Sy = A. retroflexus C. Linnaeus var. powellii (S. Watson) J. Boivin
A. retroflexus C. Linnaeus
Sy = A. retroflexus var. salicifolius I. M. Johnston
A. rudis J. Sauer
Ma = Acnida tamariscinus *auct.*, non (T. Nuttall) A. Wood
Ma = Amaranthus tamariscina *auct.*, non T. Nuttall
A. scleropoides E. Uline & W. Bray
A. spinosus C. Linnaeus
A. wrightii S. Watson

BLUTAPARON C. Rafinesque-Schmaltz
B. vermiculare (C. Linnaeus) J. Mears
Sy = Caraxeron vermiculare (C. Linnaeus) C. Rafinesque-Schmaltz
Sy = Cruzeta vermicularis (C. Linnaeus) G. Maza
Sy = Lithophila vermicularis (C. Linnaeus) E. Uline *ex* C. Millspaugh
Sy = Philoxerus vermicularis (C. Linnaeus) R. Brown *ex* J. E. Smith

CELOSIA C. Linnaeus
C. argentea C. Linnaeus
Sy = C. margaritacea C. Linnaeus
C. cristata C. Linnaeus **[cultivated]**
Sy = C. argentea C. Linnaeus var. cristata (C. Linnaeus) K. E. O. Kuntze
C. nitida M. H. Vahl
Sy = C. texana G. Scheele
C. palmeri S. Watson

FROELICHIA C. Moench
F. arizonica J. Thornber *ex* P. Standley
F. drummondii C. Moquin-Tandon
F. floridana (T. Nuttall) C. Moquin-Tandon var. ***campestris*** (J. K. Small) M. Fernald
Sy = F. campestris J. K. Small
F. floridana (T. Nuttall) C. Moquin-Tandon var. ***floridana***
F. gracilis (W. Hooker) C. Moquin-Tandon
Sy = F. braunii P. Standley
Sy = Oplotheca gracilis W. Hooker
F. interrupta (C. Linnaeus) C. Moquin-Tandon
Sy = F. texana J. Coulter & E. Fisher
Sy = Gomphrena interrupta C. Linnaeus

GOMPHRENA C. Linnaeus
G. caespitosa J. Torrey
Sy = G. viridis E. Wooton & P. Standley
Sy = Xeraea caespitosa K. E. O. Kuntze
G. globosa C. Linnaeus **[cultivated]**
G. haagenna J. Klotzsch
Sy = G. tuberifera J. Torrey

G. nealleyi J. Coulter & E. Fisher
G. nitida J. Rothrock
Sy = Xeraea nitida K. E. O. Kuntze
G. serrata C. Linnaeus
Sy = G. decumbens N. von Jacquin
Sy = G. dispersa P. Standley

GOSSYPIANTHUS W. Hooker
G. lanuginosus (J. Poiret) C. Moquin-Tandon var. ***lanuginosus***
Sy = G. lanuginosus var. sheldonii E. Uline & W. Bray
Sy = G. rigidiflorus W. Hooker
Sy = G. sheldonii (E. Uline & W. Bray) J. K. Small
Sy = Guilleminea lanuginosa (J. Poiret) J. Hooker
Sy = G. lanuginosa var. rigidiflora (W. Hooker) J. Mears
Sy = G. lanuginosa var. sheldonii (E. Uline & W. Bray) J. Mears
G. lanuginosus (J. Poiret) C. Moquin-Tandon var. ***tenuiflorus*** (W. Hooker) J. Mears *ex* J. Henrickson
Sy = G. tenuiflorus W. Hooker
Sy = Guilleminea lanuginosa var. tenuiflora (W. Hooker) J. Mears

GUILLEMINEA K. Kunth
G. densa (F. von Humboldt & A. Bonpland *ex* C. von Willdenow) C. Moquin-Tandon var. ***aggregata*** E. Uline & W. Bray
G. densa (F. von Humboldt & A. Bonpland *ex* C. von Willdenow) C. Moquin-Tandon var. ***densa***
Sy = Brayulinea densa (F. von Humboldt & A. Bonpland *ex* C. von Willdenow) J. K. Small
Sy = Illecebrum densum F. von Humboldt & A. Bonpland *ex* C. von Willdenow

IRESINE P. Browne, *nomen conservandum*
I. diffusa F. von Humboldt & A. Bonpland *ex* C. von Willdenow
Sy = I. canescens F. von Humboldt & A. Bonpland *ex* C. von Willdenow
Sy = I. celosia C. Linnaeus
Sy = I. celosioides C. Linnaeus
Sy = I. elongata F. von Humboldt & A. Bonpland *ex* C. von Willdenow
I. heterophylla P. Standley
Sy = I. celosioides C. Linnaeus var. obtusifolia J. Coulter
Sy = I. paniculata (C. Linnaeus) J. Poiret var. obtusifolia J. Coulter *ex* E. Uline & W. Bray
I. leptoclada (J. Hooker) J. Henrickson & S. Sundberg
Sy = Dicraurus leptocladus J. Hooker
I. palmeri (S. Watson) P. Standley
Sy = Hebanthe palmeri S. Watson
I. rhizomatosa P. Standley
Ma = I. celosioides *auct.*, non C. Linnaeus

TIDESTROMIA P. Standley
T. carnosa (J. Steyermark) I. M. Johnston
Sy = T. lanuginosa (T. Nuttall) P. Standley var. carnosa (J. Steyermark) V. Cory
T. gemmata I. M. Johnston
T. lanuginosa (T. Nuttall) P. Standley
Sy = Achyranthes lanuginosa T. Nuttall
Sy = Alternanthera lanuginosa C. Moquin-Tandon
Sy = Cladothrix lanuginosa T. Nuttall *ex* C. Moquin-Tandon
T. oblongifolia (S. Watson) P. Standley
Sy = Cladothrix oblongifolia S. Watson
T. suffruticosa (J. Torrey) P. Standley
Sy = Alternanthera suffruticosa J. Torrey
Sy = Cladothrix suffruticosa (J. Torrey) G. Bentham & J. Hooker

ANACARDIACEAE, *nomen conservandum*

COTINUS P. Miller
C. coggygria J. Scopoli [**cultivated**]
Sy = Rhus cotinus C. Linnaeus
C. obovatus C. Rafinesque-Schmaltz
Sy = C. americanus T. Nuttall

PISTACIA C. Linnaeus
P. chinensis A. von Bunge [**cultivated**]
P. mexicana K. Kunth
Sy = P. texana W. Swingle

RHUS C. Linnaeus
R. aromatica W. Aiton var. ***serotina*** (E. Greene) A. Rehder
Sy = R. aromatica subsp. serotina (E. Greene) R. Brooks
Sy = R. trilobata T. Nuttall var. serotina (E. Greene) F. Barkley
Sy = Schmaltzia serotina E. Greene
R. copallinum C. Linnaeus var. ***latifolia*** H. Engler
R. glabra C. Linnaeus
Sy = R. borealis E. Greene

Sy = R. calophylla E. Greene
Sy = R. cismontana E. Greene
Sy = R. glabra var. cismontana (E. Greene) F. Daniels
Sy = R. glabra var. laciniata É. Carrière
Sy = R. glabra var. occidentalis J. Torrey

R. lanceolata (A. Gray) N. Britton
Sy = R. copallinum C. Linnaeus var. lanceolata A. Gray
Sy = Schmaltzia lanceolata (A. Gray) J. K. Small

R. microphylla G. Engelmann *ex* A. Gray
Sy = Schmaltzia microphylla (G. Engelmann) J. K. Small

R. trilobata T. Nuttall var. ***pilosissima*** G. Engelmann
Sy = R. aromatica subsp. pilosissima (G. Engelmann) W. A. Weber
Sy = R. aromatica var. mollis W. Ashe
Sy = R. aromatica var. pilosissima (G. Engelmann) L. Shinners
Sy = R. trilobata var. malacophylla (E. Greene) P. Munz
Sy = Schmaltzia emoryi E. Greene
Sy = S. trilobata var. pilosissima (G. Engelmann) F. Barkley

R. trilobata T. Nuttall var. ***trilobata***
Sy = R. aromatica W. Aiton subsp. flabelliformis (L. Shinners) R. Brooks
Sy = R. aromatica var. flabelliformis L. Shinners
Sy = R. aromatica var. trilobata (T. Nuttall) A. Gray
Sy = Schmaltzia trilobata (T. Nuttall) J. K. Small

R. virens F. Lindheimer *ex* A. Gray var. ***choriophylla*** (E. Wooton & P. Standley) L. Benson
Sy = R. choriophylla E. Wooton & P. Standley
Sy = Schmaltzia choriophylla (E. Wooton & P. Standley) F. Barkley

R. virens F. Lindheimer *ex* A. Gray var. ***virens***
Sy = R. virens subsp. virens

SCHINUS C. Linnaeus

S. longifolius (J. Lindley) C. Spegazzini **[cultivated]**

S. molle C. Linnaeus **[cultivated]**

S. terebinthifolius G. Raddi var. ***raddianus*** G. Engelmann **[cultivated]**

S. terebinthifolius G. Raddi var. ***rhoifolius*** (K. von Martius) G. Engelmann **[cultivated]**

TOXICODENDRON P. Miller

T. diversilobum (J. Torrey & A. Gray) E. Greene var. ***pubescens*** P. Miller
Sy = Rhus diversiloba J. Torrey & A. Gray
Sy = R. toxicarium R. A. Salisbury
Sy = R. toxicodendron C. Linnaeus
Sy = Toxicodendron quercifolium (A. Michaux) E. Greene
Sy = T. toxicarium (R. A. Salisbury) W. Gillis
Sy = T. toxicodendron (C. Linnaeus) N. Britton

T. radicans (C. Linnaeus) K. E. O. Kuntze subsp. ***eximium*** (E. Greene) W. Gillis
Sy = Rhus toxicodendron C. Linnaeus var. eximia (E. Greene) J. McNair
Sy = Toxicodendron radicans var. eximium (E. Greene) F. Barkley

T. radicans (C. Linnaeus) K. E. O. Kuntze subsp. ***negundo*** (E. Greene) W. Gillis
Sy = T. radicans var. negundo (E. Greene) J. Reveal

T. radicans (C. Linnaeus) K. E. O. Kuntze subsp. ***pubens*** (G. Engelmann *ex* S. Watson) W. Gillis
Sy = T. radicans var. pubens (G. Engelmann *ex* S. Watson) J. Reveal

T. radicans (C. Linnaeus) K. E. O. Kuntze subsp. ***radicans***
Sy = Rhus radicans C. Linnaeus
Sy = R. radicans var. littoralis (E. Mearns) C. Deam
Sy = R. radicans var. malacotrichcarpa (A. Moore) M. Fernald
Sy = Toxicodendron vulgare P. Miller

T. radicans (C. Linnaeus) K. E. O. Kuntze subsp. ***verrucosum*** (G. Scheele) W. Gillis
Sy = Rhus verrucosa G. Scheele
Sy = Toxicodendron radicans var. verrucosum (G. Scheele) F. Barkley

T. rydbergii (J. K. Small *ex* P. Rydberg) E. Greene
Sy = Rhus radicans C. Linnaeus var. rydbergii (J. K. Small *ex* P. Rydberg) A. Rehder
Sy = R. radicans var. vulgaris (A. Michaux) A. P. de Candolle
Sy = R. toxicodendron C. Linnaeus var. vulgaris A. Michaux
Sy = Toxicodendron desertorum J. Lunell
Sy = T. radicans (C. Linnaeus) K. E. O. Kuntze var. rydbergii (J. K. Small *ex* P. Rydberg) D. S. Erskine

T. vernix (C. Linnaeus) K. E. O. Kuntze
Sy = Rhus vernix C. Linnaeus

ANNONACEAE, *nomen conservandum*

ASIMINA M. Adanson

A. parviflora (A. Michaux) M. Dunal

A. triloba (C. Linnaeus) M. Dunal

APIACEAE, *nomen conservandum*

ALETES J. Coulter & J. Rose
A. acaulis (J. Torrey) J. Coulter & J. Rose
A. filifolius M. Mathias, L. Constance, & W. Theobald

AMMI C. Linnaeus
A. majus C. Linnaeus
A. visnaga (C. Linnaeus) J. de Lamarck
Sy = Visnaga daucoides J. Gaertner

AMMOSELINUM J. Torrey & A. Gray
A. butleri (G. Engelmann *ex* S. Watson) J. Coulter & J. Rose
A. popei J. Torrey & A. Gray

ANETHUM C. Linnaeus
A. graveolens C. Linnaeus [**cultivated**]

ANTHRISCUS C. Persoon, *nomen conservandum*
A. cerefolium (C. Linnaeus) G. F. Hoffmann [**cultivated**]
Sy = Cerefolium cerefolium (C. Linnaeus) H. Schinz & A. Thellung

APIUM C. Linnaeus
A. graveolens C. Linnaeus var. **dulce** (P. Miller) A. P. de Candolle [**cultivated**]

BERULA C. Bessey *ex* W. Koch
B. erecta (W. Hudson) F. Coville
Sy = B. erecta var. incisa (J. Torrey) A. Cronquist
Sy = B. incisa (J. Torrey) G. N. Jones
Sy = B. pusilla M. Fernald
Sy = Siella erects (W. Hudson) M. Pimenov
Sy = Sium erectum W. Hudson

BIFORA G. F. Hoffmann, *nomen conservandum*
B. americana G. Bentham & W. Hooker *ex* S. Watson

BOWLESIA H. Ruiz López & J. Pavón
B. incana H. Ruiz López & J. Pavón
Sy = B. septentrionalis J. Coulter & J. Rose

BUPLEURUM C. Linnaeus
B. lancifolium J. Hornemann
Sy = B. protractum G. F. Hoffmann & J. Link
Sy = B. subovatum J. Link
B. rotundifolium C. Linnaeus

CENTELLA C. Linnaeus
C. asiatica (C. Linnaeus) I. Urban

CHAEROPHYLLUM C. Linnaeus
C. tainturieri W. Hooker var. **dasycarpum** W. Hooker *ex* S. Watson
Sy = C. dasycarpum (W. Hooker *ex* S. Watson) T. Nuttall *ex* J. K. Small
C. tainturieri W. Hooker var. **tainturieri**
Sy = C. procumbens (C. Linnaeus) H. von Crantz var. tainturieri (W. Hooker) J. Coulter & J. Rose
Sy = C. tainturieri var. floridanum J. Coulter & J. Rose
Sy = C. texanum J. Coulter & J. Rose

CICLOSPERMUM M. Lagasca y Segura
C. leptophyllum (C. Persoon) T. A. Sprague *ex* N. Britton & Percy Wilson
Sy = Apium leptophyllum (C. Persoon) F. von Mueller *ex* G. Bentham
Sy = A. tenuifolium (C. Moench) A. Thellung *ex* G. Hegi

CICUTA C. Linnaeus
C. maculata C. Linnaeus var. **maculata**
Sy = C. curtissii J. Coulter & J. Rose
Sy = C. maculata var. curtissii (J. Coulter & J. Rose) M. Fernald
Sy = C. mexicana J. Coulter & J. Rose

CONIUM C. Linnaeus
C. maculatum C. Linnaeus

CORIANDRUM C. Linnaeus
C. sativum C. Linnaeus [**cultivated**]

CRYPTOTAENIA A. P. de Candolle, *nomen conservandum*
C. canadensis (C. Linnaeus) A. P. de Candolle
Sy = Deringa canadensis (C. Linnaeus) K. E. O. Kuntze

CUMINUM C. Linnaeus
C. cyminum C. Linnaeus

CYMOPTERUS C. Rafinesque-Schmaltz
C. acaulis (F. Pursh) C. Rafinesque-Schmaltz var. **fendleri** (A. Gray) S. Goodrich
Sy = C. fendleri A. Gray
C. bulbosus A. Nelson

Sy = Phellopterus bulbosus (A. Nelson) J. Coulter & J. Rose
Sy = P. utahensis (M. E. Jones) E. Wooton & P. Standley
C. macrorhizus S. Buckley
Sy = Phellopterus macrorhizus (S. Buckley) J. Coulter & J. Rose
C. montanus T. Nuttall *ex* J. Torrey & A. Gray
Sy = Phellopterus montanus T. Nuttall *ex* J. Torrey & A. Gray
C. multinervatus (J. Coulter & J. Rose) I. Tidestrom
Sy = Phellopterus multinervatus J. Coulter & J. Rose

CYNOSCIADIUM A. P. de Candolle
C. digitatum A. P. de Candolle

DAUCOSMA G. Engelmann & A. Gray *ex* A. Gray
D. laciniatum G. Engelmann & A. Gray
Sy = Ptilimnium laciniatum (G. Engelmann & A. Gray) K. E. O. Kuntze

DAUCUS C. Linnaeus
D. carota C. Linnaeus
D. pusillus A. Michaux

ERYNGIUM C. Linnaeus
E. diffusum J. Torrey
E. heterophyllum G. Engelmann
Sy = E. wrightii A. Gray
E. hookeri W. Walpers
E. integrifolium T. Walter
Sy = E. ludovicianum T. Morong
E. leavenworthii J. Torrey & A. Gray
E. nasturtiifolium A. L. de Jussieu *ex* F. Delaroche
E. prostratum T. Nuttall *ex* A. P. de Candolle
Sy = E. prostratum var. disjunctum M. Fernald
E. yuccifolium A. Michaux
Sy = E. synchaetum (J. Coulter & J. Rose) J. Coulter & J. Rose
Sy = E. yuccifolium var. synchaetum A. Gray *ex* J. Coulter & J. Rose

EURYTAENIA J. Torrey & A. Gray
E. hinckleyi M. Mathias & L. Constance
E. texana J. Torrey & A. Gray

FOENICULUM P. Miller
F. vulgare P. Miller [**cultivated**]
Sy = Falcaria sioides (A. W. Wibel) P. Ashcerson
Sy = Foeniculum foeniculum (C. Linnaeus) G. Karsten

HYDROCOTYLE C. Linnaeus
H. bonariensis P. Commerson *ex* J. de Lamarck
H. ranunculoides C. von Linné
H. umbellata C. Linnaeus
H. verticillata C. Thunberg var. ***triradiata*** (A. Richard) M. Fernald
Sy = H. australis J. Coulter & J. Rose
Sy = H. canbyi J. Coulter & J. Rose
Sy = H. prolifera A. Kellogg
H. verticillata C. Thunberg var. ***verticillata***

LILAEOPSIS E. Greene
L. chinensis (C. Linnaeus) K. E. O. Kuntze
Sy = L. lineata (A. Michaux) E. Greene

LIMNOSCIADIUM M. Mathias & L. Constance
L. pinnatum (A. P. de Candolle) M. Mathias & L. Constance
Sy = Cynosciadium pinnatum A. P. de Candolle
L. pumilum (G. Engelmann & A. Gray) M. Mathias & L. Constance
Sy = Cynosciadium pumilum (G. Engelmann & A. Gray) J. Coulter & J. Rose

LOMATIUM C. Rafinesque-Schmaltz
L. foeniculaceum (T. Nuttall) J. Coulter & J. Rose subsp. ***daucifolium*** (J. Torrey & A. Gray) W. Theobald
Sy = Cogswellia daucifolia (J. Torrey & A. Gray) M. E. Jones
Sy = Lomatium daucifolium (J. Torrey & A. Gray) J. Coulter & J. Rose
Sy = L. foeniculaceum var. daucifolium (J. Torrey & A. Gray) A. Cronquist

OSMORHIZA C. Rafinesque-Schmaltz, *nomen conservandum*
O. longistylis (J. Torrey) A. P. de Candolle
Sy = O. aristata (C. Thunberg) T. Makino & Y. Yabe var. longistylis (J. Torrey) J. Boivin
Sy = O. longistylis (J. Torrey) A. P. de Candolle var. brachycoma S. F. Blake
Sy = O. longistylis var. villicaulis M. Fernald
Sy = Washingtonia longistylis (J. Torrey) N. Britton
* ***O. mexicana*** A. Grisebach subsp. ***bipatriata*** (L. Constance & Ren-hwa Shan) [**TOES: III**]
Sy = O. bipatriata L. Constance & Ren-hwa Shan

OXYPOLIS C. Rafinesque-Schmaltz
O. filiformis (T. Walter) N. Britton
O. rigidior (C. Linnaeus) C. Rafinesque-Schmaltz

Sy = O. longifolia (F. Pursh) J. K. Small
Sy = O. rigidior var. ambigua (T. Nuttall) B. Robinson
Sy = O. rigidior var. longifolia (F. Pursh) N. Britton
Sy = O. turgida J. K. Small
O. ternata (T. Nuttall) A. A. Heller

PASTINACA C. Linnaeus
P. sativa C. Linnaeus var. ***sativa*** [**cultivated**]

PETROSELINUM J. Hill
P. crispum (P. Miller) C. Nyman *ex* A. Hill [**cultivated**]
Sy = Apium petroselinum C. Linnaeus

POLYTAENIA A. P. de Candolle
P. nuttallii A. P. de Candolle
Sy = Pleiotaenia nuttallii (A. P. de Candolle) J. Coulter & J. Rose
P. texana (J. Coulter & J. Rose) M. Mathias & L. Constance
Sy = Pleiotaenia nuttallii (A. P. de Candolle) J. Coulter & J. Rose var. texana J. Coulter & J. Rose
Sy = Polytaenia nuttallii A. P. de Candolle var. texana J. Coulter & J. Rose

PSEUDOCYMOPTERUS J. Coulter & J. Rose
P. longiradiatus M. Mathias, L. Constance, & W. Theobald
P. montanus (A. Gray) J. Coulter & J. Rose
Sy = Cymopterus lemmonii (J. Coulter & J. Rose) R. Dorn
Sy = Pseudocymopterus tidestromii J. Coulter & J. Rose
Sy = Thaspium montanum A. Gray

PTILIMNIUM C. Rafinesque-Schmaltz
P. capillaceum (A. Michaux) C. Rafinesque-Schmaltz
P. costatum (S. Elliott) C. Rafinesque-Schmaltz
P. nuttallii (A. P. de Candolle) N. Britton
Sy = Discopleura nuttallii A. P. de Candolle
P. × texense J. Coulter & J. Rose [***capillaceum* × *nuttallii***]

SANICULA C. Linnaeus
S. canadensis C. Linnaeus var. ***canadensis***
Sy = S. canadensis var. floridana (E. Bicknell) H. Wolff
Sy = S. floridana E. Bicknell
S. canadensis C. Linnaeus var. ***grandis*** M. Fernald
S. odorata (C. Rafinesque-Schmaltz) K. Pryer & L. Phillippe
Sy = S. gregaria E. Bicknell
Sy = Triclinium odoratum C. Rafinesque-Schmaltz
S. smallii E. Bicknell

SCANDIX C. Linnaeus
S. pecten-veneris C. Linnaeus

SIUM C. Linnaeus
S. suave T. Walter
Sy = S. cicutaefolium F. von Schrank
Sy = S. floridanum J. K. Small
Sy = S. suave var. floridanum (J. K. Small) C. F. Reed

SPERMOLEPIS C. Rafinesque-Schmaltz
S. divaricata (T. Walter) C. Rafinesque-Schmaltz *ex* N. Séringe
Sy = Daucus divaricata T. Walter
S. echinata (T. Nuttall *ex* A. P. de Candolle) A. A. Heller
Sy = Apium echinatum (T. Nuttall *ex* A. P. de Candolle) G. Bentham & J. Hooker *ex* S. Watson
Sy = Leptocaulis echinatus T. Nuttall *ex* A. P. de Candolle
S. inermis (T. Nuttall *ex* A. P. de Candolle) M. Mathias & L. Constance
Sy = S. patens (T. Nuttall *ex* A. P. de Candolle) B. Robinson

TAENIDIA (J. Torrey & A. Gray) C. Drude
T. integerrima (C. Linnaeus) C. Drude

TAUSCHIA D. von Schlechtendal, *nomen conservandum*
T. texana A. Gray

THASPIUM T. Nuttall
T. barbinode (A. Michaux) T. Nuttall
Sy = T. barbinode var. angustifolium J. Coulter & J. Rose
Sy = T. chapmanii (J. Coulter & J. Rose) J. K. Small
T. trifoliatum (C. Linnaeus) A. Gray
Sy = T. aureum (C. Linnaeus) N. Britton var. trifoliatum (C. Linnaeus) J. Coulter & J. Rose

TORILIS M. Adanson
T. arvensis (W. Hudson) J. Link
T. nodosa (C. Linnaeus) J. Gaertner

TREPOCARPUS T. Nuttall *ex* A. P. de Candolle
T. aethusae T. Nuttall *ex* A. P. de Candolle

ZIZIA W. Koch
Z. aurea (C. Linnaeus) W. Koch

APOCYNACEAE, *nomen conservandum*

ALLAMANDA C. Linnaeus
A. cathartica C. Linnaeus [**cultivated**]
Sy = A. cathartica var. grandiflora (J. Aublet) L. Bailey & C. Raffill
Sy = A. schottii J. Pohl
A. neriifolia W. Hooker [**cultivated**]

AMSONIA T. Walter
A. ciliata T. Walter var. ***tenuifolia*** (C. Rafinesque-Schmaltz) R. Woodson
Sy = A. ciliata var. filifolia R. Woodson
A. ciliata T. Walter var. ***texana*** (A. Gray) J. Coulter
Sy = A. texana (A. Gray) A. A. Heller
A. illustris R. Woodson
A. longiflora J. Torrey var. ***longiflora***
A. longiflora J. Torrey var. ***salpignantha*** (R. Woodson) S. McLaughlin
Sy = A. salpignantha R. Woodson
A. palmeri A. Gray
Sy = A. hirtella P. Standley
Sy = A. hirtella var. pogonosepala (R. Woodson) I. L. Wiggins
Sy = A. pogonosepala R. Woodson
Sy = A. standleyi R. Woodson
A. repens L. Shinners
A. tabernaemontana T. Walter var. ***salicifolia*** (F. Pursh) R. Woodson
Sy = A. salicifolia (F. Pursh) R. Woodson
Sy = A. tabernaemontana var. gattingeri R. Woodson
A. tabernaemontana T. Walter var. ***tabernaemontana***
Sy = A. amsonia (C. Linnaeus) N. Britton
Sy = A. glaberrima R. Woodson
* ***A. tharpii*** R. Woodson [**TOES: IV**]
A. tomentosa J. Torrey & J. Frémont var. ***stenophylla*** T. Kearney & R. Peebles
Sy = A. arenaria P. Standley

APOCYNUM C. Linnaeus
A. androsaemifolium C. Linnaeus subsp. ***androsaemifolium***
Sy = A. ambigens E. Greene
Sy = A. androsaemifolium var. glabrum J. Macoun
Sy = A. androsaemifolium var. griseum (E. Greene) A. Béguinot & R. N. Belosersky
Sy = A. androsaemifolium var. incanum A. P. de Candolle
Sy = A. pumilum (A. Gray) E. Greene var. rhomboideum (E. Greene) A. Béguinot & R. N. Belosersky
Sy = A. scopulorum E. Greene *ex* P. Rydberg
A. cannabinum C. Linnaeus
Sy = A. cannabinum var. angustifolium (E. Wooton) N. Holmgren
Sy = A. cannabinum var. glaberrimum A. P. de Candolle
Sy = A. cannabinum var. greeneanum (A. Béguinot & R. N. Belosersky) A. Woods
Sy = A. cannabinum var. nemorale (G. Miller) M. Fernald
Sy = A. cannabinum var. pubescens J. Mitchell *ex* R. Brown
Sy = A. cannabinum var. suksdorfii (E. Greene) A. Béguinot & R. N. Belosersky
A. sibiricum N. von Jacquin
Sy = A. cannabinum C. Linnaeus var. hypericifolium A. Gray
Sy = A. hypericifolium W. Aiton
Sy = A. salignum E. Greene
Sy = Cynopaema hypericifolium J. Lunell
A. × floribundum E. Greene [***androsaemifolium* × *cannabinum***]
Sy = A. abditum E. Greene
Sy = A. cinereum A. A. Heller
Sy = A. lividum E. Greene
Sy = A. medium E. Greene var. floribundum (E. Greene) R. Woodson

BEAUMONTIA N. Wallich
B. grandiflora N. Wallich [**cultivated**]

CARISSA C. Linnaeus, *nomen conservandum*
C. macrocarpa (C. Ecklon) A. P. de Candolle [**cultivated**]
Sy = C. grandiflora (E. Meyer) A. P. de Candolle

CATHARANTHUS G. Don
C. roseus (C. Linnaeus) G. Don [**cultivated**]
Sy = Ammocallis rosea (C. Linnaeus) J. K. Small

Sy = Lochnera rosea (C. Linnaeus) H. G. L. Reichenbach *ex* E. Spach
Sy = Vinca rosea C. Linnaeus

DAMNACANTHUS K. von Gaertner
D. indicus K. von Gaertner [**cultivated**]
Sy = Carissa spinarum C. Loddiges *ex* A. L. de Candolle

HAPLOPHYTON A. P. de Candolle
H. cimicidum A. P. de Candolle
Sy = H. cimicidum var. crooksii L. Benson
Sy = H. crooksii (L. Benson) L. Benson

MACROSIPHONIA J. Müller of Aargau
M. hypoleuca (G. Bentham) J. Müller of Aargau
Sy = Echites hypoleuca G. Bentham
M. lanuginosa (M. Martens & H. Galeotti) W. Hemsley var. ***macrosiphon*** (J. Torrey) J. Henrickson
Sy = M. macrosiphon (J. Torrey) A. A. Heller

MANDEVILLA J. Lindley
M. boliviensis (J. Hooker) R. Woodson [**cultivated**]
M. splendens (J. Hooker) R. Woodson [**cultivated**]
M. × ***amabilis*** J. Backhouse & J. Backhouse f. [***splendens*** × unknown] [**cultivated**]

NERIUM C. Linnaeus
N. oleander C. Linnaeus [**cultivated**]

PLUMERIA C. Linnaeus
P. alba C. Linnaeus [**cultivated**]
P. rubra C. Linnaeus [**cultivated**]

TABERNAEMONTANA C. Plumier *ex* C. Linnaeus
T. africana A. P. de Candolle [**cultivated**]
Sy = T. grandiflora W. Hooker
T. divaricata (C. Linnaeus) R. Brown *ex* J. J. Römer & J. A. Schultes [**cultivated**]
Sy = Ervatamia coronaria (L.) O. Stapf
Sy = Tabernaemontana coronaria (C. Linnaeus) C. von Willdenow

THEVETIA M. Adanson, *nomen conservandum*
T. peruviana (C. Persoon) K. Schumann [**cultivated**]
Sy = Cerbera peruviana C. Persoon
Sy = C. thevetia C. Linnaeus
Sy = Thevetia neriifolia A. L. de Jussieu *ex* E. von Steudel

TRACHELOSPERMUM C. Lemaire
T. asiaticum (P. von Siebold & J. Zuccarini) T. Nakai [**cultivated**]
Sy = Malouetia asiatica P. von Siebold & J. Zuccarini
T. difforme (T. Walter) A. Gray
T. jasminoides (J. Lindley) C. Lemaire [**cultivated**]
Sy = Rhynchospermum jasminoides J. Lindley

VINCA C. Linnaeus
V. major C. Linnaeus [**cultivated**]
Sy = V. major var. variegata J. Loudon
V. minor C. Linnaeus [**cultivated**]

AQUIFOLIACEAE, *nomen conservandum*

ILEX C. Linnaeus
I. ambigua (A. Michaux) J. Torrey
Sy = I. beadlei W. Ash
Sy = I. buswellii J. K. Small
Sy = I. caroliniana W. Trelease *ex* J. K. Small
Sy = I. montana J. Torrey & A. Gray *ex* A. Gray var. beadlei (W. Ashe) M. Fernald
I. aquifolium C. Linnaeus [**cultivated**]
I. cassine C. Linnaeus var. ***latifolia*** W. Aiton
I. coriacea (F. Pursh) A. Chapman
I. cornuta J. Lindley & J. Paxton [**cultivated**]
I. crenata C. Thunberg [**cultivated**]
Sy = Celastrus adenophylla F. Miquel
I. decidua T. Walter
Sy = I. curtissii (M. Fernald) J. K. Small
Sy = I. decidua var. curtissii M. Fernald
I. glabra (C. Linnaeus) A. Gray
I. longipes A. Chapman *ex* W. Trelease
Sy = I. decidua T. Walter var. longipes (A. Chapman *ex* W. Trelease) H. Ahles
Sy = I. longipes var. hispuda C. Lundell
I. montana J. Torrey & A. Gray *ex* A. Gray
Sy = I. ambigua (A. Michaux) J. Torrey var. montana (J. Torrey & A. Gray *ex* A. Gray) H. Ahles
Sy = I. ambigua var. monticola (A. Wood) R. Wunderlin & J. Poppleton
Sy = I. amelanchier M. Curtis *ex* A. Chapman var. monticola A. Wood
Sy = I. monticola A. Gray
I. myrtifolia T. Walter
Sy = I. cassine C. Linnaeus var. myrtifolia (T. Walter) C. Sargent
I. opaca D. Solander var. ***opaca***
I. verticillata (C. Linnaeus) A. Gray
Sy = I. bronxensis N. Britton

Sy = I. fastigiata E. Bicknell
Sy = I. verticillata var. cyclophylla B. Robinson
Sy = I. verticillata var. fastigiata (E. Bicknell) M. Fernald
Sy = I. verticillata var. padifolia (C. von Willdenow) J. Torrey & A. Gray *ex* S. Watson
Sy = I. verticillata var. tenuifolia (J. Torrey) S. Watson
I. vomitoria W. Aiton
I. × ***attentuata*** W. Ashe [***cassine*** × ***opaca***] [**cultivated**]

ARALIACEAE, *nomen conservandum*

ARALIA C. Linnaeus
A. racemosa C. Linnaeus subsp. ***bicrenata*** (E. Wooton & P. Standley) S. Welsh & N. Atwood
Sy = A. bicrenata E. Wooton & P. Standley
A. spinosa C. Linnaeus

DIZGOTHECA N. E. Brown
D. elegantissima N. E. Brown [**cultivated**]

× **FATSHEDERA** A. Guillaumin [***Fatsia*** × ***Hedera***]
× ***F. lizei*** (P. Cochet) A. Guillaumin [***Fatsia japonica*** cv. "Moseri" × ***Hedera helix hibernica***] [**cultivated**]

FATSIA J. Decaisne & J. Planchon
F. japonica (C. Thunberg) J. Decaisne & J. Planchon [**cultivated**]
Sy = Aralia japonica C. Thunberg
Sy = A. sieboldii K. Koch

HEDERA C. Linnaeus
H. canariensis C. von Willdenow [**cultivated**]
H. colchica (K. Koch) K. Koch [**cultivated**]
H. helix C. Linnaeus var. ***helix*** [**cultivated**]

TETRAPANAX (K. Koch) K. Koch
T. papyriferus (W. Hooker) K. Koch [**cultivated**]
Sy = Aralia papyrifera W. Hooker

ARISTOLOCHIACEAE, *nomen conservandum*

ARISTOLOCHIA C. Linnaeus
A. coryi I. M. Johnston
Sy = A. brevipes var. acuminata S. Watson
A. erecta C. Linnaeus
Sy = A. longiflora G. Engelmann & A. Gray
A. littoralis L. Parodi [**cultivated**]
Sy = A. elegans M. T. Masters
A. macrophylla J. de Lamarck [**cultivated**]
Sy = A. durior J. Hill
Sy = Isotrema macrophyllum (J. Hill) C. F. Reed
A. pentandra N. von Jacquin
Sy = A. marshii P. Standley
A. reticulata N. von Jacquin
A. serpentaria C. Linnaeus
Sy = A. convolvulacea J. K. Small
Sy = A. hastata T. Nuttall
Sy = A. nashii T. Kearney
Sy = A. serpentaria var. hastata (T. Nuttall) P. Duchartre
Sy = A. serpentaria var. nashii (T. Kearney) H. Ahles
Sy = Endodeca serpentaria (C. Linnaeus) C. Rafinesque-Schmaltz var. hastata (T. Nuttall) C. F. Reed
A. tomentosa J. Sims
Sy = Isotrema tomentosa (J. Sims) H. Huber
A. wrightii B. Seemann

ASCLEPIADACEAE, *nomen conservandum*

ASCLEPIAS C. Linnaeus
A. amplexicaulis J. E. Smith
A. arenaria J. Torrey
A. asperula (J. Decaisne) R. Woodson subsp. ***asperula***
A. asperula (J. Decaisne) R. Woodson subsp. ***capricornu*** (R. Woodson) R. Woodson
Sy = A. asperula var. decumbens (T. Nuttall) L. Shinners
Sy = A. capricornu R. Woodson
Sy = A. capricornu subsp. occidentalis R. Woodson
Sy = A. decumbens (T. Nuttall) J. Decaisne
Sy = Ascerates decumbens (T. Nuttall) J. Decaisne
Sy = Asclepiodora decumbens (T. Nuttall) J. Decaisne
A. brachystephana G. Engelmann *ex* J. Torrey
A. curassavica C. Linnaeus
Sy = A. bicolor C. Moench
A. emoryi (E. Greene) A. Vail
Sy = Podostemma emoryi E. Greene
A. engelmanniana R. Woodson
A. glaucescens K. Kunth
Sy = A. elata G. Bentham
A. incarnata C. Linnaeus subsp. ***incarnata***
A. incarnata C. Linnaeus subsp. ***pulchra*** (F. Ehrhart *ex* C. von Willdenow) R. Woodson

Sy = A. incarnata var. neoscotica M. Fernald
Sy = A. incarnata var. pulchra (F. Ehrhart *ex* C. von Willdenow) C. Persoon
Sy = A. pulchra F. Ehrhart *ex* C. von Willdenow
A. *involucrata* G. Engelmann *ex* J. Torrey
A. *lanceolata* T. Walter
Sy = A. lanceolata var. paupercula (A. Michaux) M. Fernald
A. *latifolia* (J. Torrey) C. Rafinesque-Schmaltz
A. *linearis* G. Scheele
Sy = A. verticillata var. linearis (G. Scheele) C. Pollard
A. *longifolia* A. Michaux subsp. ***longifolia***
Sy = Acerates delticola J. K. Small
Sy = A. floridana (J. de Lamarck) A. Hitchcock
Sy = A. longifolia (A. Michaux) S. Elliott
A. *macrotis* J. Torrey
A. *nummularia* J. Torrey
A. *obovata* S. Elliott
Sy = A. viridiflora C. Rafinesque-Schmaltz var. obovata (S. Elliott) J. Torrey
A. *oenotheroides* A. von Chamisso & D. von Schlechtendal
Sy = A. brevicornus G. Scheele
Sy = A. lindheimeri G. Engelmann & A. Gray
Sy = A. wrightii E. Greene
Sy = Podostemma lindheimeri (G. Engelmann & A. Gray) E. Greene
A. *perennis* T. Walter
* **A. *prostrata*** W. Blackwell **[TOES: IV]**
A. *pumila* (A. Gray) A. Vail
Sy = A. verticillata C. Linnaeus var. pumila A. Gray
A. *purpurascens* C. Linnaeus
A. *quinquedentata* A. Gray
A. *rubra* C. Linnaeus
Sy = A. laurifolia A. Michaux
A. *scaposa* A. Vail
A. *speciosa* J. Torrey
Sy = A. curvipes A. Nelson
Sy = A. douglasii W. Hooker
A. *sperryi* R. Woodson
A. *stenophylla* A. Gray
Sy = Acerates angustifolia (T. Nuttall) J. Decaisne
Sy = Polyotus angustifolius T. Nuttall
A. *subverticillata* (A. Gray) A. Vail
Sy = A. verticillata C. Linnaeus var. subverticillata A. Gray
Ma = A. galioides *auct.*, non K. Kunth: American authors
A. *texana* A. A. Heller
A. *tomentosa* S. Elliott
A. *tuberosa* C. Linnaeus subsp. ***interior*** R. Woodson
Sy = A. tuberosa subsp. terminalis R. Woodson
Sy = A. tuberosa var. interior (R. Woodson) L. Shinners
A. *variegata* C. Linnaeus
Sy = Biventraria variegata (C. Linnaeus) J. K. Small
A. *verticillata* C. Linnaeus
A. *viridiflora* C. Rafinesque-Schmaltz
Sy = Acerates viridiflora (C. Rafinesque-Schmaltz) F. Pursh *ex* A. Eaton
Sy = A. viridiflora var. ivesii N. Britton
Sy = A. viridiflora var. linearis A. Gray
Sy = Asclepias viridiflora var. lanceolata J. Torrey
Sy = A. viridiflora var. linearis (A. Gray) M. Fernald
A. *viridis* T. Walter
Sy = Asclepiodora viridis (T. Walter) A. Gray

CYNANCHUM C. Linnaeus
C. *angustifolium* C. Persoon
Sy = C. palustre (F. Pursh) A. A. Heller
Sy = Lyonia palustris (F. Pursh) J. K. Small
Sy = Seutera palustris (F. Pursh) A. Vail
C. *barbigerum* (G. Scheele) L. Shinners
Sy = Metastelma barbigerum G. Scheele
C. *laeve* (A. Michaux) C. Persoon
Sy = Ampelamus albidus (T. Nuttall) N. Britton
Sy = Enslenia albida T. Nuttall
Sy = Gonolobus laevis A. Michaux
C. *maccartii* L. Shinners var. ***maccartii***
Sy = Metastelma palmeri S. Watson
Ma = Cynanchum palmeri *auct.*, non S. F. Blake: *sensu* L. Shinners
C. *pringlei* (A. Gray) J. Henrickson
Sy = C. barbigerum var. breviflorum L. Shinners
Sy = Metastelma pringlei A. Gray
C. *racemosa* (N. von Jacquin) N. von Jacquin var. ***unifarium*** (G. Scheele) E. Sundell
Sy = C. palmeri (S. Watson) S. F. Blake
Sy = C. unifarium (G. Scheele) R. Woodson
Sy = C. watsonianum R. Woodson
Sy = Rouliniella unifaria (G. Scheele) A. Vail

MATELEA J. Aublet
M. *biflora* (C. Rafinesque-Schmaltz) R. Woodson
Sy = Gonolobus biflorus C. Rafinesque-Schmaltz
M. *brevicoronata* (B. Robinson) R. Woodson
Sy = Gonolobus parviflorus var. brevicoronatus B. Robinson
M. *cynanchoides* (G. Engelmann) R. Woodson
Sy = Gonolobus cynanchoides G. Engelmann
M. *decipiens* (E. Alexander) R. Woodson

Sy = Gonolobus decipiens (E. Alexander) L. Perry
Sy = Odontostephana decipiens E. Alexander
M. edwardsensis D. Correll
M. gonocarpos (T. Walter) L. Shinners
Sy = Gonolobus gonocarpos (T. Walter) L. Perry
Sy = Vincetoxicum gonocarpos T. Walter
M. parviflora (J. Torrey) R. Woodson
M. parvifolia (J. Torrey) R. Woodson
Sy = Gonolobus californicus W. Jepson
Sy = G. parvifolius J. Torrey
Sy = Lachnostoma hastulatum A. Gray
Sy = Vincetoxicum hastulatum A. A. Heller
Sy = V. parvifolium (J. Torrey) A. A. Heller
M. producta (J. Torrey) R. Woodson
Sy = Gonolobus productus J. Torrey
Sy = Vincetoxicum productum (J. Torrey) A. Vail
* ***M. radiata*** D. Correll [**TOES: III**]
M. reticulata (G. Engelmann *ex* A. Gray) R. Woodson
Sy = Gonolobus reticulatus G. Engelmann & A. Gray
Sy = Vincetoxicum reticulatum (G. Engelmann) A. A. Heller
M. sagittifolia (A. Gray) R. Woodson
Sy = Gonolobus sagittifolius A. Gray
Sy = Matelea woodsonii L. Shinners
* ***M. texensis*** D. Correll [**TOES: III**]

PERIPLOCA C. Linnaeus
P. graeca C. Linnaeus [**cultivated**]

SARCOSTEMMA R. Brown
S. clausum (N. von Jacquin) J. J. Römer & J. A. Schultes
Sy = Funastrum clausum (N. von Jacquin) F. Schlechter
S. crispum G. Bentham
Sy = Funastrum crispum (G. Bentham) F. Schlechter
Sy = Philibertella crispa (G. Bentham) A. Vail
Sy = Sarcostemma lobata U. Waterfall
S. cynanchoides J. Decaisne subsp. ***cynanchoides***
Sy = Funastrum cynanchoides (J. Decaisne) F. Schlechter
Sy = F. cynanchoides var. subtruncatum (B. Robinson & M. Fernald) J. Macbride
Sy = Philibertella cynanchoides (J. Decaisne) A. Vail
S. cynanchoides J. Decaisne subsp. ***hartwegii*** (A. Vail) R. Holm
Sy = Funastrum heterophyllum (G. Engelmann *ex* J. Torrey) P. Standley
Sy = F. lineare (J. Decaisne) J. Macbride
Sy = Philibertella heterophylla (G. Engelmann) T. Cockerell
Sy = Philibertia heterophylla (G. Engelmann *ex* J. Torrey) W. Jepson
Sy = Sarcostemma cynanchoides var. hartwegii (A. Vail) L. Shinners
S. torreyi (A. Gray) R. Woodson

ASTERACEAE, *nomen conservandum*

ACANTHOSPERMUM F. von Schrank
A. australe (P. Löfling) K. E. O. Kuntze
Sy = Melampodium australe P. Löfling

ACHILLEA C. Linnaeus
A. filipendulina J. de Lamarck [**cultivated**]
A. millefolium C. Linnaeus subsp. ***lanulosa*** (T. Nuttall) C. Piper var. ***lanulosa***
Sy = A. millefolium var. occidentalis A. P. de Candolle
Sy = A. occidentalis (A. P. de Candolle) C. Rafinesque-Schmaltz *ex* P. Rydberg

ACMELLA L. C. Richard *ex* C. Persoon
A. oppositifolia (J. de Lamarck) R. Jansen var. ***repens*** (T. Walter) R. Jansen
Sy = A. repens (T. Walter) L. C. Richard
Sy = Spilanthes americana (J. Mutis) G. Hieronymus var. repens (T. Walter) T. Moore

ACOURTIA G. Don
A. nana (A. Gray) J. Reveal & R. King
Sy = Perezia nana A. Gray
A. runcinata (M. Lagasca y Segura *ex* D. Don) B. L. Turner
Sy = Perezia runcinata (M. Lagasca y Segura *ex* D. Don) M. Lagasca y Segura *ex* A. Gray
A. wrightii (A. Gray) J. Reveal & R. King
Sy = Perezia arizonica A. Gray
Sy = P. wrightii A. Gray

ACROPTILON A.-H. Cassini
A. repens (C. Linnaeus) A. P. de Candolle
Sy = Centaurea repens C. Linnaeus

AGERATINA E. Spach
A. altissima (C. Linnaeus) R. King & H. Robinson var. ***altissima***
Sy = Eupatorium rugosum M. Houttuyn
A. havanensis (K. Kunth) R. King & H. Robinson
Sy = Eupatorium havanense K. Kunth
Sy = E. texense (J. Torrey & A. Gray) P. Rydberg

A. herbacea (A. Gray) R. King & H. Robinson
Sy = Eupatorium herbaceum (A. Gray) E. Greene
A. rothrockii (A. Gray) R. King & H. Robinson
Sy = Eupatorium rothrockii A. Gray
A. wrightii (A. Gray) R. King & H. Robinson
Sy = Eupatorium wrightii A. Gray

AGERATUM C. Linnaeus
A. corymbosum A. Zuccagni
Sy = A. guatemalense M. F. Johnson
Sy = Coelestina corymbosa (A. Zuccagni) A. P. de Candolle

AMBLYOLEPIS A. P. de Candolle
A. setigera A. P. de Candolle
Sy = Helenium setigerum (A. P. de Candolle) N. Britton & H. Rusby

AMBROSIA C. Linnaeus
A. acanthicarpa W. Hooker
Sy = Franseria acanthicarpa (W. Hooker) F. Coville
Sy = F. hookeriana T. Nuttall
Sy = Gaertneria acanthicarpa (W. Hooker) N. Britton
Sy = G. hookeriana K. E. O. Kuntze
A. artemisiifolia C. Linnaeus
Sy = A. artemisiifolia var. elatior (C. Linnaeus) M. Descourtilz
Sy = A. elatior C. Linnaeus
Sy = A. elatior var. artemisiifolia O. Farwell
A. bidentata A. Michaux
* ***A. cheiranthifolia*** A. Gray [**TOES: IV**]
A. confertiflora A. P. de Candolle
Sy = A. fruticosa A. P. de Candolle
Sy = Franseria confertiflora (A. P. de Candolle) P. Rydberg
A. coronopifolia J. Torrey & A. Gray
Sy = A. psilostachya A. P. de Candolle var. coronopifolia (J. Torrey & A. Gray) O. Farwell
A. grayi (A. Nelson) L. Shinners
Sy = Franseria tomentosa A. Gray
Sy = Gaertneria grayi A. Nelson
Sy = G. tomentosa (A. Gray) K. E. O. Kuntze
A. psilostachya A. P. de Candolle
Sy = A. psilostachya var. californica (P. Rydberg) S. F. Blake
Sy = A. psilostachya var. lindheimeriana (G. Scheele) J. Blankinship
Sy = A. rugelii P. Rydberg
Ma = A. cuamanensis *auct.*, non K. Kunth: Hatch et al. 1990
A. trifida C. Linnaeus var. ***texana*** G. Scheele
Sy = A. trifida var. aptera (A. P. de Candolle) K. E. O. Kuntze
A. trifida C. Linnaeus var. ***trifida***
Sy = A. integrifolia G. H. Muhlenberg *ex* C. von Willdenow
Sy = A. trifida var. integrifolia (G. H. Muhlenberg *ex* C. von Willdenow) J. Torrey & A. Gray

AMPHIACHYRIS T. Nuttall
A. amoena (L. Shinners) O. Solbrig
Sy = Xanthocephalum amoenum L. Shinners
Sy = X. amoenum var. intermedium L. Shinners
A. dracunculoides (A. P. de Candolle) T. Nuttall
Sy = Brachyris dracunculoides A. P. de Candolle
Sy = Gutierrezia dracunculoides (A. P. de Candolle) S. F. Blake
Sy = Xanthocephalum dracunculoides (A. P. de Candolle) L. Shinners

ANTENNARIA J. Gaertner
A. marginata E. Greene
Sy = A. dioica (C. Linnaeus) J. Gaertner var. marginata (E. Greene) W. Jepson
A. parlinii M. Fernald subsp. ***fallax*** (E. Greene) R. Bayer & G. Stebbins
Sy = A. ambigens (E. Greene) M. Fernald
Sy = A. fallax E. Greene
Sy = A. parlinii var. farwellii (E. Greene) J. Boivin
Sy = A. plantaginifolia (C. Linnaeus) L. C. Richard var. ambigens (E. Greene) A. Cronquist

ANTHEMIS C. Linnaeus
A. cotula C. Linnaeus
Sy = Maruta cotula (C. Linnaeus) A. P. de Candolle
A. nobilis C. Linnaeus [**cultivated**]

APHANOSTEPHUS A. P. de Candolle
A. pilosus S. Buckley
A. ramosissimus A. P. de Candolle var. ***humilis*** (G. Bentham) B. L. Turner & A. Birdsong
Sy = A. humilis (G. Bentham) A. Gray
Sy = Leucopsidium humile G. Bentham
A. ramosissimus A. P. de Candolle var. ***ramosissimus***
Sy = A. arizonicus A. Gray
Sy = Egletes ramosissima (A. P. de Candolle) A. Gray

A. riddellii J. Torrey & A. Gray
Sy = A. perennis E. Wooton & P. Standley
A. skirrhobasis (A. P. de Candolle) W. Trelease var. ***kidderi*** (S. F. Blake) B. L. Turner
Sy = A. kidderi S. F. Blake
A. skirrhobasis (A. P. de Candolle) W. Trelease var. ***skirrhobasis***
Sy = A. arkansanus (A. P. de Candolle) A. Gray
Sy = Keerlia skirrhobasis A. P. de Candolle
A. skirrhobasis (A. P. de Candolle) W. Trelease var. ***thalassius*** L. Shinners

ARNOGLOSSUM C. Rafinesque-Schmaltz
A. ovatum (T. Walter) H. Robinson
Sy = Cacalia lanceolata T. Nuttall
Sy = C. ovata T. Walter
A. plantagineum C. Rafinesque-Schmaltz
Sy = Cacalia plantaginea (C. Rafinesque-Schmaltz) L. Shinners

ARTEMISIA C. Linnaeus
A. abrotanum C. Linnaeus [**cultivated**]
Sy = A. procera C. von Willdenow
A. absinthium C. Linnaeus [**cultivated**]
A. annua C. Linnaeus [**cultivated**]
A. bigelovii A. Gray
A. campestris C. Linnaeus subsp. ***caudata*** (A. Michaux) H. Hall & F. Clements
Sy = A. campestris var. caudata (A. Michaux) E. Palmer & J. Steyermark
Sy = A. caudata A. Michaux
Sy = A. forwoodii S. Watson
A. carruthii A. Wood *ex* J. Carruthers
Sy = A. carruthii var. wrightii (A. Gray) S. F. Blake
Sy = A. kansana N. Britton
Sy = A. wrightii A. Gray
A. dracunculus C. Linnaeus
Sy = A. dracunculina S. Watson
Sy = A. dracunculoides F. Pursh
Sy = A. dracunculoides var. dracunculina S. F. Blake
Sy = A. dracunculus subsp. glauca (P. von Pallas *ex* C. von Willdenow) H. Hall & F. Clements
Sy = A. dracunculus var. glauca (P. von Pallas *ex* C. von Willdenow) W. Besser
Sy = A. glauca C. Allioni *ex* C. von Willdenow
Sy = A. glauca var. dracunculoides B. Bush
Sy = Oligosporus glaucus P. Poljakov
A. filifolia J. Torrey
Sy = Oligosporus filifolius P. Poljakov
A. frigida C. von Willdenow
A. ludoviciana T. Nuttall subsp. ***mexicana*** (C. von Willdenow *ex* K. Sprengel) D. Keck var. ***albula*** (E. Wooton) L. Shinners
Sy = A. albula E. Wooton & P. Standley var. americana (C. Bessey) M. Fernald
Sy = A. albula var. gnaphalodes (T. Nuttall) J. Torrey & A. Gray
Sy = A. gnaphalodes T. Nuttall
Sy = A. rhizomata A. Nelson
Sy = A. silvicola G. Osterhout
A. ludoviciana T. Nuttall subsp. ***mexicana*** (C. von Willdenow *ex* K. Sprengel) D. Keck var. ***mexicana*** (C. von Willdenow *ex* K. Sprengel) M. Fernald
Sy = A. ludoviciana var. mexicana (C. von Willdenow *ex* K. Sprengel) A. Gray
Sy = A. mexicana C. von Willdenow *ex* K. Sprengel
Sy = A. neomexicana E. Greene *ex* P. Rydberg
A. pontica C. Linnaeus [**cultivated**]
A. stelleriana W. Besser [**cultivated**]
Sy = A. stelleriana var. sachalinensis T. Nakai
Sy = A. stelleriana var. vesiculosa A. Franchet & P. A. L. Savatier
A. vulgaris C. Linnaeus [**cultivated**]

ASTRANTHIUM T. Nuttall
A. integrifolium (A. Michaux) T. Nuttall subsp. ***ciliatum*** (C. Rafinesque-Schmaltz) D. DeJong
Sy = A. integrifolium var. ciliatum (C. Rafinesque-Schmaltz) E. L. Larsen
Sy = A. integrifolium var. triflorum (C. Rafinesque-Schmaltz) L. Shinners
A. robustum (L. Shinners) D. DeJong
Sy = A. integrifolium (A. Michaux) T. Nuttall var. robustum L. Shinners

BACCHARIS C. Linnaeus, *nomen conservandum*
B. bigelovii A. Gray
B. brachyphylla A. Gray
B. emoryi A. Gray
B. halimifolia C. Linnaeus
Sy = B. halimifolia var. angustior A. P. de Candolle
B. havardii A. Gray
B. neglecta N. Britton
B. pteronioides A. P. de Candolle
Sy = Aplopappus ramulosus A. P. de Candolle
Sy = Baccharis ramulosa (A. P. de Candolle) A. Gray
B. salicifolia (H. Ruiz López & J. Pavón) C. Persoon
Sy = B. glutinosa C. Persoon

Sy = B. viminea A. P. de Candolle
Sy = Molina salicifolia H. Ruiz López & J. Pavón
B. *salicina* J. Torrey & A. Gray
Sy = B. salicifolia T. Nuttall
B. *sarothroides* A. Gray
B. *texana* (J. Torrey & A. Gray) A. Gray
Sy = Aplopappus linearifolius S. Buckley
Sy = Linosyris texana J. Torrey & A. Gray
B. *wrightii* A. Gray

BAHIA M. Lagasca y Segura
B. *absinthifolia* G. Bentham var. ***absinthifolia***
B. *absinthifolia* G. Bentham var. ***dealbata*** (A. Gray) A. Gray
Sy = B. dealbata A. Gray
Sy = Picradeniopsis dealbata (A. Gray) E. Wooton & P. Standley
B. *bigelovii* A. Gray
B. *dissecta* (A. Gray) N. Britton
Sy = Amauria dissecta A. Gray
Sy = Amauriopsis dissecta (A. Gray) P. Rydberg
Sy = Bahia chrysanthemoides A. Gray
Sy = Eriophyllum chrysanthemoides K. E. O. Kuntze
Sy = Villanova dissecta (A. Gray) P. Rydberg
B. *pedata* A. Gray
Sy = Eriophyllum pedatum (A. Gray) K. E. O. Kuntze
Sy = Schkuhria pedata A. Gray

BAILEYA W. Harvey & A. Gray *ex* J. Torrey
B. *multiradiata* W. Harvey & A. Gray *ex* J. Torrey
Sy = B. multiradiata var. nudicaulis A. Gray
Sy = B. multiradiata var. thurberi (E. Greene) T. Kittell
Sy = B. perennis (A. Nelson) P. Rydberg
Sy = B. pleniradiata W. Harvey & A. Gray *ex* J. Torrey var. multiradiata T. Kearney

BEBBIA E. Greene
B. *juncea* (G. Bentham) E. Greene
Sy = B. aspera A. Nelson
Sy = B. juncea var. aspera E. Greene
Sy = Carpheporus junceus G. Bentham

BERLANDIERA A. P. de Candolle
B. *lyrata* G. Bentham var. ***lyrata***
Sy = B. incisa J. Torrey & A. Gray
B. *pumila* (A. Michaux) T. Nuttall
Sy = B. dealbata (J. Torrey & A. Gray) J. K. Small
Sy = B. tomentosa (F. Pursh) T. Nuttall var. dealbata J. Torrey & A. Gray
B. *texana* A. P. de Candolle
B. × *betonicifolia* (W. Hooker) J. K. Small [***pumila* × *texana***]
Sy = B. texana A. P. de Candolle var. betonicifolia (W. Hooker) W. Hooker & A. Gray

BIDENS C. Linnaeus
B. *aristosa* (A. Michaux) N. Britton
Sy = B. aristosa var. fritchey M. Fernald
Sy = B. aristosa var. mutica (A. Gray) A. Gattinger
Sy = B. aristosa var. retrorsa (E. Sherff) R. Wunderlin
Sy = B. polylepis S. F. Blake
B. *bigelovii* A. Gray
B. *bipinnata* C. Linnaeus var. ***biternatoides*** E. Sherff
B. *discoidea* (J. Torrey & A. Gray) N. Britton
B. *frondosa* C. Linnaeus
Sy = B. frondosa var. anomala T. Porter *ex* M. Fernald
Sy = B. frondosa var. caudata E. Sherff
Sy = B. frondosa var. pallida K. Wiegand
Sy = B. frondosa var. puberula K. Wiegand
Sy = B. frondosa var. stenodonta M. Fernald & H. St. John
B. *laevis* (C. Linnaeus) N. Britton, E. Sterns, & J. Poggenburg
Sy = Helianthus laevis C. Linnaeus
B. *leptocephala* E. Sherff
B. *mitis* (A. Michaux) E. Sherff
Sy = B. mitis var. leptophylla (T. Nuttall) J. K. Small
B. *odorata* A. Cavanilles var. ***odorata***
Sy = B. pilosa C. Linnaeus var. bimucronata (P. Turczaninow) K. H. Schultz-Bipontinus
B. *pilosa* C. Linnaeus
Sy = B. pilosa var. minor (K. von Blume) E. Sherff

BIGELOWIA A. P. de Candolle, *nomen conservandum*
B. *nuttallii* L. Anderson
Ma = B. virgata *auct.,* non (T. Nuttall) A. P. de Candolle: Correll & Johnston 1970

BOLTONIA L'Héritier de Brutelle
B. *asteroides* (C. Linnaeus) L'Héritier de Brutelle var. ***asteroides***
Sy = B. asteroides var. glastifolia (J. Hill) M. Fernald
Sy = Matricaria glastifolia J. Hill
B. *diffusa* S. Elliott var. ***diffusa***

BORRICHIA M. Adanson
B. frutescens (C. Linnaeus) A. P. de Candolle
Sy = B. frutescens var. angustifolia A. P. de Candolle
Sy = Buphthalmium frutescens C. Linnaeus

BRICKELLIA S. Elliott, *nomen conservandum*
* *B. baccharidea* A. Gray [**TOES: V**]
Sy = Coleosanthus baccharideus (A. Gray) K. E. O. Kuntze
B. brachyphylla (A. Gray) A. Gray var. *brachyphylla*
Sy = Coleosanthus brachyphyllus (A. Gray) K. E. O. Kuntze
* *B. brachyphylla* (A. Gray) A. Gray var. *hinckleyi* (P. Standley) L. Flyr [**TOES: V**]
Sy = B. hinckleyi P. Standley
* *B. brachyphylla* (A. Gray) A. Gray var. *terlinguensis* L. Flyr [**TOES: V**]
B. californica (J. Torrey & A. Gray) A. Gray var. *californica*
Sy = B. californica var. reniformis (A. Gray) B. Robinson
Sy = B. californica var. tener B. Robinson
Sy = B. tenera A. Gray
Sy = B. wrightii A. Gray
Sy = Bulbostylis californica J. Torrey & A. Gray
Sy = Coleosanthus albicaulis P. Rydberg
Sy = C. axillaris E. Greene
Sy = C. californicus K. E. O. Kuntze
Sy = C. tener K. E. O. Kuntze
B. conduplicata (B. Robinson) B. Robinson
B. coulteri A. Gray
B. cylindracea A. Gray & G. Engelmann
B. dentata (A. P. de Candolle) K. H. Schultz-Bipontinus
Sy = B. riddellii (J. Torrey & A. Gray) A. Gray
B. eupatorioides (C. Linnaeus) L. Shinners var. *chlorolepis* (E. Wooton & P. Standley) B. L. Turner
Sy = B. chlorolepis (E. Wooton & P. Standley) L. Shinners
Sy = Kuhnia chlorolepis E. Wooton & P. Standley
B. eupatorioides (C. Linnaeus) L. Shinners var. *corymbulosa* (J. Torrey & A. Gray) L. Shinners
Sy = Kuhnia eupatorioides C. Linnaeus var. corymbulosa J. Torrey & A. Gray
B. eupatorioides (C. Linnaeus) L. Shinners var. *eupatorioides*
Sy = Kuhnia eupatorioides C. Linnaeus
Sy = K. glutinosa S. Elliott
B. eupatorioides (C. Linnaeus) L. Shinners var. *gracillima* (A. Gray) B. L. Turner
Sy = B. leptophylla (G. Scheele) L. Shinners
Sy = Kuhnia eupatorioides C. Linnaeus var. gracillima A. Gray
B. eupatorioides (C. Linnaeus) L. Shinners var. *texana* (L. Shinners) L. Shinners
Sy = Kuhnia eupatorioides C. Linnaeus var. texana L. Shinners
B. grandiflora (W. Hooker) T. Nuttall
Sy = B. grandiflora var. minor A. Gray
Sy = B. grandiflora var. petiolaris A. Gray
Sy = Coleosanthus congestus A. Nelson
Sy = C. garrettii A. Nelson
Sy = C. grandiflorus (W. Hooker) K. E. O. Kuntze
Sy = C. minor F. Daniels
Sy = C. umbellatus E. Greene
B. laciniata A. Gray
Sy = Coleosanthus laciniatus (A. Gray) K. E. O. Kuntze
B. parvula A. Gray
B. venosa (E. Wooton & P. Standley) B. Robinson
Sy = Coleosanthus venosus E. Wooton & P. Standley
B. veronicifolia (K. Kunth) A. Gray var. *petrophila* (B. Robinson) B. Robinson
* *B. viejensis* L. Flyr [**TOES: V**]

CALYCOSERIS A. Gray
C. wrightii A. Gray

CALYPTOCARPUS C. Lessing
C. vialis C. Lessing
Sy = Synedrella vialis (C. Lessing) A. Gray

CARDUUS C. Linnaeus
C. acanthoides C. Linnaeus
C. nutans C. Linnaeus subsp. *macrocephalus* (R. Desfontaines) C. Nyman
Sy = C. macrocephalus R. Desfontaines
Sy = C. nutans var. macrocephalus (R. Desfontaines) J. Boivin
C. tenuiflorus W. Curtis

CARMINATIA J. Mociño *ex* A. P. de Candolle
C. tenuflora A. P. de Candolle

CARPHOCHAETE A. Gray
C. bigelovii A. Gray

CARTHAMUS C. Linnaeus
C. lanatus C. Linnaeus subsp. *lanatus*

CENTAUREA C. Linnaeus
*C. **americana*** T. Nuttall
*C. **cineraria*** C. Linnaeus [**cultivated**]
*C. **cyanus*** C. Linnaeus
Sy = Leucacantha cyanus (C. Linnaeus) J. Nieuwland & J. Lunell
*C. **melitensis*** C. Linnaeus
*C. **moschata*** C. Linnaeus [**cultivated**]
Sy = Amberboa moschata (C. Linnaeus) A. P. de Candolle
*C. **solstitialis*** C. Linnaeus

CHAETOPAPPA A. P. de Candolle
*C. **asteroides*** T. Nuttall *ex* A. P. de Candolle var. ***asteroides***
*C. **asteroides*** T. Nuttall *ex* A. P. de Candolle var. ***grandis*** L. Shinners
*C. **bellidifolia*** (A. Gray & G. Engelmann) L. Shinners
Sy = Bourdonia bellidifolia (A. Gray & G. Engelmann) E. Greene
Sy = Keerlia bellidifolia A. Gray & G. Engelmann
*C. **bellioides*** (A. Gray) L. Shinners
Sy = C. bellioides (A. Gray) L. Shinners var. hirticaulis L. Shinners
Sy = Diplostelma bellioides A. Gray
*C. **effusa*** (A. Gray) L. Shinners
Sy = Bourdonia effusa (A. Gray) E. Greene
Sy = Keerlia effusa A. Gray
*C. **ericoides*** (J. Torrey) G. Nesom
Sy = Aster arenosus (A. A. Heller) S. F. Blake
Sy = A. bellus S. F. Blake
Sy = A. ericifolius var. tenuis A. Gray
Sy = A. hirtifolius S. F. Blake
Sy = A. leucelene S. F. Blake
Sy = Chrysopsis ericoides J. Torrey & E. James
Sy = Diplopappus ericoides J. Torrey & A. Gray
Sy = D. ericoides var. hirtella A. Gray
Sy = Leucelene arenosa A. A. Heller *ex* P. Rydberg
Sy = L. ericoides (J. Torrey) E. Greene
* *C. **hersheyi*** S. F. Blake [**TOES: V**]
*C. **imberbis*** (A. Gray) G. Nesom
Sy = C. asteroides T. Nuttall *ex* A. P. de Candolle var. imberbis A. Gray
*C. **parryi*** A. Gray

CHAPTALIA E. Ventenat, *nomen conservandum*
*C. **texana*** E. Greene
Sy = C. carduacea E. Greene
Sy = C. leonina E. Greene
Sy = C. nutans (C. Linnaeus) K. Polák var. texana (E. Greene) A. Burkart
Sy = C. petrophila E. Greene
*C. **tomentosa*** E. Ventenat
Sy = C. integrifolia (A. Michaux) T. Nuttall
Sy = Tussilago integrifolia A. Michaux

CHLORACANTHA G. Nesom, Y. Suh, D. Morgan, S. Sundberg, & B. Simpson
*C. **spinosa*** (G. Bentham) G. Nesom var. ***spinosa***
Sy = Aster spinosus G. Bentham
Sy = Erigeron ortegae S. F. Blake var. spinosa (G. Bentham) S. Sundberg
Sy = Leucosyris spinosa (G. Bentham) E. Greene

CHROMOLAENA A. P. de Candolle
*C. **ivifolia*** (C. Linnaeus) R. King & H. Robinson
Sy = Eupatorium ivifolium C. Linnaeus
Sy = Osmia ivifolia (C. Linnaeus) K. H. Schultz-Bipontinus
*C. **odorata*** (C. Linnaeus) R. King & H. Robinson
Sy = Eupatorium odoratum C. Linnaeus
Sy = Osmia odorata (C. Linnaeus) K. H. Schultz-Bipontinus

CHRYSACTINIA A. Gray
*C. **mexicana*** A. Gray
Sy = Pectis taxifolia E. Greene

CHRYSANTHEMUM C. Linnaeus
*C. **leucanthemum*** C. Linnaeus
Sy = C. leucanthemum var. pinnatifidum H. Lecoq & M. Lamotte
Sy = Leucanthemum leucanthemum P. Rydberg
Sy = L. vulgare J. de Lamarck
*C. **maximum*** L. Ramond de Carbonnière [**cultivated**]
*C. **morifolium*** T. d'Audibert de Ramatuelle [**cultivated**]

CHRYSOPSIS (T. Nuttall) S. Elliott, *nomen conservandum*
*C. **mariana*** (C. Linnaeus) S. Elliott
Sy = C. mariana var. macradenia M. Fernald
Sy = Heterotheca mariana (C. Linnaeus) L. Shinners
Sy = Inula mariana C. Linnaeus
*C. **pilosa*** T. Nuttall
Sy = Heterotheca pilosa (T. Nuttall) L. Shinners
*C. **texana*** G. Nesom
Sy = Bradburia hirtella J. Torrey & A. Gray

CHRYSOTHAMNUS T. Nuttall
*C. **pulchellus*** (A. Gray) E. Greene subsp. ***baileyi*** (E. Wooton & P. Standley) H. Hall & F. Clements

Sy = C. baileyi E. Wooton & P. Standley
Sy = C. pulchellus var. baileyi (E. Wooton & P. Standley) S. F. Blake
C. ***pulchellus*** (A. Gray) E. Greene subsp. ***pulchellus***
Sy = Bigelowia pulchella A. Gray
Sy = Chrysothamnus pulchellus subsp. elatior (P. Standley) H. Hall & F. Clements
Sy = Linosyris pulchella A. Gray
C. ***spathulatus*** L. Anderson
Sy = C. viscidiflorus (W. Hooker) T. Nuttall var. ludens L. Shinners

CICHORIUM C. Linnaeus
C. ***intybus*** C. Linnaeus

CIRSIUM P. Miller
C. ***altissimum*** (C. Linnaeus) J. Hill
Sy = Carduus altissimus C. Linnaeus
Sy = Cirsium iowense (L. Pammel) M. Fernald
C. ***carolinianum*** (T. Walter) M. Fernald & B. Schubert
Sy = Carduus carolinianus T. Walter
C. ***engelmannii*** P. Rydberg
Sy = C. terraenigrae L. Shinners
C. ***horridulum*** A. Michaux var. ***elliottii*** J. Torrey & A. Gray
Sy = C. horridulum var. vittatum (J. K. Small) R. Long
C. ***muticum*** A. Michaux
Sy = Carduus muticus (A. Michaux) C. Persoon
Sy = Cirsium muticum var. monticola M. Fernald
Sy = C. muticum var. subpinnatifidum (N. Britton) M. Fernald
C. ***ochrocentrum*** A. Gray
Sy = Carduus ochrocentrus A. Gray
Sy = Cnicus ochrocentrus (A. Gray) A. Gray
C. ***texanum*** S. Buckley
Sy = Carduus austrinus J. K. Small
Sy = C. texanum var. stenolepis L. Shinners
C. ***turneri*** B. Warnock
C. ***undulatum*** (T. Nuttall) K. Sprengel var. ***undulatum***
Sy = Carduus undulatus T. Nuttall
Sy = Cirsium helleri (J. K. Small) V. Cory
Sy = C. megacephalum (A. Gray) T. Cockerell
Sy = C. undulatum var. megacephalum (A. Gray) M. Fernald
Sy = C. undulatum var. megacephalus A. Gray
Sy = Cnicus undulatus (T. Nuttall) A. Gray
C. ***vulgare*** (G. Savi) M. Tenore
Sy = Carduus lanceolatus C. Linnaeus
Sy = C. vulgaris G. Savi
Sy = Cirsium lanceolatum (C. Linnaeus) J. Scopoli, non J. Hill
Sy = Cnicus lanceolatus C. von Willdenow

CLAPPIA A. Gray
C. ***suaedifolia*** A. Gray

CNICUS C. Linnaeus
C. ***benedictus*** C. Linnaeus
Sy = Centaurea benedicta (C. Linnaeus) C. Linnaeus
Sy = Cirsium pugnax C. Sommier & E. Levier

CONOCLINIUM A. P. de Candolle
C. ***betonicifolium*** (P. Miller) R. King & H. Robinson
Sy = Eupatorium betonicifolium P. Miller
C. ***coelestinum*** (C. Linnaeus) A. P. de Candolle
Sy = Eupatorium coelestinum C. Linnaeus
C. ***greggii*** (A. Gray) J. K. Small
Sy = C. dissectum A. Gray
Sy = Eupatorium dissectum A. Gray
Sy = E. greggii A. Gray

CONYZA C. Lessing, *nomen conservandum*
C. ***bonariensis*** (C. Linnaeus) A. Cronquist var. ***bonariensis***
Sy = Conyzella linifolia E. Greene
Sy = Erigeron bonariensis C. Linnaeus
Sy = E. linifolium C. von Willdenow
Sy = Leptilon bonariense (C. Linnaeus) J. K. Small
Sy = L. linifolium J. K. Small
Sy = Marsea bonariensis V. Badillo
C. ***canadensis*** (C. Linnaeus) A. Cronquist var. ***canadensis***
Sy = Erigeron canadensis C. Linnaeus
Sy = Leptilon canadense (C. Linnaeus) N. Britton
C. ***canadensis*** (C. Linnaeus) A. Cronquist var. ***glabrata*** (A. Gray) A. Cronquist
Sy = Erigeron canadensis C. Linnaeus var. glabratus A. Gray
C. ***canadensis*** (C. Linnaeus) A. Cronquist var. ***pusilla*** (T. Nuttall) A. Cronquist
Sy = Erigeron pusillus T. Nuttall
Sy = Leptilon pusillum (T. Nuttall) N. Britton
C. ***ramosissima*** A. Cronquist
Sy = Erigeron divaricatus A. Michaux

COREOPSIS C. Linnaeus
C. ***auriculata*** C. Linnaeus [**cultivated**]
C. ***basalis*** (A. Dietrich) S. F. Blake
Sy = Calliopsis basalis A. Dietrich

Sy = Coreopsis drummondii (D. Don) J. Torrey & A. Gray
C. ***grandiflora*** T. Hogg *ex* R. Sweet var. ***grandiflora***
Sy = C. grandiflora var. pilosa E. Sherff
C. ***grandiflora*** T. Hogg *ex* R. Sweet var. ***harveyana*** (A. Gray) E. Sherff
Sy = C. heterolepis E. Sherff
C. ***grandiflora*** T. Hogg *ex* R. Sweet var. ***longipes*** (W. Hooker) J. Torrey & A. Gray
Sy = C. longipes W. Hooker
C. ***intermedia*** E. Sherff
C. ***lanceolata*** C. Linnaeus
Sy = C. lanceolata var. villosa A. Michaux
C. ***linifolia*** T. Nuttall
Sy = C. gladiata T. Walter var. linifolia (T. Nuttall) A. Cronquist
C. ***nuecensis*** A. A. Heller
C. ***nuecensoides*** E. B. Smith
C. ***pubescens*** S. Elliott var. ***pubescens***
C. ***tinctoria*** T. Nuttall var. ***similis*** (F. Boynton) H. Parker
Sy = C. similis F. Boynton
C. ***tinctoria*** T. Nuttall var. ***tinctoria***
Sy = Calliopsis cardaminifolia A. P. de Candolle
Sy = C. tinctoria A. P. de Candolle
Sy = Coreopsis cardaminifolia (A. P. de Candolle) J. Torrey & A. Gray
Sy = C. stenophylla F. Boynton
Sy = C. tinctoria var. imminuta E. Sherff
Sy = Diplosastera tinctoria I. Tausch
C. ***tripteris*** C. Linnaeus var. ***deamii*** P. Standley
Sy = C. tripteris var. intercedens P. Standley
Sy = C. tripteris var. smithii E. Sherff
Sy = C. tripteris var. subrhomboidea E. Sherff
C. ***verticillata*** C. Linnaeus **[cultivated]**
C. ***wrightii*** (A. Gray) H. Parker
Sy = C. basalis (A. Dietrich) S. F. Blake var. wrightii (A. Gray) S. F. Blake
Sy = C. drummondii (D. Don) J. Torrey & A. Gray var. wrightii A. Gray

COSMOS A. Cavanilles
C. ***bipinnatus*** A. Cavanilles **[cultivated]**
C. ***parviflorus*** (N. von Jacquin) C. Persoon **[cultivated]**
Sy = Coreopsis parviflora N. von Jacquin
C. ***sulphureus*** A. Cavanilles **[cultivated]**

COTULA C. Linnaeus
C. ***australis*** (F. Sieber) J. Hooker

CREPIS C. Linnaeus
C. ***capillaris*** (C. Linnaeus) C. Wallroth
C. ***pulchra*** C. Linnaeus
C. ***runcinata*** (E. James) J. Torrey & A. Gray var. ***runcinata***
Sy = Crepidium runcinatum T. Nuttall
Sy = Hieracium runcinatum E. James
Sy = Psilochaenia runcinata A. Löve & D. Löve
C. ***setosa*** H. Hallier
C. ***zacintha*** (C. Linnaeus) E. Babcock

CROPTILON C. Rafinesque-Schmaltz
C. ***divaricatum*** (T. Nuttall) C. Rafinesque-Schmaltz
Sy = Haplopappus divaricatus (T. Nuttall) J. Torrey & A. Gray
C. ***hookerianum*** (J. Torrey & A. Gray) H. House var. ***graniticum*** (E. B. Smith) E. B. Smith
Sy = C. divaricatum (T. Nuttall) C. Rafinesque-Schmaltz var. graniticum (E. B. Smith) L. Shinners
C. ***hookerianum*** (J. Torrey & A. Gray) H. House var. ***hookerianum***
Sy = C. divaricatum (T. Nuttall) C. Rafinesque-Schmaltz var. hookerianum (J. Torrey & A. Gray) L. Shinners
C. ***hookerianum*** (J. Torrey & A. Gray) H. House var. ***validum*** (P. Rydberg) E. B. Smith
Sy = Haplopappus validus (C. Rydberg) V. Cory
C. ***rigidifolium*** (E. B. Smith) E. B. Smith
Sy = C. divaricatum (T. Nuttall) C. Rafinesque-Schmaltz var. hirtellum (L. Shinners) L. Shinners
Sy = Haplopappus rigidifolius E. B. Smith

DAHLIA A. Cavanilles
D. ***pinnata*** A. Cavanilles **[cultivated]**
Sy = Dahlia rosea A. Cavanilles

DICHAETOPHORA A. Gray
D. ***campestris*** A. Gray

DICRANOCARPUS A. Gray
D. ***parviflorus*** A. Gray
Sy = D. dicranocarpus (A. Gray) E. Wooton & P. Standley
Sy = Heterosperma dicranocarpum A. Gray
Sy = Wootonia parviflora E. Greene

DOELLINGERIA C. Nees von Esenbeck
D. ***sericocarpoides*** J. K. Small
Sy = Aster sericocarpoides (J. K. Small) K. Schumann
Sy = A. umbellatus P. Miller var. latifolius A. Gray
Sy = Doellingeria umbellata (P. Miller) C. Nees von Esenbeck var. latifolia (A. Gray) H. House

DRACOPIS A.-H. Cassini
D. *amplexicaulis* (M. H. Vahl) A.-H. Cassini
Sy = Rudbeckia amplexicaulis M. H. Vahl

DYSODIOPSIS (A. Gray) P. Rydberg
D. *tagetoides* (J. Torrey & A. Gray) P. Rydberg
Sy = Dyssodia tagetoides J. Torrey & A. Gray
Sy = Hymenatherum tagetoides (J. Torrey & A. Gray) A. Gray

DYSSODIA E. Ventenat
D. *acerosa* A. P. de Candolle
Sy = Aciphyllaea acerosa (A. P. de Candolle) A. Gray
Sy = Dyssodia fusca A. Nelson
Sy = Hymenatherum acerosum A. Gray
Sy = Thymophylla acerosa (A. P. de Candolle) J. Strother
D. *aurea* (A. Gray) A. Nelson var. ***aurea***
Sy = Hymenatherum aureum A. Gray
Sy = Lowellia aurea A. Gray
Sy = Thymophylla aurea (A. Gray) E. Greene *ex* N. Britton
D. *aurea* (A. Gray) A. Nelson var. ***polychaeta*** (A. Gray) M. C. Johnston
Sy = D. polychaeta (A. Gray) B. Robinson
Sy = Hymenatherum polychaetum A. Gray
Sy = Thymophylla aurea (A. Gray) E. Greene *ex* N. Britton var. polychaeta (A. Gray) J. Strother
Sy = T. polychaeta A. Gray
D. *micropoides* (A. P. de Candolle) L. Loesener
Sy = Thymophylla micropoides (A. P. de Candolle) J. Strother
D. *papposa* (E. Ventenat) A. Hitchcock
Sy = Boebera chrysanthemoides C. von Willdenow
Sy = B. papposa (E. Ventenat) P. Rydberg
Sy = Dyssodia chrysanthemoides M. Lagasca y Segura
Sy = Tagetes papposa E. Ventenat
D. *pentachaeta* (A. P. de Candolle) B. Robinson var. ***belenidium*** (A. P. de Candolle) J. Strother
Sy = D. belenidium (A. P. de Candolle) G. Macloskie
Sy = D. thurberi (A. Gray) B. Robinson
Sy = Hymenatherum thurberi A. Gray
Sy = Thymophylla pentachaeta (A. P. de Candolle) J. K. Small var. belenidium (A. P. de Candolle) J. Strother
Sy = T. pentachaeta (A. P. de Candolle) J. K. Small var. hartwegii (A. Gray) J. Strother
D. *pentachaeta* (A. P. de Candolle) B. Robinson var. ***hartwegii*** (A. Gray) J. Strother
Sy = D. hartwegii (A. Gray) B. Robinson
Sy = D. pentachaeta subsp. hartwegii (A. Gray) J. Strother
Sy = Hymenatherum hartwegii A. Gray
Sy = Thymophylla hartwegii (A. Gray) E. Wooton & P. Standley
D. *pentachaeta* (A. P. de Candolle) B. Robinson var. ***pentachaeta***
Sy = Thymophylla pentachaeta (A. P. de Candolle) J. K. Small var. pentachaeta
D. *pentachaeta* (A. P. de Candolle) B. Robinson var. ***puberula*** (P. Rydberg) J. Strother
Sy = Thymophylla pentachaeta (A. P. de Candolle) J. K. Small var. puberula (P. Rydberg) J. Strother
D. *setifolia* (M. Lagasca y Segura) B. Robinson var. ***radiata*** (A. Gray) J. Strother
Sy = D. greggii (A. Gray) B. Robinson
Sy = D. setifolia (M. Lagasca y Segura) B. Robinson var. greggii (A. Gray) M. C. Johnston
Sy = Thymophylla greggii A. Gray
Sy = T. setifolia M. Lagasca y Segura var. radiata (A. Gray) J. Strother
D. *tenuiloba* (A. P. de Candolle) B. Robinson var. ***tenuiloba***
Sy = Thymophylla tenuiloba (A. P. de Candolle) J. K. Small var. tenuiloba
D. *tenuiloba* (A. P. de Candolle) B. Robinson var. ***texana*** (V. Cory) J. Strother
Sy = D. texana V. Cory
Sy = Thymophylla tenuiloba var. texana (V. Cory) J. Strother
D. *tenuiloba* (A. P. de Candolle) B. Robinson var. ***treculii*** (A. Gray) J. Strother
Sy = D. treculii (A. Gray) B. Robinson
Sy = Thymophylla tenuiloba (A. P. de Candolle) J. K. Small var. treculii (A. Gray) J. Strother
D. *tenuiloba* (A. P. de Candolle) B. Robinson var. ***wrightii*** (A. Gray) J. Strother
Sy = D. tenuiloba subsp. wrightii (A. Gray) J. Strother
Sy = D. wrightii (A. Gray) B. Robinson
Sy = Thymophylla tenuiloba (A. P. de Candolle) J. K. Small var. wrightii (A. Gray) J. Strother
D. *tephroleuca* S. F. Blake
* Sy = Thymophylla tephroleuca (S. F. Blake) J. Strother **[TOES: I]**

ECHINACEA C. Moench
E. *angustifolia* A. P. de Candolle var. ***angustifolia***

Sy = Brauneria angustifolia (A. P. de Candolle) A. A. Heller
Sy = Echinacea pallida (T. Nuttall) var. angustifolia (A. P. de Candolle) A. Cronquist
E. angustifolia A. P. de Candolle var. ***strigosa*** R. L. McGregor
Sy = E. pallida (T. Nuttall) T. Nuttall var. strigosa (R. L. McGregor) K. Gandhi
E. atrorubens T. Nuttall
E. pallida (T. Nuttall) T. Nuttall
Sy = Brauneria pallida (T. Nuttall) A. Cronquist
Sy = Rudbeckia pallida T. Nuttall
E. paradoxa (J. Norton) N. Britton var. ***neglecta*** R. L. McGregor
E. purpurea (C. Linnaeus) C. Moench
Sy = Brauneria purpurea (C. Linnaeus) N. Britton
Sy = Echinacea purpurea var. arkansana J. Steyermark
E. sanguinea T. Nuttall
Sy = E. pallida (T. Nuttall) T. Nuttall var. sanguinea (T. Nuttall) K. Gandhi & R. D. Thomas

ECLIPTA C. Linnaeus, *nomen conservandum*
E. prostrata (C. Linnaeus) C. Linnaeus
Sy = E. alba (C. Linnaeus) J. Hasskarl
Sy = E. erecta C. Linnaeus
Sy = Verbesina alba C. Linnaeus

EGLETES A.-H. Cassini
E. viscosa (C. Linnaeus) C. Lessing
Sy = Cotula viscosa C. Linnaeus

ELEPHANTOPUS C. Linnaeus
E. carolinianus E. Raeuschel
E. nudatus A. Gray
E. tomentosus C. Linnaeus
Sy = E. nudicaulis J. Poiret

EMILIA A.-H. Cassini
E. fosbergii D. Nicolson

ENCELIA M. Adanson
E. scaposa (A. Gray) A. Gray
Sy = E. scaposa var. stenophylla L. Shinners
Sy = Simsia scaposa A. Gray

ENGELMANNIA A. Gray *ex* T. Nuttall
E. pinnatifida A. Gray *ex* T. Nuttall

ERECHTITES C. Rafinesque-Schmaltz
E. hieraciifolia (C. Linnaeus) C. Rafinesque-Schmaltz *ex* A. P. de Candolle var. ***hieraciifolia***
Sy = E. hieraciifolia var. intermedia M. Fernald

ERICAMERIA T. Nuttall
E. laricifolia (A. Gray) L. Shinners
Sy = Haplopappus laricifolius A. Gray
E. nauseosa (P. von Pallas *ex* F. Pursh) G. Nesom & G. Baird var. ***bigelovii*** (A. Gray) G. Nesom & G. Baird
Sy = Chrysothamnus bigelovii (A. Gray) E. Greene
Sy = C. nauseosus (P. von Pallas *ex* F. Pursh) N. Britton subsp. bigelovii (A. Gray) H. Hall & F. Clements
Sy = C. nauseosus var. bigelovii (A. Gray) H. Hall
Sy = C. nauseosus var. glareosus (M. E. Jones) H. Hall
Sy = Linosyris bigelovii A. Gray
E. nauseosa (P. von Pallas *ex* F. Pursh) G. Nesom & G. Baird var. ***glabrata*** (A. Gray) G. Nesom & G. Baird
Sy = Chrysothamnus graveolens (T. Nuttall) E. Greene
Sy = C. nauseosus (P. von Pallas *ex* F. Pursh) N. Britton subsp. graveolens (T. Nuttall) C. Piper
Sy = C. nauseosus var. glabratus (A. Gray) A. Cronquist
Sy = C. nauseosus var. graveolens (T. Nuttall) C. Piper *ex* H. Hall
Sy = C. nauseosus var. petrophilus A. Cronquist
E. nauseosa (P. von Pallas *ex* F. Pursh) G. Nesom & G. Baird var. ***texensis*** (L. Anderson) G. Nesom & G. Baird
* Sy = Chrysothamnus nauseosus (P. von Pallas *ex* F. Pursh) N. Britton subsp. texensis L. Anderson [**TOES: V**]

ERIGERON C. Linnaeus
E. annuus (C. Linnaeus) C. Persoon
Sy = Aster annuus C. Linnaeus
Sy = Erigeron annuus var. discoideus (F. Victorin & J. Rousseau) A. Cronquist
E. bellidiastrum T. Nuttall var. ***arenarius*** (E. Greene) G. Nesom
E. bellidiastrum T. Nuttall var. ***bellidiastrum***
Sy = E. eastwoodiae E. Wooton & P. Standley

E. bellidiastrum T. Nuttall var. ***robustus*** A. Cronquist
E. bigelovii A. Gray
E. colomexicanus A. Nelson
Sy = E. cinereum A. Gray
Sy = E. divergens J. Torrey & A. Gray var. cinereus A. Gray
E. divergens J. Torrey & A. Gray var. ***divergens***
Sy = E. divaricatum T. Nuttall
Sy = E. incomptus A. Gray
Sy = E. lavandulaceus E. Greene
Sy = E. wootonii P. Rydberg
E. eximius E. Greene
Sy = E. eldensis E. Greene
Sy = E. superbus E. Greene *ex* P. Rydberg
E. flagellaris A. Gray
Sy = E. macdougallii A. A. Heller
Sy = E. senilis E. Wooton & P. Standley
Sy = E. tonsus E. Wooton & P. Standley
E. geiseri L. Shinners
E. modestus A. Gray
Sy = E. lobatus var. warnockii L. Shinners
E. philadelphicus C. Linnaeus
Sy = E. philadelphicus var. scaturicola (M. Fernald) M. Fernald
Sy = Tessenia philadelphia J. Lunell
E. procumbens (W. Houstoun *ex* P. Miller) G. Nesom
Sy = Aster procumbens M. Houstoun *ex* P. Miller
Sy = Erigeron myrionactis J. K. Small
E. pulchellus A. Michaux var. ***pulchellus***
E. strigosus G. H. Muhlenberg *ex* C. von Willdenow var. ***beyrichii*** (F. von Fischer & C. von Meyer) J. Torrey & A. Gray *ex* A. Gray
Sy = E. ramosus (T. Walter) N. Britton, E. Sterns, & J. Poggenburg var. beyrichii (F. von Fischer & C. von Meyer) A. Gray
E. strigosus G. H. Muhlenberg *ex* C. von Willdenow var. ***strigosus***
Sy = Doronicum ramosum T. Walter
Sy = Erigeron annuum (C. Linnaeus) C. Persoon var. ramosum N. Hylander
Sy = E. ramosus N. Britton, E. Sterns, & J. Poggenburg
Sy = E. strigosus var. discoideus J. Robbins *ex* A. Gray
Sy = E. strigosus var. eligulatus A. Cronquist
Sy = E. traversii L. Shinners
Sy = Stenactis ramosa (T. Walter) K. Domin
Sy = Tessenia ramosa (T. Walter) J. Lunell
E. tenellus A. P. de Candolle
E. tenuis J. Torrey & A. Gray
E. versicolor (J. Greenman) G. Nesom
Sy = Achaetogeron chihuahuensis E. L. Larsen *ex* S. F. Blake
Sy = A. versicolor J. Greenman
Sy = Erigeron geiseri L. Shinners var. calcicola L. Shinners
Sy = E. gilensis E. Wooton & P. Standley
* Sy = E. mimegletes L. Shinners [**TOES: V**]

EUPATORIUM C. Linnaeus
E. altissimum C. Linnaeus
E. capillifolium (J. de Lamarck) J. K. Small
E. compositifolium T. Walter
E. fistulosum J. Barratt
Sy = Eupatoriadelphus fistulosus (J. Barratt) R. King & H. Robinson
E. glaucescens S. Elliott
Sy = E. cuneifolium C. von Willdenow
E. hyssopifolium C. Linnaeus var. ***calcaratum*** M. Fernald & B. Schubert
E. lancifolium (J. Torrey & A. Gray) J. K. Small
Sy = E. semiserratum A. P. de Candolle var. lancifolium J. Torrey & A. Gray
E. leucolepis (A. P. de Candolle) J. Torrey & A. Gray var. ***leucolepis***
E. mohrii E. Greene
E. perfoliatum C. Linnaeus var. ***perfoliatum***
E. recurvans J. K. Small
E. rotundifolium C. Linnaeus var. ***rotundifolium***
E. rotundifolium C. Linnaeus var. ***scabridum*** (S. Elliott) A. Gray
Sy = E. scabridum S. Elliott
E. serotinum A. Michaux

EURYBIA (A.-H. Cassini) S. F. Gray
E. hemispherica (E. Alexander) G. Nesom
Sy = Aster gattingeri E. Alexander
Sy = A. hemisphericus E. Alexander
Sy = A. paludosus D. Solander *ex* W. Aiton subsp. hemisphericus (E. Alexander) A. Cronquist
Sy = A. pedionomus E. Alexander
Sy = A. verutifolius E. Alexander
Sy = Heleastrum hemisphericum (E. Alexander) L. Shinners

EURYOPS (A.-H. Cassini) A.-H. Cassini
E. pectinatus A.-H. Cassini [**cultivated**]

EUTHAMIA T. Nuttall *ex* A.-H. Cassini
E. gymnospermoides E. Greene
Sy = E. camporum E. Greene

Sy = E. pulverulenta E. Greene
Sy = Solidago graminifolia (C. Linnaeus) R. A. Salisbury var. media (E. Greene) S. Harris
E. leptocephala (J. Torrey & A. Gray) E. Greene
Sy = Solidago leptocephala J. Torrey & A. Gray

EVAX J. Gaertner
E. candida (J. Torrey & A. Gray) A. Gray
Sy = Filago candida (J. Torrey & A. Gray) L. Shinners
E. prolifera T. Nuttall *ex* A. P. de Candolle
Sy = Filago nuttallii L. Shinners
Sy = F. prolifera (A. P. de Candolle) N. Britton
E. verna C. Rafinesque-Schmaltz
Sy = E. multicaulis A. P. de Candolle var. drummondii (J. Torrey & A. Gray) A. Gray
Sy = E. verna var. drummondii (J. Torrey & A. Gray) J. Kartesz & K. Gandhi
Sy = Filago verna (C. Rafinesque-Schmaltz) L. Shinners var. drummondii (J. Torrey & A. Gray) L. Shinners

FACELIS A.-H. Cassini
F. retusa (J. de Lamarck) K. H. Schultz-Bipontinus

FILAGO C. Linnaeus
F. californica T. Nuttall
Sy = Oglifa californica (T. Nuttall) P. Rydberg

FLAVERIA A. L. de Jussieu
F. brownii A. M. Powell
Ma = F. oppositifolia *auct.*, non (A. P. de Candolle) P. Rydberg: Correll & Johnston 1970
F. campestris J. R. Johnston
F. chlorifolia A. Gray
F. trinervia (K. Sprengel) C. Mohr
Sy = F. repanda M. Lagasca y Segura
Sy = Oedera trinervia K. Sprengel

FLEISCHMANNIA K. H. Schultz-Bipontinus
F. incarnata (T. Walter) R. King & H. Robinson
Sy = Eupatorium incarnatum T. Walter

FLORESTINA A.-H. Cassini
F. tripteris A. P. de Candolle
Sy = Palafoxia tripteris (A. P. de Candolle) L. Shinners
Sy = P. tripteris var. brevis L. Shinners

FLOURENSIA A. P. de Candolle
F. cernua A. P. de Candolle
Sy = Helianthus cernua G. Bentham & J. Hooker

FLYRIELLA R. King & H. Robinson
* ***F. parryi*** (A. Gray) R. King & H. Robinson [**TOES: V**]
Sy = Brickellia shineri M. E. Jones
Sy = Eupatorium parryi A. Gray

GAILLARDIA Fougeroux de Bondaroy
G. aestivalis (T. Walter) H. Rock var. ***aestivalis***
Sy = G. fastigiata E. Greene
Sy = G. lanceolata A. Michaux
G. aestivalis (T. Walter) H. Rock var. ***flavovirens*** (C. Mohr) A. Cronquist
Sy = G. lutea E. Greene
* ***G. aestivalis*** (T. Walter) H. Rock var. ***winkleri*** (A. Gray) B. L. Turner [**TOES: V**]
Sy = G. lutea E. Greene var. winkleri V. Cory
G. amblyodon J. Gay
G. coahuilensis B. L. Turner
Ma = G. mexicana *auct.*, non A. Gray
G. multiceps E. Greene var. ***microcephala*** B. L. Turner
G. pinnatifida J. Torrey
Sy = G. crassa P. Rydberg
Sy = G. crassifolia A. Nelson & J. Macbride
Sy = G. globosa A. Nelson
Sy = G. gracilis A. Nelson
Sy = G. linearis P. Rydberg
Sy = G. mearnsii P. Rydberg
G. pulchella Fougeroux de Bondaroy var. ***australis*** B. L. Turner & M. Whalen
G. pulchella Fougeroux de Bondaroy var. ***pulchella***
Sy = G. pulchella var. picta (R. Sweet) A. Gray
G. suavis (G. Engelmann & A. Gray) N. Britton & H. Rusby
Sy = Agassizia suavis A. Gray & G. Engelmann
Sy = Gaillardia trinervata J. K. Small

GALINSOGA H. Ruiz López & J. Pavón
G. parviflora A. Cavanilles
Sy = G. parviflora var. semicalva A. Gray
Sy = G. semicalva (A. Gray) H. St. John & D. White

GAMOCHAETA H. Weddell
G. americana (P. Miller) H. Weddell
Sy = G. spicata (J. de Lamarck) A. Cabrera
Sy = Gnaphalium americanum P. Miller
Sy = G. purpureum C. Linnaeus var. americanum (P. Miller) F. Klatt
Sy = G. spicatum J. de Lamarck
G. falcata (J. de Lamarck) A. Cabrera
Sy = Gnaphalium falcatum J. de Lamarck
Sy = G. purpureum C. Linnaeus var. falcatum (J. de Lamarck) J. Torrey & A. Gray

G. pensilvanica (C. von Willdenow) A. Cabrera
Sy = Gnaphalium pensilvanicum C. von Willdenow
Sy = G. peregrinum M. Fernald
G. purpurea (C. Linnaeus) A. Cabrera
Sy = Gnaphalium purpureum C. Linnaeus
G. sphacilata (K. Kunth) A. Cabrera
Sy = Gnaphalium sphacilatum K. Kunth

GOCHNATIA K. Kunth
G. hypoleuca (A. P. de Candolle) A. Gray subsp. ***hypoleuca***

GRINDELIA C. von Willdenow
G. adenodonta (J. Steyermark) G. Nesom
Sy = G. microcephala A. P. de Candolle var. adenodonta J. Steyermark
G. arizonica A. Gray var. ***neomexicana*** (E. Wooton & P. Standley) G. Nesom
Sy = G. neomexicana E. Wooton & P. Standley
Sy = G. scabra E. Greene var. neomexicana (E. Wooton & P. Standley) J. Steyermark
G. grandiflora W. Hooker
G. havardii J. Steyermark
G. lanceolata T. Nuttall var. ***lanceolata***
Sy = G. littoralis J. Steyermark
Sy = G. texana G. Scheele var. lanceolata (T. Nuttall) L. Shinners
G. lanceolata T. Nuttall var. ***texana*** (G. Scheele) L. Shinners
Sy = G. texana G. Scheele
G. microcephala A. P. de Candolle
G. nuda A. Wood var. ***aphanactis*** (P. Rydberg) G. Nesom
Sy = G. aphanactis P. Rydberg
Sy = G. pinnatifida E. Wooton & P. Standley
G. nuda A. Wood var. ***nuda***
Sy = G. squarrosa (F. Pursh) M. Dunal var. nuda (A. Wood) A. Gray
* ***G. oölepis*** S. F. Blake **[TOES: V]**
G. papposa G. Nesom & Y. Suh
Sy = Haplopappus ciliatus (T. Nuttall) A. P. de Candolle
Sy = Prionopsis ciliata (T. Nuttall) T. Nuttall
G. pusilla (J. Steyermark) G. Nesom
Sy = G. microcephala A. P. de Candolle var. pusilla J. Steyermark
G. squarrosa (F. Pursh) M. Dunal var. ***squarrosa***
Sy = Donia squarrosa F. Pursh

GUTIERREZIA M. Lagasca y Segura
G. microcephala (A. P. de Candolle) A. Gray
Sy = Brachyris microcephala A. P. de Candolle
Sy = Gutierrezia glomerella E. Greene
Sy = G. lucida (E. Greene) E. Greene
Sy = G. sarothrae (F. Pursh) N. Britton & H. Rusby var. microcephala (A. P. de Candolle) L. Benson
Sy = Xanthocephalum lucidum E. Greene
Sy = X. microcephalum (A. P. de Candolle) L. Shinners
G. sarothrae (F. Pursh) N. Britton & H. Rusby
Sy = Brachyachyris euthamiae K. Sprengel
Sy = Brachyris euthamiae T. Nuttall
Sy = Gutierrezia diversifolia E. Greene
Sy = G. euthamiae J. Torrey & A. Gray
Sy = G. filifolia E. Greene
Sy = G. furfuracea E. Greene
Sy = G. goldmanii E. Greene
Sy = G. juncea E. Greene
Sy = G. lepidota E. Greene
Sy = G. linearifolia M. Lagasca y Segura
Sy = G. linearis P. Rydberg
Sy = G. linoides E. Greene
Sy = G. longifolia E. Greene
Sy = G. longipappa S. F. Blake
Sy = G. pomariensis (S. Welsh) S. Welsh
Sy = G. sarothrae var. pomariensis (S. Welsh) S. Welsh
Sy = G. tenuis E. Greene
Sy = Solidago sarothrae F. Pursh
Sy = Xanthocephalum sarothrae (F. Pursh) L. Shinners
Sy = X. tenue (E. Greene) L. Shinners
G. sphaerocephala A. Gray
Sy = G. eriocarpa A. Gray
Sy = Xanthocephalum sphaerocephalum (A. Gray) L. Shinners
Sy = X. sphaerocephalum var. eriocarpum (A. Gray) L. Shinners
G. texana (A. P. de Candolle) J. Torrey & A. Gray var. ***glutinosa*** (S. Schauer) M. Lane
Sy = G. glutinosa S. Schauer
Sy = Hemiachyris glutinosa S. Schauer
G. texana (A. P. de Candolle) J. Torrey & A. Gray var. ***texana***
Sy = Hemiachyris texana A. P. de Candolle
Sy = Xanthocephalum texanum (A. P. de Candolle) L. Shinners

GYMNOSPERMA C. Lessing
G. glutinosum (K. Sprengel) C. Lessing
Sy = Selloa glutinosa K. Sprengel
Sy = Xanthocephalum glutinosum (K. Sprengel) L. Shinners

GYMNOSTYLES A. L. de Jussieu
G. *stolonifera* (F. Brotero) T. Tutin
Sy = Soliva stolonifera (F. Brotero) J. Loudon

HAPLOESTHES A. Gray
H. greggii A. Gray var. ***texana*** (J. Coulter) I. M. Johnston
Sy = Haplopappus texanus J. Coulter

HEDYPNOIS P. Miller
H. cretica (C. Linnaeus) Dumont de Courset
Sy = H. rhagadioloides var. cretica (C. Linnaeus) G. Hegi

HELENIUM C. Linnaeus
H. amarum (C. Rafinesque-Schmaltz) H. Rock var. ***amarum***
H. amarum (C. Rafinesque-Schmaltz) H. Rock var. ***badium*** (A. Gray *ex* S. Watson) U. Waterfall
Sy = H. badium (A. Gray) E. Greene
H. autumnale C. Linnaeus var. ***autumnale***
Sy = Heleastrum autumnale K. E. O. Kuntze
Sy = Helenium autumnale var. canaliculatum (J. de Lamarck) J. Torrey & A. Gray
H. drummondii H. Rock
H. elegans A. P. de Candolle var. ***amphilobum*** (A. Gray) M. Bierner
Sy = H. amphilobum A. Gray
H. elegans A. P. de Candolle var. ***elegans***
H. flexuosum C. Rafinesque-Schmaltz
H. linifolium P. Rydberg
H. microcephalum A. P. de Candolle var. ***oöclinium*** (A. Gray) M. Bierner
Sy = H. oöclinium A. Gray
H. quadridentatum J. de Houtton Labillardière

HELIANTHUS C. Linnaeus
H. angustifolius C. Linnaeus
Sy = H. angustifolius var. planifolius M. Fernald
H. annuus C. Linnaeus
Sy = H. annuus subsp. jaegeris (C. Heiser) C. Heiser
Sy = H. annuus subsp. lenticularis (D. Douglas *ex* J. Lindley) T. Cockerell
Sy = H. annuus subsp. texanus C. Heiser
Sy = H. annuus var. lenticularis (D. Douglas *ex* J. Lindley) J. Steyermark
Sy = H. annuus var. macrocarpus (A. P. de Candolle) T. Cockerell
Sy = H. annuus var. texanus (C. Heiser) L. Shinners
Sy = H. aridus P. Rydberg
Sy = H. lenticularis D. Douglas *ex* J. Lindley
H. argophyllus J. Torrey & A. Gray
H. ciliaris A. P. de Candolle
H. debilis T. Nuttall subsp. ***cucumerifolius*** (J. Torrey & A. Gray) C. Heiser
Sy = H. cucumerifolius J. Torrey & A. Gray
Sy = H. debilis var. cucumerifolius (J. Torrey & A. Gray) A. Gray
H. debilis T. Nuttall subsp. ***silvestris*** C. Heiser
Sy = H. debilis var. silvestris (C. Heiser) A. Cronquist
H. grosseserratus M. Martens
Sy = H. grosseserratus subsp. maximus R. Long
Sy = H. grosseserratus var. hypoleucus A. Gray
Sy = H. instabilis E. Watson
H. hirsutus C. Rafinesque-Schmaltz var. ***stenophyllus*** J. Torrey & A. Gray
Sy = H. hirsutus var. trachyphyllus J. Torrey & A. Gray
H. laciniatus A. Gray
Sy = H. crenatus R. C. Jackson
Sy = H. heiseri R. C. Jackson
H. maximilianii H. A. Schrader
Sy = H. dalyi N. Britton
H. mollis J. de Lamarck
Sy = H. mollis var. cordatus S. Watson
* ***H. neglectus*** C. Heiser [**TOES: V**]
H. niveus (G. Bentham) T. Brandegee subsp. ***canescens*** (A. Gray) C. Heiser
Sy = H. canescens (A. Gray) C. Heiser
Sy = H. canus (N. Britton) E. Wooton & P. Standley
Sy = H. petiolaris T. Nuttall var. canescens A. Gray
H. occidentalis J. Riddell subsp. ***occidentalis***
Sy = H. dowellianus M. Curtis
Sy = H. occidentalis var. dowellianus (M. Curtis) J. Torrey & A. Gray
H. occidentalis J. Riddell subsp. ***plantagineus*** (J. Torrey & A. Gray) L. Shinners
Sy = H. occidentalis var. plantagineus J. Torrey & A. Gray
* ***H. paradoxus*** C. Heiser [**TOES: III**]
H. pauciflorus T. Nuttall subsp. ***pauciflorus***
Sy = H. laetiflorus C. Persoon var. rigidus (A.-H Cassini) M. Fernald
Sy = H. rigidus (A.-H. Cassini) R. Desfontaines
H. pauciflorus T. Nuttall subsp. ***subrhomboideus*** (P. Rydberg) O. Spring & E. Schilling
Sy = H. laetiflorus C. Persoon var. subrhomboideus (P. Rydberg) M. Fernald
Sy = H. pauciflorus var. subrhomboideus (P. Rydberg) M. Fernald

Sy = H. rigidus (A.-H. Cassini) R. Desfontaines subsp. laetiflorus (P. Rydberg) C. Heiser
Sy = H. rigidus subsp. subrhomboideus (P. Rydberg) C. Heiser
Sy = H. rigidus var. subrhomboideus (P. Rydberg) A. Cronquist
Sy = H. subrhomboideus P. Rydberg
H. petiolaris T. Nuttall
Sy = H. petiolaris subsp. fallax C. Heiser
* ***H. praecox*** G. Engelmann & A. Gray subsp. ***hirtus*** (C. Heiser) C. Heiser [**TOES: V**]
Sy = H. debilis T. Nuttall subsp. hirtus C. Heiser
H. praecox G. Engelmann & A. Gray subsp. ***praecox***
Sy = H. debilis T. Nuttall subsp. praecox (G. Engelmann & A. Gray) C. Heiser
H. praecox G. Engelmann & A. Gray subsp. ***runyonii*** (C. Heiser) C. Heiser
Sy = H. debilis T. Nuttall subsp. runyonii C. Heiser
Sy = H. praecox var. runyonii (C. Heiser) B. L. Turner
H. salicifolius A. Dietrich
Sy = H. filiformis J. K. Small
Sy = H. orgyalis A. P. de Candolle
H. simulans E. Watson
H. strumosus C. Linnaeus
Sy = H. montanus E. Watson
Sy = H. saxicola J. K. Small
H. tuberosus C. Linnaeus [**cultivated**]
Sy = H. neomexicanus E. Wooton & P. Standley
Sy = H. tomentosus A. Michaux
Sy = H. tuberosus var. subcanescens A. Gray

HELIOPSIS C. Persoon
H. helianthoides (C. Linnaeus) R. Sweet var. ***gracilis*** (T. Nuttall) K. Gandhi & R. D. Thomas
Sy = H. gracilis T. Nuttall
H. helianthoides (C. Linnaeus) R. Sweet var. ***scabra*** (M. Dunal) M. Fernald
Sy = H. helianthoides subsp. scabra (M. Dunal) T. Fisher
Sy = H. scabra M. Dunal
H. parvifolia A. Gray

HETEROSPERMA A. Cavanilles
H. pinnatum A. Cavanilles
Sy = H. tagetinum A. Gray

HETEROTHECA A.-H. Cassini
H. canescens (A. P. de Candolle) L. Shinners
Sy = Chrysopsis berlandiera E. Greene
Sy = C. villosa (F. Pursh) T. Nuttall *ex* A. P. de Candolle var. canescens (A. P. de Candolle) A. Gray
Sy = Haplopappus canescens A. P. de Candolle
H. fulcrata (E. Greene) L. Shinners var. ***arizonica*** J. Semple
H. fulcrata (E. Greene) L. Shinners var. ***senilis*** (E. Wooton & P. Standley) J. Semple
Sy = Chrysopsis senilis E. Wooton & P. Standley
H. stenophylla (A. Gray) L. Shinners
Sy = Chrysopsis hispida A. P. de Candolle var. stenophylla A. Gray
Sy = C. stenophylla (A. Gray) E. Greene
Sy = C. villosa (F. Pursh) T. Nuttall *ex* A. P. de Candolle var. stenophylla (A. Gray) A. Gray
H. subaxillaris (J. de Lamarck) N. Britton & H. Rusby
Sy = H. chrysopsidis A. P. de Candolle
Sy = H. latifolia S. Buckley
Sy = H. latifolia var. arkansana B. Wagenknecht
Sy = H. latifolia var. macgregoris B. Wagenknecht
Sy = H. subaxillaris var. latifolia (S. Buckley) K. Gandhi & R. D. Thomas
Sy = H. subaxillaris var. petiolaris H. Benke
Sy = H. subaxillaris var. procumbens B. Wagenknecht
Sy = H. subaxillaris var. psammophila (B. Wagenknecht) K. Gandhi
H. villosa (F. Pursh) L. Shinners var. ***villosa***
Sy = Amellus villosus F. Pursh
Sy = Chrysopsis foliosa T. Nuttall
Sy = C. nitidula E. Wooton & P. Standley
Sy = C. villosa (F. Pursh) T. Nuttall *ex* A. P. de Candolle
Sy = Heterotheca villosa var. angustifolia (P. Rydberg) V. Harms
Sy = H. villosa var. foliosa (T. Nuttall) V. Harms
Sy = H. villosa var. pedunculata (E. Greene) V. Harms *ex* J. Semple
H. viscida (A. Gray) V. Harms
Sy = Chrysopsis villosa (F. Pursh) T. Nuttall *ex* A. P. de Candolle var. viscida A. Gray
Sy = C. viscida (A. Gray) E. Greene

HIERACIUM C. Linnaeus
H. fendleri K. H. Schultz-Bipontinus
Sy = Chlorocrepis fendleri W. A. Weber
Sy = Crepis ambigua A. Gray, non G. Balbis
Sy = Heteropleura fendleri (K. H. Schultz-Bipontinus) P. Rydberg
H. gronovii C. Linnaeus
Sy = H. gronovii var. foliosum A. Michaux
H. wrightii (A. Gray) B. Robinson & J. Greenman

HYMENOCLEA J. Torrey & A. Gray *ex* A. Gray
H. monogyra J. Torrey & A. Gray *ex* A. Gray

HYMENOPAPPUS L'Héritier de Brutelle
H. artemisiifolius A. P. de Candolle var. **artemisiifolius**
Sy = H. scabiosaeus L'Héritier de Brutelle var. artemisiifolius (A. P. de Candolle) K. Gandhi & R. D. Thomas
H. artemisiifolius A. P. de Candolle var. **riograndensis** B. L. Turner
Sy = H. scabiosaeus L'Héritier de Brutelle var. riograndensis (B. L. Turner) K. Gandhi
* **H. biennis** B. L. Turner [**TOES: V**]
H. carrizoanus B. L. Turner
H. filifolius W. Hooker var. **cinereus** (P. Rydberg) I. M. Johnston
Sy = H. arenosus A. A. Heller
Sy = H. cinereus P. Rydberg
H. flavescens A. Gray var. **canotomentosus** A. Gray
Sy = H. robustus E. Greene
H. flavescens A. Gray var. **flavescens**
Sy = H. fisheri E. Wooton & P. Standley
Sy = Rothia flavescens (A. Gray) K. E. O. Kuntze
H. scabiosaeus L'Héritier de Brutelle var. **corymbosus** (J. Torrey & A. Gray) B. L. Turner
Sy = H. corymbosus J. Torrey & A. Gray
H. scabiosaeus L'Héritier de Brutelle var. **scabiosaeus**
Sy = H. carolinensis (J. de Lamarck) T. Porter
H. tenuifolius F. Pursh
Sy = Rothia tenuifolia K. E. O. Kuntze

HYMENOTHRIX A. Gray
H. wislizenii A. Gray
H. wrightii A. Gray

HYMENOXYS A.-H. Cassini
H. odorata A. P. de Candolle
H. richardsonii (W. Hooker) T. Cockerell var. **floribunda** (A. Gray) K. Parker
Sy = H. olivacea T. Cockerell
* **H. texana** (J. Coulter & J. Rose) T. Cockerell [**TOES: I**]

HYPOCHAERIS C. Linnaeus
H. brasiliensis (C. Lessing) A. Grisebach var. **tweedyi** (W. Hooker & G. Arnott) J. G. Baker
Sy = H. tweedyi W. Hooker & G. Arnott
H. microcephala (K. H. Schultz-Bipontinus) A. Cabrera var. **albiflora** (K. E. O. Kuntze) A. Cabrera
H. radicata C. Linnaeus

IONACTIS E. Greene
I. linariifolius (C. Linnaeus) E. Greene
Sy = Aster linariifolius C. Linnaeus

ISOCARPHA R. Brown
I. oppositifolia (C. Linnaeus) A.-H. Cassini var. **achyranthes** (A. P. de Candolle) D. Keil & T. Stuessy

ISOCOMA T. Nuttall
I. coronopifolia (A. Gray) E. Greene
Sy = I. coronopifolia (A. Gray) E. Greene var. pedicellata (E. Greene) G. Nesom
Sy = I. pedicellata E. Greene
I. drummondii (J. Torrey & A. Gray) E. Greene
Sy = Haplopappus drummondii (J. Torrey & A. Gray) S. F. Blake
Sy = Isocoma megalantha L. Shinners
I. pluriflora (J. Torrey & A. Gray) E. Greene
Sy = I. wrightii (A. Gray) P. Rydberg

IVA C. Linnaeus
I. ambrosiifolia (A. Gray) A. Gray
Sy = Cyclachaena ambrosiifolia (A. Gray) G. Bentham & J. Hooker
Sy = Euphrosyne ambrosiifolia A. Gray
I. angustifolia T. Nuttall *ex* A. P. de Candolle
I. annua C. Linnaeus var. **annua**
Sy = I. ciliata C. von Willdenow var. ciliata
I. annua C. Linnaeus var. **caudata** (J. K. Small) R. Jackson
Sy = I. caudata J. K. Small
I. axillaris F. Pursh
Sy = I. axillaris subsp. robustior (W. Hooker) I. J. Bassett
Sy = I. axillaris var. robustior W. Hooker
I. dealbata A. Gray
I. frutescens C. Linnaeus subsp. **frutescens**
I. imbricata T. Walter
I. texensis R. R. Johnson
Sy = I. angustifolia T. Nuttall *ex* A. P. de Candolle var. latior L. Shinners
I. xanthifolia T. Nuttall
Sy = Cyclachaena xanthifolia (T. Nuttall) J. Fresentius
Sy = Euphrosyne xanthifolia A. Gray

JEFEA J. Strother
J. brevifolia (A. Gray) J. Strother
Sy = Zexmenia brevifolia A. Gray

KOANOPHYLLON M. Arruda da Cámara *ex* R. King & H. Robinson
K. solidaginifolia (A. Gray) R. King & H. Robinson
Sy = Eupatorium solidaginifolium A. Gray

KRIGIA J. von Schreber, *nomen conservandum*
K. cespitosa (C. Rafinesque-Schmaltz) K. Chambers
Sy = K. gracilis (A. P. de Candolle) L. Shinners
Sy = K. oppositifolia C. Rafinesque-Schmaltz
K. dandelion (C. Linnaeus) T. Nuttall
Sy = Cynthia dandelion (C. Linnaeus) A. P. de Candolle
K. occidentalis T. Nuttall
Sy = Cymbia occidentalis (T. Nuttall) P. Standley
K. virginica (C. Linnaeus) C. von Willdenow
Sy = Hyoseris virginica C. Linnaeus
K. wrightii (A. Gray) K. Chambers *ex* Ki-Joong Kim

LACTUCA C. Linnaeus
L. canadensis C. Linnaeus var. **_canadensis_**
Sy = L. canadensis var. integrifolia (J. Bigelow) J. Torrey & A. Gray
L. canadensis C. Linnaeus var. **_latifolia_** K. E. O. Kuntze
L. floridana (C. Linnaeus) J. Gaertner var. **_floridana_**
Sy = Mulgedium floridanum (C. Linnaeus) A. P. de Candolle
L. floridana (C. Linnaeus) J. Gaertner var. **_villosa_** (N. von Jacquin) A. Cronquist
Sy = L. villosa N. von Jacquin
Sy = Mulgedium villosum (N. von Jacquin) J. K. Small
L. graminifolia A. Michaux var. **_graminifolia_**
Sy = L. graminifolia var. arizonica R. McVaugh
L. hirsuta G. H. Muhlenberg *ex* T. Nuttall var. **_albiflora_** (J. Torrey & A. Gray) L. Shinners
L. hirsuta G. H. Muhlenberg *ex* T. Nuttall var. **_sanguinea_** (J. Bigelow) M. Fernald
L. intybacea N. von Jacquin
L. ludoviciana (T. Nuttall) J. Riddell
L. sativa C. Linnaeus [**cultivated**]
L. serriola C. Linnaeus
Sy = L. integrata (J. Grenier & D. Godron) A. Nelson
Sy = L. scariola C. Linnaeus var. integrifolia (C. Bogenhard) G. Beck
Sy = L. scariola var. integrata J. Grenier & D. Godron
Sy = L. virosa C. Linnaeus var. integrifolia A. Gray
L. tatarica (C. Linnaeus) C. von Meyer var. **_pulchella_** (F. Pursh) A. Breitung
Sy = L. tatarica subsp. pulchella (F. Pursh) G. Stebbins
Sy = L. tatarica var. heterophylla (T. Nuttall) J. Boivin

LAENNECIA A.-H. Cassini
L. coulteri (A. Gray) G. Nesom
Sy = Conyza coulteri A. Gray
Sy = Conyzella coulteri E. Greene
Sy = Eschenbachia coulteri (A. Gray) P. Rydberg
L. filaginoides (A. P. de Candolle) G. Nesom
Sy = Conyza filaginoides (A. P. de Candolle) G. Hieronymus
L. sophiifolia (K. Kunth) G. Nesom
Sy = Conyza sophiifolia K. Kunth

LEPIDOSPARTUM (A. Gray) A. Gray
* **_L. burgessii_** B. L. Turner [**TOES: V**]

LIATRIS J. Gaertner *ex* J. von Schreber, *nomen conservandum*
L. acidota G. Engelmann & A. Gray
Sy = Lacinaria acidota (G. Engelmann & A. Gray) K. E. O. Kuntze
L. aspera A. Michaux var. **_aspera_**
Sy = Lacinaria aspera (A. Michaux) A. Greene
L. aspera A. Michaux var. **_salutans_** (J. Lunell) L. Shinners
* **_L. bracteata_** L. Gaiser [**TOES: V**]
* **_L. cymosa_** (H. Ness) K. Schumann [**TOES: V**]
L. earlei (T. Walter) K. Schumann
Ma = L. squarrulosa *auct.*, non A. Michaux
L. elegans (T. Walter) A. Michaux var. **_carizzana_** L. Gaiser
L. elegans (T. Walter) A. Michaux var. **_elegans_**
Sy = Lacinaria elegans (T. Walter) K. E. O. Kuntze
L. lancifolia (E. Greene) T. Kittell
Sy = Lacinaria lancifolia E. Greene
L. mucronata A. P. de Candolle
L. punctata W. Hooker var. **_nebraskana_** L. Gaiser
L. punctata W. Hooker var. **_punctata_**
Sy = Lacinaria punctata (W. Hooker) K. E. O. Kuntze
L. pycnostachya A. Michaux var. **_lasiophylla_** L. Shinners
L. pycnostachya A. Michaux var. **_pycnostachya_**
Sy = Lacinaria pycnostachya (A. Michaux) K. E. O. Kuntze

L. scariosa (C. Linnaeus) C. von Willdenow var. ***nieuwlandii*** J. Lunell **[cultivated]**
Sy = L. novoae-angilae (J. Lunell) L. Shinners var. nieuwlandii (J. Lunell) L. Shinners
Sy = L. × nieuwlandii (J. Lunell) L. Gaiser
L. spicata (C. Linnaeus) C. von Willdenow **[cultivated]**
Sy = Lacinaria spicata (C. Linnaeus) K. E. O. Kuntze
L. squarrosa (C. Linnaeus) A. Michaux var. ***alabamensis*** (E. Alexander) L. Gaiser
Sy = L. glabrata P. Rydberg var. alabamensis (E. Alexander) L. Shinners
L. squarrosa (C. Linnaeus) A. Michaux var. ***glabrata*** (P. Rydberg) L. Gaiser
Sy = L. glabrata P. Rydberg
L. squarrosa (C. Linnaeus) A. Michaux var. ***hirsuta*** (P. Rydberg) L. Gaiser
Sy = L. hirsuta P. Rydberg
* ***L. tenuis*** L. Shinners **[TOES: V]**

LINDHEIMERA A. Gray & G. Engelmann
L. texana A. Gray & G. Engelmann

LYGODESMIA D. Don
L. juncea (F. Pursh) D. Don *ex* W. Hooker
Sy = L. juncea var. racemosa J. Lunell
Sy = Prenanthes juncea F. Pursh
L. ramosissima J. Greenman
L. texana (J. Torrey & A. Gray) E. Greene
Sy = L. aphylla (T. Nuttall) A. P. de Candolle var. texana J. Torrey & A. Gray

MACHAERANTHERA C. Nees von Esenbeck
M. annua (P. Rydberg) L. Shinners
Sy = Haplopappus annuus (P. Rydberg) V. Cory
* ***M. aurea*** (A. Gray) L. Shinners **[TOES: V]**
Sy = Haplopappus aureus A. Gray
Sy = Sideranthus aureus (A. Gray) J. K. Small
M. blephariphylla (A. Gray) L. Shinners
Sy = Haplopappus blephariphyllus A. Gray
M. gracilis (T. Nuttall) L. Shinners
Sy = Haplopappus gracilis (T. Nuttall) A. Gray
M. gypsophila B. L. Turner
M. linearis E. Greene
Sy = Aster linearis (E. Greene) V. Cory
Sy = Machaeranthera canescens (F. Pursh) A. Gray subsp. glabra (A. Gray) B. L. Turner var. glabra
Sy = M. canescens var. viridis A. Gray
Sy = M. fremontii P. Rydberg
Sy = M. linearis E. Greene var. nebraskana B. L. Turner
M. parviflora A. Gray
Sy = Aster parviflorus A. Gray
Sy = A. parvulus S. F. Blake
Sy = A. tanacetifolius K. Kunth var. pygmaeus A. Gray
Sy = Machaeranthera pygmaea (A. Gray) E. Wooton & P. Standley
M. phyllocephala (A. P. de Candolle) L. Shinners
Sy = Haplopappus phyllocephalus A. P. de Candolle
Sy = Machaeranthera phyllocephala var. megacephala (G. Nash) L. Shinners
M. pinnatifida (W. Hooker) L. Shinners subsp. ***pinnatifida*** var. ***chihuahuanus*** B. L. Turner & R. Hartman
Sy = Haplopappus spinulosus (F. Pursh) A. P. de Candolle var. chihuahuana (B. L. Turner & R. Hartman) K. Gandhi
M. pinnatifida (W. Hooker) L. Shinners subsp. ***pinnatifida*** var. ***glaberrima*** (P. Rydberg) B. L. Turner & R. Hartman
Sy = Haplopappus spinulosus (F. Pursh) A. P. de Candolle subsp. glaberrimus (P. Rydberg) H. Hall
Sy = H. spinulosus var. glaberrimus (P. Rydberg) S. F. Blake
M. pinnatifida (W. Hooker) L. Shinners subsp. ***pinnatifida*** var. ***pinnatifida***
Sy = Aster pinnatifidus (W. Hooker) K. E. O. Kuntze
Sy = Haplopappus spinulosus (F. Pursh) A. P. de Candolle
Sy = Machaeranthera texensis (R. Jackson) L. Shinners
M. tanacetifolia (K. Kunth) C. Nees von Esenbeck
Sy = Aster chrysanthemoides C. von Willdenow *ex* K. Sprengel
Sy = A. tanacetifolius K. Kunth
Sy = Dieteria coronopifolia T. Nuttall
Sy = Machaeranthera coronopifolia (T. Nuttall) A. Nelson
Sy = M. parthenium E. Greene
M. viscida (E. Wooton & P. Standley) R. Hartman
Sy = Haplopappus havardii U. Waterfall
Sy = Machaeranthera havardii (U. Waterfall) L. Shinners

MALACOTHRIX A. P. de Candolle
M. fendleri A. Gray

MARSHALLIA J. von Schreber, *nomen conservandum*
M. caespitosa T. Nuttall *ex* A. P. de Candolle var. ***caespitosa***

M. caespitosa T. Nuttall *ex* A. P. de Candolle var. ***signata*** C. Beadle & F. Boynton
M. graminifolia (T. Walter) J. K. Small var. ***cynanthera*** (S. Elliott) C. Beadle & F. Boynton
Sy = M. graminifolia subsp. tenuifolia (C. Rafinesque-Schmaltz) L. Watson
Sy = M. tenuifolia C. Rafinesque-Schmaltz

MATRICARIA C. Linnaeus
M. recutita C. Linnaeus [**cultivated**]
Sy = Chamomilla chamomilla (C. Linnaeus) P. Rydberg
Sy = C. recutita (C. Linnaeus) S. Rauschert
Sy = Matricaria chamomilla C. Linnaeus var. coronata (J. Gay) E. Cosson & Germain de Saint-Pierre
Sy = M. suaveolens C. Linnaeus

MELAMPODIUM C. Linnaeus
M. cinereum A. P. de Candolle var. ***cinereum***
M. cinereum A. P. de Candolle var. ***hirtellum*** T. Stuessy
M. cinereum A. P. de Candolle var. ***ramosissimum*** (A. P. de Candolle) A. Gray
M. leucanthum J. Torrey & A. Gray
M. strigosum T. Stuessy

MICROSERIS D. Don
M. lindleyi (A. P. de Candolle) A. Gray
Sy = Calais lindleyi A. P. de Candolle
Sy = C. linearifolia A. P. de Candolle
Sy = Hyponema glaucum W. Hooker
Sy = Microseris linearifolia (A. P. de Candolle) K. H. Schultz-Bipontinus
Sy = Uropappus lindleyi T. Nuttall
Sy = U. linearifolia T. Nuttall

MIKANIA C. von Willdenow, *nomen conservandum*
M. cordifolia (C. von Linné) C. von Willdenow
Sy = Cacalia cordifolia C. von Linné
M. scandens (C. Linnaeus) C. von Willdenow
Sy = M. scandens var. pubescens (T. Nuttall) J. Torrey & A. Gray

NICOLLETIA A. Gray
N. edwardsii A. Gray

NOTHOCALAIS (A. Gray) E. Greene
N. cuspidata (F. Pursh) E. Greene
Sy = Agoseris cuspidata (F. Pursh) C. Rafinesque-Schmaltz
Sy = Microseris cuspidata (F. Pursh) K. H. Schultz-Bipontinus

OLIGONEURON J. K. Small
O. nitidum (J. Torrey & A. Gray) J. K. Small
Sy = Solidago nitida J. Torrey & A. Gray
O. rigidum (C. Linnaeus) J. K. Small
Sy = Aster rigidus (C. Linnaeus) K. E. O. Kuntze
Sy = Oligoneuron canescens P. Rydberg
Sy = O. grandiflorum (C. Rafinesque-Schmaltz) J. K. Small
Sy = Solidago grandiflora C. Rafinesque-Schmaltz
Sy = S. rigida C. Linnaeus
Sy = S. rigida subsp. humilis (J. Poiret) S. Heard & J. Semple
Sy = S. rigida var. glabrata E. Braun
Sy = S. rigida var. humilis T. Porter
Sy = S. rigida var. laevicaulis L. Shinners
Sy = S. rigida var. magna W. Clute
Sy = S. rigida var. microcephala A. P. de Candolle

ONOPORDUM C. Linnaeus
O. acanthium C. Linnaeus

PALAFOXIA M. Lagasca y Segura
P. callosa (T. Nuttall) J. Torrey & A. Gray
Sy = Othake callosa (T. Nuttall) B. Bush
Sy = O. tenuifolium C. Rafinesque-Schmaltz
Sy = Palafoxia callosa var. bella (V. Cory) L. Shinners
Sy = Polypteris callosa (T. Nuttall) A. Gray
Sy = Stevia callosa T. Nuttall
P. hookeriana J. Torrey & A. Gray var. ***hookeriana***
P. hookeriana J. Torrey & A. Gray var. ***minor*** L. Shinners
P. reverchonii (B. Bush) V. Cory
P. riograndensis V. Cory
P. rosea (B. Bush) V. Cory var. ***macrolepis*** (P. Rydberg) B. L. Turner & M. Morris
Sy = Othake macrolepis P. Rydberg
Sy = O. texanum B. Bush var. macrolepis (P. Rydberg) G. R. Ammerman
Sy = Palafoxia macrolepis (P. Rydberg) V. Cory
Sy = P. texana A. P. de Candolle var. macrolepis (P. Rydberg) L. Shinners
Sy = Polypteris macrolepis (P. Rydberg) V. Cory
P. rosea (B. Bush) V. Cory var. ***rosea***
P. sphacelata (T. Nuttall *ex* J. Torrey) V. Cory
Sy = Othake sphacelata (T. Nuttall *ex* J. Torrey) P. Rydberg

Sy = Polypteris sphacelata (T. Nuttall *ex* J. Torrey) W. Trelease
Sy = Stevia sphacelata T. Nuttall
P. texana A. P. de Candolle var. ***ambigua*** (L. Shinners) B. L. Turner & M. Morris
Sy = P. rosea (B. Bush) V. Cory var. ambigua L. Shinners
P. texana A. P. de Candolle var. ***texana***
Sy = P. rosea (B. Bush) V. Cory var. papposa L. Shinners

PARTHENIUM C. Linnaeus
P. argentatum A. Gray
Sy = P. lloydii H. Bartlett
P. confertum A. Gray var. ***divaricatum*** R. Rollins
P. confertum A. Gray var. ***lyratum*** (A. Gray) R. Rollins
Sy = P. hysterophorus C. Linnaeus var. lyratum A. Gray
Sy = P. lyratum (A. Gray) A. Gray
P. confertum A. Gray var. ***microcephalum*** R. Rollins
P. hysterophorus C. Linnaeus
Sy = P. lobatum S. Buckley
P. incanum K. Kunth
Sy = P. ramosissimum A. P. de Candolle
P. integrifolium C. Linnaeus var. ***hispidum*** (C. Rafinesque-Schmaltz) J. Mears
Sy = P. hispidum C. Rafinesque-Schmaltz
P. integrifolium C. Linnaeus var. ***integrifolium***

PECTIS C. Linnaeus
P. angustifolia J. Torrey var. ***fastigiata*** (A. Gray) D. Keil
Sy = P. texana V. Cory
P. angustifolia J. Torrey var. ***tenella*** (A. P. de Candolle) D. Keil
Sy = P. tenella A. P. de Candolle
P. cylindrica (M. Fernald) P. Rydberg
Sy = P. prostrata A. Cavanilles var. cylindrica M. Fernald
P. filipes W. Harvey & A. Gray var. ***subnuda*** M. Fernald
P. longipes A. Gray
P. papposa W. Harvey & A. Gray var. ***grandis*** D. Keil
P. prostrata A. Cavanilles
Sy = Chthonia prostrata A.-H. Cassini
Sy = Pectis prostrata var. urceolata (P. Rydberg) M. Fernald
Sy = P. urceolata P. Rydberg

PERICOME A. Gray
P. caudata A. Gray
Sy = P. caudata var. glandulosa (G. Goodman) H. Harrington
Sy = P. glandulosa G. Goodman

PERITYLE G. Bentham
P. aglossa A. Gray
Sy = Laphamia aglossa (A. Gray) W. Hemsley
Sy = Leptopharynx aglossa (A. Gray) P. Rydberg
P. angustifolia (A. Gray) L. Shinners
Sy = Laphamia angustifolia L. Garay
Sy = L. angustifolia subsp. laciniata (A. Gray) W. Niles
P. bisetosa (J. Torrey *ex* A. Gray) L. Shinners var. ***appressa*** A. M. Powell
* **P. bisetosa** (J. Torrey *ex* A. Gray) L. Shinners var. ***bisetosa*** **[TOES: V]**
Sy = Laphamia bisetosa J. Torrey *ex* A. Gray
* **P. bisetosa** (J. Torrey *ex* A. Gray) L. Shinners var. ***scalaris*** A. M. Powell **[TOES: V]**
P. cinerea (A. Gray) A. M. Powell
Sy = Laphamia cinerea A. Gray
Sy = Pappothrix cinerea (A. Gray) P. Rydberg
P. dissecta (J. Torrey) A. Gray
Sy = Laphamia dissecta J. Torrey
Sy = Leptopharynx dissecta (J. Torrey) P. Rydberg
P. fosteri A. M. Powell
* **P. huecoënsis** A. M. Powell **[TOES: V]**
P. lindheimeri (A. Gray) L. Shinners var. ***halimifolia*** (A. Gray) A. M. Powell
Sy = Laphamia halimifolia A. Gray
Sy = Perityle halimifolia (A. Gray) L. Shinners
P. lindheimeri (A. Gray) L. Shinners var. ***lindheimeri***
Sy = Laphamia halimifolia subsp. lindheimeri (A. Gray) W. Niles
Sy = L. lindheimeri A. Gray
Sy = Perityle rotundata (P. Rydberg) L. Shinners
P. microglossa G. Bentham var. ***microglossa***
Sy = P. spilanthoides (K. H. Schultz-Bipontinus) P. Rydberg
P. parryi A. Gray
Sy = Laphamia parryi (A. Gray) G. Bentham
Sy = Leptopharynx trisecta P. Rydberg
P. quinqueflora (J. Steyermark) L. Shinners
Sy = Laphamia quinqueflora J. Steyermark
Sy = Pappothrix quinqueflora (J. Steyermark) M. Hammack
P. rupestris (A. Gray) L. Shinners var. ***albiflora*** A. M. Powell

P. rupestris (A. Gray) L. Shinners var. **rupestris**
Sy = Laphamia rupestris A. Gray
P. vaseyi J. Coulter
* **P. vitreomontana** B. Warnock [**TOES**: **V**]
* **P. warnockii** A. M. Powell [**TOES**: **V**]

PICRADENIOPSIS P. Rydberg *ex* N. Britton
P. woodhousii (A. Gray) P. Rydberg
Sy = Bahia woodhousii (A. Gray) A. Gray

PINAROPAPPUS C. Lessing
P. parvus S. F. Blake
P. roseus (C. Lessing) C. Lessing var. **roseus**

PITYOPSIS T. Nuttall
P. graminifolia (A. Michaux) T. Nuttall var. **graminifolia**
Sy = Chrysopsis graminifolia (A. Michaux) S. Elliott
Sy = Heterotheca graminifolia (A. Michaux) L. Shinners
P. graminifolia (A. Michaux) T. Nuttall var. **tenuifolia** (J. Torrey) J. Semple & F. Bowers
Sy = Heterotheca graminifolia (A. Michaux) L. Shinners var. microcephala (J. K. Small) A. Cronquist
Sy = H. graminifolia var. tenuifolia (J. Torrey) K. Gandhi & R. D. Thomas
P. oligantha (A. Chapman *ex* J. Torrey & A. Gray) J. K. Small
Sy = Chrysopsis oligantha (A. Chapman *ex* J. Torrey) V. Harms
Sy = Heterotheca oligantha (A. Chapman) V. Harms

PLUCHEA A.-H. Cassini
P. camphorata (C. Linnaeus) A. P. de Candolle
Sy = P. petiolata A.-H. Cassini
Sy = P. viscida (C. Rafinesque-Schmaltz) H. House
P. foetida (C. Linnaeus) A. P. de Candolle var. **foetida**
Sy = P. foetida var. imbricata T. Kearney
Sy = P. tenuifolia J. K. Small
P. purpurascens (O. Swartz) A. P. de Candolle
Sy = Conyza purpurascens O. Swartz
Sy = Placus purpurascens (O. Swartz) M. Gómez
Ma = Pluchea odorata *auct.*, non (C. Linnaeus) A.-H. Cassini var. odorata: American authors
P. rosea R. Godfrey
P. sericea (T. Nuttall) F. Coville
Sy = Berthelotia sericea (T. Nuttall) P. Rydberg
Sy = Eremohylema sericea A. Nelson
Sy = Polypappus sericeus T. Nuttall
Sy = Tessaria sericea (T. Nuttall) L. Shinners

POROPHYLLUM M. Adanson
P. gracile G. Bentham
Sy = P. caesium E. Greene
Sy = P. junciforme E. Greene
Sy = P. putidum A. Nelson
Sy = P. vaseyi E. Greene
P. greggii A. Gray
P. ruderale (N. von Jacquin) A.-H. Cassini subsp. **macrocephalum** (A. P. de Candolle) R. R. Johnson
Sy = P. latifolium L. Bentham
Sy = P. macrophyllum A. P. de Candolle
Sy = P. ruderale var. macrocephalum (A. P. de Candolle) A. Cronquist
P. scoparium A. Gray
Sy = P. fruticulosum P. Rydberg

PRENANTHELLA P. Rydberg
P. exigua (A. Gray) P. Rydberg
Sy = Lygodesmia exigua (A. Gray) A. Gray

PRENANTHES C. Linnaeus
P. altissima C. Linnaeus
Sy = Nabalus altissimus (C. Linnaeus) W. Hooker
Sy = Prenanthes altissima var. cinnamomea M. Fernald
* **P. barbata** (J. Torrey & A. Gray) W. Milstead [**TOES**: **V**]
Sy = P. serpentaria F. Pursh var. barbata (J. Torrey & A. Gray) A. Gray

PSATHYROTES A. Gray
P. scaposa A. Gray
Sy = Pseudobartlettia scaposa (A. Gray) P. Rydberg

PSEUDOCLAPPIA P. Rydberg
P. arenaria P. Rydberg
Sy = Clappia suaedifolia E. Wooton & P. Standley, non A. Gray
* **P. watsonii** A. M. Powell & B. L. Turner [**TOES**: **V**]

PSEUDOGNAPHALIUM M. E. Kirpichnikov
P. brachypterum (A. P. de Candolle) A. Anderberg
Sy = Gnaphalium brachypterum A. P. de Candolle

P. canescens (A. P. de Candolle) A. Anderberg
Sy = Gnaphalium canescens A. P. de Candolle
Sy = G. wrightii A. Gray
Sy = Pseudognaphalium canescens (A. P. de Candolle) W. A. Weber
P. helleri (N. Britton) A. Anderberg
Sy = Gnaphalium helleri N. Britton
Sy = G. helleri N. Britton var. micradenium (C. Weatherby) W. Mahler
Sy = G. obtusifolium C. Linnaeus var. helleri (N. Britton) S. F. Blake
Sy = G. obtusifolium var. micradenium C. Weatherby
P. leucocephalum (A. Gray) A. Anderberg
Sy = Gnaphalium leucocephalum A. Gray
P. obtusifolium (C. Linnaeus) O. Hilliard & B. D. Burtt
Sy = Gnaphalium obtusifolium C. Linnaeus
Sy = G. saxicola N. Fassett
P. stramineum (K. Kunth) A. Anderberg
Sy = Gnaphalium chilense K. Sprengel
Sy = G. proximum E. Greene
Sy = G. stramineum K. Kunth
Sy = Pseudognaphalium stramineum (K. Kunth) W. A. Weber
P. viscosum (K. Kunth) A. Anderberg
Sy = Gnaphalium decurrens E. Ives, non C. Linnaeus
Sy = G. gracile K. Kunth
Sy = G. leptophyllum A. P. de Candolle
Sy = G. viscosum K. Kunth
Sy = Pseudognaphalium viscosum (K. Kunth) W. A. Weber
Ma = Gnaphalium macounii *auct.*, non E. Greene: American authors

PSEUDOGYNOXYS (J. Greenman) A. Cabrera
P. chenopodioides (K. Kunth) A. Cabrera var. ***chenopodioides*** **[cultivated]**
Sy = Senecio confusus J. Britten

PSILACTIS A. Gray
P. asteroides A. Gray
Sy = Aster boltoniae E. Greene
Sy = Machaeranthera boltoniae (E. Greene) B. L. Turner & D. Horne
Sy = Psilactis leptos L. Shinners
P. brevilingulata K. H. Schultz-Bipontinus *ex* W. Hemsley
Sy = Aster brevilingulata (K. H. Schultz-Bipontinus *ex* W. Hemsley) R. McVaugh
Sy = Machaeranthera brevilingulata (K. H. Schultz-Bipontinus *ex* W. Hemsley) B. L. Turner & D. Horne
* ***P. heterocarpa*** (R. Hartman & M. Lane) D. Morgan **[TOES: V]**
Sy = Machaeranthera heterocarpa R. Hartman & M. Lane
P. tenuis S. Watson
Sy = Machaeranthera tenuis (S. Watson) B. L. Turner & D. Horne

PSILOSTROPHE A. P. de Candolle
P. gnaphalioides A. P. de Candolle
P. tagetina (T. Nuttall) E. Greene var. ***cerifera*** (A. Nelson) B. L. Turner
Sy = P. villosa P. Rydberg
P. tagetina (T. Nuttall) E. Greene var. ***tagetina***
Sy = P. lanata P. Rydberg
Sy = P. tagetina var. grandiflora (P. Rydberg) C. Heiser
Sy = P. tagetina var. lanata A. Nelson
Sy = Riddellia tagetinae (T. Nuttall) E. Greene

PTEROCAULON S. Elliott
P. virgatum (C. Linnaeus) A. P. de Candolle
Sy = Gnaphalium virgatum C. Linnaeus

PYRRHOPAPPUS A. P. de Candolle, *nomen conservandum*
P. carolinianus (T. Walter) A. P. de Candolle
Sy = P. carolinianus var. georgianus (L. Shinners) H. Ahles
Sy = P. georgianus L. Shinners
Sy = Sitilias caroliniana (T. Walter) C. Rafinesque-Schmaltz
P. grandiflorus (T. Nuttall) T. Nuttall
Sy = P. scaposus A. P. de Candolle
P. pauciflorus (D. Don) A. P. de Candolle
Sy = P. geiseri L. Shinners
Sy = P. multicaulis A. P. de Candolle
Sy = P. multicaulis var. geiseri (L. Shinners) D. Northington
Sy = P. rothrockii A. Gray
Sy = Sitilias multicaulis (A. P. de Candolle) E. Greene

RAFINESQUIA T. Nuttall, *nomen conservandum*
R. neomexicana A. Gray
Sy = Nemoseris neomexicana (A. Gray) E. Greene

RATIBIDA C. Rafinesque-Schmaltz
R. columnifera (T. Nuttall) E. Wooton & P. Standley
Sy = Lepachys columnifera (T. Nuttall) P. Rydberg
Sy = Obeliscaria pulcherrima A. P. de Candolle
Sy = Ratibida columnaris (J. Sims) D. Don
Sy = R. columnaris var. pulcherrima (A. P. de Candolle) D. Don
Sy = Rudbeckia columnaris J. Sims
Sy = R. columnifera T. Nuttall
R. peduncularis (J. Torrey & A. Gray) J. Barnhart var. ***peduncularis***
Sy = Lepachys peduncularis J. Torrey & A. Gray
R. tagetes (E. James) J. Barnhart
Sy = R. peduncularis (J. Torrey & A. Gray) J. Barnhart var. tagetes (E. James) J. Barnhart
Sy = Rudbeckia tagetes E. James

RUDBECKIA C. Linnaeus
R. fulgida W. Aiton var. ***fulgida***
Sy = R. acuminata F. Boynton & C. Beadle
Sy = R. coryi L. Shinners
Sy = R. foliosa F. Boynton & C. Beadle
Sy = R. fulgida var. umbrosa (F. Boynton & C. Beadle) A. Cronquist
Sy = R. palustris H. Eggert *ex* F. Boynton & C. Beadle
Sy = R. spathulata A. Michaux
Sy = R. tenax F. Boynton & C. Beadle
R. grandiflora (D. Don) J. F. Gmelin *ex* A. P. de Candolle var. ***alismifolia*** (J. Torrey & A. Gray) A. Cronquist
Sy = R. alismifolia J. Torrey & A. Gray
R. grandiflora (D. Don) J. F. Gmelin *ex* A. P. de Candolle var. ***grandiflora***
Sy = Centrocarpa grandiflora D. Don
R. hirta C. Linnaeus var. ***angustifolia*** (T. Moore) R. Perdue
Sy = R. bicolor T. Nuttall
Sy = R. divergens T. Moore
Sy = R. floridana T. Moore var. angustifolia T. Moore
R. hirta C. Linnaeus var. ***pulcherrima*** O. Farwell
Sy = R. hirta var. corymbifera M. Fernald
Sy = R. hirta var. lanceolata (G. Bischoff) E. Core
Sy = R. hirta var. sericea (T. Moore) M. Fernald
Sy = R. hirta var. serotina (T. Nuttall) E. Core
Sy = R. longipes T. Moore
Sy = R. sericea T. Moore
Sy = R. serotina T. Nuttall
Sy = R. serotina var. corymbifera (M. Fernald) M. Fernald & B. Schubert
Sy = R. serotina var. lanceolata (G. Bischoff) M. Fernald & B. Schubert
Sy = R. serotina var. sericea (T. Moore) M. Fernald & B. Schubert
R. laciniata C. Linnaeus var. ***laciniata***
Sy = R. laciniata var. hortensis L. Bailey
R. maxima T. Nuttall
R. missouriensis G. Engelmann *ex* C. Boynton & C. Beadle
Sy = R. fulgida W. Aiton var. missouriensis (G. Engelmann *ex* C. Boynton & C. Beadle) A. Cronquist
R. nitida T. Nuttall var. ***texana*** R. Perdue
Sy = R. texana (R. Perdue) Cox & L. Urbatsch
* ***R. scabrifolia*** L. E. Brown [**TOES: V**]
R. subtomentosa F. Pursh
R. triloba C. Linnaeus var. ***trilobata***

SANTOLINA C. Linnaeus
S. chamaecyparissus C. Linnaeus [**cultivated**]
S. viridis J. Poiret [**cultivated**]
Sy = S. virens P. Miller

SANVITALIA J. de Lamarck
S. abertii A. Gray
S. angustifolia A. Gray
S. ocymoides A. P. de Candolle

SARTWELLIA A. Gray
S. flaveriae A. Gray

SCHKUHRIA A. Roth, *nomen conservandum*
S. anthemoidea (A. P. de Candolle) J. Coulter var. ***anthemoidea***
S. anthemoidea (A. P. de Candolle) J. Coulter var. ***wrightii*** (A. Gray) C. Heiser
Sy = S. wislizeni A. Gray var. wrightii (A. Gray) S. F. Blake
Sy = S. wrightii A. Gray
S. multiflora W. Hooker & G. Arnott
Sy = Bahia neomexicana (A. Gray) A. Gray
Sy = Baileya neomexicana (A. Gray) A. Gray
S. wislizenii A. Gray var. ***wislizenii***

SCLEROCARPUS N. von Jacquin
S. uniserialis (G. Bentham) W. Hemsley var. ***austrotexanus*** B. L. Turner
S. uniserialis (G. Bentham) W. Hemsley var. ***uniserialis***
Sy = Gymnopsis uniserialis W. Hooker
Sy = Sclerocarpus major J. K. Small

SCORZONERA C. Linnaeus
- ***S. laciniata*** C. Linnaeus
  - Sy = Podospermum laciniatum (C. Linnaeus) A. P. de Candolle

SENECIO C. Linnaeus
- ***S. ampullaceus*** W. Hooker
- ***S. cineraria*** A. P. de Candolle [**cultivated**]
  - Sy = S. bicolor (C. von Willdenow) A. Todaro var. cineraria (A. P. de Candolle) A. Chater
- ***S. douglasii*** A. P. de Candolle
  - Sy = S. douglasii var. monoënsis W. Jepson
  - Sy = S. douglasii var. tularensis P. Munz
  - Sy = S. flaccidus C. Lessing var. douglasii B. L. Turner & T. Barkley
  - Sy = S. flaccidus var. monoënsis B. L. Turner & T. Barkley
  - Sy = S. monoënsis E. Greene
- ***S. flaccidus*** C. Lessing
  - Sy = S. douglasii subsp. longilobus (G. Bentham) W. A. Weber
  - Sy = S. douglasii var. jamesii R. Ediger *ex* D. Correll & M. C. Johnston
  - Sy = S. douglasii var. longilobus (G. Bentham) L. Benson
  - Sy = S. filifolius T. Nuttall
  - Sy = S. filifolius var. jamesii J. Torrey & A. Gray
  - Sy = S. longilobus G. Bentham
- ***S. glabellus*** J. Poiret
- ***S. millelobatus*** P. Rydberg
- ***S. multicapitatus*** J. Greenman *ex* P. Rydberg
  - Sy = S. spartioides J. Torrey & A. Gray var. multicapitatus (J. Greenman *ex* P. Rydberg) S. Welsh
- ***S. neomexicanus*** A. Gray var. ***neomexicanus***
  - Sy = Packera neomexicana (A. Gray) W. A. Weber
  - Sy = Senecio neomexicana var. griffithsii J. Greenman
- ***S. obovatus*** G. H. Muhlenberg *ex* C. von Willdenow
  - Sy = S. obovatus var. elliottii (J. Torrey & A. Gray) M. Fernald
  - Sy = S. obovatus var. rotundus N. Britton
- ***S. parryi*** A. Gray
- ***S. plattensis*** T. Nuttall
- ***S. riddellii*** J. Torrey & A. Gray
  - Sy = S. riddellii var. parksii V. Cory
  - Sy = S. spartioides J. Torrey & A. Gray var. fremontii (J. Torrey & A. Gray) J. Greenman
  - Sy = S. spartioides var. parksii (V. Cory) L. Shinners
  - Sy = S. spartioides var. riddellii (J. Torrey & A. Gray) J. Greenman
- ***S. salignus*** A. P. de Candolle
  - Sy = Barkleyanthus salicifolius (K. Kunth) H. Robinson & P. Bretting
- ***S. spartioides*** J. Torrey & A. Gray
  - Sy = S. spartioides var. granularis B. Maguire & A. Holmgren *ex* A. Cronquist
  - Sy = S. toiyabensis S. Welsh & S. Goodrich
- ***S. tampicanus*** A. P. de Candolle
  - Sy = S. greggii P. Rydberg
  - Sy = S. imparipinnatus F. Klatt
- ***S. tomentosus*** A. Michaux
  - Sy = S. alabamensis N. Britton *ex* J. K. Small
- ***S. tridenticulatus*** P. Rydberg
  - Sy = Packera tridenticulata (P. Rydberg) W. A. Weber & A. Löve
- ***S. vulgaris*** C. Linnaeus
- ***S. wootonii*** E. Greene
  - Sy = S. anacletus E. Greene
  - Sy = S. microdontus (A. Gray) A. A. Heller, non Baker
  - Sy = S. toluccanus A. P. de Candolle var. microdontus A. Gray

SHINNERSOSERIS A. S. Tomb
- ***S. rostrata*** (A. Gray) A. S. Tomb
  - Sy = Lygodesmia rostrata (A. Gray) A. Gray

SILPHIUM C. Linnaeus
- ***S. albiflorum*** A. Gray
- ***S. asteriscus*** C. Linnaeus
  - Sy = S. asteriscus var. scabrum T. Nuttall
  - Sy = S. gatesii C. Mohr
  - Sy = S. scaberrimum S. Elliott
- ***S. gracile*** A. Gray
  - Sy = S. simpsonii E. Greene var. wrightii L. Perry
- ***S. integrifolium*** A. Michaux var. ***deamii*** L. Perry
- ***S. integrifolium*** A. Michaux var. ***integrifolium***
  - Sy = S. laevigatum F. Pursh
- ***S. integrifolium*** A. Michaux var. ***laeve*** J. Torrey & A. Gray
  - Sy = S. speciosum T. Nuttall
- ***S. laciniatum*** C. Linnaeus var. ***robinsonii*** L. Perry
- ***S. radula*** T. Nuttall
  - Sy = S. asperrimum W. Hooker

SILYBUM M. Adanson, *nomen conservandum*
- ***S. marianum*** (C. Linnaeus) J. Gaertner
  - Sy = Carduus marianus C. Linnaeus
  - Sy = Mariana mariana (C. Linnaeus) J. Hill

SIMSIA C. Persoon
S. *calva* (G. Engelmann & A. Gray) A. Gray
S. *lagasciformis* A. P. de Candolle
Sy = Encelia exaristata A. Gray
Sy = Simsia exaristata A. Gray var. exaristata
Sy = S. exaristata var. perplexa S. F. Blake

SMALLANTHUS K. Mackenzie *ex* J. K. Small
S. *uvedalia* (C. Linnaeus) K. Mackenzie *ex* J. K. Small
Sy = Osteospermum uvedalia C. Linnaeus
Sy = Polymnia uvedalia (C. Linnaeus) C. Linnaeus
Sy = P. uvedalia var. densipilis S. F. Blake

SOLIDAGO C. Linnaeus
S. *altiplanities* C. E. Taylor & J. R. Taylor
S. *arguta* W. Aiton var. ***boottii*** (W. Hooker) E. Palmer & J. Steyermark
Sy = S. arguta subsp. boottii (W. Hooker) G. Morton
Sy = S. boottii W. Hooker
Sy = S. dispersa J. K. Small
S. *auriculata* R. Shuttleworth *ex* S. F. Blake
Sy = S. amplexicaulis J. Torrey & A. Gray *ex* A. Gray, non M. Martens
Sy = S. notabilis K. Mackenzie
S. *caesia* C. Linnaeus var. ***caesia***
Sy = S. axillaris F. Pursh
Sy = S. caesia var. axillaris (F. Pursh) A. Gray
S. *caesia* C. Linnaeus var. ***curtisii*** (J. Torrey & A. Gray) A. Wood
Sy = S. caesia var. hispida A. Wood
Sy = S. curtisii J. Torrey & A. Gray
Sy = S. curtisii var. pubens (M. Curtis) A. Gray
Sy = S. lancifolia J. Torrey & A. Gray
Sy = S. monticola J. Torrey & A. Gray
Sy = S. pubens M. Curtis
S. *canadensis* C. Linnaeus var. ***gilvocanescens*** P. Rydberg
Sy = Doria gilvocanescens (P. Rydberg) J. Lunell
Sy = Solidago altissima C. Linnaeus var. gilvocanescens (P. Rydberg) J. Semple
Sy = S. canadensis subsp. gilvocanescens (P. Rydberg) A. Löve and D. Löve
Sy = S. gilvocanescens (P. Rydberg) B. Smyth
Sy = S. pruinosa E. Greene
S. *canadensis* C. Linnaeus var. ***lepida*** (A. P. de Candolle) A. Cronquist
S. *canadensis* C. Linnaeus var. ***scabra*** J. Torrey & A. Gray
Sy = S. altissima C. Linnaeus
Sy = S. altissima var. pluricephala M. C. Johnston
Sy = S. altissima var. procera (W. Aiton) M. Fernald
Sy = S. hirsutissima P. Miller
Sy = S. lunellii P. Rydberg
S. *gigantea* W. Aiton
Sy = Doria pitcheri J. Lundell
Sy = Solidago gigantea subsp. serotina (K. E. O. Kuntze) J. McNeill
Sy = S. gigantea var. leiophylla M. Fernald
Sy = S. gigantea var. pitcheri (T. Nuttall) L. Shinners
Sy = S. gigantea var. serotina (K. E. O. Kuntze) A. Cronquist
Sy = S. gigantea var. shinnersii J. Beaudry
Sy = S. pitcheri T. Nuttall
Sy = S. serotina W. Aiton
Sy = S. serotinoides A. Löve & D. Löve
S. *juliae* G. Nesom
Sy = S. altissima C. Linnaeus var. canescens (A. Gray) M. C. Johnston
Sy = S. canadensis C. Linnaeus var. canescens A. Gray
S. *ludoviciana* (A. Gray) J. K. Small
Sy = S. boottii W. Hooker var. ludoviciana A. Gray
S. *missouriensis* T. Nuttall var. ***fasciculata*** J. Holzinger
Sy = S. glaberrima M. Martens
Sy = S. glaberrima var. mortiura (G. Scheele) E. Palmer & J. Steyermark
Sy = S. missouriensis var. glaberrima (M. Martens) C. Rosendahl & A. Cronquist
S. *mollis* F. Bartling var. ***angustata*** L. Shinners
S. *mollis* F. Bartling var. ***mollis***
S. *nemoralis* W. Aiton var. ***longipetiolata*** (K. Mackenzie & B. Bush) E. Palmer & J. Steyermark
Sy = S. decemflora A. P. de Candolle
Sy = S. longipetiolata K. Mackenzie & B. Bush
Sy = S. nemoralis subsp. decemflora (A. P. de Candolle) Brammall
Sy = S. nemoralis subsp. longipetiolata (K. Mackenzie & B. Bush) G. Douglas
Sy = S. nemoralis var. decemflora (A. P. de Candolle) M. Fernald
Sy = S. pulcherrima A. Nelson
S. *nemoralis* W. Aiton var. ***nemoralis***
Sy = S. nemoralis subsp. haleana (M. Fernald) G. Douglas
Sy = S. nemoralis var. haleana M. Fernald

S. odora W. Aiton var. ***odorata***
Sy = S. suaveolens J. Schöpf
S. patula G. H. Muhlenberg *ex* C. von Willdenow var. ***strictula*** J. Torrey & A. Gray
Sy = S. arguta W. Aiton var. strigosa (J. K. Small) J. Steyermark
Sy = S. salicina S. Elliott
Sy = S. strigosa J. K. Small
S. petiolaris W. Aiton
Sy = S. angusta J. Torrey & A. Gray
Sy = S. lindheimeriana G. Scheele
Sy = S. milleriana K. Mackenzie
Sy = S. petiolaris var. angusta (J. Torrey & A. Gray) A. Gray
Sy = S. petiolaris var. wardii (N. Britton) M. Fernald
Sy = S. wardii N. Britton
S. radula T. Nuttall var. ***radula***
Sy = S. pendula J. K. Small
Sy = S. radula T. Nuttall var. laeta (E. Greene) M. Fernald
Sy = S. rotundifolia A. P. de Candolle
Sy = S. scaberrima J. Torrey & A. Gray
S. radula T. Nuttall var. ***stenolepis*** M. Fernald
S. rugosa P. Miller var. ***aspera*** (W. Aiton) M. Fernald
Sy = S. aspera W. Aiton
Sy = S. celtidifolia J. K. Small
Sy = S. drummondii J. Torrey & A. Gray
Sy = S. rugosa subsp. aspera (W. Aiton) A. Cronquist
Sy = S. rugosa var. celtidifolia (J. K. Small) M. Fernald
S. rugosa P. Miller var. ***rugosa***
Sy = S. rugosa subsp. rugosa
S. sempervirens C. Linnaeus
Sy = S. angustifolia S. Elliott
Sy = S. mexicana C. Linnaeus
Sy = S. sempervirens var. mexicana (C. Linnaeus) M. Fernald
S. simplex K. Kunth var. ***simplex***
Sy = S. neomexicana (A. Gray) E. Wooton & P. Standley
Sy = S. spathulata A. P. de Candolle var. neomexicana (A. Gray) A. Cronquist
S. speciosa T. Nuttall var. ***rigidiuscula*** J. Torrey & A. Gray
Sy = S. rigidiuscula (J. Torrey & A. Gray) T. Porter
Sy = S. speciosa var. angustata J. Torrey & A. Gray
S. stricta W. Aiton
S. tortifolia S. Elliott
S. ulmifolia G. H. Muhlenberg *ex* C. von Willdenow var. ***microphylla*** A. Gray
Sy = S. delicatula J. K. Small
Sy = S. helleri J. K. Small
Sy = S. microphylla (A. Gray) G. Engelmann *ex* J. K. Small
S. ulmifolia G. H. Muhlenberg *ex* C. von Willdenow var. ***ulmifolia***
S. velutina A. P. de Candolle
Sy = S. arizonica (A. Gray) E. Wooton & P. Standley
Sy = S. californica T. Nuttall var. nevadensis A. Gray
Sy = S. canadensis C. Linnaeus var. arizonica A. Gray
Sy = S. howellii E. Wooton & P. Standley
Sy = S. sparsiflora A. Gray
Sy = S. trinervata E. Greene
Sy = S. velutina var. nevadensis (A. Gray) C. E. Taylor & J. R. Taylor
S. wrightii A. Gray var. ***adenophora*** S. F. Blake
Sy = S. bigelovii A. Gray
S. wrightii A. Gray var. ***wrightii***
Sy = S. bigelovii var. wrightii A. Gray

SOLIVA H. Ruiz López & J. Pavón
S. mutisii K. Kunth
Sy = S. anthemifolia (A. L. de Jussieu) R. Brown
S. pterosperma (A. L. de Jussieu) C. Lessing

SONCHUS C. Linnaeus
S. arvensis C. Linnaeus var. ***arvensis***
S. arvensis C. Linnaeus var. ***glabrescens*** J. C. Günther
Sy = S. arvensis subsp. uliginosus (F. Marschall von Bieberstein) H. Neumayer
Sy = S. uliginosus F. Marschall von Bieberstein
S. asper (C. Linnaeus) J. Hill
Sy = S. oleraceus C. Linnaeus var. asper C. Linnaeus
S. oleraceus C. Linnaeus

STEPHANOMERIA T. Nuttall, *nomen conservandum*
S. exigua T. Nuttall subsp. ***exigua***
Sy = Hemiptilon bigelovii A. Gray
Sy = Lygodesmia bigelovii (A. Gray) L. Shinners
Sy = L. exigua (A. Gray) A. Gray
Sy = Prenanthella exigua P. Rydberg
Sy = Ptiloria bigelovii (A. Gray) E. Wooton & P. Standley
Sy = P. exigua (T. Nuttall) E. Greene

Sy = Stephanomeria exigua subsp. pentachaeta (A. P. de Candolle) A. Eaton
Sy = S. minima M. E. Jones
S. pauciflora (J. Torrey) A. Nelson
Sy = Lygodesmia pauciflora (J. Torrey) L. Shinners
Sy = Ptiloria cinera S. F. Blake
Sy = P. lygodesmoides A. A. Heller
Sy = P. pauciflora (J. Torrey) C. Rafinesque-Schmaltz
Sy = Stephanomeria cinera B. Blake
Sy = S. lygodesmoides M. E. Jones *ex* L. F. Henderson
Sy = S. pauciflora var. parishii P. Munz
Sy = S. runcinata T. Nuttall var. parishii W. Jepson
S. tenuifolia (C. Rafinesque-Schmaltz) H. Hall var. ***tenuifolia***
Sy = Lygodesmia minor W. Hooker
Sy = L. tenuifolia (J. Torrey) L. Shinners
Sy = Ptiloria minor (W. Hooker) T. Nuttall
Sy = P. tenuifolia (J. Torrey) C. Rafinesque-Schmaltz
Sy = Stephanomeria minor W. Hooker
Sy = S. wrightii A. Gray

STEVIA A. Cavanilles
S. ovata C. von Willdenow var. ***texana*** J. Grashoff
Sy = S. rhombifolia K. Kunth
S. serrata A. Cavanilles
Sy = S. serrata var. haplopappa B. Robinson
Sy = S. serrata var. ivifolia (C. von Willdenow) B. Robinson

STEVIOPSIS R. King & H. Robinson
S. fendleri (A. Gray) B. L. Turner
Sy = Brickellia fendleri A. Gray
Sy = Brickelliastrum fendleri (A. Gray) R. King & H. Robinson
Sy = Coleosanthus ambigens E. Greene
Sy = Eupatorium fendleri (A. Gray) A. Gray

STOKESIA L'Héritier de Brutelle
S. laevis (J. Hill) E. Greene **[cultivated]**
Sy = Carthamus laevis J. Hill

STYLOCLINE T. Nuttall
S. micropoides A. Gray

SYMPHYOTRICHUM C. Nees von Esenbeck
S. divaricatum (T. Nuttall) G. Nesom
Sy = Aster divaricatus (T. Nuttall) J. Torrey & A. Gray
Sy = A. neomexicanus E. Wooton & P. Standley
Sy = A. subulatus A. Michaux var. ligulatus L. Shinners
Sy = Tripolium divaricatum T. Nuttall
S. drummondii (J. Lindley) G. Nesom var. ***parviceps*** (L. Shinners) G. Nesom
Sy = Aster drummondii J. Lindley subsp. parviceps (L. Shinners) A. G. Jones
Sy = A. drummondii var. parviceps (L. Shinners) A. G. Jones
Sy = A. texanus E. Burgess var. parviceps L. Shinners
S. drummondii (J. Lindley) G. Nesom var. ***texanum*** (E. Burgess) G. Nesom
Sy = Aster drummondii J. Lindley subsp. texanus (E. Burgess) A. G. Jones
Sy = A. drummondii var. texanus (E. Burgess) A. G. Jones
Sy = A. texanus E. Burgess var. texanus
S. dumosum (C. Linnaeus) G. Nesom var. ***dumosum***
Sy = Aster coridifolius A. Michaux
Sy = A. dumosus C. Linnaeus var. coridifolius (A. Michaux) J. Torrey & A. Gray
Sy = A. dumosus var. dumosus
Sy = A. dumosus var. gracilentus J. Torrey & A. Gray
S. dumosum (C. Linnaeus) G. Nesom var. ***subulifolium*** (J. Torrey & A. Gray) G. Nesom
Sy = Aster dumosus C. Linnaeus var. subulifolius J. Torrey & A. Gray
S. ericoides (C. Linnaeus) G. Nesom var. ***ericoides***
Sy = Aster ericoides C. Linnaeus var. ericoides
Sy = A. multiflorus D. Solander *ex* W. Aiton
Sy = A. polycephalus P. Rydberg
Sy = Lasallea ericoides (C. Linnaeus) J. Semple & L. Brouillet
Sy = Virgulus ericoides (C. Linnaeus) J. Reveal & C. Keener
S. ericoides (C. Linnaeus) G. Nesom var. ***prostratum*** (K. E. O. Kuntze) G. Nesom
Sy = Aster ericoides C. Linnaeus var. prostratus (K. E. O. Kuntze) S. F. Blake
Sy = A. hebecladus A. P. de Candolle
Sy = A. multiflorus D. Solander *ex* W. Aiton var. prostratus K. E. O. Kuntze
Sy = A. scoparius A. P. de Candolle, non C. Nees von Esenbeck

S. eulae (L. Shinners) G. Nesom
Sy = Aster eulae L. Shinners
Sy = A. × eulae L. Shinners [lanceolatus × praealtus]
S. falcatum (J. Lindley) G. Nesom var. ***commutatum*** (J. Torrey & A. Gray) G. Nesom
Sy = Aster commutatus A. Gray
Sy = A. falcatus J. Lindley subsp. commutatus (J. Torrey & A. Gray) A. G. Jones var. commutatus (J. Torrey & A. Gray) A. G. Jones
Sy = A. multiflorus D. Solander *ex* W. Aiton var. commutatus J. Torrey & A. Gray
S. falcatum (J. Lindley) G. Nesom var. ***crassulum*** (P. Rydberg) G. Nesom
Sy = Aster crassulus P. Rydberg
Sy = A. falcatus J. Lindley subsp. commutatus (J. Torrey & A. Gray) A. G. Jones var. crassulus (P. Rydberg) A. Cronquist
Sy = A. falcatus var. crassulus (P. Rydberg) A. Cronquist
S. fendleri (A. Gray) G. Nesom
Sy = Aster fendleri A. Gray
Sy = Virgulus fendleri (A. Gray) J. Reveal & C. Keener
S. laeve (C. Linnaeus) A. Löve & D. Löve var. ***geyeri*** (A. Gray) G. Nesom
Sy = Aster geyeri (A. Gray) T. J. Howell
Sy = A. laevis C. Linnaeus var. geyeri A. Gray
* Sy = A. laevis var. guadalupensis A. G. Jones [TOES: V]
S. lanceolatum (C. von Willdenow) G. Nesom subsp. ***hesperium*** (A. Gray) G. Nesom var. ***hesperium*** (A. Gray) G. Nesom
Sy = Aster foliaceus J. Lindley *ex* A. P. de Candolle var. hesperius (A. Gray) W. Jepson
Sy = A. hesperius A. Gray
Sy = A. laetevirens E. Greene
Sy = A. lanceolatus C. von Willdenow subsp. hesperius (A. Gray) J. Semple & J. Chmielewski
Sy = A. wootonii (E. Greene) E. Greene
Sy = Symphyotrichum hesperium (A. Gray) A. Löve & D. Löve
S. lanceolatum (C. von Willdenow) G. Nesom subsp. ***lanceolatum*** var. ***lanceolatum***
Sy = Aster lanceolatus C. von Willdenow subsp. lanceolatus
S. lateriflorum (C. Linnaeus) A. Löve & D. Löve var. ***flagellare*** (L. Shinners) G. Nesom
Sy = Aster lateriflorus (C. Linnaeus) N. Britton var. flagellaris L. Shinners
Sy = A. lateriflorus (C. Linnaeus) N. Britton var. indutus L. Shinners
S. oblongifolium (T. Nuttall) G. Nesom
Sy = Aster diffusus W. Aiton var. thyrsoides A. Gray
Sy = A. kumleinii R. Fries *ex* A. Gray
Sy = A. lateriflorus (C. Linnaeus) N. Britton var. thyrsoides (A. Gray) E. Sheldon
Sy = A. missouriensis N. Britton var. thyrsoides (A. Gray) K. Wiegand
Sy = A. oblongifolius T. Nuttall var. angustatus L. Shinners
Sy = A. oblongifolius var. oblongifolius
Sy = A. oblongifolius var. rigidulus A. Gray
Sy = A. pantotrichus S. F. Blake
Sy = A. pantotrichus var. thyrsoides (A. Gray) S. F. Blake
Sy = A. tradescantii C. Linnaeus var. thyrsoides (A. Gray) J. Boivin
Sy = Lasallea oblongifolia (T. Nuttall) J. Semple & L. Brouillet
Sy = Virgulus oblongifolius (T. Nuttall) J. Reveal & C. Keener
Sy = V. oblongifolius var. angustatus (L. Shinners) J. Reveal & C. Keener
S. ontarione (K. Wiegand) G. Nesom
Sy = Aster ontarionis K. Wiegand
S. oölentangiense (J. Riddell) G. Nesom var. ***oölentagiense***
Sy = Aster azureus J. Lindley
Sy = A. oölentangiensis J. Riddell var. laevicaulis (M. Fernald) A. G. Jones
Sy = A. oölentangiensis var. oölentangiensis
S. oölentangiense (J. Riddell) G. Nesom var. ***poaceum*** (E. Burgess) G. Nesom
Sy = Aster azureus J. Lindley var. poaceus (E. Burgess) M. Fernald
Sy = A. oölentangiensis J. Riddell var. poaceus (E. Burgess) A. G. Jones
Sy = A. poaceus E. Burgess
Sy = A. vernalis G. Engelmann *ex* E. Burgess
S. patens (W. Aiton) G. Nesom var. ***gracile*** (W. Hooker) G. Nesom
Sy = Aster patens W. Aiton var. gracilis W. Hooker
Sy = Virgulus patens (W. Aiton) J. Reveal & C. Keener var. gracilis (W. Hooker) R. Reveal & C. Keener
S. patens (W. Aiton) G. Nesom var. ***patens***
Sy = Aster patens W. Aiton var. patens

Sy = Lasallea patens (W. Aiton) J. Semple & L. Brouillet
Sy = Virgulus patens (W. Aiton) J. Reveal & C. Keener
S. patens (W. Aiton) G. Nesom var. ***patentissimum*** (J. Lindley *ex* A. P. de Candolle) G. Nesom
Sy = Aster continuus J. K. Small
Sy = A. patens W. Aiton var. patentissimus (J. Lindley *ex* A. P. de Candolle) J. Torrey & A. Gray
Sy = A. patentissimus J. Lindley *ex* A. P. de Candolle
Sy = A. subsessilis E. Burgess
Sy = Virgulus patens (W. Aiton) J. Reveal & C. Keener var. patentissimus (J. Lindley *ex* A. P. de Candolle) J. Reveal & C. Keener
S. praealtum (J. Poiret) G. Nesom var. ***praealtum***
Sy = Aster coerulescens A. P. de Candolle
Sy = A. nebraskensis N. Britton
Sy = A. praealtus J. Poiret var. coerulescens (A. P. de Candolle) A. G. Jones
Sy = A. praealtus var. imbricatior K. Wiegand
Sy = A. praealtus var. nebraskensis (N. Britton) K. Wiegand
Sy = A. praealtus var. praealtus
S. praealtum (J. Poiret) G. Nesom var. ***subasperum*** (J. Lindley) G. Nesom
Sy = Aster praealtus J. Poiret var. subasper (J. Lindley) K. Wiegand
Sy = A. subasper J. Lindley
S. praealtum (J. Poiret) G. Nesom var. ***texicola*** (K. Wiegand) G. Nesom
Sy = A. praealtus J. Poiret var. texicola K. Wiegand
S. pratense (C. Rafinesque-Schmaltz) G. Nesom
Sy = Aster phyllolepis J. Torrey & A. Gray
Sy = A. pratensis C. Rafinesque-Schmaltz
Sy = A. sericeus E. Ventenat var. microphyllus A. P. de Candolle
Sy = Lasallea sericea (E. Ventenat) E. Greene var. pratensis (C. Rafinesque-Schmaltz) J. Semple & L. Brouillet
Sy = Virgulus pratensis (C. Rafinesque-Schmaltz) J. Reveal & C. Keener
S. puniceum (C. Linnaeus) A. Löve & D. Löve var. ***scabricaule*** (L. Shinners) G. Nesom
Sy = Aster puniceus C. Linnaeus subsp. elliottii (J. Torrey & A. Gray) A. G. Jones var. scabricaulis (L. Shinners) A. G. Jones
* Sy = A. puniceus var. scabricaulis (L. Shinners) A. G. Jones [**TOES**: **V**]
Sy = A. scabricaulis L. Shinners
S. racemosum (S. Elliott) G. Nesom var. ***subdumosum*** (K. Wiegand) G. Nesom
Sy = Aster fragilis C. von Willdenow var. subdumosus (K. Wiegand) A. G. Jones
Sy = A. vimineus J. de Lamarck var. subdumosus K. Wiegand
S. sericeum (E. Ventenat) G. Nesom
Sy = Aster sericeus E. Ventenat
Sy = Lasallea sericea (E. Ventenat) E. Greene
Sy = Virgulus sericeus (E. Ventenat) J. Reveal & C. Keener
S. squamatum (K. Sprengel) G. Nesom
Sy = Aster asteroides (L. Colla) H. Rusby
Sy = A. squamatus (K. Sprengel) G. Hieronymus
Sy = A. subtropicos T. Morong
Sy = A. subulatus A. Michaux var. sandwicensis (A. Gray) A. G. Jones
Sy = Baccharis asteroides L. Colla
Sy = Conyza squamata K. Sprengel
Sy = Conyzanthus squamatus (K. Sprengel) S.Tamamschjan
S. subulatum (A. Michaux) G. Nesom
Sy = Aster subulatus A. Michaux
Sy = Mesoligus subulatus (A. Michaux) C. Rafinesque-Schmaltz
S. tenuifolium (C. Linnaeus) G. Nesom
Sy = Aster tenuifolius C. Linnaeus

TAGETES C. Linnaeus
T. fragrantissima M. Sessé y Lacasta & J. Mociño
T. lemmonii A. Gray [**cultivated**]
Sy = T. alamensis P. Rydberg
Sy = T. palmeri A. Gray
T. lucida A. Cavanilles [**cultivated**]
Sy = T. anethina M. Sessé y Lacasta & J. Mociño
Sy = T. florida R. Sweet
T. micrantha A. Cavanilles

TAMAULIPA R. King & H. Robinson
T. azurea (A. P. de Candolle) R. King & H. Robinson
Sy = Eupatorium azureum A. P. de Candolle

TANACETUM C. Linnaeus
T. parthenium (C. Linnaeus) K. H. Schultz-Bipontinus [**cultivated**]
Sy = Chrysanthemum parthenium (C. Linnaeus) J. Bernharti
Sy = Matricaria parthenium C. Linnaeus

TARAXACUM G. Weber *ex* F. Wiggers, *nomen conservandum*
T. laevigatum (C. von Willdenow) A. P. de Candolle

Sy = Leontodon erythrospermum K. von Eichwald
Sy = L. laevigatus C. von Willdenow
Sy = Taraxacum erythrospermum (A. Andrzejowski *ex* W. Besser) N. Britton
Sy = T. officinale G. Weber *ex* F. Wiggers var. erythrospermum J. Weiss
T. officinale G. Weber *ex* F. Wiggers
Sy = Leontodon taraxacum C. Linnaeus
Sy = L. vulgare J. de Lamarck
Sy = Taraxacum mexicanum A. P. de Candolle
Sy = T. vulgare F. von Schrank

TETRAGONOTHECA C. Linnaeus
T. ludoviciana (J. Torrey & A. Gray) A. Gray *ex* H. Hall
Sy = Halea ludoviciana J. Torrey & A. Gray
T. repanda (S. Buckley) J. K. Small
Sy = Halea repanda S. Buckley
T. texana G. Engelmann & A. Gray *ex* A. Gray

TETRANEURIS E. Greene
T. acaulis (F. Pursh) E. Greene var. ***acaulis***
Sy = Actinea acaulis (F. Pursh) K. Sprengel
Sy = A. acaulis (F. Pursh) T. Nuttall
Sy = Actinella depressa J. Torrey & A. Gray var. pygmaea A. Gray
Sy = Hymenoxys acaulis (F. Pursh) K. Parker
T. linearifolia (W. Hooker) E. Greene var. ***arenicola*** M. Bierner
T. linearifolia (W. Hooker) E. Greene var. ***linearifolia***
Sy = Actinella linearifolia (W. Hooker) J. Torrey & A. Gray
Sy = Hymenoxys linearifolia W. Hooker
Sy = Tetraneuris oblongifolia E. Greene
T. scaposa (A. P. de Candolle) E. Greene var. ***argyrocaulon*** (K. Parker) K. Parker
Sy = Hymenoxys scaposa var. argyrocaulon K. Parker
T. scaposa (A. P. de Candolle) E. Greene var. ***linearis*** (T. Nuttall) K. Parker
Sy = Actinella scaposa (A. P. de Candolle) T. Nuttall var. linearis T. Nuttall
Sy = Hymenoxys scaposa (A. P. de Candolle) E. Greene var. linearis (T. Nuttall) K. Parker
T. scaposa (A. P. de Candolle) E. Greene var. ***scaposa***
Sy = Actinea scaposa (A. P. de Candolle) K. E. O. Kuntze
Sy = Actinella glabra T. Nuttall
Sy = A. scaposa (A. P. de Candolle) T. Nuttall
Sy = Cephalophora scaposa A. P. de Candolle
Sy = Hymenoxys glabra (T. Nuttall) L. Shinners
Sy = H. scaposa (A. P. de Candolle) K. Parker
Sy = H. scaposa var. glabra (T. Nuttall) K. Parker
Sy = Tetraneuris glabra (T. Nuttall) E. Greene
Sy = T. glabriuscula E. Greene
Sy = T. scaposa var. glabra (T. Nuttall) K. Parker
Sy = T. stenophylla P. Rydberg
T. scaposa (A. P. de Candolle) E. Greene var. ***villosa*** (L. Shinners) L. Shinners
Sy = Hymenoxys scaposa var. villosa L. Shinners
T. turneri (K. Parker) K. Parker
Sy = Hymenoxys turneri K. Parker

THELECHITONIA J. Cuatrecasas
T. trilobata (C. Linnaeus) H. Robinson & J. Cuatrecasas [**cultivated**]
Sy = Complaya trilobata (C. Linnaeus) J. Strother
Sy = Silphium trilobatum C. Linnaeus
Sy = Wedelia brasiliensis (K. Sprengel) S. F. Blake
Sy = W. paludosa A. P. de Candolle
Sy = W. trilobata (C. Linnaeus) A. Hitchcock

THELESPERMA C. Lessing
T. ambiguum A. Gray
Sy = T. fraternum L. Shinners
Sy = T. megapotamicum (K. Sprengel) K. E. O. Kuntze var. ambiguum (A. Gray) L. Shinners
T. burridgeanum (E. von Regel, F. Körnicke, & L. Rach) S. F. Blake
T. curvicarpum T. Melchert
T. filifolium (W. Hooker) A. Gray var. ***filifolium***
Sy = T. trifidum (J. Poiret) N. Britton
T. filifolium (W. Hooker) A. Gray var. ***intermedium*** (P. Rydberg) L. Shinners
Sy = T. intermedium P. Rydberg
T. flavodiscum (L. Shinners) B. L. Turner
T. longipes A. Gray
T. megapotamicum (K. Sprengel) K. E. O. Kuntze
Sy = Bidens gracilis J. Torrey, non T. Nuttall
Sy = B. megapotamica K. Sprengel
Sy = Cosmidium gracile J. Torrey & A. Gray
Sy = Isostigma megapotamicum E. Sherff
Sy = Thelesperma gracile (J. Torrey) A. Gray
T. nuecense B. L. Turner
T. simplicifolium A. Gray
Sy = T. subsimplicifolium A. Gray

THUROVIA J. Rose
* ***T. triflora*** J. Rose [**TOES: V**]
Sy = Gutierrezia triflora (J. Rose) M. Lane

TITHONIA R. Desfontaines *ex* A. L. de Jussieu
T. rotundifolia (P. Miller) S. F. Blake **[cultivated]**
Sy = Tagetes rotundifolia P. Miller

TOWNSENDIA W. Hooker
T. annua J. Beaman
T. exscapa (J. Richardson) T. Porter
Sy = Aster exscapus J. Richardson
Sy = Townsendia intermedia P. Rydberg
Sy = T. sericea W. Hooker
T. texensis E. L. Larsen

TRAGOPOGON C. Linnaeus
T. dubius J. Scopoli
Sy = T. dubius subsp. major (N. von Jacquin) F. Vollmann
Sy = T. major N. von Jacquin
T. porrifolius C. Linnaeus

TRICHOCORONIS A. Gray
T. rivularis A. Gray
Sy = Shinnersia rivularis (A. Gray) R. King & H. Robinson
T. wrightii (J. Torrey & A. Gray) A. Gray var. ***wrightii***
Sy = Ageratum wrightii J. Torrey & A. Gray

TRIXIS P. Browne
T. californica A. Kellogg var. ***californica***
Sy = T. angustifolia A. P. de Candolle var. latiuscula A. Gray
Sy = T. suffruticosa S. Watson
T. inula H. von Crantz
Sy = Inula trixis C. Linnaeus
Sy = Trixis frutescens P. Browne *ex* K. Sprengel
Sy = T. radialis (C. Linnaeus) K. E. O. Kuntze

VARILLA A. Gray
V. texana A. Gray

VERBESINA C. Linnaeus
V. alternifolia (C. Linnaeus) N. Britton *ex* T. Kearney
Sy = Actinomeris alternifolia (C. Linnaeus) A. P. de Candolle
Sy = Ridan alternifolia (C. Linnaeus) N. Britton
V. encelioides (A. Cavanilles) G. Bentham & J. Hooker *ex* A. Gray var. ***encelioides***
Sy = Ximenesia encelioides A. Cavanilles
V. helianthoides A. Michaux
Sy = Actinomeris helianthoides (A. Michaux) T. Nuttall
V. lindheimeri B. Robinson & J. Greenman
V. microptera A. P. de Candolle
Sy = V. microptera var. mollissima B. Robinson & J. Greenman
Sy = V. texana S. Buckley
V. nana (A. Gray) B. Robinson & J. Greenman
Sy = Wootonella nana (A. Gray) P. Standley
Sy = Ximenesia nana (A. Gray) L. Shinners
V. oreophila E. Wooton & P. Standley
V. virginica C. Linnaeus var. ***virginica***
Sy = Phaethusa virginica (C. Linnaeus) N. Britton

VERNONIA J. von Schreber, *nomen conservandum*
V. baldwinii J. Torrey subsp. ***baldwinii***
V. baldwinii J. Torrey subsp. ***interior*** (J. K. Small) E. Faust
Sy = V. interior J. K. Small
V. gigantea (T. Walter) W. Trelease *ex* Branner & F. Coville subsp. ***gigantea***
Sy = V. altissima T. Nuttall
Sy = V. altissima var. lilacina W. Clute
Sy = V. altissima var. taeniotricha S. F. Blake
V. larsenii B. King & S. B. Jones
Sy = V. lindheimeri A. Gray & G. Engelmann *ex* A. Gray var. leucophylla E. L. Larsen
V. lindheimeri A. Gray & G. Engelmann *ex* A. Gray
V. marginata (J. Torrey) C. Rafinesque-Schmaltz
Sy = V. altissima J. Torrey var. marginata J. Torrey
Sy = V. marginata var. tenuifolia (J. K. Small) L. Shinners
V. missurica C. Rafinesque-Schmaltz
Sy = V. aborigina H. Gleason
V. texana (A. Gray) J. K. Small
V. × vulturina L. Shinners [***baldwinii × marginata***]

VIGUIERA K. Kunth
V. cordifolia A. Gray
V. dentata (A. Cavanilles) K. Sprengel var. ***dentata***
Sy = Helianthus dentata A. Cavanilles
Sy = Viguiera texana J. Torrey & A. Gray
V. longifolia (B. Robinson & J. Greenman) S. F. Blake
Sy = Gymnolomia longifolia B. Robinson & J. Greenman
Sy = G. multiflora G. Bentham & J. Hooker *ex* W. Hemsley var. annua M. E. Jones
Sy = Heliomeris annua (M. E. Jones) T. Cockerell
Sy = H. longifolia (B. Robinson & J. Greenman) T. Cockerell
Sy = H. longifolia (B. Robinson & J. Greenman) T. Cockerell var. annua (M. E. Jones) H. Yates
Sy = Viguiera annua (M. E. Jones) S. F. Blake

V. phenax S. F. Blake
Sy = V. ludens (L. Shinners) M. C. Johnston
V. stenoloba S. F. Blake var. ***chihuahuensis*** M. Butterwick
Sy = Heliomeris tenuifolia A. Gray
V. stenoloba S. F. Blake var. ***stenoloba***

WEDELIA N. von Jacquin, *nomen conservandum*
W. texana (A. Gray) B. L. Turner
Ma = W. hispida *auct.*, non K. Kunth: Texas authors
Ma = Zexmenia hispida *auct.*, non (K. Kunth) A. Gray *ex* J. K. Small: Texas authors

XANTHISMA A. P. de Candolle
X. texanum A. P. de Candolle subsp. ***drummondii*** (J. Torrey & A. Gray) J. Semple
Sy = X. texanum var. drummondii (J. Torrey & A. Gray) A. Gray
X. texanum A. P. de Candolle subsp. ***texanum*** var. ***orientale*** J. Semple
X. texanum A. P. de Candolle subsp. ***texanum*** var. ***texanum***

XANTHIUM C. Linnaeus
X. spinosum C. Linnaeus
Sy = Acanthoxanthium spinosum (C. Linnaeus) J. Fourreau
Sy = Xanthium spinosum var. inerme C. P. Bélanger
X. strumarium C. Linnaeus var. ***canadense*** (P. Miller) J. Torrey & A. Gray
Sy = X. acerosum E. Greene
Sy = X. campestre E. Greene
Sy = X. canadense P. Miller
Sy = X. cavanillesii J. Schouw
Sy = X. cenchroides C. Millspaugh & E. Sherff
Sy = X. commune N. Britton
Sy = X. echinatum J. Murray
Sy = X. glanduliferum E. Greene
Sy = X. italicum E. Morris
Sy = X. macounii N. Britton
Sy = X. oligacanthum C. Piper
Sy = X. oviforme C. Wallroth
Sy = X. pensylvanicum C. Wallroth
Sy = X. saccharatum C. Wallroth
Sy = X. speciosum T. Kearney
Sy = X. strumarium subsp. italicum (G. L. Moretti) D. Löve
Sy = X. strumarium var. oviforme (C. Wallroth) M. Peck
Sy = X. strumarium var. pensylvanicum (C. Wallroth) M. Peck
Sy = X. varians E. Greene
X. strumarium C. Linnaeus var. ***glabratum*** (A. P. de Candolle) A. Cronquist
Sy = X. americanum T. Walter
Sy = X. calvum C. Millspaugh & E. Sherff
Sy = X. chasei M. Fernald
Sy = X. chinense P. Miller
Sy = X. curvescens C. Millspaugh & E. Sherff
Sy = X. cylindraceum C. Millspaugh & E. Sherff
Sy = X. echinellum E. Greene
Sy = X. globosum G. H. Shull
Sy = X. inflexum K. Mackenzie & B. Bush
Sy = X. macrocarpon A. P. de Candolle var. glabratum A. P. de Candolle
Sy = X. orientale C. Linnaeus
Sy = X. strumarium C. Linnaeus var. wootonii (T. Cockerell) W. Martin & C. Hutchins
Sy = X. wootonii T. Cockerell

XANTHOCEPHALUM C. von Willdenow
X. gymnospermoides (A. Gray) G. Bentham & J. Hooker
Sy = Grindelia gymnospermoides (A. Gray) J. Ruffin
Sy = Gutierrezia gymnospermoides A. Gray

XYLORHIZA T. Nuttall
X. wrightii (A. Gray) E. Greene
Sy = Aster wrightii A. Gray
Sy = Machaeranthera wrightii (A. Gray) A. Cronquist & D. Keck
Sy = Townsendia wrightii (A. Gray) A. Gray

XYLOTHAMIA G. Nesom, Y. Suh, D. Morgan, & B. Simpson
X. palmeri (A. Gray) G. Nesom
Sy = Aster palmeri A. Gray
Sy = Ericameria austrotexana M. C. Johnston
Sy = Isocoma palmeri (A. Gray) L. Shinners
X. triantha (S. F. Blake) G. Nesom
Sy = Ericameria triantha (S. F. Blake) L. Shinners
Sy = Haplopappus trianthus S. F. Blake

YOUNGIA A.-H. Cassini
Y. japonica (C. Linnaeus) A. P. de Candolle
Sy = Crepis japonica (C. Linnaeus) G. Bentham
Sy = Youngia japonica subsp. elstonii (B. P. G. Hochreutiner) G. Stebbins

ZINNIA C. Linnaeus, *nomen conservandum*
Z. acerosa (A. P. de Candolle) A. Gray
Sy = Crassina acerosa K. E. O. Kuntze
Sy = Diplothrix acerosa A. P. de Candolle
Sy = Zinnia pumila A. Gray
Z. anomala A. Gray
Sy = Crassina anomala (A. Gray) K. E. O. Kuntze
Z. grandiflora T. Nuttall
Sy = Crassina grandiflora (T. Nuttall) K. E. O. Kuntze
Z. peruviana (C. Linnaeus) C. Linnaeus
Sy = Crassina multiflora (C. Linnaeus) K. E. O. Kuntze
Sy = Zinnia multiflora C. Linnaeus
Z. violacea A. Cavanilles **[cultivated]**
Sy = Crassina elegans (N. von Jacquin) K. E. O. Kuntze
Sy = Zinnia elegans N. von Jacquin

AVICENNIACEAE, *nomen conservandum*

AVICENNIA C. Linnaeus
A. germinans (C. Linnaeus) C. Linnaeus
Sy = A. nitida N. von Jacquin
Sy = Bontia germinans C. Linnaeus

BALSAMINACEAE, *nomen conservandum*

IMPATIENS C. Linnaeus
I. balsamina C. Linnaeus **[cultivated]**
I. capensis N. Meerburg

BASELLACEAE, *nomen conservandum*

ANREDERA A. L. de Jussieu
A. baselloides (K. Kunth) H. Baillon
Sy = Boussingaultia baselloides K. Kunth
A. cordifolia (M. Tenore) C. van Steenis **[cultivated]**
Sy = Boussingaultia gracilis J. Miers
A. leptostachys (C. Moquin-Tandon) C. van Steenis
Sy = Boussingaultia leptostachya C. Moquin-Tandon
A. vesicaria (J. de Lamarck) K. von Gaertner
Sy = Basella vesicaria J. de Lamarck
Ma = Anredera scandens *auct.*, non (C. Linnaeus) C. Moquin-Tandon

BATACEAE, *nomen conservandum*

BATIS P. Browne
B. maritima C. Linnaeus

BERBERIDACEAE, *nomen conservandum*

BERBERIS C. Linnaeus
B. aquifolium F. Pursh var. **aquifolium** **[cultivated]**
Sy = Mahonia aquifolium (F. Pursh) T. Nuttall
Sy = Odostemon aquifolium (F. Pursh) P. Rydberg
B. aquifolium F. Pursh var. **repens** (J. Lindley) H. Scoggan
Sy = B. nana E. Greene
Sy = B. repens J. Lindley
Sy = Mahonia repens (J. Lindley) G. Don
Sy = Odostemon repens (J. Lindley) T. Cockerell
B. bealei R. Fortune **[cultivated]**
Sy = Mahonia bealei (R. Fortune) É. Carrière
B. haematocarpa E. Wooton
Sy = Mahonia haematocarpa (E. Wooton) F. Fedde
Sy = Odostemon haematocarpus (E. Wooton) A. A. Heller
B. julianae C. Schneider **[cultivated]**
B. swaseyi S. Buckley
Sy = Mahonia swaseyi (S. Buckley) F. Fedde
B. thunbergii A. P. de Candolle **[cultivated]**
B. trifoliata M. Moricand
Sy = Mahonia trifoliata (M. Moricand) F. Fedde
Sy = M. trifoliata var. glauca (I. M. Johnston) M. C. Johnston *ex* J. de Houtton Labillardière
Sy = Odostemon trifoliolatus (M. Moricand) A. A. Heller
B. × gladwynensis E. Anderson [**verruculosa** × **julianiae**] **[cultivated]**

NANDINA C. Thunberg
N. domestica C. Thunberg **[cultivated]**

PODOPHYLLUM C. Linnaeus
P. peltatum C. Linnaeus

BETULACEAE, *nomen conservandum*

ALNUS P. Miller
A. serrulata (J. Dryander *ex* W. Aiton) C. von Willdenow
Sy = A. serrulata var. subelliptica M. Fernald
Sy = Betula serrulata J. Dryander

BETULA C. Linnaeus
B. nigra C. Linnaeus

CARPINUS C. Linnaeus
C. ***caroliniana*** T. Walter subsp. ***caroliniana***
Sy = C. americana A. Michaux

OSTRYA J. Scopoli, *nomen conservandum*
* ***O. chisosensis*** D. Correll [**TOES**: V]
O. knowltonii F. Coville
Sy = O. baileyi J. Rose
O. virginiana (P. Miller) K. Koch
Sy = O. virginiana var. lasia M. Fernald

BIGNONIACEAE, *nomen conservandum*

BIGNONIA C. Linnaeus, *nomen conservandum*
B. capreolata C. Linnaeus
Sy = Anisostichus capreolata (C. Linnaeus) E. Bureau
Sy = A. crucigera (C. Linnaeus) E. Bureau *ex* J. K. Small

CAMPSIS J. de Loureiro, *nomen conservandum*
C. radicans (C. Linnaeus) B. Seemann *ex* E. Bureau
Sy = Bignonia radicans C. Linnaeus
Sy = Tecoma radicans (C. Linnaeus) A. L. de Jussieu
C. × ***tagliabuana*** (R. de Visiani) A. Rehder [***grandiflora*** × ***radicans***] [**cultivated**]

CATALPA J. Scopoli
C. bignonioides T. Walter
Sy = C. catalpa (C. Linnaeus) G. Karsten
C. speciosa (J. Warder) J. Warder *ex* G. Engelmann
Sy = C. bignonioides T. Walter var. speciosa R. Barneby

CHILOPSIS D. Don
C. linearis (A. Cavanilles) R. Sweet subsp. ***arcuata*** (F. Fosberg) J. Henrickson
C. linearis (A. Cavanilles) R. Sweet subsp. ***linearis***
Sy = Bignonia linearis A. Cavanilles
Sy = Chilopsis linearis var. glutinosa (G. Engelmann) F. Fosberg
Sy = C. linearis var. originaria F. Fosberg
Sy = C. saligna D. Don

JACARANDA A. L. de Jussieu
J. mimosifolia D. Don [**cultivated**]
Ma = J. acutifolia *auct.*, non F. von Humboldt & A. Bonpland: Texas authors

MACFADYENA A. P. de Candolle
M. unguis-cati (C. Linnaeus) A. Gentry [**cultivated**]
Sy = Batocydia unguis-cati (C. Linnaeus) K. von Martius *ex* N. Britton
Sy = Bignonia unguis-cati C. Linnaeus
Sy = Doxantha unguis-cati (C. Linnaeus) J. Miers

PSEUDOCALYMMA A. Sampaio & J. Kuhlmann
P. alliaceum (J. de Lamarck) N. Sandwith [**cultivated**]

PYROSTEGIA K. Presl
P. venusta (J. Ker-Gawler) J. Miers [**cultivated**]
Sy = Bignonia venusta J. Ker-Gawler

TABEBUIA B. A. Gomes (1769–1823) *ex* A. P. de Candolle
T. argentea N. Britton [**cultivated**]

TECOMA A. L. de Jussieu
T. australis R. Brown [**cultivated**]
Sy = Pandorea pandorana A. L. de Jussieu *ex* E. von Steudel
T. capensis (C. Thunberg) J. Lindley [**cultivated**]
Sy = Bignonia capensis C. Thunberg
Sy = Tecomaria capensis (C. Thunberg) E. Spach
T. jasminoides J. Lindley [**cultivated**]
T. stans (C. Linnaeus) A. L. de Jussieu *ex* K. Kunth var. ***angustata*** A. Rehder
Sy = Stenolobium incisum J. Rose & P. Standley
Sy = Tecoma incisa I. M. Johnston
Sy = T. tronadora (L. Loesener) I. M. Johnston
Ma = T. stans *auct.*, non (C. Linnaeus) A. L. de Jussieu *ex* K. Kunth var. stans: Texas authors

BIXACEAE, *nomen conservandum*

AMOREUXIA J. Mociño & M. Sessé y Lacasta *ex* A. P. de Candolle
* ***A. wrightii*** A. Gray [**TOES**: V]

BIXA C. Linnaeus
B. orellana C. Linnaeus [**cultivated**]

BOMBACACEAE, *nomen conservandum*

CHORISIA K. Kunth
C. speciosa A. de Saint-Hilaire [**cultivated**]

PSEUDOBOMBAX A. Dugand
P. ellipticum (K. Kunth) A. Dugand [**cultivated**]
Sy = Bombax ellipticum K. Kunth

Sy = Pachira fatuosa (J. Mociño & M. Sessé y Lacasta *ex* A. P. de Candolle) J. Decaisne

BORAGINACEAE, *nomen conservandum*

AMSINCKIA J. Lehmann, *nomen conservandum*
A. intermedia F. von Fischer & C. von Meyer
Sy = A. intermedia var. echinata (A. Gray) I. L. Wiggins
Sy = A. media E. Krause
Sy = Benthamia intermedia G. Druce
A. lycopsoides J. Lehmann
Sy = Benthamia lycopsoides (J. Lehmann) J. Lindley *ex* G. Druce
A. menziesii (J. Lehmann) A. Nelson & J. Macbride
Sy = A. debilis A. Brand
Sy = A. idahoënsis M. E. Jones
Sy = A. kennedyi W. Suksdorf
Sy = A. micrantha W. Suksdorf
Sy = Benthamia micrantha G. Druce
Sy = Echium menziesii J. Lehmann

ANCHUSA C. Linnaeus
A. azurea P. Miller
Sy = A. italica A. Retzius

ANTIPHYTUM A. P. de Candolle & C. Meisner
A. floribundum (J. Torrey) A. Gray
Sy = Eritrichium floribundum J. Torrey
A. heliotropioides A. L. de Candolle

BORAGO C. Linnaeus
B. officinalis C. Linnaeus [**cultivated**]

CORDIA C. Linnaeus
C. boissieri A. L. de Candolle
C. podocephala J. Torrey
C. sebestena C. Linnaeus [**cultivated**]
Sy = Sebesten sebestena (C. Linnaeus) N. Britton *ex* J. K. Small

CRYPTANTHA J. Lehmann *ex* G. Don
C. albida (K. Kunth) I. M. Johnston
Sy = Myosotis albida K. Kunth
C. angustifolia (J. Torrey) E. Greene
Sy = C. inaequata I. M. Johnston
Sy = Cryptanthe angustifolia E. Greene
Sy = Eremocarya angustifolia J. Torrey
Sy = Eritrichium angustifolium J. Torrey
Sy = Krynitzkia angustifolia (J. Torrey) A. Gray
C. barbigera (A. Gray) E. Greene
Sy = Cryptanthe barbigera E. Greene
Sy = Eritrichium barbigera A. Gray
Sy = Krynitzkia barbigera A. Gray
Sy = K. mixta M. E. Jones
C. cinerea (E. Greene) A. Cronquist var. ***cinerea***
Sy = C. jamesii (J. Torrey) E. Payson var. cinerea (E. Greene) E. Payson
Sy = C. jamesii var. multicaulis (J. Torrey) E. Payson
Sy = C. jamesii var. setosa (M. E. Jones) I. M. Johnston *ex* I. Tidestrom
Sy = Hemisphaerocarya cinerea A. Brand
Sy = Oreocarya cinerea E. Greene
Sy = O. multicaulis E. Greene var. cinerea J. Macbride
Sy = O. suffruticosa (C. Piper) E. Greene var. cinerea E. Payson
C. cinerea (E. Greene) A. Cronquist var. ***jamesii*** A. Cronquist
Sy = C. jamesii (J. Torrey) E. Payson var. disticha (A. Eastwood) E. Payson
Sy = C. jamesii var. jamesii
Sy = Eritrichium jamesii J. Torrey
Sy = E. multicaule J.Torrey
Sy = Hemisphaerocarya suffruticosa E. Greene
Sy = Krynitzkia jamesii (J. Torrey) A. Gray
Sy = Myosotis jamesii J. Torrey
Sy = M. suffruticosa J. Torrey, non C. Piper
Sy = Oreocarya suffruticosa (J. Torrey) E. Greene
C. cinerea (E. Greene) A. Cronquist var. ***laxa*** (J. Macbride) L. Higgins
Sy = C. jamesii (J. Torrey) E. Payson var. laxa (J. Macbride) E. Payson
* ***C. crassipes*** I. M. Johnston [**TOES: I**]
C. crassisepala (J. Torrey & A. Gray) E. Greene var. ***crassisepala***
Sy = Cryptanthe crassisepala E. Greene
Sy = C. dicarpa A. Nelson
Sy = Eritrichium crassisepalum J. Torrey & A. Gray
Sy = Krynitzkia crassisepala (J. Torrey & A. Gray) A. Gray
C. crassisepala (J. Torrey & A. Gray) E. Greene var. ***elachantha*** I. M. Johnston
C. mexicana (T. Brandegee) I. M. Johnston
C. micrantha (J. Torrey) I. M. Johnston var. ***micrantha***
Sy = Eremocarya micrantha (J. Torrey) E. Greene
Sy = Eritrichium micranthum J. Torrey
Sy = Krynitzkia micrantha A. Gray
C. minima P. Rydberg
C. oblata (M. E. Jones) E. Payson

Sy = Krynitzkia oblata M. E. Jones
Sy = Oreocarya hispidissima E. Wooton & P. Standley
Sy = O. oblata (M. E. Jones) J. Macbride
C. palmeri (A. Gray) E. Payson
Sy = C. coryi I. M. Johnston
Sy = Krynitzkia palmeri A. Gray
Sy = Oreocarya palmeri (A. Gray) E. Greene
* ***C. paysonii*** (J. Macbride) I. M. Johnston **[TOES: V]**
Sy = Oreocarya paysonii J. Macbride
C. pterocarya (J. Torrey) E. Greene var. ***cycloptera*** (E. Greene) J. Macbride
Sy = C. cycloptera E. Greene
Sy = Krynitzkia cycloptera E. Greene
C. pusilla (J. Torrey & A. Gray) E. Greene
Sy = Eritrichium pusillum J. Torrey & A. Gray
Sy = Krynitzkia pusilla (J. Torrey & A. Gray) A. Gray
C. texana (A. L. de Candolle) E. Greene
Sy = Eritrichium texanum A. L. de Candolle

CYNOGLOSSUM C. Linnaeus
C. amabile O. Stapf & J. R. Drummond **[cultivated]**
C. virginianum C. Linnaeus var. ***virginianum***
C. zeylanicum (J. Hornemann) C. Thunberg *ex* J. Lehmann

ECHIUM C. Linnaeus
E. vulgare C. Linnaeus

EHRETIA P. Browne
E. anacua (M. Terán & J. Berlandier) I. M. Johnston

HACKELIA P. M. Opiz
H. besseyi (P. Rydberg) J. Gentry
Sy = H. grisea (E. Wooton & P. Standley) I. M. Johnston
H. floribunda (J. Lehmann) I. M. Johnston
Sy = Echinospermum deflexum J. Lehmann var. floribundum (J. Lehmann) S. Watson
Sy = E. floribundum J. Lehmann
Sy = Lappula floribunda (J. Lehmann) E. Greene
H. pinetorum (E. Greene *ex* A. Gray) I. M. Johnston var. ***pinetorum***
H. virginiana (C. Linnaeus) I. M. Johnston
Sy = Lappula virginiana (C. Linnaeus) E. Greene
Sy = Myosotis virginiana C. Linnaeus

HELIOTROPIUM C. Linnaeus
H. amplexicaule M. H. Vahl
H. angiospermum J. Murray
Sy = H. parviflorum C. Linnaeus
Sy = Schobera angiosperma (J. Murray) N. Britton
H. confertifolium (J. Torrey) J. Torrey *ex* A. Gray
H. convolvulaceum (T. Nuttall) A. Gray
Sy = Batschia albiflora C. Rafinesque-Schmaltz
Sy = Euploca albiflora I. M. Johnston
Sy = E. albiflora var. californica W. Jepson & R. Hoover
Sy = E. convolvulacea subsp. californica L. Abrams
Sy = E. convolvulacea T. Nuttall subsp. convolvulacea
Sy = E. grandiflora J. Torrey
Sy = Heliotropium californicum E. Greene
Sy = H. convolvulaceum var. californicum (E. Greene) I. M. Johnston
H. curassavicum C. Linnaeus var. ***curassavicum***
Sy = H. curassavicum var. xerophilum (T. Cockerell) A. Nelson & J. Macbride
Sy = H. xerophilum T. Cockerell
H. curassavicum C. Linnaeus var. ***obovatum*** A. P. de Candolle
Sy = H. spathulatum P. Rydberg
H. europaeum C. Linnaeus
H. glabriusculum (J. Torrey) A. Gray
Sy = Heliophytum glabriusculum J. Torrey
H. greggii J. Torrey
H. indicum C. Linnaeus
Sy = Tiaridium indicum (C. Linnaeus) J. Lehmann
H. molle (J. Torrey) I. M. Johnston
H. procumbens P. Miller var. ***procumbens***
H. racemosum J. Rose & P. Standley
Sy = H. convolvulaceum (T. Nuttall) A. Gray var. racemosum (J. Rose & P. Standley) I. M. Johnston
H. tenellum (T. Nuttall) J. Torrey
Sy = Lithococca tenella (T. Nuttall) J. K. Small
Sy = Lithospermum tenellum T. Nuttall
H. texanum I. M. Johnston
H. torreyi I. M. Johnston
Sy = H. angustifolium J. Torrey

LAPPULA C. Moench
L. occidentalis (S. Watson) E. Greene var. ***cupulata*** (A. Gray) L. Higgins
Sy = L. redowskii (J. Hornemann) E. Greene var. cupulata (A. Gray) M. E. Jones
Sy = L. redowskii var. texana (G. Scheele) A. Brand
Sy = L. texana (G. Scheele) N. Britton

L. occidentalis (S. Watson) E. Greene var. ***occidentalis***
Sy = L. echinata J.-E. Gilibert var. occidentalis (S. Watson) J. Boivin
Sy = L. redowskii (J. Hornemann) E. Greene var. occidentalis (S. Watson) P. Rydberg
L. squarrosa (A. Retzius) C. Dumortier
Sy = L. echinata J.-E. Gilibert
Sy = L. squarrosa var. erecta (A. Nelson) R. Dorn

LITHOSPERMUM C. Linnaeus
L. arvense C. Linnaeus
Sy = Buglossoides arvense (C. Linnaeus) I. M. Johnston
L. calycosum (J. Macbride) I. M. Johnston
L. caroliniense (T. Walter *ex* J. F. Gmelin) C. MacMillan var. ***caroliniense***
Sy = Anonymos caroliniensis T. Walter
Sy = Batschia caroliniensis T. Walter *ex* J. F. Gmelin
Sy = Lithospermum bejariense A. L. de Candolle
L. cobrense E. Greene
L. confine I. M. Johnston
L. incisum J. Lehmann
Sy = Batschia decumbens T. Nuttall, non E. Ventenat
Sy = B. longiflora F. Pursh
Sy = Cyphorima angustifolia J. Nieuwland
Sy = Lithospermum angustifolium A. Michaux
Sy = L. breviflorum G. Engelmann & A. Gray
Sy = L. linearifolium J. Goldie
Sy = L. mandanensis A. P. de Candolle
Sy = L. oblongum E. Greene
L. matamorense A. P. de Candolle
L. mirabile J. K. Small
Sy = L. longiflorum A. L. de Candolle var. mirabile (J. K. Small) A. Brand
L. multiflorum J. Torrey *ex* A. Gray
L. parksii I. M. Johnston var. ***parksii***
L. parksii I. M. Johnston var. ***rugulosum*** I. M. Johnston
L. tuberosum F. Rugel *ex* A. P. de Candolle
L. viride E. Greene

MYOSOTIS C. Linnaeus
M. macrosperma G. Engelmann
Sy = M. verna T. Nuttall var. macrosperma (G. Engelmann) A. Chapman
M. verna T. Nuttall

OMPHALODES P. Miller
O. aliena A. Gray

ONOSMODIUM A. Michaux
O. bejariense A. P. de Candolle *ex* A. L. de Candolle subsp. ***bejariense*** var. ***bejariense***
Sy = O. molle A. Michaux subsp. bejariense (A. P. de Candolle *ex* A. L. de Candolle) T. Cochrane
Sy = O. molle var. bejariense (A. P. de Candolle *ex* A. L. de Candolle) A. Cronquist
Ma = O. hispidissimum K. Mackenzie
O. bejariense A. P. de Candolle *ex* A. L. de Candolle subsp. ***bejariense*** var. ***occidentale*** (K. Mackenzie) B. L. Turner
Sy = O. molle A. Michaux subsp. occidentale (K. Mackenzie) J. Boivin
Sy = O. molle A. Michaux subsp. occidentale (K. Mackenzie) T. Cochrane
Sy = O. molle var. occidentale (K. Mackenzie) I. M. Johnston
Sy = O. occidentale K. Mackenzie
* ***O. helleri*** J. K. Small [**TOES: V**]

PECTOCARYA A. P. de Candolle *ex* C. Meisner
P. heterocarpa (I. M. Johnston) I. M. Johnston
Sy = P. penicillata (W. Hooker *ex* G. Arnott) A. P. de Candolle var. heterocarpa I. M. Johnston
P. platycarpa (P. Munz & I. M. Johnston) P. Munz & I. M. Johnston
Sy = P. gracilis I. M. Johnston var. platycarpa P. Munz & I. M. Johnston
Sy = P. linearis (H. Ruiz López & J. Pavón) A. P. de Candolle var. platycarpa (P. Munz & I. M. Johnston) A. Cronquist

SYMPHYTUM C. Linnaeus
S. officinale C. Linnaeus [**cultivated**]

TIQUILIA C. Persoon
T. canescens (A. P. de Candolle) A. Richardson var. ***canescens***
Sy = Coldenia canescens A. P. de Candolle
Sy = Stegnocarpus canescens (A. P. de Candolle) J. Torrey
T. gossypina (E. Wooton & P. Standley) A. Richardson
Sy = Coldenia gossypina (E. Wooton & P. Standley) I. M. Johnston
Sy = Eddya gossypina E. Wooton & P. Standley
T. greggii (J. Torrey & A. Gray) A. Richardson
Sy = Coldenia greggii (J. Torrey & A. Gray) A. Gray
Sy = Ptilocalyx greggii J. Torrey & A. Gray

T. hispidissima (J. Torrey & A. Gray) A. Richardson
Sy = Coldenia hispidissima (J. Torrey & A. Gray) A. Gray
Sy = Eddya hispidissima J. Torrey & A. Gray
T. mexicana (S. Watson) A. Richardson
Sy = Coldenia mexicana S. Watson
Sy = C. mexicana var. tomentosa (S. Watson) I. M. Johnston

TOURNEFORTIA C. Linnaeus
T. volubilis C. Linnaeus
Sy = Messerschmidia candida M. Martens & H. Galeotti
Sy = Myriopus volubilis (C. Linnaeus) A. Bertoloni *ex* K. Sprengel
Sy = Tournefortia candida W. Walpers
Sy = T. floribunda K. Kunth
Sy = T. velutina K. Kunth

BRASSICACEAE, *nomen conservandum*

ARABIDOPSIS G. Heynhold, *nomen conservandum*
A. thaliana (C. Linnaeus) G. Heynhold
Sy = Arabis thaliana C. Linnaeus
Sy = Sisymbrium thalianum (C. Linnaeus) J. Gay & J. Monnard

ARABIS C. Linnaeus
A. canadensis C. Linnaeus
A. fendleri (S. Watson) E. Greene var. ***fendleri***
Sy = A. holboellii J. Hornemann var. fendleri S. Watson
Sy = Boechera fendleriana (S. Watson) W. A. Weber
A. perennans S. Watson
Sy = A. ermophila E. Greene
A. petiolaris (A. Gray) A. Gray
Sy = Erysimum petiolare (A. Gray) K. E. O. Kuntze
Sy = Streptanthus petiolaris A. Gray

ARMORACIA P. Gaertner, B. Meyer, & J. Scherbius, *nomen conservandum*
A. lacustris (A. Gray) I. Al-Shehbaz & V. Bates
Sy = A. aquatica (A. Eaton) K. Wiegand
Sy = Nasturtium lacustre A. Gray
Sy = Neobeckia aquatica (A. Eaton) E. Greene
A. rusticana P. Gaertner, B. Meyer, & J. Scherbius [**cultivated**]
Sy = A. armoracia (C. Linnaeus) N. Britton
Sy = A. lapathifolia J.-E. Gilibert
Sy = Cochlearia armoracia C. Linnaeus
Sy = Radicula armoracia (C. Linnaeus) B. Robinson
Sy = Rorippa armoracia (C. Linnaeus) A. S. Hitchcock

AURINIA N. Desvaux
A. saxtilis (C. Linnaeus) N. Desvaux [**cultivated**]
Sy = Alyssum saxatile C. Linnaeus

BRASSICA C. Linnaeus
B. caulorapa F. Pasquale [**cultivated**]
B. juncea (C. Linnaeus) V. Czernajew [**cultivated**]
Sy = B. juncea var. crispifolia L. Bailey
Sy = B. juncea var. japonica (C. Thunberg) L. Bailey
Sy = Sinapis juncea C. Linnaeus
B. nigra (C. Linnaeus) W. Koch
Sy = Sinapis nigra C. Linnaeus
B. oleracea C. Linnaeus var. ***acephala*** A. P. de Candolle [**cultivated**]
B. oleracea C. Linnaeus var. ***botrytis*** C. Linnaeus [**cultivated**]
B. oleracea C. Linnaeus var. ***capitata*** C. Linnaeus [**cultivated**]
B. oleracea C. Linnaeus var. ***gemmifera*** J. Zenker [**cultivated**]
B. oleracea C. Linnaeus var. ***italica*** J. von Plenck [**cultivated**]
B. oleracea C. Linnaeus var. ***oleracea***
B. rapa C. Linnaeus var. ***chinensis*** (C. Linnaeus) S. Kitamura [**cultivated**]
Sy = B. chinensis C. Linnaeus
B. rapa C. Linnaeus var. ***rapa*** [**cultivated**]
Sy = B. campestris C. Linnaeus
Sy = B. campestris var. rapa (C. Linnaeus) R. Hartman
Sy = B. rapa subsp. campestris (C. Linnaeus) A. Clapham
Sy = B. rapa subsp. olifera A. P. de Candolle
Sy = B. rapa subsp. sylvestris E. Janchen
Sy = B. rapa var. campestris (C. Linnaeus) W. Koch
B. tournefortii A. Gouan
Sy = B. tournefortii var. sisymbrioides (F. von Fischer) A. Grossheim

CAKILE P. Miller
C. constricta J. Rodman
C. geniculata (B. Robinson) C. Millspaugh
Sy = C. lanceolata (C. von Willdenow) O. E. Schulz var. geniculata (B. Robinson) L. Shinners

Sy = C. maritima J. Scopoli var. geniculata B. Robinson
C. ***lanceolata*** (C. von Willdenow) O. E. Schulz subsp. ***pseudoconstricta*** J. Rodman

CAMELINA H. von Crantz
C. ***microcarpa*** A. Andrzejowski *ex* A. P. de Candolle
Sy = C. sativa (C. Linnaeus) H. von Crantz subsp. microcarpa (A. P. de Candolle) E. Schmid
C. ***rumelica*** J. Velenovský

CAPSELLA F. Medikus, *nomen conservandum*
C. ***bursa-pastoris*** (C. Linnaeus) F. Medikus
Sy = C. bursa-pastoris (C. Linnaeus) N. Britton
Sy = C. bursa-pastoris (C. Linnaeus) F. Medikus var. bifida F. Crépin
Sy = Thlaspi bursa-pastoris C. Linnaeus

CARDAMINE C. Linnaeus
C. ***concatenata*** (A. Michaux) O. Swartz
Sy = C. laciniata (C. von Willdenow) A. Wood
Sy = Dentaria laciniata G. H. Muhlenberg *ex* C. von Willdenow
C. ***hirsuta*** C. Linnaeus
C. ***macrocarpa*** T. Brandegee var. ***texana*** R. Rollins
C. ***parviflora*** C. Linnaeus var. ***arenicola*** (N. Britton) O. E. Schulz
Sy = C. arenicola N. Britton
C. ***pensylvanica*** G. H. Muhlenberg *ex* C. von Willdenow
Sy = C. pensylvanica var. brittoniana O. Farwell
C. ***rhomboidea*** (C. Persoon) A. P. de Candolle
Sy = Arabis bulbosa J. von Schreber *ex* G. H. Muhlenberg
Sy = Cardamine bulbosa (J. von Schreber *ex* G. H. Muhlenberg) N. Britton, E. Sterns, & J. Poggenburg

CARDARIA N. Desvaux
C. ***draba*** (C. Linnaeus) N. Desvaux
Sy = Lepidium draba C. Linnaeus

CHORISPORA R. Brown *ex* A. P. de Candolle, *nomen conservandum*
C. ***tenella*** (P. von Pallas) A. P. de Candolle
Sy = Raphanus tenella P. von Pallas

COELOPHRAGMUS O. E. Schulz
C. ***auriculatus*** (A. Gray) O. E. Schulz
Sy = Sisymbrium auriculatum A. Gray

CONRINGIA L. Heister *ex* P. Fabricius
C. ***orientalis*** (C. Linnaeus) A. Andrzejowski
Sy = Brassica orientalis C. Linnaeus

CORONOPUS J. Zinn, *nomen conservandum*
C. ***didymus*** (C. Linnaeus) J. E. Smith
Sy = Carara didyma (C. Linnaeus) N. Britton
Sy = Lepidium didymum C. Linnaeus

DESCURAINIA P. Webb & S. Berthelot, *nomen conservandum*
D. ***incana*** (J. Bernhardi *ex* F. von Fischer & C. von Meyer) R. Dorn subsp. ***incana***
Sy = D. incana var. major (W. Hooker) R. Dorn
Sy = D. richardsonii (R. Sweet) O. E. Schulz var. incisa (G. Engelmann) L. Detling
Sy = Sisymbrium incisum G. Engelmann
D. ***incana*** (J. Bernhardi *ex* F. von Fischer & C. von Meyer) R. Dorn subsp. ***viscosa*** (P. Rydberg) J. Kartesz & K. Gandhi
Sy = D. incana var. viscosa (P. Rydberg) R. Dorn
Sy = D. richardsonii (R. Sweet) O. E. Schulz subsp. viscosa (P. Rydberg) L. Detling
Sy = D. richardsonii var. viscosa (P. Rydberg) J. Blankinship
Sy = Sophia viscosa P. Rydberg
D. ***pinnata*** (T. Walter) N. Britton subsp. ***brachycarpa*** (J. Richardson) L. Detling
Sy = D. brachycarpa (J. Richardson) O. E. Schulz
Sy = Sisymbrium brachycarpum J. Richardson
Sy = S. pinnata var. brachycarpa (J. Richardson) M. Fernald
D. ***pinnata*** (T. Walter) N. Britton subsp. ***glabra*** (E. Wooton & P. Standley) L. Detling
Sy = D. pinnata (T. Walter) N. Britton var. glabra (E. Wooton & P. Standley) L. Shinners
Sy = Sophia glabra E. Wooton & P. Standley
D. ***pinnata*** (T. Walter) N. Britton subsp. ***ochroleuca*** (E. Wooton) L. Detling
Sy = D. pinnata (T. Walter) N. Britton var. ochroleuca (E. Wooton) L. Shinners
Sy = Sophia ochroleuca E. Wooton
D. ***pinnata*** (T. Walter) N. Britton subsp. ***pinnata***
Sy = Erysimum pinnatum T. Walter
Sy = Sisymbrium pinnatum (T. Walter) E. Greene
Sy = Sophia pinnata (T. Walter) T. Howell
D. ***sophia*** (C. Linnaeus) P. Webb *ex* K. Prantl
Sy = Sisymbrium sophia C. Linnaeus
Sy = Sophia sophia (C. Linnaeus) N. Britton

DIMORPHOCARPA (C. Rafinesque-Schmaltz) R. Rollins
D. ***candicans*** (C. Rafinesque-Schmaltz) R. Rollins
Sy = D. palmeri (E. Payson) R. Rollins
Sy = Dithyrea palmeri (E. Payson) R. Rollins
Sy = D. wislizenii G. Engelmann var. palmeri E. Payson
Sy = Iberis candicans C. Rafinesque-Schmaltz

D. wislizenii (G. Engelmann) R. Rollins
Sy = Dithyrea wislizenii G. Engelmann
Sy = D. wislizenii var. griffithsii (E. Wooton & P. Standley) E. Payson

DIPLOTAXIS A. P. de Candolle
D. muralis (C. Linnaeus) A. P. de Candolle
Sy = Sisymbrium murale C. Linnaeus
D. tenuifolia (C. Linnaeus) A. P. de Candolle
Sy = Sisymbrium tenuifolium C. Linnaeus

DRABA C. Linnaeus
D. brachycarpa T. Nuttall *ex* J. Torrey & A. Gray
Sy = D. brachycarpa (T. Nuttall *ex* J. Torrey & A. Gray) E. Greene
D. cuneifolia T. Nuttall *ex* J. Torrey & A. Gray var. ***cuneifolia***
Sy = D. cuneifolia var. foliosa R. Mohlenbrock & J. Voigt
Sy = D. cuneifolia var. helleri (J. K. Small) O. E. Schulz
Sy = D. cuneifolia var. leiocarpa O. E. Schulz
D. cuneifolia T. Nuttall *ex* J. Torrey & A. Gray var. ***integrifolia*** S. Watson
Sy = D. sonorae E. Greene var. integrifolia (S. Watson) O. E. Schulz
D. platycarpa J. Torrey & A. Gray
Sy = D. cuneifolia T. Nuttall *ex* J. Torrey & A. Gray var. platycarpa (J. Torrey & A. Gray) S. Watson
D. reptans (J. de Lamarck) M. Fernald
Sy = Arabis reptans J. de Lamarck
Sy = Draba coloradensis P. Rydberg
Sy = D. micrantha T. Nuttall
Sy = D. reptans subsp. stellifera (O. E. Schulz) L. Abrams
Sy = D. reptans var. micrantha (T. Nuttall) M. Fernald
Sy = D. reptans var. stellifera (O. E. Schulz) C. Hitchcock
* ***D. standleyi*** J. Macbride & E. Payson [**TOES: V**]
Sy = D. chrysantha S. Watson var. gilgiana (E. Wooton & P. Standley) O. E. Schulz
Sy = D. gilgiana E. Wooton & P. Standley, non R. Muschler

DRYOPETALON A. Gray
D. runcinatum A. Gray var. ***runcinatum***
Sy = Coelophragmus umbrosus (B. Robinson) O. E. Schulz
Sy = Sisymbrium umbrosum B. Robinson

ERUCA P. Miller
E. vesicaria (C. Linnaeus) A. Cavanilles subsp. ***sativa*** (P. Miller) A. Thellung
Sy = E. sativa P. Miller

ERUCASTRUM K. Presl
E. gallicum (C. von Willdenow) O. E. Schulz
Sy = Brassica erucastrum C. Linnaeus

ERYSIMUM C. Linnaeus
E. asperum (T. Nuttall) A. P. de Candolle
Sy = Cheiranthus asper T. Nuttall
Sy = Cheirinia arida E. Greene
Sy = C. aspera (T. Nuttall) P. Rydberg
Sy = Erysimum asperum var. dolichocarpum O. E. Schulz
E. capitatum (D. Douglas *ex* W. Hooker) E. Greene var. ***capitatum***
Sy = Cheiranthus capitatus D. Douglas *ex* W. Hooker
Sy = Erysimum asperum (T. Nuttall) A. P. de Candolle var. capitatum (D. Douglas *ex* W. Hooker) J. Boivin
Sy = E. elatum T. Nuttall
Sy = E. wheeleri J. Rothrock
E. repandum C. Linnaeus
Sy = Cheirinia repanda (C. Linnaeus) J. Link

HALIMOLOBOS I. Tausch
H. diffusa (A. Gray) O. E. Schulz var. ***diffusa***
Sy = Sisymbrium diffusum A. Gray

IBERIS C. Linnaeus
I. amara C. Linnaeus [**cultivated**]
I. sempervirens C. Linnaeus [**cultivated**]
I. umbellata C. Linnaeus [**cultivated**]

IODANTHUS J. Torrey & A. Gray *ex* E. von Steudel
I. pinnatifidus (A. Michaux) E. von Steudel
Sy = Hesperis pinnatifidus A. Michaux
Sy = Thelypodium pinnatifidum (A. Michaux) S. Watson

LEAVENWORTHIA J. Torrey
L. aurea J. Torrey var. ***texana*** (W. Mahler) R. Rollins
* Sy = L. texana W. Mahler [**TOES: V**]

LEPIDIUM C. Linnaeus
L. alyssoides A. Gray var. ***alyssoides***
Sy = L. alyssoides var. minus A. Thellung
Sy = L. alyssoides var. polycarpum A. Thellung
Sy = L. alyssoides var. streptocarpum A. Thellung

Sy = L. eastwoodiae E. Wooton
Sy = L. montanum T. Nuttall *ex* J. Torrey & A. Gray subsp. alyssoides (A. Gray) C. Hitchcock
Sy = L. montanum var. alyssoides (A. Gray) M. E. Jones
Sy = Nasturtium alyssoides (A. Gray) K. E. O. Kuntze

L. alyssoides A. Gray var. **angustifolium** (C. Hitchcock) R. Rollins
Sy = L. montanum T. Nuttall *ex* J. Torrey & A. Gray subsp. angustifolium (C. Hitchcock) C. Hitchcock
Sy = L. montanum var. angustifolium C. Hitchcock

L. austrinum J. K. Small
Sy = L. austrinum var. orbiculare A. Thellung
Sy = L. lasiocarpum T. Nuttall *ex* J. Torrey & A. Gray var. obiculare (A. Thellung) C. Hitchcock

L. densiflorum H. A. Schrader var. ***densiflorum***
Sy = L. apetalum C. Linnaeus
Sy = L. neglectum A. Thellung

L. lasiocarpum T. Nuttall *ex* J. Torrey & A. Gray var. ***wrightii*** (A. Gray) C. Hitchcock
Sy = L. lasiocarpum subsp. wrightii (A. Gray) A. Thellung
Sy = L. lasiocarpum var. rotundum C. Hitchcock
Sy = L. nelsonii L. O. Williams
Sy = L. wrightii A. Gray

L. latifolium C. Linnaeus

L. oblongum J. K. Small var. ***oblongum***
Sy = L. greenei A. Thellung

L. ruderale C. Linnaeus
Sy = L. texanum S. Buckley
Sy = L. virginicum C. Linnaeus subsp. texanum (S. Buckley) A. Thellung

L. sordidum A. Gray
Sy = L. granulare J. Rose

L. thurberi E. Wooton

L. virginicum C. Linnaeus var. ***medium*** (E. Greene) C. Hitchcock
Sy = L. medium E. Greene

L. virginicum C. Linnaeus var. ***virginicum***
Sy = L. virginicum subsp. euvirginicum A. Thellung
Sy = L. virginicum var. linearifolium O. Farwell

LESQUERELLA S. Watson

L. angustifolia (T. Nuttall *ex* J. Torrey & A. Gray) S. Watson
Sy = L. longifolia V. Cory
Sy = Vesicaria angustifolia T. Nuttall *ex* J. Torrey & A. Gray

L. argyraea (A. Gray) S. Watson var. ***argyraea***
Sy = Alyssum argyraeum (A. Gray) K. E. O. Kuntze
Sy = Vesicaria argyraea A. Gray

L. auriculata (G. Engelmann & A. Gray) S. Watson
Sy = Alyssum auriculatum (G. Engelmann & A. Gray) K. E. O. Kuntze
Sy = Vesicaria auriculata G. Engelmann & A. Gray

L. densiflora (A. Gray) S. Watson
Sy = Alyssum densiflora (A. Gray) K. E. O. Kuntze
Sy = Vesicaria densiflora A. Gray

L. engelmannii (A. Gray) S. Watson
Sy = Alyssum engelmannii (A. Gray) K. E. O. Kuntze
Sy = Vesicaria engelmannii A. Gray
Sy = V. pulchella K. Kunth & P. C. Bouché

L. fendleri (A. Gray) S. Watson
Sy = L. stenophylla (A. Gray) P. Rydberg
Sy = Vesicaria fendleri A. Gray

L. gordonii (A. Gray) S. Watson var. ***gordonii***
Sy = Alyssum gordonii (A. Gray) K. E. O. Kuntze
Sy = Vesicaria gordonii A. Gray

L. gracilis (W. Hooker) S. Watson subsp. ***gracilis***
Sy = Alyssum gracile (W. Hooker) K. E. O. Kuntze
Sy = Lesquerella polyantha (D. von Schlechtendal) J. K. Small
Sy = Vesicaria gracilis W. Hooker
Sy = V. polyantha D. von Schlechtendal

L. gracilis (W. Hooker) S. Watson subsp. ***nuttallii*** (J. Torrey & A. Gray) R. Rollins & E. Shaw
Sy = Alyssum nuttallii (J. Torrey & A. Gray) K. E. O. Kuntze
Sy = A. repandum (T. Nuttall *ex* J. Torrey & A. Gray) K. E. O. Kuntze
Sy = Lesquerella gracilis var. repanda (T. Nuttall *ex* J. Torrey & A. Gray) E. Payson
Sy = L. nuttallii (J. Torrey & A. Gray) S. Watson
Sy = L. repanda (T. Nuttall *ex* J. Torrey & A. Gray) S. Watson
Sy = Vesicaria nuttallii J. Torrey & A. Gray
Sy = V. repanda J. Torrey & A. Gray

L. grandiflora (W. Hooker) S. Watson
Sy = Alyssum grandiflorum (W. Hooker) K. E. O. Kuntze
Sy = Vesicaria brevistyla J. Torrey & A. Gray
Sy = V. grandiflora W. Hooker

L. lasiocarpa (W. Hooker *ex* A. Gray) S. Watson var. ***berlandieri*** (A. Gray) E. Payson

Sy = L. lasiocarpa subsp. berlandieri (A. Gray) R. Rollins & E. Shaw var. hispida (S. Watson) R. Rollins & E. Shaw
Sy = L. lasiocarpa var. ampla R. Rollins
Sy = L. lasiocarpa var. hispida S. Watson
Sy = Synthlipsis berlandieri A. Gray

L. lasiocarpa (W. Hooker *ex* A. Gray) S. Watson var. ***lasiocarpa***
Sy = Alyssum lasiocarpum (W. Hooker *ex* A. Gray) K. E. O. Kuntze
Sy = Vesicaria lasiocarpa W. Hooker *ex* A. Gray

L. lindheimeri (A. Gray) S. Watson
Sy = Alyssum lindheimeri (A. Gray) K. E. O. Kuntze
Sy = Lesquerella gracilis (W. Hooker) S. Watson var. pilosa C. Lundell
Sy = Vesicaria lindheimeri A. Gray

L. mcvaughiana R. Rollins

L. ovalifolia P. Rydberg *ex* N. Britton subsp. ***ovalifolia***
Sy = L. engelmannii (A. Gray) S. Watson subsp. ovalifolia (P. Rydberg *ex* N. Britton) C. Clark
Sy = L. ovata E. Greene

* ***L. pallida*** (J. Torrey & A. Gray) S. Watson **[TOES: I]**
Sy = Alyssum pallidum (J. Torrey & A. Gray) K. E. O. Kuntze
Sy = Vesicaria grandiflora W. Hooker var. pallida J. Torrey & A. Gray
Sy = V. pallidum (J. Torrey & A. Gray) K. E. O. Kuntze

L. purpurea (A. Gray) S. Watson
Sy = L. purpurea subsp. foliosa (R. Rollins) R. Rollins & A. Shaw
Sy = L. purpurea var. albiflora A. Gray
Sy = L. purpurea var. foliosa R. Rollins
Sy = Vesicaria purpurea A. Gray

L. recurvata (G. Engelmann *ex* A. Gray) S. Watson
Sy = Alyssum recurvatum (G. Engelmann *ex* A. Gray) K. E. O. Kuntze
Sy = Vesicaria recurvata G. Engelmann *ex* A. Gray

L. sessilis (S. Watson) J. K. Small
Sy = L. gordonii (A. Gray) S. Watson var. sessilis S. Watson
Sy = L. gracilis (W. Hooker) S. Watson var. sessilis S. Watson

* ***L. thamnophila*** R. Rollins & E. Shaw **[TOES: IV]**

L. valida E. Greene
Sy = L. lepidota V. Cory

LOBULARIA N. Desvaux, *nomen conservandum*

L. maritima (C. Linnaeus) N. Desvaux **[cultivated]**
Sy = Alyssum maritimum (C. Linnaeus) J. de Lamarck
Sy = Clypeola maritima C. Linnaeus
Sy = Konga maritima (C. Linnaeus) R. Brown

MANCOA H. Weddell, *nomen conservandum*

M. pubens (A. Gray) R. Rollins
Sy = Capsella pubens (A. Gray) S. Watson
Sy = Hymenolobos pubens A. Gray
Sy = Poliophyton pubens (A. Gray) O. E. Schulz

MATTHIOLA R. Brown, *nomen conservandum*

M. incana (C. Linnaeus) R. Brown
Sy = Cheiranthus incanus C. Linnaeus

M. longipetala (E. Ventenat) A. P. de Candolle
Sy = Cheiranthus longipetalus E. Ventenat
Sy = Matthiola bicornis J. Sibthorp & J. E. Smith
Sy = M. longipetala subsp. bicornis (J. Sibthorp & J. E. Smith) P. Ball

MYAGRUM C. Linnaeus

M. perfoliatum C. Linnaeus

NERISYRENIA E. Greene

N. camporum (A. Gray) E. Greene
Sy = Greggia camporum A. Gray
Sy = Parrasia camporum (A. Gray) E. Greene

N. linearifolium (S. Watson) E. Greene var. ***linearifolium***
Sy = Greggia camporum A. Gray var. linearifolia (S. Watson) M. E. Jones
Sy = Parrasia linearifolia (S. Watson) E. Greene

PENNELLIA J. Nieuwland

P. longifolia (G. Bentham) R. Rollins
Sy = Streptanthus longifolius G. Bentham
Sy = Thelypodium longifolium (G. Bentham) S. Watson

P. micrantha (A. Gray) J. Nieuwland
Sy = Streptanthus micranthus A. Gray
Sy = Thelypodium micrantha (A. Gray) S. Watson

RAPHANUS C. Linnaeus

R. sativus C. Linnaeus **[cultivated]**
Sy = R. raphanistrum C. Linnaeus var. sativus (C. Linnaeus) G. Beck

RAPISTRUM H. von Crantz, *nomen conservandum*

R. rugosum (C. Linnaeus) C. Allioni
Sy = Myagrum rugosum C. Linnaeus

RORIPPA J. Scopoli
R. curvipes E. Greene var. **_truncata_** (W. Jepson) R. Rollins
Sy = Radicula integra (P. Rydberg) A. A. Heller
Sy = Rorippa integra P. Rydberg
Sy = R. obtusa (T. Nuttall *ex* J. Torrey & A. Gray) N. Britton var. integra (P. Rydberg) J. Marie-Victorin
R. nasturtium-aquaticum (C. Linnaeus) A. von Hayek
Sy = Nasturtium officinale R. Brown
Sy = Radicula nasturtium-aquaticum (C. Linnaeus) N. Britton & A. Rendle
Sy = Rorippa nasturtium H. Rusby
Sy = Sisymbrium nasturtium-aquaticum C. Linnaeus
R. palustris (C. Linnaeus) W. Besser subsp. **_fernaldiana_** (F. Butters & E. Abbe) B. E. Jonsell
Sy = R. islandica (G. Oeder) V. von Borbás subsp. fernaldiana (F. Butters & E. Abbe) O. Hultén
Sy = R. islandica var. fernaldiana F. Butters & E. Abbe
Sy = R. palustris var. fernaldiana (F. Butters & E. Abbe) R. Stuckey
* **_R. ramosa_** R. Rollins [**TOES**: **V**]
R. sessilifora (T. Nuttall *ex* J. Torrey & A. Gray) A. Hitchcock
Sy = Nasturtium sessiliflorum T. Nuttall
Sy = Radicula sessiliflora (T. Nuttall) E. Greene
R. sinuata (T. Nuttall *ex* J. Torrey & A. Gray) A. Hitchcock
Sy = Nasturtium sinuatum T. Nuttall
Sy = Radicula sinuata (T. Nuttall) E. Greene
R. sphaerocarpa (A. Gray) N. Britton
Sy = Nasturtium sphaerocarpum A. Gray
Sy = Radicula sphaerocarpa (A. Gray) E. Greene
Sy = Rorippa obtusa (T. Nuttall) N. Britton var. sphaerocarpa (A. Gray) V. Cory
R. tenerrima E. Greene
Sy = Radicula tenerrima (E. Greene) E. Greene
R. teres (A. Michaux) R. Stuckey
Sy = Cardamine teres A. Michaux
Sy = Rorippa obtusa (T. Nuttall) N. Britton
Sy = R. walteri (S. Elliott) E. Greene

SCHENOCRAMBE E. Greene
S. linearifolia (A. Gray) R. Rollins
Sy = Hesperidanthus linearifolius (A. Gray) P. Rydberg
Sy = Sisymbrium linearifolium (A. Gray) E. Payson
Sy = Thelepodiopsis linearifolius (A. Gray) I. Al-Shehbaz

SELENIA T. Nuttall
S. aurea T. Nuttall
Sy = S. aptera (S. Watson) J. K. Small
Sy = S. aurea var. aptera S. Watson
S. dissecta J. Torrey & A. Gray
Sy = S. mexicana P. Standley
S. grandis R. Martin
Sy = S. oinosepala J. Steyermark
S. jonesii V. Cory var. **_jonesii_**
S. jonesii V. Cory var. **_obovata_** R. Rollins

SIBARA E. Greene
S. grisea R. Rollins
S. viereckii (O. E. Schulz) R. Rollins
Sy = Arabis runcinata S. Watson, non J. de Lamarck
Sy = Sibara runcinata (S. Watson) R. Rollins
S. virginica (C. Linnaeus) R. Rollins
Sy = Arabis virginica (C. Linnaeus) J. Poiret
Sy = Cardamine virginica C. Linnaeus

SINAPSIS C. Linnaeus
S. alba C. Linnaeus
Sy = Brassica alba G. Rabenhorst
Sy = B. hirta C. Linnaeus
S. arvensis C. Linnaeus
Sy = Brassica arvensis (C. Linnaeus) G. Rabenhorst
Sy = B. kaber (A. P. de Candolle) L. Wheeler

SISYMBRIUM C. Linnaeus
S. altissimum C. Linnaeus
Sy = Hesperis altissima (C. Linnaeus) K. E. O. Kuntze
Sy = Norta altissima (C. Linnaeus) N. Britton
S. irio C. Linnaeus
Sy = Erysimum irio (C. Linnaeus) O. Farwell
Sy = Norta irio (C. Linnaeus) N. Britton
S. officinale (C. Linnaeus) J. Scopoli
Sy = Erysimum officinale C. Linnaeus
Sy = Sisymbrium officinale var. leiocarpum A. P. de Candolle
S. orientale C. Linnaeus
Sy = S. columnae N. von Jacquin
S. polyceratium C. Linnaeus
Sy = Chamaeplium polyceratium (C. Linnaeus) C. Wallroth

STANLEYA T. Nuttall
S. *pinnata* (F. Pursh) N. Britton var. ***integrifolia*** (C. James *ex* J. Torrey) R. Rollins
Sy = S. integrifolia E. James *ex* J. Torrey
Sy = S. pinnatifida J. Torrey var. integrifolia B. Robinson
S. *pinnata* (F. Pursh) N. Britton var. ***pinnata***
Sy = Cleome pinnata F. Pursh
Sy = Stanleya arcuata P. Rydberg
Sy = S. canescens P. Rydberg
Sy = S. fruticosa T. Nuttall
Sy = S. glauca P. Rydberg
Sy = S. heterophylla T. Nuttall *ex* J. Torrey & A. Gray
Sy = S. pinnatifida T. Nuttall

STREPTANTHUS T. Nuttall
* ***S. bracteatus*** A. Gray [**TOES: III**]
Sy = Erysimum bracteatum (A. Gray) K. E. O. Kuntze
S. carinatus C. Wright *ex* A. Gray subsp. ***arizonicus*** (S. Watson) A. Kruckeberg, J. Rodman, & R. Worthington
Sy = Disaccanthus arizonicus (S. Watson) E. Greene
Sy = D. luteus E. Greene
Sy = Streptanthus arizonicus S. Watson
Sy = S. arizonicus var. luteus T. Kearney & H. Peebles
S. carinatus C. Wright *ex* A. Gray subsp. ***carinatus***
Sy = Disaccanthus validus E. Greene
Sy = Streptanthus validus (E. Greene) V. Cory
* ***S. cutleri*** V. Cory [**TOES: V**]
S. hyacinthoides W. Hooker
Sy = Erysimum hyacinthoides (W. Hooker) K. E. O. Kuntze
Sy = Euklisia hyacinthoides (W. Hooker) J. K. Small
Sy = Icianthus atratus E. Greene
Sy = I. glabrifolius (S. Buckley) E. Greene
Sy = I. hyacinthoides (W. Hooker) E. Greene
Sy = Steptanthus glabrifolius S. Buckley
S. maculatus T. Nuttall subsp. ***maculatus***
Sy = Erysimum maculatum (T. Nuttall) K. E. O. Kuntze
S. platycarpus A. Gray
Sy = Erysimum platycarpum (A. Gray) K. E. O. Kuntze
* ***S. sparsiflorus*** R. Rollins [**TOES: V**]

SYNTHLIPSIS A. Gray
S. greggii A. Gray var. ***greggii***

THELEPODIOPSIS P. Rydberg
T. purpusii (T. Brandegee) R. Rollins
Sy = Sisymbrium purpusii (T. Brandegee) O. E. Schulz
Sy = Thelypodium purpusii T. Brandegee
T. shinnersii (M. C. Johnston) R. Rollins
Sy = Sisymbrium shinnersii M. C. Johnston
Sy = Thelypodium shinnersii (M. C. Johnston) R. Rollins
T. vaseyi (S. Watson *ex* B. Robinson) R. Rollins
Sy = Sisymbrium vaseyi S. Watson *ex* B. Robinson

THELYPODIUM S. Endlicher
* ***T. tenue*** R. Rollins [**TOES: V**]
T. texanum (V. Cory) R. Rollins
Sy = Stanleyella texana V. Cory
T. wrightii A. Gray subsp. ***wrightii***
Sy = Stanleyella wrightii (A. Gray) P. Rydberg

THLASPI C. Linnaeus
T. arvense C. Linnaeus
Sy = Teruncius arvensis (C. Linnaeus) J. Lunell
T. montanum C. Linnaeus var. ***fendleri*** (A. Gray) P. Holmgren
Sy = T. fendleri A. Gray
Sy = T. prolixum A. Nelson
Sy = T. stipitatum A. Nelson

BUDDLEJACEAE, *nomen conservandum*

BUDDLEJA C. Linnaeus
B. alternifolia C. Maximowicz [**cultivated**]
B. davidii A. Franchet [**cultivated**]
B. lindleyana R. Fortune *ex* J. Lindley
Sy = Adenoplea lindleyana (R. Fortune *ex* J. Lindley) J. K. Small
B. marrubiifolia G. Bentham subsp. ***marrubiifolia***
B. racemosa J. Torrey subsp. ***incana*** (J. Torrey) E. Norman
Sy = B. racemosa var. incana J. Torrey
B. racemosa J. Torrey subsp. ***racemosa***
B. scordioides K. Kunth
Sy = B. scordioides var. capitata S. Watson
B. sessiliflora K. Kunth
Sy = B. barbata K. Kunth
Sy = B. melliodora K. Kunth
Sy = B. pringlei A. Gray
Sy = B. pseudoverticillata M. Martens & H. Galeotti
Sy = B. simplex F. Kränzlin
Sy = B. verticillata K. Kunth

EMORYA J. Torrey
  E. suaveolens J. Torrey

POLYPREMUM C. Linnaeus
  P. procumbens C. Linnaeus

BUXACEAE, *nomen conservandum*

BUXUS C. Linnaeus
  B. microphylla P. von Siebold & J. Zuccarini [**cultivated**]
  B. sempervirens C. Linnaeus [**cultivated**]

PACHYSANDRA A. Michaux
  P. terminalis P. von Siebold & J. Zuccarini [**cultivated**]

SARCOCOCCA J. Lindley
  S. humilis O. Stapf [**cultivated**]

CABOMBACEAE, *nomen conservandum*

BRASENIA J. von Schreber
  B. schreberi J. F. Gmelin
    Sy = B. peltata F. Pursh

CABOMBA J. Aublet
  C. caroliniana A. Gray var. ***caroliniana***
    Sy = C. caroliniana var. pulcherrima R. Harper

CACTACEAE, *nomen conservandum*

ACANTHOCEREUS (G. Engelmann *ex* A. Berger) N. Britton & J. Rose
  A. tetragonus (C. Linnaeus) E. Hummel
    Sy = A. floridanus J. K. Small
    Sy = A. pentagonus N. Britton & J. Rose
    Sy = Cereus pentagonus (C. Linnaeus) A. Haworth

ANCISTROCACTUS (K. Schumann) N. Britton & J. Rose
  A. brevihamatus (G. Engelmann) N. Britton & J. Rose
    Sy = Echinocactus brevihamatus G. Engelmann
  A. sheeri (J. Salm-Reifferscheid-Dyck) N. Britton & J. Rose
    Sy = Echinocactus megarhizus J. Rose
    Sy = E. sheeri J. Salm-Reifferscheid-Dyck
    Sy = Sclerocactus scheeri (J. Salm-Reifferscheid-Dyck) N. Taylor
* ***A. tobuschii*** (W. Marshall) C. Backeberg [**TOES: I**]
    Sy = Echinocactus tobushii D. Weniger

ARIOCARPUS M. Scheidweiler
  A. fissuratus (G. Engelmann) K. Schumann var. ***fissuratus***
    Sy = Anhalonium engelmannii C. Lemaire
    Sy = Mammillaria fissurata G. Engelmann

ASTROPHYTUM C. Lemaire
* ***A. asterias*** (J. Zuccarini) C. Lemaire [**TOES: I**]
    Sy = Echinocactus asterias J. Zuccarini

CORYPHANTHA (G. Engelmann) C. Lemaire, *nomen conservandum*
* ***C. albicolumnaria*** (J. Hester) A. Zimmerman [**TOES: V**]
    Sy = C. strobiliformis (H. Poselger) C. Orcutt var. durispina (L. Quehl) L. Benson
    Sy = C. strobiliformis var. orcuttii (F. Bödecker) L. Benson
    Sy = Escobaria albicolumnaria J. Hester
    Sy = E. orcuttii F. Bödecker
    Sy = Mammillaria albicolumnaria (J. Hester) D. Weniger
* ***C. chaffeyi*** (N. Britton & J. Rose) F. Fosberg [**TOES: V**]
    Sy = C. dasyacantha (G. Engelmann) C. Orcutt var. chaffeyi N. Britton & J. Rose
    Sy = Escobaria chaffeyi N. Britton & J. Rose
    Sy = E. dasyacantha (G. Engelmann) N. Britton & J. Rose var. chaffeyi (N. Britton & J. Rose) N. Taylor
* ***C. dasyacantha*** (G. Engelmann) C. Orcutt var. ***dasyacantha*** [**TOES: V**]
    Sy = Escobaria dasyacantha (G. Engelmann) N. Britton & J. Rose var. dasyacantha
    Sy = Mammillaria dasyacantha G. Engelmann
  C. dasyacantha (G. Engelmann) C. Orcutt var. ***varicolor*** (E. Tiegel) L. Benson
    Sy = Escobaria tuberculosa (G. Engelmann) N. Britton & J. Rose
    Sy = E. tuberculosa (G. Engelmann) S. Brack & K. Heil
    Sy = Mammillaria tuberculosa G. Engelmann
    Sy = M. varicolor (E. Tiegel) D. Weniger
    Ma = Coryphantha strobiliformis *auct.*, non (H. Poselger) C. Orcutt
* ***C. duncanii*** (J. Hester) L. Benson [**TOES: V**]
    Sy = Escobaria dasyacantha (G. Engelmann) N. Britton & J. Rose var. duncanii (J. Hester) N. Taylor
    Sy = E. duncanii (J. Hester) C. Backeberg
    Sy = Mammillaria duncanii (J. Hester) D. Weniger

*C. **echinus*** (G. Engelmann) N. Britton & J. Rose
Sy = C. cornifera (A. P. de Candolle) C. Lemaire var. echinus (G. Engelmann) L. Benson
Sy = C. pectinata (G. Engelmann) N. Britton & J. Rose
Sy = Mammillaria echinus G. Engelmann
Sy = M. pectinata G. Engelmann
* *C. **hesteri*** Y. Wright **[TOES: V]**
Sy = Escobaria hesteri (Y. Wright) F. Buxbaum
Sy = Mammillaria hesteri (Y. Wright) D. Weniger
*C. **macromeris*** (G. Engelmann) C. Lemaire var. ***macromeris***
Sy = Mammillaria macromeris G. Engelmann
* *C. **macromeris*** (G. Engelmann) C. Lemaire var. ***runyonii*** (N. Britton & J. Rose) L. Benson **[TOES: IV]**
Sy = C. runyonii N. Britton & J. Rose
Sy = Mammillaria runyonii (N. Britton & J. Rose) V. Cory
* *C. **minima*** R. O. Baird **[TOES: I]**
Sy = C. nellieae L. Croizat
Sy = Escobaria minima (R. O. Baird) D. Hunt
Sy = E. nellieae (L. Croizat) C. Backeberg
Sy = Mammillaria nellieae (L. Croizat) L. Croizat
*C. **missouriensis*** (R. Sweet) N. Britton & J. Rose var. ***caespitosa*** (G. Engelmann) L. Benson
Sy = C. similis (G. Engelmann) N. Britton & J. Rose
Sy = Escobaria missouriensis (R. Sweet) D. Hunt var. similis (G. Engelmann) N. Taylor
Sy = Mammillaria similis G. Engelmann
Sy = Neomammillaria similis (G. Engelmann) N. Britton & J. Rose *ex* P. Rydberg
*C. **missouriensis*** (R. Sweet) N. Britton & J. Rose var. ***missouriensis***
Sy = Escobaria missouriensis (R. Sweet) D. Hunt var. missouriensis
Sy = Mammillaria missouriensis R. Sweet
Sy = Neobesseya missouriensis (R. Sweet) N. Britton & J. Rose
* *C. **ramillosa*** L. Cutak **[TOES: II]**
Sy = Mammillaria ramillosa (L. Cutak) D. Weniger
*C. **robertii*** A. Berger
Sy = Escobaria emskoetteriana (L. Quehl) J. Borg
Sy = Mammillaria robertii (A. Berger) D. Weniger
*C. **scheeri*** (F. Mühlenpfordt) C. Lemaire var. ***scheeri***
Sy = C. muhlenpfordtii (H. Poselger) N. Britton & J. Rose
Sy = Mammillaria sheeri F. Mühlenpfordt
*C. **scheeri*** (F. Mühlenpfordt) C. Lemaire var. ***uncinata*** L. Benson
*C. **scheeri*** (F. Mühlenpfordt) C. Lemaire var. ***valida*** (G. Engelmann) L. Benson
* *C. **sneedii*** (N. Britton & J. Rose) A. Berger var. ***sneedii*** **[TOES: I]**
* Sy = Escobaria guadalupensis S. Brack & K. Heil **[TOES: V]**
Sy = E. sneedii N. Britton & J. Rose var. sneedii
Sy = Mammillaria sneedii (N. Britton & J. Rose) V. Cory
* *C. **sulcata*** (G. Engelmann) N. Britton & J. Rose var. ***nickelsiae*** (K. Brandegee) L. Benson **[TOES: V]**
Sy = Mammillaria nickelsiae K. Brandegee
*C. **sulcata*** (G. Engelmann) N. Britton & J. Rose var. ***sulcata***
Sy = C. missouriensis (R. Sweet) N. Britton & J. Rose var. robustior (G. Engelmann) L. Benson
Sy = Mammillaria sulcata G. Engelmann
*C. **vivipara*** (T. Nuttall) N. Britton & J. Rose var. ***neomexicana*** (G. Engelmann) C. Backeberg
Sy = C. neomexicana (G. Engelmann) N. Britton & J. Rose
Sy = Escobaria vivipara (T. Nuttall) F. Buxbaum var. neomexicana (G. Engelmann) F. Buxbaum
Sy = Mammillaria neomexicana (G. Engelmann) G. Engelmann var. neomexicana
*C. **vivipara*** (T. Nuttall) N. Britton & J. Rose var. ***radiosa*** (G. Engelmann) C. Backeberg
Sy = C. fragrans J. Hester
Sy = C. radiosa (G. Engelmann) P. Rydberg
Sy = Escobaria vivipara (T. Nuttall) F. Buxbaum var. radiosa (G. Engelmann) D. Hunt
Sy = Mammillaria vivipara (T. Nuttall) A. Haworth var. radiosa G. Engelmann
*C. **vivipara*** (T. Nuttall) N. Britton & J. Rose var. ***vivipara***
Sy = Escobaria vivipara (T. Nuttall) F. Buxbaum var. vivipara
Sy = Mammillaria vivipara (T. Nuttall) A. Haworth
Sy = M. vivipara var. borealis G. Engelmann

ECHINOCACTUS J. Link & C. Otto
*E. **horizonthalonius*** C. Lemaire var. ***horizonthalonius***
Sy = E. horizonthalonius var. moelleri F. A. Haage
*E. **texensis*** C. Höpffer
Sy = Homalocephala texensis (C. Höpffer) N. Britton & J. Rose

ECHINOCEREUS G. Engelmann

E. berlandieri (G. Engelmann) F. A. Haage
Sy = Cereus berlandieri G. Engelmann
Sy = Echinocereus blanckii H. Poselger var. berlandieri (G. Engelmann) C. Backeberg
Sy = E. poselgerianus A. Linke

* ***E. chisoënsis*** W. Marshall var. ***chisoënsis*** **[TOES: II]**
Sy = E. reichenbachii (Terscheck *ex* W. Walpers) F. A. Haage var. chisoënsis (W. Marshall) L. Benson

E. chloranthus (G. Engelmann) F. A. Haage var. ***chloranthus***
Sy = Cereus chloranthus G. Engelmann

* ***E. chloranthus*** (G. Engelmann) F. A. Haage var. ***neocapillus*** D. Weniger **[TOES: V]**

E. chloranthus (G. Engelmann) F. A. Haage var. ***russanthus*** (D. Weniger) A. Lambert *ex* W. Rowley
Sy = E. russanthus D. Weniger

E. coccineus G. Engelmann var. ***arizonicus*** (J. Rose *ex* C. Orcutt) A. Ferguson
Ma = E. triglochidiatus G. Engelmann var. neomexicanus *auct.*, non (P. Standley) P. Standley *ex* H. Marshall

E. coccineus G. Engelmann var. ***coccineus***
Sy = Cereus conoideus G. Engelmann & J. Bigelow
Sy = Echinocereus coccineus (G. Engelmann) G. Engelmann var. melanacanthus G. Engelmann
Sy = E. coccineus var. cylindricus G. Engelmann
Sy = Mammillaria aggregata G. Engelmann

E. coccineus G. Engelmann var. ***gurneyi*** (L. Benson) K. Heil & S. Brack
Sy = E. triglochidiatus G. Engelmann var. gurneyi L. Benson

E. coccineus G. Engelmann var. ***paucispinus*** (G. Engelmann) A. Ferguson
Sy = Cereus roemeri F. Mühlenpfordt
Sy = C. roemeri G. Engelmann
Sy = Echinocereus paucispinus (G. Engelmann) G. Engelmann *ex* T. Rümpler
Sy = E. roemeri (G. Engelmann) T. Rümpler
Sy = E. triglochidiatus G. Engelmann var. paucispinus (G. Engelmann) L. Benson

E. dasyacanthus G. Engelmann
Sy = E. pectinatus (M. Scheidweiler) G. Engelmann var. dasyacanthus (G. Engelmann) N. Taylor
Sy = E. pectinatus var. neomexicanus (J. Coulter) L. Benson

E. davisii A. Houghton
* Sy = E. viridiflorus G. Engelmann var. davisii (A. Houghton) H. Marshall **[TOES: I]**

E. enneacanthus G. Engelmann var. ***dubius*** (G. Engelmann) L. Benson
Sy = Cereus dubius G. Engelmann
Sy = Echinocereus dubius (G. Engelmann) T. Rümpler

E. enneacanthus G. Engelmann var. ***enneacanthus***
Sy = E. enneacanthus var. brevispinus (W. Moore) L. Benson

E. fendleri (G. Engelmann) F. Seitz var. ***fendleri***
Sy = Cereus fendleri G. Engelmann
Sy = C. fendleri var. pauperculus G. Engelmann
Sy = Echinocereus albiflorus W. Weingart

E. fendleri (G. Engelmann) F. Seitz var. ***rectispinus*** (R. Peebles) L. Benson
Sy = E. rectispinus R. Peebles

E. papillosus A. Linke *ex* T. Rümpler var. ***angusticeps*** (E. Clover) W. Marshall
Sy = E. angusticeps E. Clover
* Sy = E. berlandieri (G. Engelmann) F. A. Haage var. angusticeps (E. Clover) L. Benson **[TOES: IV]**
Sy = E. blankii (H. Poselger) H. Poselger *ex* T. Rümpler var. angusticeps (E. Clover) L. Benson

E. papillosus A. Linke *ex* T. Rümpler var. ***papillosus***
Sy = E. berlandieri (G. Engelmann) G. Engelmann *ex* T. Rümpler var. papillosus (J. Link) L. Benson

E. pectinatus (M. Scheidweiler) G. Engelmann var. ***wenigeri*** L. Benson

E. pentalophus (A. P. de Candolle) C. Lemaire
Sy = Cereus pentalophus A. P. de Candolle
Sy = C. procumbens G. Engelmann
Sy = C. propinguus J. Salm-Reifferscheid-Dyck *ex* F. Otto
Sy = Echinocereus pentalophus var. procumbens (G. Engelmann) P. Fournier
Sy = E. procumbens (G. Engelmann) C. Lemaire

E. pseudopectinatus (N. Taylor) N. Taylor
Ma = E. pectinatus (M. Scheidweiler) G. Engelmann var. minor *auct.*, non (G. Engelmann) L. Benson

* ***E. reichenbachii*** (Terscheck *ex* W. Walpers) F. A. Haage var. ***albertii*** L. Benson **[TOES: I]**

E. reichenbachii (Terscheck *ex* W. Walpers) F. A. Haage var. ***baileyi*** (J. Rose) N. Taylor
Sy = E. albispinus B. Lahman
Sy = E. baileyi J. Rose var. albispinus (B. Lahman) C. Backeberg

Sy = E. bailey var. baileyi
Sy = E. reichenbachii var. albispinus (B. Lahman) L. Benson

E. reichenbachii (Terscheck *ex* W. Walpers) F. A. Haage var. ***fitchii*** (N. Britton & J. Rose) L. Benson
Sy = E. fitchii N. Britton & J. Rose

E. reichenbachii (Terscheck *ex* W. Walpers) F. A. Haage var. ***perbellus*** (N. Britton & J. Rose) L. Benson
Sy = E. perbellus N. Britton & J. Rose
Sy = E. reichenbachii var. caespitosus (G. Engelmann) G. Engelmann

E. reichenbachii (Terscheck *ex* W. Walpers) F. A. Haage var. ***reichenbachii***
Sy = Cereus caespitosus G. Engelmann
Sy = Echinocactus reichenbachii Terscheck *ex* W. Walpers
Sy = Echinocereus caespitosus (G. Engelmann) G. Engelmann
Sy = E. caespitosus var. minor G. Engelmann
Sy = E. caespitosus var. purpureus D. Weniger
Sy = E. purpureus B. Lahman

E. stramineus (G. Engelmann) F. Seitz
Sy = Cereus stramineus G. Engelmann
Sy = Echinocereus enneacanthus G. Engelmann var. stramineus (G. Engelmann) L. Benson

* ***E. viridiflorus*** G. Engelmann var. ***correllii*** L. Benson **[TOES: V]**

E. viridiflorus G. Engelmann var. ***cylindricus*** (G. Engelmann) G. Engelmann *ex* T. Rümpler
Sy = Cereus viridiflorus (G. Engelmann) G. Engelmann var. cylindricus G. Engelmann
Sy = Echinocereus chloranthus (G. Engelmann) F. A. Haage var. cylindricus (G. Engelmann) N. Taylor

E. viridiflorus G. Engelmann var. ***viridiflorus***
Sy = E. viridiflorus var. standleyi (N. Britton & J. Rose) C. Orcutt *ex* D. Weniger

* ***E.*** × ***lloydii*** N. Britton & J. Rose [***coccineus*** × ***dasyacanthus***] **[TOES: I]**

ECHINOMASTUS N. Britton & J. Rose

E. intertextus (G. Engelmann) N. Britton & J. Rose var. ***dasyacanthus*** (G. Engelmann) C. Backeberg
Sy = Echinocactus intertextus G. Engelmann var. dasyacanthus G. Engelmann
Sy = Neolloydia intertexta (G. Engelmann) L. Benson var. dasyacantha (G. Engelmann) L. Benson
Sy = Sclerocactus intertextus (G. Engelmann) N. Taylor var. dasyacanthus (G. Engelmann) N. Taylor

E. intertextus (G. Engelmann) N. Britton & J. Rose var. ***intertextus***
Sy = Echinocactus intertextus G. Engelmann
Sy = Neolloydia intertexta (G. Engelmann) L. Benson var. intertexta
Sy = Sclerocactus intertextus (G. Engelmann) N. Taylor var. intertextus

E. mariposensis J. Hester
Sy = Echinocactus mariposensis (J. Hester) D. Weniger
* Sy = Neolloydia mariposensis (J. Hester) L. Benson **[TOES: II]**
Sy = Sclerocactus mariposensis (J. Hester) N. Taylor

E. warnockii (L. Benson) C. Glass & R. Foster
Sy = Echinocactus warnockii (L. Benson) D. Weniger
Sy = Neolloydia warnockii L. Benson
Sy = Sclerocactus warnockii (L. Benson) N. Taylor

EPITHELANTHA N. Britton & J. Rose

E. bokei L. Benson
Sy = E. micromeris (G. Engelmann) A. Weber *ex* N. Britton & J. Rose var. bokei (L. Benson) C. Glass & R. Foster

E. micromeris (G. Engelmann) A. Weber *ex* N. Britton & J. Rose var. ***micromeris***
Sy = Mammillaria micromeris G. Engelmann

FEROCACTUS N. Britton & J. Rose

F. hamatacanthus (F. Mühlenpfordt) N. Britton & J. Rose var. ***hamatacanthus***
Sy = Echinocactus hamatacanthus F. Mühlenpfordt

F. hamatacanthus (F. Mühlenpfordt) N. Britton & J. Rose var. ***sinuatus*** (F. Dietrich) L. Benson
Sy = Echinocactus sinuatus A. Dietrich

* ***F. wislizenii*** (G. Engelmann) N. Britton & J. Rose **[TOES: V]**
Sy = Echinocactus wislizenii G. Engelmann

GLANDULICACTUS C. Backeberg

G. uncinatus (G. Engelmann) C. Backeberg var. ***wrightii*** C. Backeberg
Sy = Ancistrocactus uncinatus (H. Galeoti) L. Benson var. wrightii (G. Engelmann) L. Benson
Sy = Echinocactus uncinatus H. Galeoti var. wrightii G. Engelmann
Sy = E. wrightii G. Engelmann *ex* T. Rümpler

Sy = Ferocactus uncinatus (H. Galeotti) N. Britton & J. Rose var. wrightii (G. Engelmann) N. Taylor
Sy = Sclerocactus uncinatus (H. Galeotti) N. Taylor var. wrightii (G. Engelmann) N. Taylor

LOPHOPHORA J. Coulter
L. williamsii (C. Lemaire *ex* J. Salm-Reifferscheid-Dyck) J. Coulter
Sy = Echinocactus rapa F. von Fischer & C. von Meyer
Sy = Lophophora williamsii var. echinata (L. Croizat) H. Bravo

MAMMILLARIA A. Haworth, *nomen conservandum*
M. grahamii G. Engelmann var. ***grahamii***
Sy = M. microcarpa G. Engelmann
Sy = M. microcarpa var. auricarpa W. Marshall
Sy = M. milleri (N. Britton & J. Rose) F. Bödecker
M. heyderi F. Mühlenpfordt var. ***hemisphaerica*** G. Engelmann
Sy = M. gummifera G. Engelmann var. hemisphaerica (G. Engelmann) L. Benson
M. heyderi F. Mühlenpfordt var. ***heyderi***
Sy = M. gummifera G. Engelmann var. applanata (G. Engelmann) L. Benson
Sy = M. heyderi var. applanata G. Engelmann
M. lasiacantha G. Engelmann
Sy = M. lasiacantha var. denudata G. Engelmann
M. meiacantha G. Engelmann
Sy = M. gummifera G. Engelmann var. meiacantha (G. Engelmann) L. Benson
Sy = M. heyderi F. Mühlenpfordt var. meiacantha (G. Engelmann) L. Benson
M. pottsii F. Scheer *ex* J. Salm-Reifferscheid-Dyck
M. prolifera (P. Miller) A. Haworth var. ***texana*** (G. Engelmann) J. Borg
Sy = M. multiceps J. Salm-Reifferscheid-Dyck
M. sphaerica A. Dietrich
Sy = M. longimamma A. P. de Candolle var. sphaerica (A. Dietrich) K. Brandegee
M. wrightii G. Engelmann var. ***wrightii***

NEOLLOYDIA N. Britton & J. Rose
N. conoidea N. Britton & J. Rose var. ***conoidea***
Sy = Cactus conoideus H. Poselger
Sy = Echinocactus conoideus (A. P. de Candolle) H. Poselger
Sy = Mammillaria conoidea A. P. de Candolle
Sy = Neolloydia conoidea var. texensis L. Kladiwa & H. Fittkau
Sy = N. texensis N. Britton & J. Rose
N. gautii L. Benson

OPUNTIA P. Miller [Note: The 74th state legislature designated the prickly pear as the ***state plant***. All members of subgenus ***Opuntia***, which have flat stems, are considered the state plant; those of subgenus ***Cylindropuntia***, with cylindrical stems, are not.]
* ***O. arenaria*** G. Engelmann [**TOES: V**]
O. atrispina D. Griffiths
* ***O. aureispina*** (S. Brack & K. Heil) D. Pinkava & B. Parfitt [**TOES: V**]
Sy = O. macrocentra G. Engelmann var. aureispina S. Brack & K. Heil
O. chisosensis (E. Anthony) D. Ferguson
Sy = O. lindheimeri G. Engelmann var. chisosensis E. Anthony
O. edwardsii V. Grant & K. Grant
O. ellisiana D. Griffiths
O. emoryi G. Engelmann
O. engelmannii J. Salm-Reifferscheid-Dyck var. ***engelmannii***
Sy = O. dillei D. Griffiths
Sy = O. phaeacantha G. Engelmann var. discata (D. Griffiths) L. Benson & D. Walkington
* ***O. engelmannii*** J. Salm-Reifferscheid-Dyck var. ***flexospina*** (D. Griffths) B. Parfitt & D. Pinkava [**TOES: IV**]
Sy = O. flexospina D. Griffiths
Sy = O. strigil G. Engelmann var. flexospina (D. Griffiths) L. Benson
O. engelmannii J. Salm-Reifferscheid-Dyck var. ***lindheimeri*** (G. Engelmann) B. Parfitt & D. Pinkava
Sy = O. engelmannii var. alta (D. Griffiths) D. Weniger
Sy = O. lindheimeri G. Engelmann var. lindheimeri
Sy = O. lindheimeri var. lehmannii L. Benson
Sy = O. lindheimeri var. tricolor (D. Griffiths) L. Benson
O. engelmannii J. Salm-Reifferscheid-Dyck var. ***linguiformis*** (D. Griffiths) B. Parfitt & D. Pinkava
Sy = O. lindheimeri G. Engelmann var. linguiformis (D. Griffiths) L. Benson
Sy = O. linguiformis D. Griffiths
O. ficus-indica (C. Linnaeus) P. Miller
Sy = Cactus ficus-indica C. Linnaeus
Sy = Opuntia megacantha J. Salm-Reifferscheid-Dyck
O. fragilis (T. Nuttall) A. Haworth var. ***fragilis***
Sy = Cactus fragilis T. Nuttall
Sy = Opuntia fragilis var. denudata K. Wiegand & C. Backeberg

O. grahamii G. Engelmann
Sy = O. schottii G. Engelmann var. grahamii (G. Engelmann) L. Benson
O. humifusa (C. Rafinesque-Schmaltz) C. Rafinesque-Schmaltz var. ***austrina*** (J. K. Small) W. Dress
Sy = O. austrina J. K. Small
Sy = O. compressa J. Macbride var. austrina (J. K. Small) L. Benson
O. humifusa (C. Rafinesque-Schmaltz) C. Rafinesque-Schmaltz var. ***humifusa***
Sy = O. compressa J. Macbride
Sy = O. compressa var. fuscoatra (G. Engelmann) D. Weniger
* ***O. imbricata*** (A. Haworth) A. P. de Candolle var. ***argentea*** M. Anthony [**TOES: V**]
O. imbricata (A. Haworth) A. P. de Candolle var. ***imbricata***
Sy = Cereus imbricata A. Haworth
Sy = Cylindropuntia imbricata (A. Haworth) F. Knuth
Sy = Opuntia imbricata var. arborescens (G. Engelmann) D. Weniger
Sy = O. imbricata var. vexans (D. Griffiths) D. Weniger
O. kleiniae A. P. de Candolle
O. kunzei J. Rose
Sy = O. standlyi G. Engelmann var. standlyi
O. leptocaulis A. P. de Candolle
O. macrocentra G. Engelmann var. ***macrocentra***
Sy = O. violacea G. Engelmann var. castetteri L. Benson
Sy = O. violacea var. macrocentra (G. Engelmann) L. Benson
Sy = O. violacea G. Engelmann var. violacea
O. macrorhiza G. Engelmann var. ***macrorhiza***
Sy = O. compressa J. Macbride var. macrorhiza (G. Engelmann) L. Benson
Sy = O. cymochila G. Engelmann & J. Bigelow
Sy = O. leptocarpa B. MacKensen
Sy = O. mackensenii J. Rose
Sy = O. plumbea J. Rose
Sy = O. roseana B. MacKensen
Sy = O. sanguinocula D. Griffiths
Sy = O. seguina C. Z. Nelson
Sy = O. stenochila G. Engelmann
Sy = O. tortispina G. Engelmann & J. Bigelow
O. macrorhiza G. Engelmann var. ***pottsii*** (J. Salm-Reifferscheid-Dyck) L. Benson
Sy = O. phaeacantha G. Engelmann var. tenuispina (G. Engelmann & J. Bigelow) D. Weniger
Sy = O. pottsii J. Salm-Reifferscheid-Dyck
O. phaeacantha G. Engelmann var. ***camanchica*** (G. Engelmann & J. Bigelow) L. Benson
Sy = O. camanchica G. Engelmann & J. Bigelow
O. phaeacantha G. Engelmann var. ***major*** G. Engelmann
Sy = O. cyclodes (G. Engelmann & J. Bigelow) J. Rose
Sy = O. phaeacantha var. brunnea (G. Engelmann) G. Engelmann
Sy = O. phaeacantha var. mojavensis (G. Engelmann) F. Fosberg
O. phaeacantha G. Engelmann var. ***phaeacantha***
Sy = O. phaeacantha var. nigricans (G. Engelmann) G. Engelmann
Sy = O. zuniensis D. Griffiths
O. polyacantha A. Haworth var. ***rufispina*** (G. Engelmann & J. Bigelow *ex* G. Engelmann) L. Benson
Sy = O. rutila T. Nuttall
O. polyacantha A. Haworth var. ***trichophora*** (G. Engelmann & J. Bigelow) J. Coulter
Sy = O. trichophora (G. Engelmann & J. Bigelow) N. Britton & J. Rose
O. pusilla (A. Haworth) T. Nuttall
Sy = Cactus pusillus A. Haworth
Sy = Opuntia drummondii R. Graham
O. rufida G. Engelmann
O. santa-rita (D. Griffiths & R. Hare) J. Rose
Sy = O. chlorotica G. Engelmann & J. Bigelow var. santa-rita D. Griffiths & R. Hare
Sy = O. violacea G. Engelmann var. santa-rita (D. Griffiths & R. Hare) L. Benson
O. schottii G. Engelmann
O. stricta (A. Haworth) A. Haworth var. ***dillenii*** (J. Ker-Gawler) L. Benson
Sy = Cactus dillenii J. Ker-Gawler
Sy = Opuntia dillenii (J. Ker-Gawler) A. Haworth
O. stricta (A. Haworth) A. Haworth var. ***stricta***
Sy = Cactus strictus A. Haworth
Sy = Opuntia macrantha L. Gibbes
O. strigil G. Engelmann
O. tunicata (J. Lehmann) J. Link & C. Otto var. ***davisii*** (G. Engelmann & J. Bigelow) L. Benson
Sy = O. davisii G. Engelmann & J. Bigelow
O. tunicata (J. Lehmann) J. Link & C. Otto var. ***tunicata***
O. × ***spinosibacca*** M. Anthony [***aureispina*** × ***phaeacantha***]
O. × ***subarmata*** D. Griffiths [***engelmannii*** × ***ficus-indica***]

PEDIOCACTUS N. Britton & J. Rose
* ***P. papyracanthus*** (G. Engelmann) L. Benson [**TOES: V**]
Sy = Sclerocactus papyracanthus (G. Engelmann) N. Taylor

PENIOCEREUS (A. Berger) N. Britton & J. Rose
P. greggii (G. Engelmann) N. Britton & J. Rose var. ***greggii***
* Sy = Cereus greggii G. Engelmann var. greggii [**TOES: V**]

SELENICEREUS (A. Berger) N. Britton & J. Rose
S. spinulosus (A. P. de Candolle) N. Britton & J. Rose
NOTE: Likely extirpated from Texas due to habitat destruction; probably never anything but rare in Texas (see Benson 1982).
Sy = Cereus spinulosus A. P. de Candolle

THELOCACTUS (K. Schumann) N. Britton & J. Rose
T. bicolor (H. Galeotti) N. Britton & J. Rose var. ***bicolor***
Sy = T. bicolor H. Galeotti *ex* H. Pfeiffer
Sy = T. bicolor var. schottii G. Engelmann
Sy = T. bicolor var. schottii (G. Engelmann) H. Krainz
* ***T. bicolor*** (H. Galeotti *ex* H. Pfeiffer) N. Britton & J. Rose var. ***flavidispinus*** C. Backeberg [**TOES: IV**]
Sy = Echinocactus flavidispinus (C. Backeberg) D. Weniger
T. setispinus (G. Engelmann) E. Anderson
Sy = Echinocactus setispinus G. Engelmann
Sy = Ferocactus setispinus (G. Engelmann) L. Benson

WILCOXIA N. Britton & J. Rose
W. poselgeri (C. Lemaire) N. Britton & J. Rose
Sy = Cereus poselgeri (C. Lemaire) J. Coulter
Sy = C. tuberosus H. Poselger
Sy = Echinocereus poselgeri C. Lemaire
Sy = Wilcoxia kroenleinii A. Cartier
Sy = W. tuberosa K. Kreuzinger

CALLITRICHACEAE, *nomen conservandum*

CALLITRICHE C. Linnaeus
C. heterophylla F. Pursh subsp. ***heterophylla***
Sy = C. anceps M. Fernald
C. nuttallii J. Torrey
C. palustris C. Linnaeus
Sy = C. palustris var. verna (C. Linnaeus) K. Fenley *ex* W. Jepson
Sy = C. verna C. Linnaeus
C. peploides T. Nuttall
Sy = C. peploides var. semialata N. Fassett
C. terrestris C. Rafinesque-Schmaltz
Sy = C. austrinii G. Engelmann
Sy = C. deflexa A. Braun

CALYCANTHACEAE, *nomen conservandum*

CALYCANTHUS C. Linnaeus, *nomen conservandum*
C. floridus C. Linnaeus [**cultivated**]

CAMPANULACEAE, *nomen conservandum*

CAMPANULA C. Linnaeus
C. rapunculoides C. Linnaeus
Sy = C. rapunculoides var. ucranica (W. Besser) K. Koch
* ***C. reverchonii*** A. Gray [**TOES: V**]
C. rotundifolia C. Linnaeus
Sy = C. petiolata A. P. de Candolle
Sy = C. rotundifolia var. arkansana A. Gray
Sy = C. rotundifolia var. intercedens (J. Witasek) O. Farwell
Sy = C. rotundifolia var. lancifolia F. Mertens & W. Koch
Sy = C. rotundifolia var. petiolata J. Henry
Sy = C. rotundifolia var. velutina A. P. de Candolle

LOBELIA C. Linnaeus
L. appendiculata A. L. de Candolle var. ***appendiculata***
L. berlandieri A. L. de Candolle var. ***berlandieri***
L. berlandieri A. L. de Candolle var. ***brachypoda*** (A. Gray) R. McVaugh
Sy = L. brachypoda (A. Gray) A. L. de Candolle *ex* J. K. Small
L. cardinalis C. Linnaeus subsp. ***cardinalis*** var. ***cardinalis***
L. cardinalis C. Linnaeus subsp. ***graminea*** (J. de Lamarck) R. McVaugh var. ***graminea***
Sy = L. cardinalis var. pseudosplendens R. McVaugh
L. cardinalis C. Linnaeus subsp. ***graminea*** (J. de Lamarck) R. McVaugh var. ***phyllostachya*** (G. Engelmann) R. McVaugh
L. cardinalis C. Linnaeus subsp. ***graminea*** (J. de Lamarck) R. McVaugh var. ***propinqua*** (J. Paxton) W. M. Bowden

Sy = L. cardinalis var. multiflora (J. Paxton) R. McVaugh
L. fenestralis A. Cavanilles
L. flaccidifolia J. K. Small
Sy = L. halei J. K. Small
L. floridana A. Chapman
L. puberula A. Michaux var. **mineolana** F. Wimmer
L. puberula A. Michaux var. **pauciflora** B. Bush
Sy = L. puberula A. Michaux subsp. pauciflora (B. Bush) W. M. Bowden
Sy = L. reverchonii B. L. Turner
L. puberula A. Michaux var. **simulans** M. Fernald
L. siphilitica C. Linnaeus var. **ludoviciana** A. L. de Candolle
L. spicata J. de Lamarck var. **scaposa** R. McVaugh

NEMACLADUS T. Nuttall
N. glanduliferus W. Jepson var. **orientalis** R. McVaugh

PLATYCODON A. P. de Candolle
P. grandiflorum (N. von Jacquin) A. P. de Candolle [**cultivated**]

TRIODANIS C. Rafinesque-Schmaltz *ex* E. Greene
T. coloradoënsis (S. Buckley) R. McVaugh
Sy = Specularia coloradoënsis (S. Buckley) J. K. Small
T. holzingeri R. McVaugh
Sy = Specularia holzingeri (R. McVaugh) M. Fernald
T. lamprosperma R. McVaugh
Sy = Specularia lamprosperma (R. McVaugh) M. Fernald
T. leptocarpa (T. Nuttall) J. Nieuwland
Sy = Specularia leptocarpa (T. Nuttall) A. Gray
T. perfoliata (C. Linnaeus) J. Nieuwland var. **biflora** (H. Ruiz López & J. Pavón) T. Bradley
Sy = Specularia biflora (H. Ruiz López & J. Pavón) F. von Fischer & C. von Meyer
Sy = Triodanis biflora (H. Ruiz López & J. Pavón) E. Greene
T. perfoliata (C. Linnaeus) J. Nieuwland var. **perfoliata**
Sy = Campanula amplexicaulis A. Michaux
Sy = C. perfoliata C. Linnaeus
Sy = Dysmicodon perfoliatum T. Nuttall
Sy = Legousia perfoliata (C. Linnaeus) N. Britton
Sy = Pentagonia perfoliata K. E. O. Kuntze
Sy = Prismatocarpus perfoliatus R. Sweet
Sy = Specularia perfoliata (C. Linnaeus) A. P. de Candolle
T. texana R. McVaugh
Sy = Specularia texana (R. McVaugh) M. Fernald

WAHLENBERGIA H. A. Schrader *ex* A. Roth, *nomen conservandum*
W. marginata (C. Thunberg) A. L. de Candolle
Sy = W. gracilis (J. G. Forster) A. P. de Candolle

CANNABACEAE, *nomen conservandum*

CANNABIS C. Linnaeus
C. sativa C. Linnaeus subsp. **indica** (J. de Lamarck) E. Small & A. Cronquist [**cultivated**]

CAPPARACEAE, *nomen conservandum*

CAPPARIS C. Linnaeus
C. incana K. Kunth

CLEOME C. Linnaeus
C. aculeata C. Linnaeus
C. gynandra C. Linnaeus
Sy = Gynandropsis gynandra (C. Linnaeus) J. Briquet
C. hassleriana R. Chodat [**cultivated**]
Ma = C. houtteana *auct.*, non D. von Schlechtendal: American authors
Ma = C. pungens *auct.*, non C. von Willdenow: American authors
Ma = C. spinosa *auct.*, non N. von Jacquin: American authors
C. multicaulis M. Sessé y Lacasta & J. Mociño *ex* A. P. de Candolle
C. serrulata F. Pursh
Sy = C. integrifolium J. Torrey & A. Gray
Sy = C. serrulata var. angusta (M. E. Jones) I. Tidestrom
Sy = Peritoma serrulatum (F. Pursh) A. P. de Candolle

CLEOMELLA A. P. de Candolle
C. angustifolia J. Torrey
C. longipes J. Torrey

KOEBERLINIA J. Zuccarini
K. spinosa J. Zuccarini var. **spinosa**

POLANISIA C. Rafinesque-Schmaltz
P. dodecandra (C. Linnaeus) A. P. de Candolle subsp. **dodecandra**
Sy = P. graveolens C. Rafinesque-Schmaltz
P. dodecandra (C. Linnaeus) A. P. de Candolle subsp. **riograndensis** H. Iltis

P. dodecandra (C. Linnaeus) A. P. de Candolle subsp. ***trachysperma*** (J. Torrey & A. Gray) H. Iltis var. ***trachysperma*** (J. Torrey & A. Gray) H. Iltis
Sy = P. trachysperma J. Torrey & A. Gray
P. erosa (T. Nuttall) H. Iltis subsp. ***breviglandulosa*** H. Iltis
P. erosa (T. Nuttall) H. Iltis subsp. ***erosa***
Sy = Cristatella erosa T. Nuttall
Sy = Polanisia erosa var. erosa
P. jamesii (J. Torrey & A. Gray) H. Iltis
Sy = Cristatella jamesii J. Torrey & A. Gray
P. uniglandulosa (A. Cavanilles) A. P. de Candolle
Sy = Cleome uniglandulosa A. Cavanilles
Sy = Polanisia dodecandra (C. Linnaeus) A. P. de Candolle subsp. uniglandulosa (A. Cavanilles) H. Iltis

WISLIZENIA Engelmann
W. refracta G. Engelmann subsp. ***refracta***
Sy = W. refracta var. melilotoides (E. Greene) I. M. Johnston

CAPRIFOLIACEAE, *nomen conservandum*

ABELIA A. Brown
A. × ***grandiflora*** (É. André) A. Rehder [***chinensis*** × ***uniflora***] [**cultivated**]

LONICERA C. Linnaeus
L. albiflora J. Torrey & A. Gray var. ***albiflora***
L. albiflora J. Torrey & A. Gray var. ***dumosa*** (A. Gray) A. Rehder
Sy = L. dumosa A. Gray
L. arizonica A. Rehder
L. fragrantissima J. Lindley & J. Paxton [**cultivated**]
Sy = Xylosteon fragrantissimum (J. Lindley & J. Paxton) J. K. Small
L. heckrottii A. Osborn [**cultivated**]
L. japonica C. Thunberg
Sy = L. japonica var. chinensis (P. Watson) J. G. Baker
Sy = Nintoöa japonica (C. Thunberg) R. Sweet
L. maackii (F. Ruprecht) C. Maximowicz [**cultivated**]
L. periclymenum C. Linnaeus [**cultivated**]
L. ruprechtiana E. von Regel [**cultivated**]
Sy = L. × muscaviensis A. Rehder
L. sempervirens C. Linnaeus var. ***sempervirens***
Sy = L. sempervirens var. minor W. Aiton
Sy = Phenianthus sempervirens (C. Linnaeus) C. Rafinesque-Schmaltz
L. tatarica C. Linnaeus [**cultivated**]

SAMBUCUS C. Linnaeus
S. canadensis C. Linnaeus var. ***canadensis***
Sy = S. canadensis var. laciniata A. Gray
S. canadensis C. Linnaeus var. ***submollis*** A. Rehder
S. mexicana K. Presl *ex* A. P. de Candolle
Sy = S. bipinnata D. von Schlechtendal & A. von Chamisso
Sy = S. canadensis C. Linnaeus var. mexicana (K. Presl) C. Sargent
Sy = S. cerulea C. Rafinesque-Schmaltz
Sy = S. cerulea var. arizonica C. Sargent
Sy = S. cerulea var. mexicana (K. Presl *ex* A. P. de Candolle) L. Benson
Sy = S. rehderana G. Schwerin

SYMPHORICARPOS Duhamel de Monceau
* ***S. guadalupensis*** D. Correll [**TOES: V**]
S. longiflorus A. Gray
Sy = S. fragrans Nelson & Kennedy
S. occidentalis W. Hooker
S. orbiculatus C. Moench
Sy = S. symphoricarpos (C. Linnaeus) C. MacMillan
S. oreophilus A. Gray var. ***oreophilus***
Sy = S. rotundifolius A. Gray var. oreophilus (A. Gray) M. E. Jones
S. palmeri G. N. Jones
S. rotundifolius A. Gray var. ***rotundifolius***

TRIOSTEUM C. Linnaeus
T. angustifolium C. Linnaeus
Sy = T. angustifolium var. eamesii K. Wiegand

VIBURNUM C. Linnaeus
V. acerifolium C. Linnaeus var. ***acerifolium***
V. acerifolium C. Linnaeus var. ***glabrescens*** A. Rehder
V. acerifolium C. Linnaeus var. ***ovatum*** (A. Rehder) W. McAtee
V. carlesii W. Hemsley [**cultivated**]
Sy = Solenolantana carlesii (W. Hemsley) T. Nakai
V. dentatum C. Linnaeus var. ***dentatum***
Sy = V. dentatum var. semitomentosum A. Michaux
V. dentatum C. Linnaeus var. ***scabrellum*** J. Torrey & A. Gray
Sy = V. dentatum var. ashei (B. Bush) W. McAtee
Sy = V. scabrellum (J. Torrey & A. Gray) A. Chapman
V. nudum C. Linnaeus var. ***cassinoides*** (C. Linnaeus) J. Torrey & A. Gray

Sy = V. cassinoides C. Linnaeus
Sy = V. nitidum W. Aiton
*V. **nudum*** C. Linnaeus var. ***nudum***
Sy = V. nudum var. angustifolium J. Torrey & A. Gray
Sy = V. nudum var. grandifolium A. Gray
Sy = V. nudum var. ovale A. Wood
Sy = V. nudum var. serotinum H. Ravenel *ex* A. Chapman
*V. **odoratissimum*** J. Ker-Gawler [**cultivated**]
*V. **opulus*** C. Linnaeus var. ***opulus*** [**cultivated**]
Sy = V. opulus var. roseum C. Linnaeus
Sy = V. roseum (C. Linnaeus) E. von Steudel
*V. **prunifolium*** C. Linnaeus
Sy = V. prunifolium var. bushii (W. Ashe) E. Palmer & J. Steyermark
Sy = V. prunifolium var. globosum G. Nash *ex* C. Schneider
*V. **rafinesquianum*** J. A. Schultes var. ***affine*** (B. Bush *ex* C. Schneider) H. House
Sy = V. affine B. Bush *ex* C. Schneider
Sy = V. australe C. Morton
Sy = V. pubescens (W. Aiton) F. Pursh var. affine (C. Schneider) A. Rehder
*V. **rafinesquianum*** J. A. Schultes var. ***rafinesquianum***
Sy = V. affine B. Bush *ex* C. Schneider var. hypomalacum S. F. Blake
*V. **rhytidophyllum*** W. Hemsley [**cultivated**]
*V. **rufidulum*** C. Rafinesque-Schmaltz
Sy = V. ferrugineum J. K. Small
Sy = V. prunifolium C. Linnaeus var. ferrugineum J. Torrey & A. Gray
Sy = V. rufidulum var. margarettiae W. Ashe
Sy = V. rufotomentosum J. K. Small
*V. **suspensum*** J. Lindley [**cultivated**]
*V. **tinus*** C. Linnaeus [**cultivated**]

WEIGELA C. Thunberg
W. florida (A. von Bunge) A. P. de Candolle [**cultivated**]
Sy = Diervilla florida (A. von Bunge) P. von Siebold & J. Zuccarini
W. japonica C. Thunberg [**cultivated**]

CARICACEAE, *nomen conservandum*

CARICA C. Linnaeus
C. papaya C. Linnaeus [**cultivated**]
Sy = C. vulgaris A. P. de Candolle

CARYOPHYLLACEAE, *nomen conservandum*

AGROSTEMMA C. Linnaeus
A. githago C. Linnaeus

ARENARIA C. Linnaeus
A. benthamii E. Fenzl *ex* J. Torrey & A. Gray
A. fendleri A. Gray var. ***fendleri***
Sy = A. fendleri var. diffusa T. Porter
Sy = Eremogone eastwoodiae (P. Rydberg) S. Ikonnikov
A. lanuginosa (A. Michaux) P. Rohrbach subsp. ***lanuginosa***
A. lanuginosa (A. Michaux) P. Rohrbach subsp. ***saxosa*** (A. Gray) B. Maguire
* ***A. livermorensis*** D. Correll [**TOES: IV**]
A. ludens L. Shinners
A. serpyllifolia C. Linnaeus subsp. ***serpyllifolia***

CERASTIUM C. Linnaeus
C. axillare D. Correll
C. brachypodum (G. Engelmann *ex* A. Gray) B. Robinson
Sy = C. brachypodum var. compactum B. Robinson
Sy = C. nutans C. Rafinesque-Schmaltz var. brachypodum G. Engelmann *ex* A. Gray
C. fontanum J. Baumgarten subsp. ***vulgare*** (C. Hartman) W. Greuter & H. Burdet
Sy = C. fontanum subsp. triviale (J. Link) J. Jalas
Sy = C. vulgatum C. Linnaeus 1762, non 1755
C. glomeratum J. Thuillier
Sy = C. glomeratum var. apetalum (B. Dumortier) E. Fenzl
Ma = C. viscosum *auct.*, non C. Linnaeus: American authors
C. nutans C. Rafinesque-Schmaltz
Sy = C. longepedunculatum G. H. Muhlenberg *ex* N. Britton
Sy = C. nutans var. obtectum T. Kearney & R. Peebles
Sy = C. nutans var. occidentale P. Boivin
C. texanum N. Britton

DIANTHUS C. Linnaeus
D. armeria C. Linnaeus [**cultivated**]
D. barbatus C. Linnaeus [**cultivated**]
Sy = D. carthusianorum C. Linnaeus
D. caryophyllus C. Linnaeus [**cultivated**]
D. chinensis C. Linnaeus [**cultivated**]
D. plumarius C. Linnaeus [**cultivated**]

DRYMARIA C. von Willdenow *ex* J. A. Schultes
D. *laxiflora* G. Bentham
D. *leptophylla* (A. Chapman & D. von Schlechtendal) E. Fenzl *ex* P. Rohrbach
Sy = D. tenella A. Gray
D. *molluginea* (M. Lagasca y Segura) D. Didrichsen
Sy = D. sperguloides A. Gray
D. *pachyphylla* E. Wooton & P. Standley

GYPSOPHILA C. Linnaeus
G. *elegans* F. Marschall von Bieberstein [**cultivated**]

LOEFLINGIA C. Linnaeus
L. *squarrosa* T. Nuttall subsp. ***squarrosa***
Sy = L. squarrosa var. texana (W. Hooker) R. Dorn
Sy = L. texana W. Hooker

MINUARTIA C. Linnaeus
M. *drummondii* (L. Shinners) J. McNeill
Sy = Arenaria drummondii L. Shinners
Sy = Stellaria nuttallii J. Torrey & A. Gray
M. *michauxii* (E. Fenzl) O. Farwell var. ***texana*** (B. Robinson) J. Mattfeld
Sy = Arenaria stricta A. Michaux subsp. texana (B. Robinson) B. Maguire
Sy = A. stricta var. texana B. Robinson
Sy = A. texana (B. Robinson) N. Britton
M. *muscorum* (N. Fassett) R. Rabler
Sy = Arenaria muriculata B. Maguire
Sy = A. patula var. robusta (J. Steyermark) B. Maguire
Sy = Minuartia muriculata (B. Maguire) J. McNeill
Sy = M. patula (A. Michaux) J. Mattfeld var. robusta (J. Steyermark) J. McNeill
M. *patula* (A. Michaux) J. Mattfeld var. ***patula***
Sy = Alsinopsis patula (A. Michaux) J. K. Small
Sy = Arenaria patula A. Michaux
Sy = Sabulina patula (A. Michaux) J. K. Small

PARONYCHIA P. Miller
* ***P. congesta*** D. Correll [**TOES: III**]
P. drummondii J. Torrey & A. Gray
Sy = P. drummondii subsp. parviflora M. Chaudhri
P. fastigiata (C. Rafinesque-Schmaltz) M. Fernald var. ***fastigiata***
P. fastigiata (C. Rafinesque-Schmaltz) M. Fernald var. ***paleacea*** M. Fernald
P. jamesii J. Torrey & A. Gray
Sy = P. jamesii var. hirsuta M. Chaudhri
Sy = P. jamesii var. parviflora M. Chaudhri
Sy = P. jamesii var. praelongifolia D. Correll
P. jonesii M. C. Johnston
P. lindheimeri G. Engelmann *ex* A. Gray var. ***lindheimeri***
Sy = P. chorizanthoides J. K. Small
* ***P. lundellorum*** B. L. Turner [**TOES: V**]
* ***P. maccartii*** D. Correll [**TOES: III**]
P. monticola V. Cory
Sy = P. nudata D. Correll
P. sessiliflora T. Nuttall
P. setacea J. Torrey & A. Gray var. ***longibracteata*** M. Chaudhri
P. setacea J. Torrey & A. Gray var. ***setacea***
P. virginica K. Sprengel
Sy = P. parksii V. Cory
Sy = P. virginica var. parksii (V. Cory) M. Chaudhri
Sy = P. virginica var. scoparia (J. K. Small) V. Cory
* ***P. wilkinsonii*** S. Watson [**TOES: V**]

PETRORHAGIA (N. Séringe) J. Link
P. dubia (C. Rafinesque-Schmaltz) G. López & A. Romo
Sy = P. velutina (G. Gussone) P. Ball & V. Heywood
Ma = P. prolifera *auct.*, non (C. Linnaeus) P. Ball & V. Heywood: American authors

POLYCARPON C. Linnaeus
P. tetraphyllum (C. Linnaeus) C. Linnaeus
Sy = Mollugo tetraphylla C. Linnaeus

PSEUDOSTELLARIA F. Pax
P. jamesiana (J. Torrey) W. A. Weber & R. Hartman
Sy = Alsine jamesiana (J. Torrey) A. A. Heller
Sy = A. jamesii (J. Torrey) J. Holzinger
Sy = Arenaria jamesiana (J. Torrey) L. Shinners
Sy = Stellaria jamesiana J. Torrey

SAGINA C. Linnaeus
S. decumbens (S. Elliott) J. Torrey & A. Gray subsp. ***decumbens***
Sy = S. decumbens var. smithii (A. Gray) S. Watson

SAPONARIA C. Linnaeus
S. officinalis C. Linnaeus
Sy = Lychnis saponaria K. Jessen

SILENE C. Linnaeus, *nomen conservandum*
S. antirrhina C. Linnaeus

Sy = S. antirrhina var. confinis M. Fernald
Sy = S. antirrhina var. depauperata P. Rydberg
Sy = S. antirrhina var. divaricata B. Robinson
Sy = S. antirrhina var. laevigata G. Engelmann & A. Gray
Sy = S. antirrhina var. subglaber G. Engelmann & A. Gray
Sy = S. antirrhina var. vaccarifolia P. Rydberg
Sy = S. dioica A. von Chamisso & D. von Schlechtendal

S. gallica C. Linnaeus [proposed *nomen conservandum* (proposal 1147)]
Sy = S. pendula C. Linnaeus
Ma = S. anglica C. Linnaeus: American authors

S. laciniata A. Cavanilles subsp. ***greggii*** (A. Gray) C. Hitchcock & B. Maguire
Sy = S. laciniata var. greggii (A. Gray) S. Watson

S. plankii C. Hitchcock & B. Maguire

* ***S. subciliata*** B. Robinson [**TOES: IV**]

SPERGULA C. Linnaeus

S. arvensis C. Linnaeus var. ***arvensis***

SPERGULARIA (C. Persoon) J. Presl & K. Presl, *nomen conservandum*

S. echinosperma (L. J. Čelakovský) P. Ascherson & K. Graebner
Sy = S. salsuginea (A. von Bunge) E. Fenzl var. bracteata B. Robinson

S. marina (C. Linnaeus) A. Grisebach
Sy = S. alata K. Wiegand
Sy = S. leiosperma (N. Kindberg) F. Schmidt
Sy = S. salina J. Presl & K. Presl
Sy = S. sparsiflora (E. Greene) A. Nelson
Sy = Tissa marina (C. Linnaeus) N. Britton

S. platensis (J. Cambessèdes) E. Fenzl
Sy = Spergula platensis (J. Cambessèdes) A. de Saint-Hilaire & A. L. de Jussieu

STELLARIA C. Linnaeus

S. cuspidata C. von Willdenow *ex* D. von Schlechtendal
Sy = Alsine cuspidata (C. von Willdenow) E. Wooton & P. Standley

S. media (C. Linnaeus) D. Villars subsp. ***media***
Sy = Alsine media C. Linnaeus
Sy = Stellaria apetala B. da Ucria *ex* J. J. Römer
Sy = S. media var. procera F. Klatt & K. Richter

S. prostrata W. Baldwin
Sy = Alsine baldwinii J. K. Small

VACCARIA N. von Wolf

V. hispanica (P. Miller) S. Rauschert
Sy = Saponaria vaccaria C. Linnaeus
Sy = Vaccaria pyramidata F. Medikus
Sy = V. segetalis C. Garcke *ex* P. Ascherson
Sy = V. vaccaria (C. Linnaeus) N. Britton
Sy = V. vulgaris N. Host

CASUARINACEAE, *nomen conservandum*

CASUARINA G. Rumphius *ex* C. Linnaeus

C. cunninghamiana F. Miquel [**cultivated**]

C. equisetifolia C. Linnaeus *ex* J. R. Forster & J. G. Forster [**cultivated**]
Sy = C. litorea C. Linnaeus *ex* F. Fosberg & M.-H. Sachet

C. glauca P. von Siebold *ex* K. Sprengel [**cultivated**]

CELASTRACEAE, *nomen conservandum*

CELASTRUS C. Linnaeus

C. scandens C. Linnaeus

EUONYMUS C. Linnaeus, *nomen conservandum*

E. alata (C. Thunberg) P. von Siebold [**cultivated**]
Sy = Celastrus alatus C. Thunberg

E. americana C. Linnaeus

E. atropurpurea N. von Jacquin var. ***atropurpurea***

E. atropurpurea N. von Jacquin var. ***cheatumii*** C. Lundell

E. fortunei (P. Turczaninow) H. Handel-Mazzetti [**cultivated**]

E. japonica C. Thunberg [**cultivated**]

E. kiautschovica L. Loesener [**cultivated**]

MAYTENUS J. Molina

M. phyllanthoides G. Bentham
Sy = M. texana C. Lundell
Sy = Tricerma crassifolium F. Liebmann
Sy = T. texana (C. Lundell) C. Lundell

MORTONIA A. Gray

M. greggii A. Gray
Sy = M. effusa P. Turczaninow

M. sempervirens A. Gray subsp. ***scabrella*** (A. Gray) B. Prigge
Sy = M. scabrella A. Gray

M. sempervirens A. Gray subsp. ***sempervirens***

PAXISTIMA C. Rafinesque-Schmaltz

P. myrsinites (F. Pursh) C. Rafinesque-Schmaltz

SCHAEFFERIA N. von Jacquin
S. cuneifolia A. Gray

CERATOPHYLLACEAE, *nomen conservandum*

CERATOPHYLLUM C. Linnaeus
C. demersum C. Linnaeus
C. muricatum A. von Chamisso subsp. **australe** (A. Grisebach) D. Les
Sy = C. australe A. Grisebach
Sy = C. echinatum A. Gray
Sy = C. floridanum N. Fassett

CHENOPODIACEAE, *nomen conservandum*

ALLENROLFEA K. E. O. Kuntze
A. occidentalis (S. Watson) K. E. O. Kuntze
Sy = A. mexicana C. Lundell
Sy = Halostachys occidentalis S. Watson
Sy = Salicornia occidentalis (S. Watson) E. Greene
Sy = Spirostachys occidentalis S. Watson

ATRIPLEX C. Linnaeus
A. acanthocarpa (J. Torrey) S. Watson subsp. **acanthocarpa**
Sy = Obione acanthocarpa J. Torrey
A. acanthocarpa (J. Torrey) S. Watson subsp. **coahuilensis** J. Henrickson
A. argentea T. Nuttall subsp. **argentea** var. **argentea**
A. argentea T. Nuttall subsp. **expansa** (S. Watson) H. Hall & F. Clements
Sy = A. argentea var. mohavensis M. E. Jones
Sy = A. expansa S. Watson
Sy = A. expansa var. mohavensis M. E. Jones
A. canescens (F. Pursh) T. Nuttall var. **canescens**
Sy = A. canescens var. laciniata S. Parish
Sy = A. canescens var. macilenta W. Jepson
Sy = A. canescens var. occidentalis (J. Torrey & J. Frémont) S. Welsh & H. Stutz
Sy = A. tetraptera (G. Bentham) P. Rydberg
Sy = Calligonum canescens F. Pursh
A. confertifolia (J. Torrey & J. Frémont) S. Watson
Sy = A. collina E. Wooton & P. Standley
Sy = A. jonesii P. Standley
Sy = A. subconferta P. Rydberg
Sy = Obione confertifolia J. Torrey & J. Frémont
A. elegans (C. Moquin-Tandon) D. Dietrich var. **elegans**
Sy = Obione elegans C. Moquin-Tandon
Sy = O. radiata J. Torrey
A. holocarpa F. von Mueller
* **A. klebergorum** M. C. Johnston **[TOES: V]**
A. matamorensis A. Nelson
Sy = A. oppositifolia S. Watson, non D. Villars
A. obovata C. Moquin-Tandon
Sy = A. greggii S. Watson
A. pentandra (N. von Jacquin) P. Standley
Sy = A. arenaria T. Nuttall
Sy = A. littoralis (N. von Jacquin) W. Fawcett & A. Rendle
A. prostrata J. Boucher de Crèvecoeur *ex* A. P. de Candolle
Sy = A. latifolia G. Wahlenberg
Sy = A. patula var. triangularis (C. von Willdenow) R. F. Thorne & S. Welsh
Sy = A. triangularis C. von Willdenow
Ma = A. patula C. Linnaeus var. hastata *auct.*, non (C. Linnaeus) A. Gray
A. rosea C. Linnaeus
Sy = A. spatiosa A. Nelson
A. saccaria S. Watson
A. semibaccata R. Brown
A. texana S. Watson
A. wardii P. Standley
A. wrightii S. Watson
Sy = A. radiata J. Coulter

BASSIA C. Allioni
B. hyssopifolia (P. von Pallas) A. Volkart
Sy = Echinopsilon hyssopifolius (P. von Pallas) C. Moquin-Tandon
Sy = Kochia hyssopifolia (P. von Pallas) H. A. Schrader

BETA C. Linnaeus
B. vulgaris C. Linnaeus var. **cicla** C. Linnaeus **[cultivated]**
B. vulgaris C. Linnaeus var. **vulgaris** **[cultivated]**

CHENOPODIUM C. Linnaeus
C. albescens J. K. Small
C. album C. Linnaeus var. **album**
Sy = C. album var. lanceolatum (G. H. Muhlenberg *ex* C. von Willdenow) E. Cosson & Germain de Saint-Pierre
Sy = C. album var. polymorphum P. Aellen
Sy = C. giganteum D. Don
Sy = C. opulifolium H. A. Schrader *ex* W. Koch & J. Ziz
Sy = C. viride C. Linnaeus
C. album C. Linnaeus var. **missouriense** (P. Aellen) I. J. Bassett & C. Crompton

Sy = C. missouriense P. Aellen
Sy = C. missouriense var. bushianum P. Aellen
C. ambrosioides C. Linnaeus var. **ambrosioides**
Sy = C. ambrosioides subsp. euambrosioides P. Aellen
Sy = C. ambrosioides var. anthelminticum (C. Linnaeus) A. Gray
Sy = C. ambrosioides var. anthelminticum (C. Linnaeus) P. Aellen
Sy = C. ambrosioides var. chilense (H. A. Schrader) C. Spegazzini
Sy = C. ambrosioides var. vagans (P. Standley) J. Howell
Sy = Teloxys ambrosioides (C. Linnaeus) W. A. Weber
C. atrovirens P. Rydberg
Sy = C. aridum A. Nelson
Sy = C. fremontii S. Watson var. atrovirens (P. Rydberg) F. Fosberg
Sy = C. wolfii P. Rydberg
C. berlandieri C. Moquin-Tandon var. **berlandieri**
Sy = C. album C. Linnaeus var. berlandieri (C. Moquin-Tandon) K. Mackenzie & B. Bush
C. berlandieri C. Moquin-Tandon var. **boscianum** (C. Moquin-Tandon) H. Wahl
Sy = C. boscianum C. Moquin-Tandon
C. berlandieri C. Moquin-Tandon var. **sinuatum** (J. Murray) H. Wahl
Sy = C. berlandieri subsp. pseudopetiolare P. Aellen
C. berlandieri C. Moquin-Tandon var. **zschackii** (J. Murray) J. Murray *ex* P. Ascherson
Sy = C. acerifolium A. Andrzejowski
Sy = C. berlandieri subsp. platyphyllum (E. Issler) Ludwig
Sy = C. berlandieri subsp. zschackii (J. Murr) A. Zobel
Sy = C. berlandieri var. farinosum (Ludwig) P. Aellen
C. botryodes J. E. Smith
Sy = Blitum chenopodioides C. Linnaeus
Sy = Chenopodium chenopodioides (C. Linnaeus) P. Aellen
Sy = C. chenopodioides var. degenianum (P. Aellen) P. Aellen
Sy = C. chenopodioides var. lengylianum (P. Aellen) P. Aellen
Sy = C. rubrum C. Linnaeus glomeratum C. Wallroth
C. botrys C. Linnaeus
Sy = Botrydium botrys (C. Linnaeus) J. K. Small
Sy = Teloxys botrys (C. Linnaeus) W. A. Weber
C. carnosulum C. Moquin-Tandon var. **patagonicum** (R. Philippi) H. Wahl
Sy = C. patagonicum R. Philippi
C. cycloides A. Nelson
C. fremontii S. Watson var. **fremontii**
C. fremontii S. Watson var. **pringlei** (P. Standley) P. Aellen
C. glaucum C. Linnaeus
Sy = C. glaucum subsp. euglaucum P. Aellen
C. graveolens C. von Willdenow
Sy = C. graveolens var. neomexicanum (P. Aellen) P. Aellen
Sy = C. incisum J. Poiret
Sy = C. incisum var. neomexicanum P. Aellen
Sy = Teloxys graveolens (C. von Willdenow) W. A. Weber
C. hians P. Standley
C. incanum (S. Watson) A. A. Heller var. **elatum** L. Crawford
C. incanum (S. Watson) A. A. Heller var. **incanum**
Sy = C. fremontii S. Watson var. incanum S. Watson
C. leptophyllum (C. Moquin-Tandon) T. Nuttall *ex* S. Watson
Sy = C. album C. Linnaeus var. leptophyllum C. Moquin-Tandon
C. murale C. Linnaeus
Sy = Atriplex muralis H. von Crantz
C. neomexicanum P. Standley var. **neomexicanum**
Sy = C. arizonicum P. Standley
C. pallescens P. Standley
C. pratericola P. Rydberg
Sy = C. desiccatum A. Nelson var. leptophylloides (J. Murray) H. Wahl
Sy = C. pratericola subsp. eupratericola P. Aellen
Sy = C. pratericola var. leptophylloides (J. Murray) P. Aellen
C. pumilio R. Brown
Sy = Teloxys pumilio (R. Brown) W. A. Weber
C. simplex (J. Torrey) C. Rafinesque-Schmaltz
Sy = C. gigantospermum P. Aellen
Sy = C. hybridum C. Linnaeus subsp. gigantospermum (P. Aellen) O. Hultén
Sy = C. hybridum var. gigantospermum (P. Aellen) E. Rouleau
Sy = C. hybridum var. simplex J. Torrey

C. standleyanum P. Aellen
Sy = C. gigantospermum P. Aellen var. standleyanum (P. Aellen) P. Aellen
Sy = C. hybridum C. Linnaeus var. standleyanum (P. Aellen) M. Fernald
C. vulvaria C. Linnaeus

CORISPERMUM C. Linnaeus
C. hyssopifolium C. Linnaeus
Sy = C. americanum T. Nuttall
Sy = C. hyssopifolium var. rubricaule W. Hooker
Sy = C. imbricatum A. Nelson
Sy = C. marginale A. Nelson
Sy = C. simplicissimum J. Lunell
Ma = C. nitidum *auct.*, non P. Kitaibel *ex* J. A. Schultes
C. orientale J. de Lamarck
Sy = C. emarginatum P. Rydberg
Sy = C. hyssopifolium C. Linnaeus var. emarginatum (P. Rydberg) J. Boivin
Sy = C. orientale var. emarginatum (P. Rydberg) J. Macbride
Sy = C. villosum P. Rydberg

CYCLOLOMA C. Moquin-Tandon
C. atriplicifolium (K. Sprengel) J. Coulter
Sy = Kochia atriplicifolia K. Sprengel
Sy = Salsola atriplicifolia K. Sprengel
Sy = S. platyphylla A. Michaux

KOCHIA A. Roth
K. americana S. Watson
Sy = K. americana var. vestita S. Watson
Sy = K. vestita (S. Watson) P. Rydberg
K. scoparia (C. Linnaeus) A. Roth *ex* H. A. Schrader
Sy = Bassia sieversiana (P. von Pallas) W. A. Weber
Sy = Kochia alata J. Bates
Sy = K. scoparia var. culta O. Farwell
Sy = K. scoparia var. pubescens E. Fenzl
Sy = K. scoparia var. subvillosa C. Moquin-Tandon
Sy = K. scoparia var. trichophylla (O. Stapf) L. Bailey

KRASCHENINNIKOVIA A. von Gueldenstaedt
K. lanata (F. Pursh) A. von Gueldenstaedt
Sy = Ceratoides lanata (F. Pursh) J. Howell
Sy = C. lanata var. ruinina S. Welsh
Sy = C. lanata var. subspinosa (P. Rydberg) J. Howell
Sy = Diotis lanata F. Pursh
Sy = Eurotia lanata (F. Pursh) C. Moquin-Tandon
Sy = E. subspinosa P. Rydberg

MONOLEPIS H. A. Schrader
M. nuttalliana (J. A. Schultes) E. Greene
Sy = Blitum nuttallianum J. A. Schultes

SALICORNIA C. Linnaeus
S. bigelovii J. Torrey
Sy = S. mucronata J. Bigelow, non M. Lagasca y Segura
S. virginica C. Linnaeus
Sy = S. depressa P. Standley

SALSOLA C. Linnaeus
S. tragus C. Linnaeus
Sy = S. australis R. Brown
Sy = S. iberica F. Sennen & C. Pau
Sy = S. kali C. Linnaeus subsp. ruthenica (M. Iljin) Soó von Bere
Sy = S. kali subsp. tenuifolia I. Tausch
Sy = S. kali subsp. tragus (C. Linnaeus) P. Aellen

SARCOBATUS C. Nees von Esenbeck
S. vermiculatus (W. Hooker) J. Torrey
Sy = Batis vermiculatus W. Hooker
Sy = Sarcobatus baileyi F. Coville
Sy = S. vermiculatus var. baileyi (F. Coville) W. Jepson

SARCOCORNIA A. Scott
S. utahensis (I. Tidestrom) A. Scott
Sy = Salicornia utahensis I. Tidestrom
Sy = Sarcocornia pacifica (P. Standley) A. Scott var. utahensis (I. Tidestrom) P. Munz

SPINACIA C. Linnaeus
S. oleracea C. Linnaeus [**cultivated**]

SUAEDA P. Forsskål *ex* J. F. Gmelin, *nomen conservandum*
S. calceoliformis (W. Hooker) C. Moquin-Tandon
Sy = Schobera occidentalis S. Watson
Sy = Suaeda americana (C. Persoon) M. Fernald
Sy = S. depressa (F. Pursh) S. Watson var. erecta S. Watson
Sy = S. maritima (C. Linnaeus) B. Dumortier var. americana (C. Persoon) J. Boivin
Sy = S. minutiflora S. Watson
Sy = S. occidentalis (S. Watson) S. Watson
Ma = S. depressa *auct.*, non (F. Pursh) S. Watson

S. conferta (J. K. Small) I. M. Johnston
Sy = Dondia conferta J. K. Small
S. linearis (S. Elliott) C. Moquin-Tandon
Sy = Dondia linearis (S. Elliott) A. A. Heller
S. mexicana (P. Standley) P. Standley
S. moquinii (J. Torrey) E. Greene
Sy = S. intermedia S. Watson
Sy = S. nigra J. Macbride
Sy = S. nigra var. ramosissima (P. Standley) P. Munz
Sy = S. torreyana S. Watson
Sy = S. torreyana var. ramosissima (P. Standley) P. Munz
S. suffrutescens S. Watson var. ***detonsa*** I. M. Johnston
* Sy = S. duripes I. M. Johnston [**TOES: III**]
Sy = S. nigrescens I. M. Johnston
Sy = S. nigrescens I. M. Johnston var. glabra I. M. Johnston
S. suffrutescens S. Watson var. ***suffrutescens***
Sy = Dondia suffrutescens (S. Watson) A. A. Heller
S. tampicensis (P. Standley) P. Standley

SUCKLEYA A. Gray
S. suckleyana (J. Torrey) P. Rydberg
Sy = Obione suckleyana J. Torrey

CISTACEAE, *nomen conservandum*

HELIANTHEMUM P. Miller
H. carolinianum (T. Walter) A. Michaux
Sy = Crocanthemum carolinianum (T. Walter) E. Spach
H. georgianum A. Chapman
Sy = Crocanthemum georgianum (A. Chapman) J. Barnhart
H. glomeratum (M. Lagasca y Segura) M. Lagasca y Segura *ex* M. Dunal
Sy = Crocanthemum glomeratum (M. Lagasca y Segura) E. Janchen
H. rosmarinifolium F. Pursh
Sy = Crocanthemum rosmarinifolium (F. Pursh) E. Janchen

LECHEA C. Linnaeus
* ***L. mensalis*** A. Hodgdon [**TOES: V**]
L. mucronata C. Rafinesque-Schmaltz
Sy = L. minor C. Linnaeus var. villosa (S. Elliott) J. Boivin
Sy = L. villosa S. Elliott
Sy = L. villosa var. macrotheca A. Hodgdon
Sy = L. villosa var. schaffneri A. Hodgdon
L. pulchella C. Rafinesque-Schmaltz var. ***pulchella***
Sy = L. leggettii N. Britton & C. Hollick
Sy = L. leggettii var. ramosissima A. Hodgdon
L. san-sabeana (S. Buckley) A. Hodgdon
L. tenuifolia A. Michaux
Sy = L. tenuifolia var. occidentalis A. Hodgdon

CLETHRACEAE, *nomen conservandum*

CLETHRA C. Linnaeus
C. alnifolia C. Linnaeus
Sy = C. alnifolia var. tomentosa (J. de Lamarck) A. Michaux
Sy = C. tomentosa J. de Lamarck

CLUSIACEAE, *nomen conservandum*

HYPERICUM C. Linnaeus
H. chinense C. Linnaeus [**cultivated**]
H. cistifolium J. de Lamarck
Sy = H. opacum J. Torrey & A. Gray
H. crux-andreae (C. Linnaeus) H. von Crantz
Sy = Ascyrum crux-andreae C. Linnaeus
Sy = A. stans A. Michaux *ex* C. von Willdenow
H. densiflorum F. Pursh
Sy = H. glomeratum J. K. Small
H. drummondii (R. Greville & W. Hooker) J. Torrey & A. Gray
Sy = Sarothra drummondii R. Greville & W. Hooker
H. fasciculatum J. de Lamarck
Sy = H. galioides var. fasciculatum (J. de Lamarck) H. Svenson
H. frondosum A. Michaux
Sy = H. aureum W. Bartram
Sy = H. splendens J. K. Small
H. galioides J. de Lamarck
Sy = H. galioides var. pallidum C. Mohr
H. gentianoides (C. Linnaeus) N. Britton, E. Sterns, & J. Poggenburg
Sy = Sarothra gentianoides C. Linnaeus
H. gymnanthum G. Engelmann & A. Gray
H. hypericoides (C. Linnaeus) H. von Crantz subsp. ***hypericoides***
Sy = Ascyrum hypericoides C. Linnaeus
Sy = A. hypericoides var. oblongifolium (E. Spach) M. Fernald
H. hypericoides (C. Linnaeus) H. von Crantz subsp. ***multicaule*** (A. Michaux *ex* C. von Willdenow) B. Robinson

Sy = Ascyrum hypericoides C. Linnaeus var. multicaule (A. Michaux) M. Fernald
Sy = Hypericum hypericoides var. multicaule (A. Michaux *ex* C. von Willdenow) F. Fosberg
H. lobocarpum A. Gattinger
Sy = H. densiflorum F. Pursh var. lobocarpum (A. Gattinger) H. Svenson
Sy = H. oklahomense E. Palmer
H. mutilum C. Linnaeus
Sy = H. mutilum var. latisepalum M. Fernald
Sy = H. mutilum var. parviflorum (C. von Willdenow) M. Fernald
H. nudiflorum A. Michaux *ex* C. von Willdenow
Sy = H. apocynifolium J. K. Small
H. patulum C. Thunberg [**cultivated**]
H. pauciflorum K. Kunth
H. perforatum C. Linnaeus
H. prolificum C. Linnaeus
Sy = H. spathulatum (E. Spach) E. von Steudel
H. pseudomaculatum B. Bush
Sy = H. punctatum J. de Lamarck var. pseudomaculatum (B. Bush) M. Fernald
H. punctatum J. de Lamarck
Sy = H. subpetiolatum E. Bicknell *ex* J. K. Small
H. setosum C. Linnaeus
H. sphaerocarpum A. Michaux
Sy = H. sphaerocarpum var. turgidum (J. K. Small) H. Svenson
Sy = H. turgidum J. K. Small

TRIADENUM C. Rafinesque-Schmaltz
T. tubulosum (T. Walter) H. Gleason
Sy = Hypericum tubulosum T. Walter
T. virginicum (C. Linnaeus) C. Rafinesque-Schmaltz
Sy = Hypericum virginicum C. Linnaeus
T. walteri (J. G. Gmelin) H. Gleason
Sy = Hypericum petiolatum T. Walter
Sy = H. tubulosum T. Walter var. walteri (J. G. Gmelin) H. Lott
Sy = H. walteri J. G. Gmelin
Sy = Triadenum petiolatum (T. Walter) N. Britton

COMBRETACEAE, *nomen conservandum*

QUISQUALIS C. Linnaeus
Q. indica C. Linnaeus [**cultivated**]

CONVOLVULACEAE, *nomen conservandum*

BONAMIA L. Aubert du Petit-Thouars, *nomen conservandum*
* ***B. ovalifolia*** (J. Torrey) H. Hallier [**TOES: V**]
Sy = Breweria ovalifolia (J. Torrey) A. Gray
B. repens (I. M. Johnston) D. Austin & G. Staples
Sy = Petrogenia repens I. M. Johnston

CALYSTEGIA R. Brown, *nomen conservandum*
C. sepium (C. Linnaeus) R. Brown subsp. ***angulata*** R. Brummitt
Sy = C. sepium var. angulata (R. Brummitt) N. Holmgren
Sy = C. sepium var. repens (C. Linnaeus) A. Gray
Sy = Convolvulus sepium var. repens (C. Linnaeus) A. Gray
C. sepium (C. Linnaeus) R. Brown var. ***sepium***
Sy = Convolvulus sepium C. Linnaeus
Sy = C. sepium var. communis R. Tryon
C. silvatica (P. Kitaibel) A. Grisebach subsp. ***fraterniflora*** (K. Mackenzie & B. Bush) R. Brummitt
Sy = C. fraterniflora (K. Mackenzie & B. Bush) R. Brummitt
Sy = C. sepium var. fraterniflora (K. Mackenzie & B. Bush) L. Shinners

CONVOLVULUS C. Linnaeus
C. arvensis C. Linnaeus
Sy = C. ambigens H. House
Sy = Strophocaulos arvensis (C. Linnaeus) J. K. Small
C. equitans G. Bentham
Sy = C. hermannioides A. Gray

CRESSA C. Linnaeus
C. nudicaulis A. Grisebach
C. truxillensis K. Kunth
Sy = C. depressa L. Goodding
Sy = C. erecta P. Rydberg
Sy = C. minima A. A. Heller
Sy = C. truxillensis var. minima (A. A. Heller) P. Munz
Sy = C. truxillensis var. vallicola (A. A. Heller) P. Munz
Sy = C. vallicola A. A. Heller

DICHONDRA J. R. Forster & J. G. Forster
D. argentea F. von Humboldt & A. Bonpland *ex* C. von Willdenow
D. brachypoda E. Wooton & P. Standley
D. carolinensis A. Michaux
Sy = D. repens J. R. Forster var. carolinensis (A. Michaux) J. Choisy

D. micrantha I. Urban
D. recurvata B. Tharp & M. C. Johnston

EVOLVULUS C. Linnaeus
E. alsinoides (C. Linnaeus) C. Linnaeus var. ***angustifolius*** J. Torrey
Sy = E. alsinoides var. acapulcensis (C. von Willdenow) S. van Ooststroom
Sy = E. alsinoides var. grisebachianus C. Meisner
Sy = E. alsinoides var. hirticaulis J. Torrey
Sy = E. alsinoides var. linifolius (C. Linnaeus) J. G. Baker
E. nuttallianus J. A. Schultes
Sy = E. pilosus T. Nuttall
E. sericeus O. Swartz var. ***sericeus***
Sy = E. sericeus var. discolor (L. Bentham) A. Gray
Sy = E. wilcoxianus H. House

IPOMOEA C. Linnaeus, *nomen conservandum*
I. alba C. Linnaeus [**cultivated**]
Sy = Calonyction aculeatum (C. Linnaeus) H. House
Sy = Ipomoea bona-nox C. Linnaeus
I. amnicola T. Morong
I. aquatica P. Forsskål [**federal noxious weed**]
I. barbatisepala A. Gray
I. batatas (C. Linnaeus) J. de Lamarck [**cultivated**]
Sy = Convolvulus batatas C. Linnaeus
I. cairica (C. Linnaeus) R. Sweet
Sy = I. cairica var. hederacea H. Hallier
I. capillacea (K. Kunth) G. Don
Sy = Convolvulus capillaceus K. Kunth
I. cardiophylla A. Gray
Ma = I. aristolochiifolia *auct.*, non (K. Kunth) G. Don
I. carnea N. von Jacquin subsp. ***fistulosa*** (K. von Martius *ex* J. Choisy) D. Austin [**cultivated**]
Sy = I. fistulosa K. von Martius *ex* J. Choisy
I. coccinea C. Linnaeus
Sy = Quamoclit coccinea (C. Linnaeus) C. Moench
I. cordatotriloba A. Dennstaedt var. ***cordatotriloba***
Sy = Convolvulus carolinus C. Linnaeus
Sy = Ipomoea trichocarpa S. Elliott var. trichocarpa
I. cordatotriloba A. Dennstaedt var. ***torreyana*** (A. Gray) D. Austin
Sy = I. trichocarpa S. Elliott var. torreyana (A. Gray) L. Shinners
Sy = I. trifida (K. E. O. Kuntze) G. Don var. torreyana A. Gray
I. costellata J. Torrey
I. cristulata H. Hallier
I. dumetorum C. von Willdenow *ex* J. J. Römer & J. A. Schultes
I. hederacea N. von Jacquin
Sy = I. hederacea var. integriuscula A. Gray
Sy = Pharbitis hederacea (C. Linnaeus) J. Choisy
I. hederifolia C. Linnaeus
Sy = I. coccinea var. hederifolia (C. Linnaeus) A. Gray
Sy = Quamoclit hederifolia (C. Linnaeus) G. Don
I. imperati (M. H. Vahl) A. Grisebach
Sy = I. stolonifera (D. Cyrillo) J. G. Gmelin
I. indica (N. Burman) E. Merrill var. ***acuminata*** (M. H. Vahl) F. Fosberg
Sy = I. acuminata (M. H. Vahl) J. J. Römer & J. A. Schultes
I. lacunosa C. Linnaeus
I. leptophylla J. Torrey
I. lindheimeri A. Gray
Sy = Pharbitis lindheimeri (A. Gray) J. K. Small
I. nil (C. Linnaeus) A. Roth
Sy = Convolvulus hederaceus C. Linnaeus
Sy = C. nil C. Linnaeus
Sy = Pharbitis nil (C. Linnaeus) J. Choisy
I. pandurata (C. Linnaeus) G. Meyer
Sy = I. pandurata var. rubescens J. Choisy
I. pes-caprae (C. Linnaeus) R. Brown subsp. ***brasiliensis*** (C. Linnaeus) S. van Ooststroom
Sy = I. brasiliensis (C. Linnaeus) R. Sweet
Sy = I. pes-caprae var. emarginata H. Hallier
I. pubescens J. de Lamarck
Sy = I. heterophylla C. Ortega
I. purpurea (C. Linnaeus) A. Roth [**cultivated**]
Sy = Convolvulus purpureus C. Linnaeus
Sy = Ipomoea purpurea var. diversifolia (J. Lindley) C. O'Donell
Sy = Pharbitis purpurea (C. Linnaeus) J. Voigt
I. quamoclit C. Linnaeus [**cultivated**]
Sy = Quamoclit quamoclit (C. Linnaeus) N. Britton
I. rupicola H. House
I. sagittata J. Poiret
I. setosa J. Ker-Gawler
Sy = I. melanotricha T. Brandegee
I. shumardiana (J. Torrey) L. Shinners
I. tenuiloba J. Torrey
Sy = I. tenuiloba var. lemmonii (A. Gray) G. Yatskievych & F. Mason
I. tricolor A. Cavanilles

I. turbinata M. Lagasca y Segura
Sy = I. muricata (C. Linnaeus) N. von Jacquin
I. violacea C. Linnaeus
Sy = I. tuba (D. von Schlechtendal) G. Don
I. wrightii A. Gray
Sy = I. heptaphylla (C. Rottbøll & C. von Willdenow) J. Voigt
Sy = I. pulchella A. Roth
Sy = I. spiralis H. House
I. × ***leucantha*** N. von Jacquin [***cordatotriloba*** × ***lacunosa***]
I. × ***multifida*** (C. Rafinesque-Schmaltz) L. Shinners [***coccinea*** × ***pinnata***] [**cultivated**]
Sy = I. sloteri (J. Nieuwland) S. van Ooststroom

JACQUEMONTIA J. Choisy
J. tamnifolia (C. Linnaeus) A. Grisebach
Sy = Ipomoea tamnifolia C. Linnaeus
Sy = Thyella tamnifolia (C. Linnaeus) C. Rafinesque-Schmaltz

MERREMIA A. Dennstaedt *ex* S. Endlicher, *nomen conservandum*
M. dissecta (N. von Jacquin) H. Hallier
Sy = Convolvulus dissectus N. von Jacquin
Sy = Ipomoea dissecta (N. von Jacquin) F. Pursh
Sy = Operculina dissecta (N. von Jacquin) H. House
M. tuberosa (C. Linnaeus) A. Rendle [**cultivated**]
Sy = Ipomoea tuberosa C. Linnaeus
Sy = Operculina tuberosa (C. Linnaeus) C. Meisner

OPERCULINA S. Manso
O. pinnatifida (K. Kunth) C. O'Donell
Sy = Ipomoea pinnatifida (K. Kunth) G. Don

STYLISMA C. Rafinesque-Schmaltz
S. aquatica (T. Walter) C. Rafinesque-Schmaltz
Sy = Bonamia aquatica (T. Walter) A. Gray
Sy = Breweria aquatica (T. Walter) A. Gray
S. humistrata (T. Walter) A. Chapman
Sy = Bonamia humistrata (T. Walter) A. Gray
Sy = Breweria humistrata (T. Walter) A. Gray
S. pickeringii (J. Torrey *ex* W. Curtis) A. Gray var. ***pattersonii*** (M. Fernald & B. Schubert) T. Myint
Sy = Breweria pickeringii (W. Curtis) A. Gray var. pattersonii M. Fernald & B. Schubert
Sy = Stylisma pattersonii (M. Fernald & B. Schubert) G. N. Jones
S. villosa (G. Nash) H. House
Sy = Bonamia villosa (G. Nash) K. Wilson
Sy = Breweria villosa G. Nash

TURBINA C. Rafinesque-Schmaltz
T. corymbosa (C. Linnaeus) C. Rafinesque-Schmaltz
Sy = Convolvulus corymbosus C. Linnaeus
Sy = Ipomoea corymbosa (C. Linnaeus) A. Roth

CORNACEAE, *nomen conservandum*

AUCUBA C. Thunberg
A. japonica C. Thunberg [**cultivated**]
Sy = Eubasis dichotoma R. A. Salisbury

CORNUS C. Linnaeus
C. drummondii C. von Meyer
Sy = C. priceae J. K. Small
C. florida C. Linnaeus subsp. ***florida***
Sy = Cynoxylon floridum (C. Linnaeus) C. Rafinesque-Schmaltz *ex* B. Jackson
C. foemina P. Miller
Sy = C. stricta J. de Lamarck
Sy = Swida foemina (P. Miller) P. Rydberg
Sy = S. stricta (J. de Lamarck) J. K. Small
C. racemosa J. de Lamarck
Sy = C. candidissima H. Marshall, non P. Miller
Sy = C. foemina P. Miller subsp. racemosa (J. de Lamarck) J. Wilson
Sy = Swida candidissima (H. Marshall) J. K. Small
Sy = Thelyerania candidissima (H. Marshall) H. N. Pojarkova

CRASSULACEAE, *nomen conservandum*

CRASSULA C. Linnaeus
C. aquatica (C. Linnaeus) J. Schoenlein
Sy = Bulliarda aquatica (C. Linnaeus) A. P. de Candolle
Sy = Tillaea aquatica C. Linnaeus
C. connata (H. Ruiz López & J. Pavón) A. Berger var. ***connata***
Sy = C. erecta (W. Hooker & G. Arnott) A. Berger
Sy = Tillaea erecta W. Hooker & G. Arnott

ECHEVERIA A. P. de Candolle
E. strictiflora A. Gray
Sy = Cotyledon strictiflora (A. Gray) J. G. Baker

KALANCHOË M. Adanson
K. delagoënsis C. Ecklon & C. Zeyher [**cultivated**]
Sy = Bryophyllum tubiflorum W. Harvey

Sy = Kalanchoë tubiflora (W. Harvey) R. Hamet
Sy = K. verticillata S. Elliott

LENOPHYLLUM J. Rose
L. ***texanum*** (J. G. Smith) J. Rose
Sy = Sedum texanum J. G. Smith
Sy = Villadia texana (J. G. Smith) J. Rose

PENTHORUM C. Linnaeus
P. ***sedoides*** C. Linnaeus subsp. ***sedoides***

SEDUM C. Linnaeus
S. acre C. Linnaeus **[cultivated]**
S. cockerellii N. Britton
Sy = Cockerellia cockerellii (N. Britton) A. Löve & D. Löve
Sy = Sedum wootonii N. Britton
* ***S. havardii*** J. Rose **[TOES: V]**
S. mexicanum N. Britton **[cultivated]**
S. moranense K. Kunth
S. nuttallianum C. Rafinesque-Schmaltz
S. parvum W. Hemsley subsp. ***nanifolium*** (H. Fröderström) R. Clausen
* Sy = S. robertsianum E. Alexander **[TOES: V]**
S. pulchellum A. Michaux
Sy = Chetyson pulchella (A. Michaux) A. Löve & D. Löve
S. reflexum C. Linnaeus **[cultivated]**
Sy = Pterosedum reflexum (C. Linnaeus) V. Grulich
S. spectabile A. Boreau **[cultivated]**
S. telephium C. Linnaeus **[cultivated]**
S. wrightii A. Gray subsp. ***priscum*** R. Clausen
S. wrightii A. Gray subsp. ***wrightii***

VILLADIA J. Rose
V. ***squamulosa*** (S. Watson) J. Rose
Sy = Cotyledon parviflora var. squamulosa S. Watson

CROSSOMATACEAE, *nomen conservandum*

GLOSSOPETALON A. Gray
G. planitierum (M. Ensign) H. St. John
Sy = Forsellesia planitierum M. Ensign
G. spinescens A. Gray var. ***spinescens***
Sy = Forsellesia spinescens (A. Gray) E. Greene
G. texense (M. Ensign) H. St. John
* Sy = Forsellesia texensis M. Ensign **[TOES: V]**

CUCURBITACEAE, *nomen conservandum*

APODANTHERA G. Arnott
A. undulata A. Gray

CAYAPONIA S. Manso, *nomen conservandum*
C. quinqueloba (C. Rafinesque-Schmaltz) L. Shinners
Sy = Arkezostis quinqueloba C. Rafinesque-Schmaltz
Sy = Cayaponia boykinii (J. Torrey & A. Gray) C. Cogniaux

CITRULLUS H. A. Schrader, *nomen conservandum*
C. colocynthis (C. Linnaeus) H. A. Schrader **[cultivated]**
Sy = C. vulgaris H. A. Schrader
Sy = Cucumis colocynthis C. Linnaeus
C. lanatus (C. Thunberg) J. Matsumura & T. Nakai var. ***citroides*** (L. Bailey) R. Mansfeld **[cultivated]**
Sy = C. vulgaris H. A. Schrader var. citroides L. Bailey

COCCINEA R. Wight & G. Arnott
C. grandis (C. Linnaeus) J. Voigt
Sy = Bryonia grandis C. Linnaeus
Ma = Coccinea cordifolia *auct.*, non (C. Linnaeus) C. Cogniaux

CUCUMIS C. Linnaeus
C. anguria C. Linnaeus var. ***anguria*** **[cultivated]**
Sy = C. anguria subsp. cubensis M. Gandoger
Sy = C. anguria subsp. jamaicensis M. Gandoger
Sy = C. angurioides J. J. Römer
Sy = C. arada C. Linnaeus *ex* C. Naudin & Müller
Sy = C. erinaceus C. Naudin & J. Huber
Sy = C. longipes J. Hooker
Sy = C. macrocarpus G. Wenderoth
C. anguria C. Linnaeus var. ***longaculeatus*** J. H. Kirkbride
C. dipsaceus C. G. Ehrenberg *ex* E. Spach
C. melo C. Linnaeus subsp. ***melo*** **[cultivated]**
Sy = C. dudaim C. Linnaeus
Sy = C. melo var. dudaim (C. Linnaeus) C. Naudin
C. sativus C. Linnaeus **[cultivated]**
Sy = C. esculentus R. A. Salisbury
Sy = C. hardwickii J. Royle
Sy = C. rumphii J. Hasskarl
Sy = C. setosus C. Cogniaux
Sy = C. spharocarpus S. Gabaev
Sy = C. vilmorinii K. Sprenger

CUCURBITA C. Linnaeus
C. digitata A. Gray
C. foetidissima K. Kunth
Sy = Cucumis perennis E. James

Sy = Cucurbita perennis (E. James) A. Gray
Sy = Pepo foetidissima (K. Kunth) N. Britton
C. maxima A. Duchesne [**cultivated**]
C. moschata (A. Duchesne *ex* J. de Lamarck) A. Duchesne *ex* J. Poiret [**cultivated**]
C. pepo C. Linnaeus [**cultivated**]
* ***C. texana*** A. Gray [**TOES: V**]

CYCLANTHERA H. A. Schrader
C. dissecta (J. Torrey & A. Gray) G. Arnott
Sy = Disclanthera dissecta J. Torrey & A. Gray

ECHINOCYSTIS J. Torrey & A. Gray
E. lobata (A. Michaux) J. Torrey & A. Gray
Sy = Micrampelis lobata (A. Michaux) E. Greene
Sy = Sicyos lobata A. Michaux

ECHINOPEPON C. Naudin
E. wrightii (A. Gray) S. Watson
Sy = Echinocystis wrightii (A. Gray) C. Cogniaux
Sy = Elaterium wrightii A. Gray

IBERVILLEA E. Greene
I. lindheimeri (A. Gray) E. Greene
Sy = Bryonia abyssinica Gouault, non J. de Lamarck
Sy = Ibervillea tenella (C. Naudin) J. K. Small
Sy = I. tripartita (C. Naudin) E. Greene
Sy = Maximowiczia lindheimeri (A. Gray) C. Cogniaux
Sy = M. tripartita (C. Naudin) C. Cogniaux
Sy = Sicydium lindheimeri A. Gray
Sy = S. tenellum C. Naudin
Sy = S. tripartitum C. Naudin
I. tenuisecta (A. Gray) J. K. Small
Sy = Maximowiczia lindheimeri (A. Gray) C. Cogniaux var. tenuisecta (A. Gray) C. Cogniaux
Sy = M. tripartita (C. Naudin) C. Cogniaux var. tenuisecta (A. Gray) S. Watson
Sy = Sicydium lindheimeri (A. Gray) E. Greene var. tenuisectum A. Gray

LAGENARIA N. Séringe
L. siceraria (J. Molina) P. Standley
Sy = Cucurbita lagenaria C. Linnaeus
Sy = Lagenaria leucantha H. Rusby
Sy = L. vulgaris N. Séringe

LUFFA P. Miller
L. aegyptiaca P. Miller
Sy = L. cylindrica (C. Linnaeus) M. Römer
Sy = Mormordica cylindrica C. Linnaeus

MELOTHRIA C. Linnaeus
M. pendula C. Linnaeus var. ***pendula***
Sy = M. chlorocarpa G. Engelmann
Sy = M. guadalupensis (K. Sprengel) C. Cogniaux
Sy = M. pendula var. chlorocarpa (G. Engelmann) C. Cogniaux

MOMORDICA C. Linnaeus
M. balsamina C. Linnaeus
M. charantia C. Linnaeus

SECHIUM P. Browne, *nomen conservandum*
S. edule (N. von Jacquin) O. Swartz [**cultivated**]

SICYOS C. Linnaeus
S. ampelophyllus E. Wooton & P. Standley
Sy = S. laciniatus C. Linnaeus var. subintigera C. Cogniaux
S. angulatus C. Linnaeus
S. glaber E. Wooton
S. laciniatus C. Linnaeus
Sy = S. laciniatus var. genuinus C. Cogniaux
S. microphyllus K. Kunth
S. parviflorus C. von Willdenow

CUSCUTACEAE, *nomen conservandum*

CUSCUTA C. Linnaeus
C. applanta G. Engelmann
C. cephalanthi G. Engelmann
Sy = C. tenuiflora G. Engelmann
Sy = Epithymum cephalanthi (G. Engelmann) J. Nieuwland & J. Lunell
Sy = Grammica cephalanthi (G. Engelmann) E. Hadac & J. Chrtek
C. compacta A. L. de Jussieu *ex* J. Choisy var. ***compacta***
C. coryli G. Engelmann
Sy = Grammica coryli (G. Engelmann) E. Hadac & J. Chrtek
C. cuspidata G. Engelmann
Sy = Grammica cuspidata (G. Engelmann) E. Hadac & J. Chrtek
C. decipiens T. Yuncker
C. exaltata G. Engelmann
C. glomerata J. Choisy
C. gronovii C. von Willdenow *ex* J. A. Schultes var. ***calyptrata*** G. Engelmann
Sy = C. calyptrata (G. Engelmann) J. K. Small
C. gronovii C. von Willdenow *ex* J. A. Schultes var. ***gronovii***

Sy = C. gronovii var. latiflora G. Engelmann
Sy = C. gronovii var. saururi (G. Engelmann) C. MacMillan
C. indecora J. Choisy var. ***indecora***
Sy = C. decora G. Englemann var. indecora G. Engelmann
Sy = C. neuropetala G. Engelmann
Sy = Grammica indecora (J. Choisy) W. A. Weber
C. indecora J. Choisy var. ***longisepala*** T. Yuncker
C. leptantha G. Engelmann
C. obtusiflora K. Kunth var. ***glandulosa*** G. Engelmann
Sy = C. glandulosa (G. Engelmann) J. K. Small
C. pentagona G. Engelmann var. ***glabrior*** (G. Engelmann) K. Gandhi, R. D. Thomas, & S. Hatch
Sy = C. arvensis H. Beyrich *ex* G. Engelmann var. glabrior (G. Engelmann) G. Engelmann
Sy = C. glabrior (G. Engelmann) T. Yuncker
Sy = C. verrucosa G. Engelmann
Sy = C. verrucosa var. glabrior G. Engelmann
C. pentagona G. Engelmann var. ***pentagona***
Sy = C. arvensis H. Beyrich *ex* G. Engelmann var. pentagona (G. Engelmann) G. Engelmann
Sy = C. campestris T. Yuncker
Sy = C. pentagona var. calycina G. Englemann
Sy = Grammica pentagona (G. Engelmann) W. A. Weber
C. pentagona G. Engelmann var. ***pubescens*** (G. Engelmann) T. Yuncker
Sy = C. arvensis H. Beyrich *ex* G. Engelmann var. pubescens G. Engelmann
Sy = C. glabrior var. pubescens (G. Engelmann) T. Yuncker
C. polygonorum G. Engelmann
C. runyonii T. Yuncker
C. squamata G. Engelmann
C. suaveolens N. Séringe
Sy = C. racemosa K. von Martius var. chiliana G. Engelmann
C. umbellata K. Kunth
Sy = C. umbellata (K. Kunth) E. Hadac & J. Chrtek
Sy = C. umbellata var. reflexa (J. Coulter) T. Yuncker

CYRILLACEAE, *nomen conservandum*

CYRILLA A. Garden *ex* C. Linnaeus
C. racemiflora C. Linnaeus

DILLENIACEAE, *nomen conservandum*

HIBBERTIA G. Andréanszky
H. volubilis G. Andréanszky [**cultivated**]
Sy = H. scandens J. Dryander

DIPSACACEAE, *nomen conservandum*

SCABIOSA C. Linnaeus
S. atropurpurea C. Linnaeus

DROSERACEAE, *nomen conservandum*

DROSERA C. Linnaeus
D. brevifolia F. Pursh
Sy = D. annua E. L. Reed
Sy = D. leucantha L. Shinners
D. capillaris J. Poiret
D. intermedia F. Hayne

EBENACEAE, *nomen conservandum*

DIOSPYROS C. Linnaeus
D. kaki C. von Linné [**cultivated**]
D. texana G. Scheele
Sy = Brayodendron texanum (G. Scheele) J. K. Small
D. virginiana C. Linnaeus
Sy = D. mosieri J. K. Small
Sy = D. virginiana var. mosieri (J. K. Small) C. Sargent
Sy = D. virginiana var. platycarpa C. Sargent
Sy = D. virginiana var. pubescens (F. Pursh) L. Dippel

ELAEAGNACEAE, *nomen conservandum*

ELAEAGNUS C. Linnaeus
E. angustifolia C. Linnaeus [**cultivated**]
E. commutata J. Bernhardi *ex* P. Rydberg [**cultivated**]
Sy = E. argentea F. Pursh, non C. Moench
E. macrophylla C. Thunberg [**cultivated**]
E. pungens C. Thunberg [**cultivated**]

ELATINACEAE, *nomen conservandum*

BERGIA C. Linnaeus
B. texana (W. Hooker) M. A. Seubert *ex* W. Walpers
Sy = Merimea texana W. Hooker

ELATINE C. Linnaeus
E. brachysperma A. Gray
Sy = E. triandra C. Schkuhr var. brachysperma (A. Gray) N. Fassett
E. triandra C. Schkuhr
Sy = E. americana (F. Pursh) G. Arnott
Sy = E. triandra var. americana (F. Pursh) N. Fassett

ERICACEAE, *nomen conservandum*

ARBUTUS C. Linnaeus
A. xalapensis K. Kunth
Sy = A. texana S. Buckley
Sy = A. xalapensis var. texana (S. Buckley) A. Gray

ARCTOSTAPHYLOS M. Adanson, *nomen conservandum*
A. pungens K. Kunth
Sy = A. pungens subsp. chaloneorum (J. Roof) J. Roof

LEUCOTHOË D. Don
L. axillaris (J. de Lamarck) D. Don [**cultivated**]
Sy = L. axillaris var. ambigens M. Fernald
Sy = L. catesbaei (T. Walter) A. Gray
Sy = L. platyphylla J. K. Small
L. fontanesiana (E. von Steudel) H. Sleumer [**cultivated**]
Sy = L. axillaris (J. de Lamarck) D. Don var. editorum (M. Fernald & B. Schubert) H. Ahles
Sy = L. editorum M. Fernald & B. Schubert
* *L. racemosa* (C. Linnaeus) A. Gray [**TOES**: **V**]
Sy = Eubotrys elongata J. K. Small
Sy = E. racemosa (C. Linnaeus) T. Nuttall
Sy = Leucothoë elongata (J. K. Small) J. K. Small
Sy = L. racemosa var. projecta M. Fernald

LYONIA T. Nuttall, *nomen conservandum*
L. ligustrina (C. Linnaeus) A. P. de Candolle var. *foliosiflora* (A. Michaux) M. Fernald
Sy = L. ligustrina var. capriifolia (P. Watson) A. P. de Candolle
Sy = L. ligustrina var. salicifolia (P. Watson) A. P. de Candolle
L. mariana (C. Linnaeus) D. Don
Sy = Neopieris mariana (C. Linnaeus) N. Britton
Sy = Xolisma mariana (C. Linnaeus) A. Rehder

PIERIS D. Don
P. japonica (C. Thunberg) D. Don [**cultivated**]
Sy = Andromeda japonica C. Thunberg

RHODODENDRON C. Linnaeus
R. canescens (A. Michaux) R. Sweet var. *subglabrum* A. Rehder
Sy = Azalea canescens A. Michaux var. subglabra (A. Rehder) J. K. Small
R. oblongifolium (J. K. Small) J. Millais
Sy = Azalea oblongifolia J. K. Small
R. obtusum (J. Lindley) J. Planchon var. *amoenum* (J. Lindley) A. Rehder [**cultivated**]
R. prinophyllum (J. K. Small) J. Millais
Sy = Azalea prinophylla J. K. Small
Sy = R. roseum (J. Loiseleur-Deslongchamps) A. Rehder
R. viscosum (C. Linnaeus) J. Torrey
Sy = Azalea viscosa C. Linnaeus
Sy = R. coryi L. Shinners
Sy = R. viscosum var. aemulans A. Rehder
Sy = R. viscosum var. montanum A. Rehder
Sy = R. viscosum var. nitidum (F. Pursh) A. Gray
Sy = R. viscosum var. serrulatum (J. K. Small) H. Ahles
Sy = R. viscosum var. tomentosum A. Rehder
R. yedoënse C. Maximowicz [**cultivated**]

VACCINIUM C. Linnaeus
V. arboreum H. Marshall
Sy = Batodendron arboreum (H. Marshall) T. Nuttall
Sy = Vaccinium arboreum var. glaucescens (E. Greene) C. Sargent
V. corymbosum C. Linnaeus
Sy = V. amoenum W. Aiton
Sy = V. arkansanum W. Ashe
Sy = V. atrococcum (A. Gray) A. A. Heller
Sy = V. corymbosum var. albiflorum (W. Hooker) M. Fernald
Sy = V. elliottii A. Chapman
Sy = V. virgatum W. Aiton
V. darrowii W. Camp
Sy = V. myrsinites J. de Lamarck var. glaucum A. Gray
V. ovatum F. Pursh var. *ovatum* [**cultivated**]
V. stamineum C. Linnaeus
Sy = Polycodium stamineum (C. Linnaeus) E. Greene
Sy = V. caesium E. Greene
Sy = V. stamineum var. candicans (J. K. Small) C. Mohr
Sy = V. stamineum var. interius (W. Ashe) E. Palmer & J. Steyermark
Sy = V. stamineum var. melanocarpum C. Mohr

Sy = V. stamineum var. neglectum (J. K. Small) C. Deam

EUPHORBIACEAE, *nomen conservandum*

ACALYPHA C. Linnaeus

A. gracilens A. Gray var. **gracilens**
Sy = A. gracilens var. delzii L. P. Miller
Sy = A. gracilens var. fraseri (J. Müller of Aargau) C. Weatherby
A. gracilens A. Gray var. **monococca** G. Engelmann *ex* A. Gray
Sy = A. gracilens subsp. monococca (G. Engelmann *ex* A. Gray) G. Webster
Sy = A. monococca (G. Engelmann *ex* A. Gray) L. Miller & K. Gandhi
A. hispida N. Burman [**cultivated**]
A. lindheimeri J. Müller of Aargau
Sy = A. lindheimeri var. major F. Pax & K. Hoffmann
A. monostachya A. Cavanilles
Sy = A. hederacea J. Torrey
A. neomexicana J. Müller of Aargau
Sy = Ricinocarpus neomexicanus K. E. O. Kuntze
A. ostryifolia J. Riddell
Sy = A. caroliniana S. Elliott
Sy = Ricinocarpus carolinianus K. E. O. Kuntze
A. poiretii K. Sprengel
A. radians J. Torrey
A. tricolor K. von Seemen [**cultivated**]
Sy = A. wilkesiana G. Argus
A. virginica C. Linnaeus var. **rhomboidea** (C. Rafinesque-Schmaltz) T. Cooperrider
Sy = A. rhomboidea C. Rafinesque-Schmaltz
A. virginica C. Linnaeus var. **virginica,** nomen conservandum

ADELIA C. Linnaeus, *nomen conservandum*

* **A. vaseyi** (J. Coulter) F. Pax & K. Hoffmann [**TOES: V**]
Sy = Euphorbia vaseyi J. Coulter

ANDRACHNE C. Linnaeus

* **A. arida** (B. Warnock & M. C. Johnston) G. Webster [**TOES: V**]
Sy = Savia arida B. Warnock & M. C. Johnston
A. phyllanthoides (T. Nuttall) J. Coulter
Sy = Savia phyllanthoides (T. Nuttall) F. Pax & K. Hoffmann

ARGYTHAMNIA P. Browne

* **A. aphoroides** J. Müller of Aargau [**TOES: V**]
Sy = Ditaxis aphoroides (J. Müller of Aargau) F. Pax
* **A. argyraea** V. Cory [**TOES: V**]
A. astroplethes J. Ingram
A. humilis (G. Engelmann & A. Gray) J. Müller of Aargau var. **humilis**
Sy = Ditaxis humilis (G. Engelmann & A. Gray) F. Pax
A. humilis (G. Engelmann & A. Gray) J. Müller of Aargau var. **laevis** (J. Torrey) L. Shinners
Sy = A. laevis J. Torrey
Sy = Ditaxis laevis (J. Torrey) A. A. Heller
A. mercurialina (T. Nuttall) J. Müller of Aargau var. **mercurialina**
Sy = Ditaxis mercurialina (T. Nuttall) J. Coulter
A. mercurialina (T. Nuttall) J. Müller of Aargau var. **pilosissima** (G. Bentham) L. Shinners
A. neomexicana J. Müller of Aargau
Sy = Ditaxis neomexicana (J. Müller of Aargau) A. A. Heller
A. simulans J. Ingram

BERNARDIA P. Miller

B. myricifolia (G. Scheele) S. Watson
Sy = B. viridis C. Millspaugh
Sy = Tyria myricifolia G. Scheele
B. obovata I. M. Johnston

CAPERONIA A. de Saint-Hilaire

C. palustris (C. Linnaeus) A. de Saint-Hilaire
Sy = Croton palustris C. Linnaeus

CNIDOSCOLUS J. Pohl

C. texanus (J. Müller of Aargau) J. K. Small
Sy = Jatropha texana J. Müller of Aargau

CODIAEUM A. L. Jussieu, *nomen conservandum*

C. variegatum (C. Linnaeus) A. L. Jussieu [**cultivated**]

CROTON C. Linnaeus

* **C. alabamensis** E. A. Smith *ex* A. Chapman var. **texensis** S. Ginzbarg [**TOES: V**]
C. argenteus C. Linnaeus
Sy = Julocroton argenteus (C. Linnaeus) D. Didrichsen
C. argyranthemus A. Michaux
C. capitatus A. Michaux var. **capitatus**
C. capitatus A. Michaux var. **lindheimeri** (G. Engelmann & A. Gray) J. Müller of Aargau

Sy = C. capitatus var. albinoides (A. Ferguson) L. Shinners
Sy = C. lindheimeri (G. Engelmann & A. Gray) G. Engelmann & A. Gray *ex* A. Wood
C. ciliatoglandulifer C. Ortega
Sy = C. penicillatus E. Ventenat
C. cortesianus K. Kunth
Sy = C. trichocarpus J. Torrey
C. coryi L. Croizat
C. dioicus A. Cavanilles
Sy = Astrogyne crotonoides G. Bentham
Sy = Croton elaeagnifolius M. Vahl
Sy = C. gracilis K. Kunth
Sy = C. neomexicana J. Müller of Aargau
C. fruticulosus G. Engelmann *ex* J. Torrey
C. glandulosus C. Linnaeus var. ***lindheimeri*** J. Müller of Aargau
C. glandulosus C. Linnaeus var. ***pubentissimus*** L. Croizat
C. glandulosus C. Linnaeus var. ***septentrionalis*** J. Müller of Aargau
Sy = C. glandulosus var. angustifolius J. Müller of Aargau
C. humilis C. Linnaeus
Sy = C. berlandieri J. Torrey
C. incanus K. Kunth
Ma = C. torreyanus *auct.*, non J. Müller of Aargau: Texas authors
C. leucophyllus J. Müller of Aargau
C. lindheimerianus G. Scheele var. ***lindheimerianus***
C. lindheimerianus G. Scheele var. ***tharpii*** M. C. Johnston
C. michauxii G. Webster
Sy = Crotonopsis linearis A. Michaux
C. monanthogynus A. Michaux
C. parksii L. Croizat
C. pottsii (J. Klotzch) J. Müller of Aargau var. ***pottsii***
Sy = C. corymbulosus G. Engelmann
Sy = C. eremophilus E. Wooton & P. Standley
* ***C. pottsii*** (J. Klotzch) J. Müller of Aargau var. ***thermophilus*** (M. C. Johnston) M. C. Johnston **[TOES: V]**
Sy = C. corymbulosus G. Engelmann var. thermophilus M. C. Johnston
C. punctatus N. von Jacquin
C. sancti-lazari L. Croizat
Sy = C. abruptus M. C. Johnston
C. soliman A. von Chamisso & D. von Schlechtendal
* ***C. suaveolens*** J. Torrey **[TOES: V]**
C. texensis (J. Klotzch) J. Müller of Aargau var. ***texensis***
Sy = C. luteovirens E. Wooton & P. Standley
Sy = Hendecaudra texensis J. Klotzsch
Sy = Oxydectes texensis K. E. O. Kuntze
C. willdenowii G. Webster
Sy = Crotonopsis elliptica C. von Willdenow

EUPHORBIA C. Linnaeus
E. acuta G. Engelmann
Sy = Chamaesyce acuta (G. Engelmann) C. Millspaugh
Sy = Euphorbia georgei R. Oudejans
E. albomarginata J. Torrey & A. Gray
Sy = Chamaesyce albomarginata (J. Torrey & A. Gray) J. K. Small
E. angusta G. Engelmann
Sy = Chamaesyce angusta (G. Engelmann) J. K. Small
E. antisyphilitica J. Zuccarini
E. arizonica G. Engelmann
Sy = Chamaesyce arizonica (G. Engelmann) J. Arthur
Sy = C. versicolor (E. Greene) J. B. S. Norton
Sy = Euphorbia versicolor E. Greene
* ***E. astyla*** G. Engelmann *ex* P. Boissier **[TOES: V]**
Sy = Chamaesyce astyla (G. Engelmann *ex* P. Boissier) C. Millspaugh
E. bicolor G. Engelmann & A. Gray
E. bifurcata G. Engelmann
E. bilobata G. Engelmann
E. bombensis (N. von Jacquin)
Sy = Chamaesyce ammannioides (K. Kunth) J. K. Small
Sy = C. bombensis (N. von Jacquin) A. Dugand
Sy = C. ingallsii J. K. Small
Sy = Euphorbia ammannioides K. Kunth
E. brachycera G. Engelmann
E. carunculata U. Waterfall
Sy = Chamaesyce carunculata (U. Waterfall) L. Shinners
E. chaetocalyx (P. Boissier) I. Tidestrom var. ***chaetocalyx***
Sy = Chamaesyce chaetocalyx (P. Boissier) E. Wooton & P. Standley
Sy = C. fendleri (J. Torrey & A. Gray) J. K. Small var. chaetocalyx (P. Boissier) L. Shinners
Sy = Euphorbia fendleri J. Torrey & A. Gray var. chaetocalyx P. Boissier
* ***E. chaetocalyx*** (P. Boissier) I. Tidestrom var. ***triligulata*** (L. Wheeler) M. C. Johnston **[TOES: V]**

Sy = Chamaesyce chaetocalyx var. triligulata (L. Wheeler) M. Mayfield
Sy = Euphorbia fendleri var. triligulata L. Wheeler

E. cinerascens G. Engelmann
Sy = Chamaesyce cinerascens (G. Engelmann) J. K. Small

E. cordifolia S. Elliott
Sy = Chamaesyce cordifolia (S. Elliott) J. K. Small

E. corollata C. Linnaeus
Sy = E. corollata var. angustifolia S. Elliott

E. cyathophora J. Murray
Sy = E. heterophylla C. Linnaeus var. cyathophora (J. Murray) A. Grisebach
Sy = Poinsettia cyathophora (J. Murray) J. Klotzsch & C. Garcke

E. dentata A. Michaux
Sy = E. dentata var. cuphosperma (G. Engelmann) M. Fernald
Sy = E. dentata var. gracillima C. Millspaugh
Sy = Poinsettia dentata (A. Michaux) J. Klotzsch & C. Garcke

E. eriantha G. Bentham

E. exstipulata G. Engelmann var. ***exstipulata***

E. exstipulata G. Engelmann var. ***lata*** B. Warnock & M. C. Johnston

E. fendleri J. Torrey & A. Gray
Sy = Chamaesyce fendleri (J. Torrey & A. Gray) J. K. Small var. fendleri

E. geyeri G. Engelmann var. ***geyeri***
Sy = Chamaesyce geyeri (G. Engelmann) J. K. Small var. geyeri

* ***E. geyeri*** var. ***wheeleriana*** B. Warnock & M. C. Johnston **[TOES: V]**
Sy = Chamaesyce geyeri var. wheeleriana (B. Warnock & M. C. Johnston) M. Mayfield

E. glyptosperma G. Engelmann
Sy = Chamaesyce glyptosperma (G. Engelmann) J. K. Small

* ***E. golondrina*** L. Wheeler **[TOES: V]**
Sy = Chamaesyce golondrina (L. Wheeler) J. K. Small

E. helioscopia C. Linnaeus
Sy = Galarhoeus helioscopius (C. Linnaeus) A. Haworth
Sy = Tithymalus helioscopius (C. Linnaeus) J. Hill

E. helleri C. Millspaugh
Sy = Tithymalus helleri (C. Millspaugh) J. K. Small

E. heterophylla C. Linnaeus
Sy = E. geniculata C. Ortega
Sy = E. prunifolia N. von Jacquin
Sy = Poinsettia heterophylla (C. Linnaeus) J. Klotzsch & C. Garcke

E. hexagona T. Nuttall *ex* K. Sprengel
Sy = Zygophyllidium hexagonum (T. Nuttall *ex* K. Sprengel) J. K. Small

E. hirta C. Linnaeus
Sy = Chamaesyce hirta (C. Linnaeus) C. Millspaugh

E. humistrata G. Engelmann
Sy = Chamaesyce humistrata (G. Engelmann) J. K. Small

E. hypericifolia C. Linnaeus
Sy = Chamaesyce hypericifolia (C. Linnaeus) C. Millspaugh

E. hyssopifolia C. Linnaeus
Sy = Chamaesyce hyssopifolia (C. Linnaeus) J. K. Small

E. indivisa (G. Engelmann) I. Tidestrom
Sy = Chamaesyce dioica (K. Kunth) C. Millspaugh
Sy = C. indivisa (G. Engelmann) C. Millspaugh

E. innocua L. Wheeler

E. jejuna M. C. Johnston & B. Warnock
Sy = Chamaesyce jejuna (M. C. Johnston & B. Warnock) L. Shinners

E. laredana C. Millspaugh
Sy = Chamaesyce laredana (C. Millspaugh) J. K. Small

E. lata G. Engelmann
Sy = Chamaesyce lata (G. Engelmann) J. K. Small

E. lathyris C. Linnaeus
Sy = Galarhoeus lathyris (C. Linnaeus) A. Haworth
Sy = Tithymalus lathyris (C. Linnaeus) J. Hill

E. longicruris G. Scheele

E. maculata C. Linnaeus
Sy = Chamaesyce maculata (C. Linnaeus) J. K. Small
Sy = C. matthewsii J. K. Small
Sy = C. supina (C. Rafinesque-Schmaltz) H. Moldenke
Sy = C. tracyi J. K. Small
Sy = Euphorbia supina C. Rafinesque-Schmaltz

E. marginata F. Pursh
Sy = Agaloma marginata (F. Pursh) A. Löve & D. Löve
Sy = Dichrophyllum marginatum (F. Pursh) J. Klotzsch & C. Garcke
Sy = Lepadena marginata (F. Pursh) J. Nieuwland

E. micromera P. Boissier
Sy = Chamaesyce micromera (P. Boissier) E. Wooton & P. Standley
E. missurica C. Rafinesque-Schmaltz
Sy = Chamaesyce missurica (C. Rafinesque-Schmaltz) L. Shinners
Sy = C. missurica var. calcicola L. Shinners
Sy = C. nuttallii (G. Engelmann) J. K. Small
Sy = C. petaloidea (G. Engelmann) J. K. Small
Sy = C. zygophylloides (P. Boissier) J. K. Small
Sy = Euphorbia missurica var. intermedia (G. Engelmann) L. Wheeler
Sy = E. missurica var. petaloidea (G. Engelmann) R. Dorn
E. nutans M. Lagasca y Segura
Sy = Chamaesyce nutans (M. Lagasca y Segura) J. K. Small
Sy = C. preslii (G. Gussone) J. Arthur
Sy = Euphorbia preslii G. Gussone
E. parryi G. Engelmann
Sy = Chamaesyce parryi (G. Engelmann) P. Rydberg
E. peplidion G. Engelmann
* ***E. perennans*** (L. Shinners) B. Warnock & M. C. Johnston [**TOES:V**]
Sy = Chamaesyce perennans L. Shinners
E. prostata W. Aiton
Sy = Chamaesyce prostrata (W. Aiton) J. K. Small
E. pubentissima A. Michaux
Sy = E. corollata C. Linnaeus var. mollis C. Millspaugh
Sy = E. corollata var. paniculata P. Boissier
E. pycnanthema G. Engelmann
Sy = Chamaesyce capitellata (G. Engelmann) C. Millspaugh
Sy = C. pycnanthema (G. Engelmann) C. Millspaugh
Sy = Euphorbia capitellata G. Engelmann
E. radians G. Bentham
Sy = Poinsettia radians (G. Bentham) J. Klotzsch & C. Garcke
E. revoluta G. Engelmann
Sy = Chamaesyce revoluta (G. Engelmann) J. K. Small
E. roemeriana G. Scheele
E. serpens K. Kunth
Sy = Chamaesyce serpens (K. Kunth) J. K. Small
E. serpyllifolia C. Persoon
Sy = Chamaesyce albicaulis (P. Rydberg) P. Rydberg
Sy = C. neomexicana (E. Greene) P. Standley
Sy = C. serpyllifolia (C. Persoon) J. K. Small subsp. serpyllifolia
Sy = Euphorbia neomexicana E. Greene
E. serrula G. Engelmann
Sy = Chamaesyce serrula (G. Engelmann) E. Wooton & P. Standley
E. setiloba G. Engelmann *ex* J. Torrey
Sy = Chamaesyce setiloba (G. Engelmann *ex* J. Torrey) C. Millspaugh *ex* S. Parish
E. simulans (L. Wheeler) B. Warnock & M. C. Johnston
Sy = Chamaesyce polycarpa (G. Bentham) C. Millspaugh *ex* S. Parish var. simulans (L. Wheeler) L. Shinners
Sy = C. simulans (L. Wheeler) M. Mayfield
Sy = Euphorbia polycarpa G. Bentham var. simulans L. Wheeler
E. spathulata J. de Lamarck
Sy = E. arkansana G. Engelmann & A. Gray
Sy = E. dictyosperma F. von Fischer & C. von Meyer
Sy = E. obtusata F. Pursh
Sy = Tithymalus dictyospermus (F. von Fischer & C. von Meyer) A. A. Heller
Sy = T. spathulatus (J. de Lamarck) W. A. Weber
E. stictospora G. Engelmann
Sy = Chamaesyce stictospora (G. Engelmann) J. K. Small var. stictospora
E. strictior J. Holzinger
Sy = Tithymalopsis strictior (J. Holzinger) E. Wooton & P. Standley
E. tetrapora G. Engelmann
E. texana (C. Millspaugh *ex* A. A. Heller) P. Boissier
Sy = Tithymalus texanus C. Millspaugh *ex* A. A. Heller
E. theriaca L. Wheeler var. ***theriaca***
Sy = Chamaesyce theriaca (L. Wheeler) L. Shinners var. theriaca
E. tirucalli C. Linnaeus [**cultivated**]
E. vellerifolia J. Klotzsch & C. Garcke
Sy = Chamaesyce vellerifolia (J. Klotzsch & C. Garcke) C. Millspaugh
E. villifera G. Scheele
Sy = Chamaesyce villifera (G. Scheele) J. K. Small
Sy = Euphorbia villifera var. nuda G. Engelmann *ex* P. Boissier
E. wrightii J. Torrey & A. Gray

JATROPHA C. Linnaeus
J. cathartica M. Terán & J. Berlandier
Sy = Mozinna cardiophylla J. Torrey
J. dioica V. de Cervantes var. ***dioica***

Sy = J. dioica var. sessiliflora W. Hooker
Sy = Mozinna spathulata C. Ortega
J. dioica V. de Cervantes var. ***graminea*** R. McVaugh
J. integerrima N. von Jacquin var. ***integerrima*** **[cultivated]**
J. macrorhiza G. Bentham var. ***septemfida*** G. Engelmann
Sy = J. arizonica I. M. Johnston

MANIHOT P. Miller
M. esculenta H. von Crantz
Sy = Jatropha manihot C. Linnaeus
Sy = Manihot manihot G. Karsten
Sy = M. utilissima J. Pohl
M. subspicata D. Rogers & S. Appan
* ***M. walkerae*** L. Croizat **[TOES: I]**

PHYLLANTHUS C. Linnaeus
P. abnormis H. Baillon var. ***abnormis***
Sy = P. drummondii J. K. Small
Sy = P. garberi J. K. Small
P. abnormis H. Baillon var. ***riograndensis*** G. Webster
P. caroliniensis T. Walter subsp. ***caroliniensis***
* ***P. ericoides*** J. Torrey **[TOES: V]**
P. niruri C. Linnaeus subsp. ***lathyroides*** (K. Kunth) G. Webster
Sy = P. lathyroides K. Kunth
P. polygonoides T. Nuttall *ex* K. Sprengel
P. pudens L. Wheeler
P. tenellus W. Roxburgh
P. urinaria C. Linnaeus

REVERCHONIA A. Gray
R. arenaria A. Gray

RICINUS C. Linnaeus
R. communis C. Linnaeus **[cultivated]**

SAPIUM P. Browne
S. sebiferum (C. Linnaeus) W. Roxburgh
Sy = Croton sebiferum C. Linnaeus
Sy = Triadica sebifera (C. Linnaeus) J. K. Small

SEBASTIANIA K. Sprengel
S. fruticosa (W. Bartram) M. Fernald
Sy = S. ligustrina (A. Michaux) J. Müller of Aargau

STILLINGIA A. Garden *ex* C. Linnaeus
S. sylvatica A. Garden *ex* C. Linnaeus subsp. ***sylvatica***
Sy = S. smallii E. Wooton & P. Standley
Sy = S. sylvatica var. salicifolia J. Torrey
S. texana I. M. Johnston
Sy = S. sylvatica A. Garden *ex* C. Linnaeus var. linearifolia (J. Torrey) J. Müller of Aargau
S. treculiana (J. Müller of Aargau) I. M. Johnston

TRAGIA C. Linnaeus
T. amblyodonta (J. Müller of Aargau) F. Pax & K. Hoffmann
T. betonicifolia T. Nuttall
T. brevispica G. Engelmann & A. Gray
T. cordata A. Michaux
Sy = T. macrocarpa C. von Willdenow
T. glanduligera F. Pax & K. Hoffmann
* ***T. nigricans*** B. Bush **[TOES: V]**
T. ramosa J. Torrey
Sy = T. nepetifolia A. Cavanilles var. leptophylla (J. Torrey) L. Shinners
Sy = T. nepetifolia var. ramosa (J. Torrey) J. Müller of Aargau
T. smallii L. Shinners
T. urens C. Linnaeus
Sy = T. linearifolia S. Elliott
T. urticifolia A. Michaux var. ***texana*** L. Shinners

FABACEAE, *nomen conservandum*

ACACIA P. Miller
A. angustissima (P. Miller) K. E. O. Kuntze var. ***chisosiana*** D. Isely
A. angustissima (P. Miller) K. E. O. Kuntze var. ***hirta*** (T. Nuttall) B. Robinson
Sy = A. hirta J. Torrey & A. Gray
A. angustissima (P. Miller) K. E. O. Kuntze var. ***texensis*** (J. Torrey & A. Gray) D. Isely
Sy = A. texensis J. Torrey & A. Gray
A. baileyana F. von Mueller **[cultivated]**
A. berlandieri G. Bentham
Sy = A. emoryana G. Bentham
Sy = A. tephroloba A. Gray
Sy = Senegalia berlandieri (G. Bentham) N. Britton & J. Rose
A. collinsii W. Safford **[cultivated]**
Sy = A. spadicigera C. Millspaugh
Sy = A. yucatanensis N. Schenck
A. constricta A. Gray var. ***constricta***
Sy = Acaciopsis constricta (A. Gray) N. Britton & J. Rose
A. constricta A. Gray var. ***paucispina*** E. Wooton & P. Standley
A. greggii A. Gray var. ***greggii***
Sy = A. greggii var. arizonica D. Isely

Sy = Senegalia greggii (A. Gray) N. Britton & J. Rose
A. greggii A. Gray var. ***wrightii*** (G. Bentham) D. Isely
Sy = A. wrightii G. Bentham *ex* A. Gray
Sy = Senegalia wrightii (G. Bentham) N. Britton & J. Rose
A. minuata (M. E. Jones) P. de Beauchamp subsp. ***minuata***
Sy = A. smallii D. Isely
Sy = Pithecellobium minutum M. E. Jones
Ma = Acacia farnesiana *auct.*, non (C. Linnaeus) C. von Willdenow: Texas authors
A. neovernicosa D. Isely
Sy = A. constricta G. Bentham var. vernicosa (P. Standley) L. Benson
Sy = A. vernicosa P. Standley
Sy = Acaciopsis vernicosa (P. Standley) N. Britton & J. Rose
A. rigidula G. Bentham
Sy = A. amentacea A. P. de Candolle
Sy = Acaciopsis rigidula (G. Bentham) N. Britton & J. Rose
A. roemeriana G. Scheele
Sy = A. malacophylla A. Gray
Sy = Senegalia roemeriana (G. Scheele) N. Britton & J. Rose
A. schaffneri (S. Watson) F. J. Hermann var. ***bravoënsis*** D. Isely
Ma = A. tortuosa *auct.*, non (C. Linnaeus) C. von Willdenow: Texas authors
A. schottii J. Torrey
Sy = Acaciopsis schottii (J. Torrey) N. Britton & J. Rose
A. sphaerocephala D. von Schlechtendal & A. von Chamisso [**cultivated**]

AESCHYNOMENE C. Linnaeus
A. indica C. Linnaeus
A. rudis G. Bentham
A. viscidula A. Michaux
Sy = Secula viscidula (A. Michaux) J. K. Small

ALBIZIA I. Durazzo
A. julibrissin I. Durazzo

ALHAGI A. Gagnebin de la Ferrière
A. maurorum F. Medikus
Sy = A. camelorum F. von Fischer
Sy = A. pseudalhagi (F. Marschall von Bieberstein) N. Desvaux *ex* B. Keller & Schaparenko

AMORPHA C. Linnaeus
A. canescens T. Nuttall *ex* F. Pursh
Sy = A. brachycarpa E. Palmer
A. fruticosa C. Linnaeus
Sy = A. fruticosa var. angustifolia (P. Watson) P. Mouillefert
Sy = A. fruticosa var. croceolanata (P. Watson) P. Watson *ex* P. Mouillefert
Sy = A. fruticosa var. emarginata F. Pursh
Sy = A. fruticosa var. oblongata E. Palmer
Sy = A. fruticosa var. occidentalis (L. Abrams) T. Kearney & R. Peebles
Sy = A. fruticosa var. tennesseënsis (R. Shuttleworth *ex* G. Kunze) E. Palmer
A. laevigata T. Nuttall
A. paniculata J. Torrey & A. Gray
A. roemeriana G. Scheele
Sy = A. texana S. Buckley
Sy = A. texana var. glabrescens E. Palmer

AMPHICARPAEA S. Elliott *ex* T. Nuttall, *nomen conservandum*
A. bracteata (C. Linnaeus) M. Fernald
Sy = A. bracteata var. comosa (C. Linnaeus) M. Fernald

APIOS P. Fabricius, *nomen conservandum*
A. americana F. Medikus
Sy = A. americana var. turrigera M. Fernald
Sy = Glycine apios C. Linnaeus

ARACHIS C. Linnaeus
A. hypogaea C. Linnaeus [**cultivated**]

ASTRAGALUS C. Linnaeus
A. allochrous A. Gray
A. amphioxys A. Gray var. ***amphioxys***
Sy = A. amphioxys var. melanocalyx (P. Rydberg) I. Tidestrom
Sy = A. crescenticarpus E. Sheldon
Sy = Xylophacos amphioxys (A. Gray) P. Rydberg
A. brazoënsis S. Buckley
A. canadensis C. Linnaeus var. ***canadensis***
Sy = A. canadensis var. carolinianus (C. Linnaeus) M. E. Jones
Sy = A. canadensis var. longilobus N. Fassett
Sy = A. oreophilus P. Rydberg
A. crassicarpus T. Nuttall var. ***berlandieri*** R. Barneby
Sy = A. neomexicanus A. P. de Candolle

A. crassicarpus T. Nuttall var. ***crassicarpus***
Sy = A. carnosus F. Pursh
Sy = A. caryocarpus J. Ker-Gawler
Sy = Geoprumnon crassicarpum (T. Nuttall) P. Rydberg *ex* J. K. Small
A. crassicarpus T. Nuttall var. ***trichocalyx*** (T. Nuttall) R. Barneby
Sy = A. mexicanus A. P. de Candolle var. trichocalyx (T. Nuttall) M. Fernald
Sy = A. trichocalyx T. Nuttall
A. distortus J. Torrey & A. Gray var. ***distortus***
Sy = Holcophacos distortus (J. Torrey & A. Gray) P. Rydberg
A. distortus J. Torrey & A. Gray var. ***engelmannii*** (E. Sheldon) M. E. Jones
Sy = A. engelmannii E. Sheldon
A. emoryanus (P. Rydberg) V. Cory var. ***emoryanus***
Sy = Hamosa emoryana P. Rydberg
A. emoryanus (P. Rydberg) V. Cory var. ***terlinguensis*** (V. Cory) R. Barneby
A. giganteus S. Watson
Sy = A. giganteus (P. von Pallas) E. Sheldon
Sy = A. giganteus S. Watson var. yaquianus (S. Watson) M. E. Jones
Sy = A. texanus E. Sheldon
Sy = A. yaquianus S. Watson
A. gracilis T. Nuttall
* ***A. gypsodes*** R. Barneby [**TOES: V**]
A. humistratus A. Gray var. ***humistratus***
Sy = A. datilensis (P. Rydberg) I. Tidestrom
Sy = Batidophaca humistrata (A. Gray) P. Rydberg
Sy = Pisophaca datilensis P. Rydberg
Sy = Tium humistratum (A. Gray) P. Rydberg
Sy = Tragacantha humistrata (A. Gray) K. E. O. Kuntze
A. leptocarpus J. Torrey & A. Gray
A. lindheimeri G. Engelmann *ex* A. Gray
A. lotiflorus W. Hooker
Sy = A. elatiocarpus E. Sheldon
Sy = A. lotiflorus var. reverchonii (A. Gray) M. E. Jones
Sy = Batidophaca lotiflora (W. Hooker) P. Rydberg
Sy = Phaca lotiflora (W. Hooker) J. Torrey & A. Gray
Sy = Tragacantha lotiflora (W. Hooker) K. E. O. Kuntze
A. missouriensis T. Nuttall var. ***missouriensis***
Sy = A. melanocarpus T. Nuttall
Sy = Tragacantha missouriensis (T. Nuttall) K. E. O. Kuntze
Sy = Xylophacos missouriensis (T. Nuttall) P. Rydberg
A. mollissimus J. Torrey var. ***bigelovii*** (A. Gray) R. Barneby
Sy = A. bigelovii A. Gray
A. mollissimus J. Torrey var. ***coryi*** I. Tidestrom
A. mollissimus J. Torrey var. ***earlei*** (E. Greene *ex* P. Rydberg) I. Tidestrom
Sy = A. earlei E. Greene *ex* P. Rydberg
* ***A. mollissimus*** J. Torrey var. ***marcidus*** (E. Greene *ex* P. Rydberg) R. Barneby (**TOES: V**)
Sy = A. bigelovii A. Gray var. marcidus (E. Greene *ex* P. Rydberg) D. Isely
Sy = A. marcidus E. Greene *ex* P. Rydberg
A. mollissimus J. Torrey var. ***mollissimus***
Sy = A. simulans T. Cockerell
Sy = Phaca villosa E. James *ex* S. Watson
A. nuttallianus A. P. de Candolle var. ***austrinus*** (J. K. Small) R. Barneby
Sy = A. austrinus (J. K. Small) O. E. Schulz
Sy = Hamosa austrina J. K. Small
A. nuttallianus A. P. de Candolle var. ***macilentus*** (J. K. Small) R. Barneby
A. nuttallianus A. P. de Candolle var. ***nuttallianus***
A. nuttallianus A. P. de Candolle var. ***trichocarpus*** J. Torrey & A. Gray
A. pictiformis R. Barneby
A. plattensis T. Nuttall
Sy = A. pachycarpus J. Torrey & A. Gray
Sy = Geoprumnon plattense (T. Nuttall) P. Rydberg
Sy = Phaca plattense (T. Nuttall) C. MacMillan
Sy = Tragacantha plattensis (T. Nuttall) K. E. O. Kuntze
A. pleianthus (L. Shinners) R. Barneby
Sy = A. nuttallianus A. P. de Candolle var. pleianthus (L. Shinners) R. Barneby
A. praelongus E. Sheldon var. ***ellisiae*** (P. Rydberg) R. Barneby
Sy = A. ellisiae (P. Rydberg) C. L. Porter
Sy = Jonesiella ellisiae P. Rydberg
A. puniceus G. Osterhout var. ***puniceus*** G. Osterhout
Sy = A. gracilentus A. Gray var. exsertus M. E. Jones
Sy = Pisophaca punicea (G. Osterhout) P. Rydberg
Sy = Xylophacos puniceus (G. Osterhout) P. Rydberg
A. racemosus F. Pursh var. ***racemosus***
Sy = Tium racemosum (F. Pursh) P. Rydberg
A. reflexus J. Torrey & A. Gray

A. soxmaniroum C. Lundell
A. waterfallii R. Barneby
A. wootonii E. Sheldon
Sy = A. allochrous A. Gray var. playanus (M. E. Jones) D. Isely
Sy = A. playanus M. E. Jones
Sy = A. triflorus A. Gray var. playanus M. E. Jones
Sy = Phaca triflora A. P. de Candolle
Sy = P. wootonii (E. Sheldon) P. Rydberg
A. wrightii A. Gray

BAPTISIA E. Ventenat
B. alba (C. Linnaeus) E. Ventenat var. ***macrophylla*** (M. Larisey) D. Isely
Sy = B. lactea (C. Rafinesque-Schmaltz) J. Thieret
Sy = B. leucantha J. Torrey & A. Gray
B. australis (C. Linnaeus) R. Brown *ex* W. T. Aiton var. ***australis***
B. australis (C. Linnaeus) R. Brown *ex* W. T. Aiton var. ***minor*** (J. Lehmann) M. Fernald
Sy = B. minor (J. Lehmann) M. Fernald
Sy = B. minor var. aberrans M. Larisey
Sy = B. texana S. Buckley
Sy = B. vespertina J. K. Small *ex* P. Rydberg
B. bracteata G. H. Muhlenberg *ex* S. Elliott var. ***laevicaulis*** (A. Gray *ex* W. Canby) D. Isely
Sy = B. laevicaulis (W. Canby) J. K. Small
Sy = B. leucophaea T. Nuttall var. laevicaulis W. Canby
B. bracteata G. H. Muhlenberg *ex* S. Elliott var. ***leucophaea*** (T. Nuttall) J. Kartesz & K. Gandhi
Sy = B. leucophaea T. Nuttall
Sy = B. leucophaea var. glabrescens M. Larisey
B. nuttalliana J. K. Small
B. shaerocarpa T. Nuttall
Sy = B. viridis M. Larisey

BAUHINIA C. Linnaeus
B. forficata J. Link [**cultivated**]
B. galpinii N. E. Brown [**cultivated**]
Sy = B. punctata K. A. Bolle, non N. von Jacquin
B. lunarioides A. Gray *ex* S. Watson
Sy = B. congesta (N. Britton & J. Rose) C. Lundell
B. macranthera G. Bentham [**cultivated**]
B. purpurea C. Linnaeus [**cultivated**]
B. tomentosa C. Linnaeus [**cultivated**]
B. variegata C. Linnaeus [**cultivated**]
B. × blakeana S. T. Dunn [***purpurea × variegata***] [**cultivated**]

BRONGNIARTIA K. Kunth
* ***B. minutifolia*** S. Watson [**TOES: V**]
Sy = B. minutifolia var. canescens S. Watson

CAESALPINIA C. Linnaeus
* ***C. brachycarpa*** (A. Gray) E. Fisher [**TOES: V**]
Sy = Hoffmannseggia brachycarpa A. Gray
C. caudata (A. Gray) E. Fisher
Sy = Hoffmannseggia caudata A. Gray
C. drepanocarpa (A. Gray) E. Fisher
Sy = Hoffmannseggia drepanocarpa A. Gray
Sy = Larrea drepanocarpa N. Britton
C. drummondii (J. Torrey & A. Gray) E. Fisher
Sy = C. texensis (E. Fisher) E. Fisher
Sy = Hoffmannseggia drummondii J. Torrey & A. Gray
C. gilliesii (W. Hooker) N. Wallich *ex* D. Dietrich
Sy = C. macrantha A. Delile
Sy = Erythrostemon gilliesii (W. Hooker) J. Link
Sy = Poinciana gilliesii N. Wallich *ex* W. Hooker
C. jamesii (J. Torrey & A. Gray) E. Fisher
Sy = Hoffmannseggia jamesii J. Torrey & A. Gray
C. mexicana A. Gray [**cultivated**]
Sy = Poinciana mexicana (A. Gray) N. Britton & J. Rose
C. parryi (E. Fisher) I. Eifert
Sy = C. melanosticta var. parryi (E. Fisher) E. Fisher
Sy = Hoffmannseggia melanosticta (J. Schauer) E. Fisher var. parryi E. Fisher
Sy = H. parryi (E. Fisher) B. L. Turner
* ***C. phyllanthoides*** P. Standley [**TOES: V**]
C. wootonii (N. Britton) I. Eifert *ex* D. Isely
Sy = C. atropunctata I. Eifert
Sy = C. melanosticta (J. Schauer) E. Fisher
Sy = Pomaria melanosticta J. Schauer

CALLIANDRA G. Bentham, *nomen conservandum*
* ***C. biflora*** B. Tharp [**TOES: V**]
C. conferta G. Bentham
C. coulteri S. Watson
C. eriophylla G. Bentham var. ***chamaedrys*** D. Isely
Sy = C. chamaedrys G. Engelmann
C. guildingii G. Bentham [**cultivated**]
C. haematocephala J. Hasskarl [**cultivated**]
C. humilis (D. von Schlechtendal) G. Bentham var. ***humilis***
Sy = Acacia humilis D. von Schlechtendal
Sy = Anneslia herbacea (A. Gray) N. Britton & J. Rose

Sy = Calliandra herbacea G. Engelmann *ex* A. Gray
C. tweedii G. Bentham [**cultivated**]

CANAVALIA A. P. de Candolle, *nomen conservandum*
C. rosea (O. Swartz) A. P. de Candolle
Sy = C. maritima (J. Aublet) L. Aubert du Petit-Thouars

CASSIA C. Linnaeus, *nomen conservandum*
C. fistula C. Linnaeus [**cultivated**]
Sy = C. fistulosa C. Linnaeus *ex* R. Long & O. Lakela
C. grandis C. von Linné [**cultivated**]
C. javanica C. Linnaeus [**cultivated**]
Sy = C. nodosa W. Hamilton
C. splendida J. Vogel [**cultivated**]

CENTROSEMA (A. P. de Candolle) G. Bentham, *nomen conservandum*
C. virginianum (C. Linnaeus) G. Bentham
Sy = Bradburya virginiana (C. Linnaeus) K. E. O. Kuntze
Sy = Centrosema virginianum var. ellipticum (A. P. de Candolle) M. Fernald

CERCIS C. Linnaeus
C. canadensis C. Linnaeus var. ***canadensis***
C. canadensis C. Linnaeus var. ***mexicana*** (J. Rose) M. Hopkins
Sy = C. mexicana J. Rose
C. canadensis C. Linnaeus var. ***texensis*** (S. Watson) M. Hopkins
Sy = C. occidentalis J. Torrey *ex* A. Gray

CHAMAECRISTA (C. Linnaeus) C. Moench
C. calycioides (A. P. de Candolle *ex* L. Colladon) E. Greene
Sy = Cassia aristellata (F. Pennell) V. Cory & H. Parks
Sy = C. calycioides A. P. de Candolle *ex* L. Colladon
C. chamaecristoides (L. Colladon) E. Greene var. ***cruziana*** (N. Britton & J. Rose) H. Irwin & R. Barneby
Sy = C. cruziana N. Britton & J. Rose
C. fasciculata (A. Michaux) E. Greene
Sy = Cassia fasciculata A. Michaux var. brachiata (C. Pollard) T. Pullen *ex* D. Isely
Sy = C. fasciculata var. depressa (C. Pollard) J. Macbride
Sy = C. fasciculata var. ferrisiae (N. Britton & J. Rose) B. L. Turner
Sy = C. fasciculata var. macrosperma M. Fernald
Sy = C. fasciculata var. puberula (E. Greene) J. Macbride
Sy = C. fasciculata var. robusta (C. Pollard) J. Macbride
Sy = C. fasciculata var. rostrata (E. Wooton & P. Standley) B. L. Turner
Sy = C. fasciculata var. tracyi (C. Pollard) J. Macbride
C. flexuosa (C. Linnaeus) E. Greene var. ***texana*** (S. Buckley) H. Irwin & R. Barneby
Sy = Cassia texana S. Buckley
Sy = Chamaecrista texana (S. Buckley) F. Pennell
C. greggii (A. Gray) C. Pollard *ex* A. A. Heller
Sy = Cassia greggii A. Gray
C. nictitans (C. Linnaeus) C. Moench subsp. ***nictitans*** var. ***leptadenia*** (J. Greenman) K. Gandhi & S. Hatch
Sy = Cassia leptadenia J. Greenman
Sy = C. nictitans var. mensalis (J. Greenman) H. Irwin & R. Barneby
Sy = Chamaecrista leptadenia (J. Greenman) T. Cockerell
C. nictitans (C. Linnaeus) C. Moench subsp. ***nictitans*** var. ***nictitans***
Sy = Cassia nictitans C. Linnaeus
Sy = C. nictitans var. hebecarpa M. Fernald
Sy = C. nictitans var. leiocarpa M. Fernald
Sy = Chamaecrista nictitans var. hebecarpa (M. Fernald) C. F. Reed
Sy = C. nictitans var. leiocarpa (M. Fernald) H. Moldenke

CHLOROLEUCON N. Britton & J. Rose
C. ebano (J. Berlandier) L. Rico
Sy = Acacia flexicaulis G. Bentham
Sy = Pithecellobium ebano (J. Berlandier) C. H. Muller
Sy = P. flexicaule (G. Bentham) J. Coulter
Sy = Samanca flexicaulis (G. Bentham) J. Macbride
Sy = Siderocarpos flexicaulis (G. Bentham) J. K. Small

CLADRASTIS C. Rafinesque-Schmaltz
C. kentukea (Dumont de Courset) V. Rudd [**cultivated**]
Sy = C. lutea (F. Michaux) K. Koch
Sy = Sophora kentukea Dumont de Courset

CLITORIA C. Linnaeus
C. mariana C. Linnaeus
Sy = Martiusia mariana (C. Linnaeus) J. K. Small
C. ternatea C. Linnaeus [**cultivated**]

COLOGANIA K. Kunth
*C. **angustifolia*** K. Kunth
Sy = C. longifolia A. Gray
*C. **pallida*** J. Rose

CORONILLA C. Linnaeus
*C. **varia*** C. Linnaeus
Sy = Securigera varia (C. Linnaeus) P. Lassen

COURSETIA A. P. de Candolle
*C. **axillaris*** J. Coulter & J. Rose

CROTALARIA C. Linnaeus
*C. **incana*** C. Linnaeus
*C. **pumila*** C. Ortega
Sy = C. lupulina K. Kunth
*C. **purshii*** A. P. de Candolle
Sy = C. purshii var. bracteolifera M. Fernald
*C. **retusa*** C. Linnaeus
*C. **sagittalis*** C. Linnaeus
Sy = C. sagittalis var. blumeriana G. Senn
Sy = C. sagittalis var. fruticosa (P. Miller) W. Fawcett & A. Rendle
*C. **spectabilis*** A. Roth
Sy = C. retzii A. Hitchcock

CYAMOPSIS A. P. de Candolle
*C. **tetragonoloba*** (C. Linnaeus) P. Taubert **[cultivated]**

DALEA C. Linnaeus, *nomen conservandum*
*D. **aurea*** T. Nuttall *ex* F. Pursh
Sy = Parosela aurea (T. Nuttall *ex* F. Pursh) N. Britton
* *D. **bartonii*** R. Barneby **[TOES: V]**
*D. **bicolor*** F. von Humboldt & A. Bonpland *ex* C. von Willdenow var. ***argyraea*** (A. Gray) R. Barneby
Sy = D. argyraea A. Gray
Sy = Parosela argyraea (A. Gray) A. A. Heller
*D. **brachystachys*** A. Gray
Sy = D. lemmonii C. Parry *ex* A. Gray
*D. **candida*** C. von Willdenow var. ***candida***
Sy = Kuhniastera candida K. E. O. Kuntze
Sy = Petalostemon candidum A. Michaux
Sy = Psoralea candida J. Poiret
*D. **candida*** C. von Willdenow var. ***oligophylla*** (J. Torrey) L. Shinners
Sy = D. oligophylla (J. Torrey) L. Shinners
Sy = Kuhniastera gracilis K. E. O. Kuntze
Sy = Petalostemon candidus A. Michaux var. oligophyllus (J. Torrey) F. J. Hermann
Sy = P. gracile T. Nuttall var. oligophyllum J. Torrey
Sy = P. oligophyllus (J. Torrey) J. Torrey *ex* B. Smyth
*D. **compacta*** K. Sprengel var. ***compacta***
Sy = Petalostemon compactus (K. Sprengel) G. Swezey
Sy = P. decumbens T. Nuttall
*D. **compacta*** K. Sprengel var. ***pubescens*** (A. Gray) R. Barneby
Sy = D. helleri L. Shinners
Sy = Petalostemon pulcherrimus (A. A. Heller) A. A. Heller
*D. **cylindriceps*** R. Barneby
Sy = Petalostemon macrostachyus J. Torrey
*D. **emarginata*** (J. Torrey & A. Gray) L. Shinners
Sy = Petalostemon emarginatus J. Torrey & A. Gray
*D. **enneandra*** T. Nuttall
Sy = D. laxiflora F. Pursh
Sy = Parosela enneandra (T. Nuttall) N. Britton
*D. **formosa*** J. Torrey
Sy = Parosela formosa (J. Torrey) A. Vail
*D. **frutescens*** A. Gray
Sy = D. frutescens var. laxa (P. Rydberg) B. L. Turner
Sy = Parosela frutescens (A. Gray) A. Vail
*D. **greggii*** A. Gray
Sy = Parosela greggii (A. Gray) A. A. Heller
*D. **hallii*** A. Gray
*D. **jamesii*** (J. Torrey) J. Torrey & A. Gray
Sy = Parosela jamesii (J. Torrey) A. Vail
Sy = Psoralea jamesii J. Torrey
*D. **lachynostachys*** A. Gray
Sy = Parosela lachnostachys (A. Gray) A. A. Heller
*D. **lanata*** K. Sprengel var. ***lanata***
Sy = Parosela lanata (K. Sprengel) N. Britton
*D. **lanata*** K. Sprengel var. ***terminalis*** (M. E. Jones) R. Barneby
Sy = D. glaberrima S. Watson
Sy = D. terminalis M. E. Jones
Sy = Parosela terminalis A. A. Heller
*D. **laniceps*** R. Barneby
*D. **lasiathera*** A. Gray
*D. **leporina*** (W. Aiton) A. Bullock
Sy = D. alba A. Michaux
Sy = D. alopecuroides C. von Willdenow
Sy = D. lagopus (A. Cavanilles) C. von Willdenow
Sy = Parosela alopecuroides (C. von Willdenow) P. Rydberg

Sy = P. dalea C. Linnaeus
Sy = Psoralea leporina W. Aiton
D. multiflora (T. Nuttall) L. Shinners
Sy = Petalostemon multiflorus T. Nuttall
D. nana J. Torrey *ex* A. Gray var. ***carnescens*** T. Kearney & R. Peebles
Sy = D. carnescens A. Bullock
Sy = D. nana var. elatior A. Gray *ex* B. L. Turner
Sy = Parosela carnescens P. Rydberg
Sy = P. rubescens A. Nelson
D. nana J. Torrey *ex* A. Gray var. ***nana***
Sy = Parosela nana (A. Gray) A. A. Heller
D. neomexicana (A. Gray) V. Cory var. ***longipila*** (P. Rydberg) R. Barneby
Sy = D. longipila (P. Rydberg) V. Cory
D. neomexicana (A. Gray) V. Cory var. ***neomexicana***
Sy = Parosela neomexicana A. A. Heller
D. obovata (J. Torrey & A. Gray) L. Shinners
Sy = Petalostemon obovatus J. Torrey & A. Gray
D. phleoides (J. Torrey & A. Gray) L. Shinners var. ***microphylla*** (J. Torrey & A. Gray) R. Barneby
Sy = D. drummondiana L. Shinners
Sy = Petalostemon microphyllus (J. Torrey & A. Gray) A. A. Heller
Sy = P. phleoides var. microphyllus (J. Torrey & A. Gray) R. Barneby
D. phleoides (J. Torrey & A. Gray) L. Shinners var. ***phleoides***
Sy = Petalostemon glandulosus J. Coulter & E. Fisher
Sy = P. phleoides J. Torrey & A. Gray
D. pogonathera A. Gray var. ***pogonathera***
Sy = Parosela pogonathera (A. Gray) A. Vail
D. pogonathera A. Gray var. ***walkerae*** (B. Tharp & F. Barkley) B. L. Turner
D. polygonoides A. Gray
Sy = D. polygonoides var. anomala (M. E. Jones) C. Morton
D. purpurea E. Ventenat var. ***arenicola*** (D. Wemple) R. Barneby
Sy = Petalostemon arenicola D. Wemple
D. purpurea E. Ventenat var. ***purpurea***
Sy = Petalostemon purpureus (E. Ventenat) P. Rydberg
Sy = P. purpureus var. mollis (P. Rydberg) J. Boivin
* ***D. reverchonii*** (S. Watson) L. Shinners **[TOES: V]**
Sy = Petalostemon reverchonii S. Watson
* ***D. sabinalis*** (S. Watson) L. Shinners **[TOES: III]**
Sy = Petalostemon sabinalis S. Watson
D. scandens (P. Miller) R. Clausen var. ***paucifolia*** (J. Coulter) R. Barneby
Sy = D. domingensis A. P. de Candolle var. paucifolia J. Coulter
Sy = D. thyrsiflora A. Gray
D. tenuifolia (A. Gray) L. Shinners
Sy = Petalostemon tenuifolius A. Gray
D. tenuis (J. Coulter) L. Shinners
Sy = Petalostemon stanfieldii J. K. Small
Sy = P. tenuis (J. Coulter) A. A. Heller
D. villosa (T. Nuttall) K. Sprengel var. ***grisea*** (J. Torrey & A. Gray) R. Barneby
Sy = Petalostemon griseus J. Torrey & A. Gray
D. villosa (T. Nuttall) K. Sprengel var. ***villosa***
Sy = Petalostemon villosus T. Nuttall
D. wrightii A. Gray
Sy = D. parrasana T. Brandegee
Sy = D. sabulicola T. Brandegee
Sy = Parosela wrightii (A. Gray) A. Vail

DELONIX C. Rafinesque-Schmaltz
D. regia (W. Bojer *ex* W. Hooker) C. Rafinesque-Schmaltz **[cultivated]**
Sy = Poinciana regia W. Bojer *ex* W. Hooker

DESMANTHUS C. von Willdenow, *nomen conservandum*
D. brevipes B. L. Turner
D. cooleyi (A. Eaton) W. Trelease
Sy = Acuan cooleyi (A. Eaton) N. Britton & J. Rose
Sy = A. jamesii (J. Torrey & A. Gray) K. E. O. Kuntze
Sy = Dalea jamesii J. Torrey & A. Gray
D. glandulosus (B. L. Turner) M. Luckow
Sy = D. virgatus (C. Linnaeus) C. von Willdenow var. glandulosus B. L. Turner
D. illinoënsis (A. Michaux) C. MacMillan *ex* B. Robinson & M. Fernald
Sy = Acuan illinoënse (A. Michaux) K. E. O. Kuntze
Sy = Mimosa illinoënsis A. Michaux
D. leptolobus J. Torrey & A. Gray
Sy = Acuan leptolobum (J. Torrey & A. Gray) K. E. O. Kuntze
D. obtusus S. Watson
Sy = Acuan fallax J. K. Small
Sy = A. obtusum (S. Watson) A. A. Heller
D. reticulatus G. Bentham
D. velutinus G. Scheele
Sy = Acuan velutinum (G. Scheele) K. E. O. Kuntze
D. virgatus (C. Linnaeus) C. von Willdenow var. ***acuminatus*** (G. Bentham) D. Isely
Sy = D. acuminatus G. Bentham

D. virgatus (C. Linnaeus) C. von Willdenow var. ***depressus*** (F. von Humboldt & A. Bonpland *ex* C. von Willdenow) B. L. Turner
Sy = Acuan depressum (F. von Humboldt & A. Bonpland *ex* C. von Willdenow) K. E. O. Kuntze
Sy = Desmanthus depressus C. von Willdenow

DESMODIUM N. Desvaux, *nomen conservandum*

D. canadense (C. Linnaeus) A. P. de Candolle
Sy = Meibomia canadensis (C. Linnaeus) K. E. O. Kuntze

D. canescens (C. Linnaeus) A. P. de Candolle
Sy = D. canescens var. hirsutum (W. Hooker) B. Robinson
Sy = D. canescens var. villosissimum J. Torrey & A. Gray
Sy = Meibomia canescens (C. Linnaeus) K. E. O. Kuntze

D. ciliare (G. H. Muhlenberg *ex* C. von Willdenow) A. P. de Candolle var. ***ciliare***
Sy = Meibomia ciliaris (G. H. Muhlenberg *ex* C. von Willdenow) S. F. Blake

D. cuspidatum (G. H. Muhlenberg *ex* C. von Willdenow) J. Loudon var. ***cuspidatum***
Sy = D. bracteosum (A. Michaux) A. P. de Candolle
Sy = D. grandiflorum A. P. de Candolle

D. fernaldii B. Schubert

D. glabellum (A. Michaux) A. P. de Candolle
Sy = Meibomia glabella (A. Michaux) K. E. O. Kuntze

D. glutinosum (G. H. Muhlenberg *ex* C. von Willdenow) A. Wood
Sy = D. acuminatum (A. Michaux) A. P. de Candolle
Sy = Meibomia acuminata (A. Michaux) S. F. Blake

D. grahamii A. Gray
Sy = Meibomia grahamii (A. Gray) K. E. O. Kuntze

D. incanum A. P. de Candolle var. ***angustifolium*** A. Grisebach
Sy = D. canum (J. F. Gmelin) H. Schinz & A. Thellung var. angustifolium (A. Grisebach) H. Léon & Brother Alain

D. laevigatum (T. Nuttall) A. P. de Candolle
Sy = Meibomia laevigata (T. Nuttall) K. E. O. Kuntze

* ***D. lindheimeri*** A. Vail **[TOES: V]**
Sy = Meibomia lindheimeri (A. Vail) A. Vail

D. lineatum A. P. de Candolle
Sy = D. lineatum var. polymorphum A. Gray
Sy = Meibomia lineata (A. P. de Candolle) K. E. O. Kuntze

D. marilandicum (C. Linnaeus) A. P. de Candolle
Sy = Meibomia marilandica (C. Linnaeus) K. E. O. Kuntze

D. neomexicanum A. Gray
Sy = D. bigelowii A. Gray
Sy = D. exiguum A. Gray
Sy = D. neomexicanum var. bigelowii (A. Gray) S. Watson
Sy = D. spirale A. P. de Candolle var. bigelowii (A. Gray) B. Robinson & J. Greenman
Sy = Meibomia neomexicana (A. Gray) K. E. O. Kuntze

D. nudiflorum (C. Linnaeus) A. P. de Candolle
Sy = Meibomia nudiflora (C. Linnaeus) K. E. O. Kuntze

D. nuttallii (A. Schindler) B. Schubert
Sy = Meibomia nuttallii A. Schindler

D. obtusum (G. H. Muhlenberg *ex* C. von Willdenow) A. P. de Candolle
Sy = D. rigidum (S. Elliott) A. P. de Candolle
Sy = Meibomia obtusa (G. H. Muhlenberg *ex* C. von Willdenow) A. Vail

D. paniculatum (C. Linnaeus) A. P. de Candolle var. ***epetiolatum*** B. Schubert

D. paniculatum (C. Linnaeus) A. P. de Candolle var. ***paniculatum***
Sy = D. dichromum L. Shinners
Sy = D. paniculatum var. angustifolium J. Torrey & A. Gray
Sy = D. paniculatum var. pubens J. Torrey & A. Gray
Sy = Meibomia paniculata (C. Linnaeus) K. E. O. Kuntze

D. pauciflorum (T. Nuttall) A. P. de Candolle
Sy = Meibomia pauciflora (T. Nuttall) K. E. O. Kuntze

D. perplexum B. Schubert
Sy = D. dillenii J. Darlington
Sy = D. paniculatum (C. Linnaeus) A. P. de Candolle var. dillenii (J. Darlington) D. Isely

D. psilophyllum D. von Schlechtendal
Sy = D. wrightii A. Gray
Sy = Meibomia psilophylla (D. von Schlechtendal) K. E. O. Kuntze

D. rotundifolium A. P. de Candolle
Sy = D. michauxii (A. Vail) F. Daniels
Sy = Meibomia rotundifolia (A. P. de Candolle) K. E. O. Kuntze

D. sessilifolium (J. Torrey) J. Torrey & A. Gray

Sy = Meibomia sessilifolia (J. Torrey) K. E. O. Kuntze
D. tortuosum (O. Swartz) A. P. de Candolle
Sy = D. purpureum (P. Miller) W. Fawcett & A. Rendle
Sy = Meibomia tortuosa (O. Swartz) K. E. O. Kuntze
D. tweedyi N. Britton
Sy = Meibomia tweedyi (N. Britton) A. Vail
D. viridiflorum (C. Linnaeus) A. P. de Candolle
Sy = Meibomia viridiflora (C. Linnaeus) K. E. O. Kuntze

DIOCLEA K. Kunth
D. multiflora (J. Torrey & A. Gray) C. Mohr
Sy = Galactia mohlenbrockii R. Maxwell

ERYTHRINA C. Linnaeus
E. crista-galli C. Linnaeus [**cultivated**]
Sy = Micropteryx crista-galli (C. Linnaeus) W. Walpers
E. herbacea C. Linnaeus
Sy = E. arborea (A. Chapman) J. K. Small

EYSENHARDTIA K. Kunth, *nomen conservandum*
* ***E. spinosa*** G. Engelmann [**TOES: V**]
E. texana G. Scheele
Sy = E. angustifolia F. Pennell

GALACTIA P. Browne
G. canescens G. Bentham
G. erecta (T. Walter) A. Vail
Sy = G. brachypoda J. Torrey & A. Gray
G. heterophylla A. Gray
Sy = G. grayi A. Vail
G. longifolia (N. von Jacquin) G. Bentham *ex* F. Hoehne
Sy = Collaea longifolia G. Bentham
G. marginalis G. Bentham
G. regularis (C. Linnaeus) N. Britton, E. Sterns, & J. Poggenburg
Sy = G. glabella A. Michaux
G. texana (G. Scheele) A. Gray
G. volubilis (C. Linnaeus) N. Britton
Sy = G. macreei M. Curtis
Sy = G. volubilis var. mississippiensis A. Vail
G. wrightii A. Gray var. ***wrightii***
Sy = G. tephrodes A. Gray

GENISTIDIUM I. M. Johnston
* ***G. dumosum*** I. M. Johnston [**TOES: V**]

GLEDITSIA C. Linnaeus
G. aquatica H. Marshall
G. triacanthos C. Linnaeus
Sy = G. triacanthos var. inermis (C. Linnaeus) C. Schneider

GLOTTIDIUM N. Desvaux
G. vesicaria (N. von Jacquin) R. Harper
Sy = Sesbania platycarpa C. Persoon
Sy = S. vesicaria (N. von Jacquin) S. Elliott

GLYCINE C. Linnaeus, *nomen conservandum*
G. max (C. Linnaeus) E. Merrill [**cultivated**]

GLYCYRRHIZA C. Linnaeus
G. lepidota F. Pursh
Sy = G. lepidota var. glutinosa (T. Nuttall) S. Watson
Sy = Liquivitia lepidota T. Nuttall

GYMNOCLADUS J. de Lamarck
G. dioica (C. Linnaeus) K. Koch

HARDENBERGIA G. Bentham
H. violacea (G. Schneevoogt) W. Stearn [**cultivated**]

HAVARDIA J. K. Small
H. pallens (G. Bentham) N. Britton & J. Rose
Sy = Pithecellobium pallens (G. Bentham) P. Standley

HEDYSARUM C. Linnaeus
H. boreale T. Nuttall subsp. ***boreale*** T. Nuttall var. ***boreale***
Sy = H. boreale var. cinerascens (P. Rydberg) R. Rollins
Sy = H. boreale var. obovatum R. Rollins
Sy = H. boreale var. pabulare (A. Nelson) R. Dorn
Sy = H. boreale var. rivulare (L. O. Williams) T. E. Northstrom
Sy = H. boreale var. utahense (P. Rydberg) R. Rollins

HOFFMANNSEGGIA A. Cavanilles, *nomen conservandum*
H. glauca (C. Ortega) I. Eifert
Sy = H. densiflora A. Gray
Sy = H. falcaria A. Cavanilles
H. oxycarpa G. Bentham *ex* A. Gray
Sy = Caesalpinia oxycarpa (G. Bentham) E. Fisher
* ***H. tenella*** B. Tharp & L. O. Williams [**TOES: I**]

INDIGOFERA C. Linnaeus
I. kirlowii C. Maximowicz [**cultivated**]
Sy = I. koreana J. Ohwi
I. lindheimeriana G. Scheele
I. miniata C. Ortega var. ***leptosepala*** (T. Nuttall *ex* J. Torrey & A. Gray) B. L. Turner
Sy = I. leptosepala T. Nuttall *ex* J. Torrey & A. Gray
I. miniata C. Ortega var. ***miniata***
I. miniata C. Ortega var. ***texana*** (S. Buckley) B. L. Turner
I. suffruticosa P. Miller

KUMMEROWIA A. Schindler
K. stipulacea (C. Maximowicz) T. Makino
Sy = Lespedeza stipulacea C. Maximowicz
K. striata (C. Thunberg) A. Schindler
Sy = Lespedeza striata (C. Thunberg) W. Hooker & G. Arnott

LATHYRUS C. Linnaeus
L. graminifolius (S. Watson) T. White
L. hirsutus C. Linnaeus
L. lanszwertii A. Kellogg var. ***leucanthus*** (P. Rydberg) R. Dorn
Sy = L. leucanthus P. Rydberg
L. latifolius C. Linnaeus [**cultivated**]
L. odoratus C. Linnaeus [**cultivated**]
L. polymorphus T. Nuttall subsp. ***incanus*** (J. G. Smith & P. Rydberg) C. Hitchcock
Sy = L. incanus (J. E. Smith & P. Rydberg) P. Rydberg
Sy = L. palustris C. Linnaeus var. graminifolius S. Watson
Sy = L. polymorphus var. incanus (J. E. Smith & P. Rydberg) R. Dorn
L. polymorphus T. Nuttall subsp. ***polymorphus*** var. ***polymorphus***
Sy = L. decaphyllus F. Pursh
L. pusillus S. Elliott
L. venosus G. H. Muhlenberg *ex* C. von Willdenow var. ***intonsus*** F. Butters & H. St. John
Sy = L. oreophilus E. Wooton & P. Standley
Sy = L. venosus subsp. arkansanus (N. Fassett) C. Hitchcock
Sy = L. venosus var. arkansanus N. Fassett
Sy = L. venosus var. meridionalis F. Butters & H. St. John

LESPEDEZA A. Michaux
L. capitata A. Michaux
Sy = L. capitata var. stenophylla C. Bissell & M. Fernald
Sy = L. capitata var. velutina (E. Bicknell) M. Fernald
Sy = L. capitata var. vulgaris J. Torrey & A. Gray
L. cuneata (Dumont de Courset) G. Don
L. hirta (C. Linnaeus) J. Hornemann subsp. ***hirta***
Sy = L. capitata A. Michaux var. calycina (A. Schindler) M. Fernald
Sy = L. hirta var. sparsiflora J. Torrey & A. Gray
L. intermedia (S. Watson) N. Britton
L. procumbens A. Michaux
L. repens (C. Linnaeus) W. Barton
Sy = Hedysarum repens C. Linnaeus
L. stuevei T. Nuttall
L. texana N. Britton
L. violacea (C. Linnaeus) C. Persoon
Sy = Hedysarum violaceum C. Linnaeus
Sy = Lespedeza frutescens (C. Linnaeus) S. Elliott
Sy = L. prairea (K. Mackenzie & B. Bush) N. Britton
L. virginica (C. Linnaeus) N. Britton
L. × ***neglecta*** K. Mackenzie & B. Bush [***stuevei*** × ***virginica***]
Sy = L. stuevei T. Nuttall var. angustifolia N. Britton, non F. Pursh

LEUCAENA G. Bentham
L. leucocephala (J. de Lamarck) H. C. de Wit
Sy = Acacia glauca C. von Willdenow
Sy = Leucaena glauca (C. von Willdenow) G. Bentham
Sy = Mimosa glauca C. Linnaeus
Sy = M. leucocephala J. de Lamarck
L. pulverulenta (D. von Schlechtendal) G. Bentham
Sy = Acacia pulverulenta D. von Schlechtendal
L. retusa G. Bentham
Sy = Caudoleucaena retusa (A. Gray) N. Britton & J. Rose

LOTUS C. Linnaeus
L. corniculatus C. Linnaeus
Sy = L. corniculatus var. arvensis (C. Schkuhr) N. Séringe *ex* A. P. de Candolle
L. humistratus E. Greene
Sy = Anisolotus brachycarpus (G. Bentham) P. Rydberg
Sy = A. trispermus (E. Greene) E. Wooton & P. Standley
Sy = Hosackia brachycarpa G. Bentham
Sy = Lotus trispermus E. Greene
L. plebeius (A. Brand) R. Barneby
Sy = L. longebracteatus P. Rydberg
Sy = L. neomexicanus P. Rydberg

Sy = L. oroboides (K. Kunth) A. Ottley var. nanus (A. Gray) D. Isely
Sy = L. oroboides var. nummularius (M. E. Jones) D. Isely
Ma = L. oroboides var. oroboides *auct.*, non (Kunth) A. Ottley: Texas authors

L. unifoliatus (W. Hooker) G. Bentham var. ***unifoliatus***
Sy = Acmispon americanum (T. Nuttall) P. Rydberg
Sy = Hosackia purshiana G. Bentham
Sy = Lotus americanus (T. Nuttall) G. Bischoff
Sy = L. purshianus F. Clements & E. Clements
Sy = L. purshianus var. glaber (T. Nuttall) P. Munz

LUPINUS C. Linnaeus [**state flower**]

L. concinnus J. Agardh subsp. ***concinnus***
Sy = L. concinnus subsp. optatus (C. P. Smith) D. Dunn
Sy = L. concinnus subsp. orcuttii (S. Watson) D. Dunn

L. havardii S. Watson

L. perennis C. Linnaeus subsp. ***gracilis*** (T. Nuttall) D. Dunn
Sy = L. gracilis T. Nuttall
Sy = L. nuttallii S. Watson
Sy = L. perennis var. gracilis (T. Nuttall) A. Chapman

L. plattensis S. Watson
Sy = L. glabratus (S. Watson) P. Rydberg
Sy = L. ornatus D. Douglas *ex* J. Lindley var. glabratus S. Watson

L. subcarnosus W. Hooker

L. texensis W. Hooker

MACROPTILIUM (G. Bentham) I. Urban

M. atropurpureum (J. Mociño & M. Sessé y Lacasta *ex* A. P. de Candolle) I. Urban
Sy = Phaseolus atropurpureus J. Mociño & M. Sessé y Lacasta *ex* A. P. de Candolle

M. gibbosifolium (C. Ortega) A. Delgado
Sy = Phaseolus heterophyllus C. von Willdenow
Sy = P. heterophyllus var. rotundifolius (A. Gray) C. Piper

MEDICAGO C. Linnaeus, *nomen conservandum*

M. arabica (C. Linnaeus) W. Hudson
Sy = M. arabica subsp. inermis P. Ricker

M. lupulina C. Linnaeus
Sy = M. lupulina var. cupaniana (G. Gussone) P. Boissier
Sy = M. lupulina var. glandulosa A. Neilreich

M. minima (C. Linnaeus) C. Linnaeus
Sy = M. minima var. compacta Neyraut
Sy = M. minima var. longiseta A. P. de Candolle
Sy = M. minima var. pubescens P. Webb

M. orbicularis (C. Linnaeus) B. Bartalini

M. polymorpha C. Linnaeus
Sy = M. hispida J. Gaertner
Sy = M. polymorpha var. brevispina (G. Bentham) C. Heyn
Sy = M. polymorpha var. ciliaris (N. Séringe) L. Shinners
Sy = M. polymorpha var. polygyra (I. Urban) L. Shinners
Sy = M. polymorpha var. tricycla (J. Grenier & D. Godron) L. Shinners
Sy = M. polymorpha var. vulgaris (G. Bentham) L. Shinners

M. sativa C. Linnaeus
Sy = M. sativa subsp. falcata (C. Linnaeus) G. Arcangeli
Sy = M. sativa subsp. sativa

MELILOTUS P. Miller

M. albus F. Medikus
Sy = Trifolium melilotus J. Georgi, non C. Linnaeus

M. indicus (C. Linnaeus) C. Allioni
Sy = M. parviflora N. Desvaux
Sy = Trifolium melilotus C. Linnaeus

M. officinalis (C. Linnaeus) J. de Lamarck
Sy = Trifolium melilotus A. von Gueldenstaedt *ex* C. von Ledebour, non C. Linnaeus

MILLETTIA R. Wight & G. Arnott

M. reticulata G. Bentham [**cultivated**]

MIMOSA C. Linnaeus

M. aculeaticarpa C. Ortega var. ***biuncifera*** (G. Bentham) R. Barneby
Sy = M. biuncifera G. Bentham
Sy = M. biuncifera var. flexuosa (G. Bentham) B. Robinson
Sy = M. biuncifera var. glabrescens A. Gray
Sy = M. biuncifera var. lindheimeri (A. Gray) B. Robinson
Sy = M. flexuosa G. Bentham
Sy = M. lindheimeri A. Gray
Sy = M. warnockii B. L. Turner
Sy = Mimosopsis biuncifera (G. Bentham) N. Britton & J. Rose
Sy = M. flexuosa (G. Bentham) N. Britton & J. Rose

M. asperata C. Linnaeus
Sy = M. pigra C. Linnaeus var. berlandieri (A. Gray) B. L. Turner
M. borealis A. Gray
Sy = M. fragrans A. Gray
M. dysocarpa G. Bentham var. ***dysocarpa***
Sy = M. dysocarpa var. wrightii (A. Gray) T. Kearney & H. Peebles
M. emoryana G. Bentham
M. hystricina (J. K. Small *ex* J. Rose) B. L. Turner
Sy = M. quadrivalvis C. Linnaeus var. hystricina (J. K. Small) R. Barneby
Sy = Schrankia hystricina (N. Britton & J. Rose) P. Standley
M. latidens (J. K. Small) B. L. Turner
Sy = M. quadrivalvis C. Linnaeus var. latidens (J. K. Small) R. Barneby
Sy = Schrankia latidens (J. K. Small) K. Schumann
M. malacophylla A. Gray
M. microphylla J. Dryander
M. nuttallii (A. P. de Candolle) B. L. Turner
Sy = Schrankia nuttallii (A. P. de Candolle *ex* N. Britton & J. Rose) P. Standley
M. roemeriana G. Scheele
Sy = M. quadrivalvis C. Linnaeus var. platycarpa (A. Gray) R. Barneby
Sy = Schrankia roemeriana (G. Scheele) J. Blankinship
M. rupertiana B. L. Turner
Sy = Leptoglottis occidentalis (E. Wooton & P. Standley) N. Britton & J. Rose
Sy = Mimosa occidentalis (E. Wooton & P. Standley) B. L. Turner
Sy = M. quadrivalvis C. Linnaeus var. occidentalis (E. Wooton & P. Standley) R. Barneby
Sy = Morongia occidentalis E. Wooton & P. Standley
Sy = Schrankia occidentalis (E. Wooton & P. Standley) P. Standley
M. strigillosa J. Torrey & A. Gray
M. texana (A. Gray) J. K. Small
Sy = M. wherryana (N. Britton) P. Standley
Sy = Mimosopsis wherryana N. Britton
Ma = Mimosa biuncifera var. lindheimeri *auct.*, non (A. Gray) B. Robinson: *sensu* B. L. Turner
M. turneri R. Barneby
Ma = M. zygophylla *auct.*, non Bentham: Texas authors

NEPTUNIA J. de Loureiro
N. lutea (M. Leavenworth) G. Bentham
Sy = N. lutea var. multipinnatifida B. L. Turner
N. pubescens G. Bentham var. ***microcarpa*** (J. Rose) D. Windler
Sy = N. palmeri N. Britton & J. Rose
N. pubescens G. Bentham var. ***pubescens***
Sy = N. pubescens var. floridana (J. K. Small) B. L. Turner
Sy = N. pubescens var. lindheimeri (B. Robinson) B. L. Turner

NISSOLIA N. von Jacquin
N. platycalyx S. Watson

ORBEXILUM C. Rafinesque-Schmaltz
O. pedunculatum (P. Miller) P. Rydberg var. ***pedunculatum***
Sy = Psoralea psoralioides (T. Walter) V. Cory var. eglandulosa (S. Elliott) F. Freeman
Sy = P. psoralioides var. pedunculata (P. Miller) K. Gandhi
O. simplex (T. Nuttall *ex* J. Torrey & A. Gray) P. Rydberg
Sy = Psoralea simplex T. Nuttall *ex* J. Torrey & A. Gray

OXYRHYNCHUS T. Brandegee
O. volubilis T. Brandegee

OXYTROPIS A. P. de Candolle, *nomen conservandum*
O. lambertii F. Pursh var. ***articulata*** (E. Greene) R. Barneby
Sy = Astragalus lambertii (F. Pursh) K. Sprengel var. abbreviatus (E. Greene) L. Shinners
O. sericea T. Nuttall var. ***sericea***
Sy = Aragallus pinetorum A. A. Heller var. pinetorum
Sy = A. pinetorum var. vegana T. Cockerell
Sy = A. sericea E. Greene
Sy = Oxytropis pinetorum (A. A. Heller) K. Schumann
Sy = O. vegana (T. Cockerell) E. Wooton & P. Standley

PARKINSONIA C. Linnaeus
P. aculeata C. Linnaeus
P. texana (A. Gray) S. Watson var. ***macra*** (I. M. Johnston) D. Isely
Sy = Cercidium macrum I. M. Johnston

P. texana (A. Gray) S. Watson var. ***texana***
Sy = Cercidium texanum A. Gray

PEDIOMELUM P. Rydberg
P. argophyllum (F. Pursh) J. Grimes
Sy = Psoralea argophylla F. Pursh
Sy = P. collina P. Rydberg
Sy = Psoralidium argophyllum (F. Pursh) P. Rydberg
P. cuspidatum (F. Pursh) P. Rydberg
Sy = Psoralea cuspidata F. Pursh
P. cyphocalyx (A. Gray) P. Rydberg
Sy = Psoralea cyphocalyx A. Gray
P. digitatum (T. Nuttall *ex* J. Torrey & A. Gray) D. Isely
Sy = P. digitatum var. parvifolium (L. Shinners) K. Gandhi & L. E. Brown
Sy = Psoralea digitata T. Nuttall *ex* J. Torrey & A. Gray
* ***P. humile*** P. Rydberg [**TOES: V**]
Sy = Psoralea humilis (P. Rydberg) J. Macbride
Sy = P. rydbergii V. Cory
P. hypogaeum (T. Nuttall *ex* J. Torrey & A. Gray) P. Rydberg var. ***hypogaeum***
Sy = Psoralea hypogaea J. Torrey & A. Gray
Sy = P. scaposa (A. Gray) J. Macbride var. breviscapa L. Shinners
P. hypogaeum (T. Nuttall *ex* J. Torrey & A. Gray) P. Rydberg var. ***scaposum*** (A. Gray) W. Mahler
Sy = P. goughae B. Tharp & F. Barkley
Sy = Psoralea hypogaea T. Nuttall *ex* J. Torrey & A. Gray subsp. scaposa (A. Gray) D. Ockendon
Sy = P. hypogaea var. scaposa (A. Gray) J. Macbride
Sy = P. scaposa (A. Gray) J. Macbride
P. hypogaeum (T. Nuttall *ex* J. Torrey & A. Gray) P. Rydberg var. ***subulatum*** (B. Bush) J. Grimes
Sy = P. subulatum (B. Bush) P. Rydberg
Sy = Psoralea subulata B. Bush
Sy = P. subulata var. minor L. Shinners
P. latestipulatum (L. Shinners) W. Mahler var. ***appressum*** (D. Ockendon) K. Gandhi & L. E. Brown
Sy = Psoralea latestipulata L. Shinners var. appressa D. Ockendon
P. latestipulatum (L. Shinners) W. Mahler var. ***latestipulatum***
Sy = Psoralea latestipulata L. Shinners
P. linearifolium (J. Torrey & A. Gray) P. Rydberg
Sy = Psoralea linearifolia J. Torrey & A. Gray
Sy = Psoralidium linearifolium (J. Torrey & A. Gray) P. Rydberg
Sy = P. linearifolium var. palodurense B. Tharp & F. Barkley
Sy = P. linearifolium var. texense B. Tharp & F. Barkley
* ***P. pentaphyllum*** (C. Linnaeus) J. Grimes [**TOES: V**]
Sy = P. trinervatum P. Rydberg
Sy = Psoralea pentaphylla C. Linnaeus
Sy = P. trinervata (P. Rydberg) P. Standley
P. reverchonii (S. Watson) P. Rydberg
Sy = Psoralea reverchonii S. Watson
P. rhombifolium (J. Torrey & A. Gray) P. Rydberg
Sy = P. brachypus P. Rydberg
Sy = P. coryi B. Tharp & F. Barkley
Sy = Psoralea brachypus (P. Rydberg) P. Standley
Sy = P. rhomifolia J. Torrey & A. Gray

PETERIA A. Gray
P. scoparia A. Gray

PHASEOLUS C. Linnaeus
P. acutifolius A. Gray var. ***tenuifolius*** A. Gray
Sy = P. tenuifolius (A. Gray) E. Wooton & P. Standley
P. angustissimus A. Gray
Sy = P. angustissimus var. latus M. E. Jones
Sy = P. dilatatus E. Wooton & P. Standley
P. coccineus C. Linnaeus [**cultivated**]
P. filiformis G. Bentham
Sy = P. wrightii A. Gray
P. lunatus C. Linnaeus [**cultivated**]
Sy = P. limensis J. Macfadyen
Sy = P. lunatus var. lunonanus L. Bailey
P. maculatus G. Scheele
Sy = P. metcalfei E. Wooton & P. Standley
P. polymorphus S. Watson
P. polystachios (C. Linnaeus) N. Britton, E. Sterns, & J. Poggenburg var. ***polystachios***
Sy = P. polystachios var. aquilonius M. Fernald
P. vulgaris C. Linnaeus [**cultivated**]
Sy = P. vulgaris var. humilis F. Alefeld

PISUM C. Linnaeus
P. sativum C. Linnaeus var. ***arvense*** (C. Linnaeus) J. Poiret [**cultivated**]
P. sativum C. Linnaeus var. ***sativum*** [**cultivated**]

PROSOPIS C. Linnaeus
P. glandulosa J. Torrey var. ***glandulosa***
Sy = Algarobia glandulosa (J. Torrey) J. Torrey & A. Gray

Sy = Neltuma glandulosa (J. Torrey) N. Britton & J. Rose
Sy = Prosopis chilensis (J. Molina) S. Stuntz var. glandulosa (J. Torrey) P. Standley
Sy = P. juliflora (O. Swartz) A. P. de Candolle var. glandulosa (J. Torrey) T. Cockerell

P. glandulosa J. Torrey var. ***torreyana*** (L. Benson) M. C. Johnston
Sy = P. glandulosa subsp. torreyana (L. Benson) E. Murray
Sy = P. juliflora (O. Swartz) A. P. de Candolle var. torreyana L. Benson
Sy = P. odorata J. Torrey & J. Frémont

P. laevigata (F. von Humboldt & A. Bonpland *ex* C. von Willdenow) M. C. Johnston

P. pubescens G. Bentham
Sy = P. emoryi J. Torrey
Sy = Strombocarpa pubescens (G. Bentham) A. Gray

P. reptans G. Bentham var. ***cinerascens*** (A. Gray) A. Burkart
Sy = P. cinerascens (A. Gray) G. Bentham
Sy = Strombocarpa cinerascens A. Gray

PSORALIDIUM P. Rydberg

P. lanceolatum (F. Pursh) P. Rydberg
Sy = Psoralea elliptica F. Pursh
Sy = P. lanceolata F. Pursh
Sy = P. laxiflora T. Nuttall
Sy = P. micrantha A. Gray
Sy = P. scabra T. Nuttall
Sy = P. stenophylla P. Rydberg
Sy = P. stenostachys P. Rydberg
Sy = Psoralidium lanceolatum var. stenophyllum (P. Rydberg) C. Toft & S. Welsh
Sy = P. lanceolatum var. stenostachys (P. Rydberg) S. Welsh

P. tenuiflorum (F. Pursh) P. Rydberg
Sy = Psoralea floribunda T. Nuttall
Sy = P. obtusiloba J. Torrey & A. Gray
Sy = P. tenuiflora F. Pursh
Sy = P. tenuiflora var. bigelovii (P. Rydberg) J. Macbride
Sy = P. tenuiflora var. floribunda (T. Nuttall *ex* J. Torrey & A. Gray) P. Rydberg
Sy = Psoralidium batesii P. Rydberg
Sy = P. bigelovii P. Rydberg
Sy = P. youngiae B. Tharp & F. Barkley

PSOROTHAMNUS P. Rydberg

P. scoparius (A. Gray) P. Rydberg
Sy = Dalea scoparia A. Gray
Sy = D. scoparia var. subrosea T. Cockerell
Sy = Parosela scoparia (A. Gray) A. A. Heller

PUERARIA A. P. de Candolle

P. montana (J. de Loureiro) E. Merrill var. ***lobata*** (C. von Willdenow) L. van der Maesen & S. Almeida
Sy = P. lobata (C. von Willdenow) J. Ohwi
Sy = P. lobata var. thomsonii (G. Bentham) L. van der Maesen
Sy = P. thunbergiana (P. von Siebold & J. Zuccarini) G. Bentham

RHYNCHOSIA J. de Loureiro, *nomen conservandum*

R. americana (W. Houstoun *ex* P. Miller) M. C. Metz
Sy = Lathyrus americanus W. Houstoun *ex* P. Miller

R. difformis (S. Elliott) A. P. de Candolle
Sy = Arcyphyllum difforme S. Elliott
Sy = Rhynchosia tomentosa (C. Linnaeus) W. Hooker G. Arnott var. volubilis (A. Michaux) J. Torrey & A. Gray

R. latifolia T. Nuttall *ex* J. Torrey & A. Gray
Sy = Dolicholus latifolius (T. Nuttall *ex* J. Torrey & A. Gray) A. Vail
Sy = Rhynchosia reticulata A. P. de Candolle var. latifolia (J. Torrey & A. Gray) K. E. O. Kuntze
Sy = R. torreyi A. Vail

R. minima (C. Linnaeus) A. P. de Candolle var. ***diminifolia*** W. Walraven

R. minima (C. Linnaeus) A. P. de Candolle var. ***minima***
Sy = Dolicholus minima (C. Linnaeus) F. Medikus
Sy = Dolichos minimus C. Linnaeus

R. reniformis (F. Pursh) A. P. de Candolle
Sy = Glycine reniformis F. Pursh
Sy = Rhynchosia tomentosa (C. Linnaeus) W. Hooker & G. Arnott var. monophylla (A. Michaux) J. Torrey & A. Gray

R. senna J. Gillies *ex* W. Hooker var. ***texana*** (J. Torrey & A. Gray) M. C. Johnston
Sy = Dolicholus texanus (J. Torrey & A. Gray) A. Vail
Sy = Rhynchosia senna var. angustifolia (A. Gray) J. Grear
Sy = R. texana J. Torrey & A. Gray

R. tomentosa (C. Linnaeus) W. Hooker & G. Arnott var. ***tomentosa***
Sy = Glycine tomentosa C. Linnaeus
Sy = Rhynchosia erecta (T. Walter) A. P. de Candolle

Sy = R. intermedia (J. Torrey & A. Gray) J. K. Small
Sy = R. tomentosa var. erecta (T. Walter) J. Torrey & A. Gray

ROBINIA C. Linnaeus
R. hispida C. Linnaeus var. ***hispida***
Sy = R. pallida W. Ashe
Sy = R. speciosa W. Ashe
R. neomexicana A. Gray var. ***neomexicana***
Sy = R. luxurians (G. Dieck) P. Rydberg
Sy = R. neomexicana var. luxurians G. Dieck
Sy = R. neomexicana var. subvelutina (P. Rydberg) T. Kearney & R. Peebles
Sy = R. rusbyi E. Wooton & P. Standley
Sy = R. subvelutina P. Rydberg
R. pseudoacacia C. Linnaeus
Sy = R. pseudoacacia var. rectissima (C. Linnaeus) O. Raber

SENNA P. Miller
S. alata (C. Linnaeus) W. Roxburgh **[cultivated]**
Sy = Cassia alata C. Linnaeus
Sy = Herpetica alata (C. Linnaeus) C. Rafinesque-Schmaltz
S. bauhinioides (A. Gray) H. Irwin & R. Barneby
Sy = Cassia bauhinioides A. Gray
Sy = C. bauhinioides var. arizonica B. Robinson *ex* J. Macbride
Sy = Earleocassia bauhinioides (A. Gray) N. Britton & J. Rose
S. bicapsularis (C. Linnaeus) W. Roxburgh **[cultivated]**
Sy = Adipera bicapsularis (C. Linnaeus) N. Britton & J. Rose *ex* N. Britton & Percy Wilson
Sy = Cassia bicapsularis C. Linnaeus
Sy = C. emarginata C. Linnaeus
S. corymbosa (J. de Lamarck) H. Irwin & R. Barneby
Sy = Adipera corymbosa (J. de Lamarck) N. Britton & J. Rose
Sy = Cassia corymbosa J. de Lamarck
S. durangensis (J. Rose) H. Irwin & R. Barneby var. ***iselyi*** H. Irwin & R. Barneby
Sy = Cassia durangensis J. Rose var. iselyi H. Irwin & R. Barneby
S. lindheimeriana (G. Scheele) H. Irwin & R. Barneby
Sy = Cassia lindheimeriana G. Scheele
Sy = Earleocassia lindheimeriana (G. Scheele) N. Britton
S. marilandica (C. Linnaeus) J. Link
Sy = Cassia marilandica C. Linnaeus
Sy = Ditremexa marilandica (C. Linnaeus) N. Britton & J. Rose
S. monozyx (H. Irwin & R. Barneby) H. Irwin & R. Barneby
S. multiglandulosa (N. von Jacquin) H. Irwin & R. Barneby **[cultivated]**
Sy = Adipera tomentosa (C. von Linné) N. Britton & J. Rose
Sy = Cassia multiglandulosa N. von Jacquin
Sy = C. tomentosa C. von Linné
S. obtusifolia (C. Linnaeus) H. Irwin & R. Barneby
Sy = Cassia obtusifolia C. Linnaeus
S. occidentalis (C. Linnaeus) J. Link
Sy = Cassia occidentalis C. Linnaeus
Sy = Ditremexa occidentalis (C. Linnaeus) N. Britton & J. Rose *ex* N. Britton & Percy Wilson
* ***S. orcuttii*** (N. Britton & J. Rose) H. Irwin & R. Barneby **[TOES: V]**
Sy = Cassia orcuttii (N. Britton & J. Rose) B. L. Turner
S. pendula (F. von Humboldt & A. Bonpland *ex* C. von Willdenow) H. Irwin & R. Barneby var. ***glabrata*** (J. Vogel) H. Irwin & R. Barneby
Sy = Cassia coluteoides L. Colladon
S. pilosior (B. Robinson *ex* J. Macbride) H. Irwin & R. Barneby
Sy = Cassia pilosior (B. Robinson *ex* J. Macbride) H. Irwin & R. Barneby
S. pumilo (A. Gray) H. Irwin & R. Barneby
Sy = Cassia pumilio A. Gray
* ***S. ripleyana*** (H. Irwin & R. Barneby) H. Irwin & R. Barneby **[TOES: V]**
Sy = Cassia ripleyana H. Irwin & R. Barneby
S. roemeriana (G. Scheele) H. Irwin & R. Barneby
Sy = Cassia roemeriana G. Scheele
Sy = Earleocassia roemeriana (G. Scheele) N. Britton
S. septemtrionalis (D. Viviani) H. Irwin & R. Barneby **[cultivated]**
Sy = Adipera laevigata (C. von Willdenow) N. Britton & J. Rose
Sy = Cassia laevigata C. von Willdenow
S. wislizenii (A. Gray) H. Irwin & R. Barneby
Sy = Cassia wislizenii A. Gray
Sy = Palmerocassia wislizenii (A. Gray) N. Britton

SESBANIA J. Scopoli, *nomen conservandum*
S. drummondii (P. Rydberg) V. Cory
Sy = Daubentonia drummondii P. Rydberg
S. exaltata (C. Rafinesque-Schmaltz) P. Rydberg *ex* A. Hill

Sy = Darwinia exaltata C. Rafinesque-Schmaltz
Sy = Sesbania macrocarpa G. H. Muhlenberg *ex* C. Rafinesque-Schmaltz
S. punicea (A. Cavanilles) G. Bentham
Sy = Daubentonia punicea (A. Cavanilles) A. P. de Candolle

SOPHORA C. Linnaeus
S. affinis J. Torrey & A. Gray
S. gypsophila B. L. Turner & A. M. Powell var. ***guadalupensis*** B. L. Turner & A. M. Powell
S. japonica C. Linnaeus [**cultivated**]
S. nuttalliana B. L. Turner
Sy = Patrinia sericea C. Rafinesque-Schmaltz
Sy = Radiusia sericea (C. Rafinesque-Schmaltz) G. Heynhold
Sy = Sophora carnosa G. Yakovlev
Sy = S. sericea T. Nuttall
Sy = Vexibia nuttalliana (B. L. Turner) W. A. Weber
S. secundiflora (C. Ortega) M. Lagasca y Segura *ex* A. P. de Candolle
Sy = Broussonetia secundiflora C. Ortega
Sy = Virgilla secundiflora (C. Ortega) A. Cavanilles
S. tomentosa C. Linnaeus var. ***occidentalis*** (C. Linnaeus) D. Isely
Sy = S. tomentosa subsp. occidentalis (C. Linnaeus) R. Brummitt

SPARTIUM C. Linnaeus
S. junceum C. Linnaeus [**cultivated**]

SPHAEROPHYSA A. P. de Candolle
S. salsula (P. von Pallas) A. P. de Candolle
Sy = Phaca salsula P. von Pallas
Sy = Swainsona salsula (P. von Pallas) P. Taubert

STROPHOSTYLES S. Elliott
S. helvula (C. Linnaeus) S. Elliott
Sy = S. helvula var. missouriensis (S. Watson) N. Britton
S. leiosperma (J. Torrey & A. Gray) C. Piper
Sy = Phaseolus leiospermus J. Torrey & A. Gray
Sy = Strophostyles pauciflora (G. Bentham) S. Watson
S. umbellata (G. H. Muhlenberg *ex* C. von Willdenow) N. Britton
Sy = S. umbellata var. paludigena M. Fernald

STYLOSANTHES O. Swartz
S. biflora (C. Linnaeus) N. Britton, E. Sterns, & J. Poggenburg
Sy = S. biflora var. hispidissima (A. Michaux) C. Pollard & C. Ball
Sy = S. floridana S. F. Blake
Sy = S. riparia T. Kearney
S. viscosa O. Swartz

TEPHROSIA C. Persoon, *nomen conservandum*
T. lindheimeri A. Gray
T. onobrychoides T. Nuttall
Sy = Cracca onobrychoides (T. Nuttall) K. E. O. Kuntze
Sy = Tephrosia angustifolia A. Featherman
Sy = T. multiflora A. Featherman
Sy = T. texana (P. Rydberg) V. Cory
T. potosina T. Brandegee
Sy = Cracca potosina (T. Brandegee) P. Standley
T. tenella A. Gray
T. virginiana (C. Linnaeus) C. Persoon
Sy = Cracca virginiana C. Linnaeus
Sy = Tephrosia latidens (J. K. Small) P. Standley
Sy = T. mohrii (P. Rydberg) R. Godfrey
Sy = T. virginiana var. glabra T. Nuttall
Sy = T. virginiana var. holosericea (T. Nuttall) J. Torrey & A. Gray

TRIFOLIUM C. Linnaeus
T. arvense C. Linnaeus var. ***arvense***
T. bejariense M. Moricand
Sy = T. macrocalyx W. Hooker
T. campestre J. von Schreber var. ***campestre***
T. carolinianum A. Michaux
Sy = T. saxicola J. K. Small
T. dubium J. Sibthorp
T. incarnatum C. Linnaeus var. ***incarnatum***
Sy = T. incarnatum var. elatius G. Gibelli & C. Belli
T. lappaceum C. Linnaeus var. ***lappaceum***
T. polymorphum J. Poiret *ex* J. de Lamarck & J. Poiret
Sy = T. amphianthum J. Torrey & A. Gray
T. pratense C. Linnaeus
Sy = T. pratense var. sativum (P. Miller) J. von Schreber
T. reflexum C. Linnaeus var. ***reflexum***
Sy = T. reflexum var. glabrum M. Lojacono
T. repens C. Linnaeus var. ***repens***
Sy = Amoria repens K. Presl
T. resupinatum C. Linnaeus
T. vesiculosum G. Savi var. ***vesiculosum***

T. willdenowii K. Sprengel
Sy = T. tridentatum J. Lindley
Sy = T. tridentatum var. aciculare (T. Nuttall) L. McDermott

VICIA C. Linnaeus
V. americana G. H. Muhlenberg *ex* C. von Willdenow subsp. ***americana***
Sy = V. americana subsp. oregana (T. Nuttall) L. Abrams
Sy = V. americana var. oregana (T. Nuttall) A. Nelson
Sy = V. americana var. truncata (T. Nuttall) J. Brewer
Sy = V. americana var. villosa (A. Kellogg) F. J. Hermann
Sy = V. dissitifolia (T. Nuttall) P. Rydberg
V. americana G. H. Muhlenberg *ex* C. von Willdenow subsp. ***minor*** (W. Hooker) C. Gunn
Sy = V. americana var. angustifolia C. Nees von Esenbeck
Sy = V. americana var. linearis (T. Nuttall) S. Watson
Sy = V. americana var. minor W. Hooker
V. caroliniana T. Walter
Sy = V. hugeri J. K. Small
V. hirsuta (C. Linnaeus) S. Gray
V. ludoviciana T. Nuttall subsp. ***leavenworthii*** (J. Torrey & A. Gray) J. Lassetter & C. Gunn
Sy = V. leavenworthii J. Torrey & A. Gray
V. ludoviciana T. Nuttall subsp. ***ludoviciana***
Sy = V. exigua T. Nuttall
Sy = V. leavenworthii J. Torrey & A. Gray var. occidentalis L. Shinners
Sy = V. ludoviciana var. laxiflora L. Shinners
Sy = V. ludoviciana var. ludoviciana
Sy = V. ludoviciana var. texana (J. Torrey & A. Gray) L. Shinners
Sy = V. producta P. Rydberg
Sy = V. texana (J. Torrey & A. Gray) J. K. Small
V. minutiflora F. Dietrich
Sy = V. micrantha T. Nuttall *ex* J. Torrey & A. Gray
Sy = V. reverchonii S. Watson
V. sativa C. Linnaeus subsp. ***nigra*** (C. Linnaeus) F. Ehrhart
Sy = V. angustifolia C. Linnaeus
Sy = V. sativa var. angustifolia (C. Linnaeus) N. Séringe
Sy = V. sativa var. nigra C. Linnaeus
Sy = V. sativa var. segetalis (J. Thuiller) N. Séringe
V. sativa C. Linnaeus subsp. ***sativa***
Sy = V. sativa var. linearis J. M. Lange
Sy = V. sativa var. sativa
V. tetrasperma (C. Linnaeus) J. von Schreber
Sy = V. tetrasperma var. tenuissima G. Druce
V. villosa A. Roth subsp. ***varia*** (N. Host) F. Corbière
Sy = V. dasycarpa M. Tenore
Sy = V. villosa var. glabrescens W. Koch
V. villosa A. Roth subsp. ***villosa***

VIGNA G. Savi, *nomen conservandum*
V. luteola (N. von Jacquin) G. Bentham
Sy = V. repens (C. Linnaeus) K. E. O. Kuntze
V. unguiculata (C. Linnaeus) W. Walpers
Sy = V. sinensis (C. Linnaeus) G. Savi

WISTERIA T. Nuttall, *nomen conservandum*
W. frutescens (C. Linnaeus) J. Poiret
Sy = W. macrostachya (J. Torrey & A. Gray) B. Robinson & M. Fernald
W. sinensis (J. Sims) A. P. de Candolle [**cultivated**]
Sy = Rehsonia sinensis (J. Sims) L. Stritch

ZORNIA J. F. Gmelin
Z. bracteata J. F. Gmelin
Z. gemella J. Vogel
Ma = Z. diphylla *auct.*, non (C. Linnaeus) C. Persoon
Z. reticulata J. E. Smith

FAGACEAE, *nomen conservandum*

CASTANEA P. Miller
C. pumila (C. Linnaeus) P. Miller var. ***pumila***
Sy = C. alnifolia T. Nuttall var. floridana C. Sargent
Sy = C. ashei (G. Sudworth) G. Sudworth
Sy = C. paucispinna W. Ashe
Sy = C. pumila var. ashei G. Sudworth
Sy = C. pumila var. margarettiae W. Ashe

FAGUS C. Linnaeus
F. grandifolia F. Ehrhart
Sy = F. grandifolia subsp. heterophylla W. Camp
Sy = F. grandifolia var. caroliniana (R. Lowden) M. Fernald & A. Rehder

QUERCUS C. Linnaeus
Q. acutissima J. Carruthers [**cultivated**]

Q. agrifolia L. Neé [**cultivated**]
Q. alba C. Linnaeus
Sy = Q. alba var. subcaerulea A. Pickens & M. Pickens
Sy = Q. alba var. subflavea A. Pickens & M. Pickens
Q. arizonica C. Sargent
Sy = Q. endemic C. H. Muller
Q. arkansana C. Sargent
Sy = Q. caput-rivuli W. Ashe
* ***Q. boyntonii*** C. Beadle [**TOES: V**]
Sy = Q. stiletto F. von Wangenheim var. boyntonii (C. Beadle) C. Sargent
Q. buckleyi K. Nixon & L. Dorr
Ma = Q. texana *auct.*, non S. Buckley: Texas authors
Q. carmenensis C. H. Muller
Q. chihuahuensis W. Trelease
* ***Q. depressipes*** W. Trelease [**TOES: V**]
Q. emoryi J. Torrey
Sy = Q. hastata F. Liebmann
Q. falcata A. Michaux var. ***falcata***
Sy = Q. digitata (H. Marshall) G. Sudworth
Sy = Q. falcata var. trilobata (A. Michaux) T. Nuttall
Q. falcata A. Michaux var. ***pagodifolia*** S. Elliott
Sy = Q. falcata var. leucophylla (W. Ashe) E. Palmer & J. Steyermark
Sy = Q. pagoda C. Rafinesque-Schmaltz
Q. gambelii T. Nuttall var. ***gambelii***
Sy = Q. novomexicana (A. P. de Candolle) P. Rydberg
Sy = Q. vreelandii P. Rydberg
* ***Q. graciliformis*** C. H. Muller [**TOES: V**]
Sy = Q. canbyi V. Cory & H. Parks
Sy = Q. graciliformis var. parvilobata C. H. Muller
Q. gravesii G. Sudworth
Sy = Q. chisosensis (C. Sargent) C. H. Muller
Sy = Q. stellipila (C. Sargent) H. Parks
Q. grisea F. Liebmann
Q. havardii P. Rydberg var. ***havardii***
* ***Q. hinckleyi*** C. H. Muller [**TOES: II**]
Q. hypoleucoides A. Camus
Sy = Q. hypoleuca G. Engelmann
Q. incana W. Bartram
Sy = Q. cinerea A. Michaux
Q. intricata W. Trelease
Q. laceyi J. K. Small
Ma = Q. glaucoides *auct.*, non M. Martens & H. Galeotti: Texas authors
Q. laurifolia A. Michaux
Sy = Q. obtusa (C. von Willdenow) W. Ashe
Ma = Q. hemisphaerica *auct.*, non W. Bartram *ex* C. von Willdenow
Q. lyrata T. Walter
Q. macrocarpa A. Michaux var. ***macrocarpa***
Sy = Q. marcocarpa var. oliviformis (F. Michaux) A. Gray
Q. marilandica O. von Münchhausen
Sy = Q. marilandica var. ashei G. Sudworth
Sy = Q. neoashei B. Bush
Q. michauxii T. Nuttall
Ma = Q. prinus *auct.*, non C. Linnaeus
Q. minima (C. Sargent) J. K. Small
Sy = Q. virginiana P. Miller var. minima C. Sargent
Q. mohriana S. Buckley *ex* P. Rydberg
Q. muhlenbergii G. Engelmann
Sy = Q. brayi J. K. Small
Sy = Q. prinoides C. von Willdenow var. acuminata (A. Michaux) H. Gleason
Q. myrsinifolia K. von Blume [**cultivated**]
Q. nigra C. Linnaeus
Sy = Q. nigra var. heterophylla (W. Aiton) W. Ashe
Q. oblongifolia J. Torrey
Q. palustris O. von Münchhausen [**cultivated**]
Q. phellos C. Linnaeus
Q. polymorpha D. von Schlechtendal & A. von Chamisso [**cultivated**]
Q. pungens F. Liebmann var. ***pungens***
Sy = Q. undulata O. Sperry
Sy = Q. undulata var. pungens (F. Liebmann) G. Engelmann
Q. pungens F. Liebmann var. ***vaseyana*** (S. Buckley) C. H. Muller
Sy = Q. sillae W. Trelease
Sy = Q. undulata O. Sperry var. vaseyana S. Buckley
Sy = Q. vaseyana S. Buckley
Q. rubra C. Linnaeus [**cultivated**]
Q. rugosa L. Neé
Sy = Q. diversicolor W. Trelease
Sy = Q. reticulata F. von Humboldt & A. Bonpland
Q. shumardii S. Buckley var. ***schneckii*** (N. Britton) C. Sargent
Sy = Q. schneckii N. Britton
Q. shumardii S. Buckley var. ***shumardii***
Sy = Q. shumardii var. macrocarpa (J. Torrey) L. Shinners
Q. similis W. Ashe
Sy = Q. ashei W. Sterrett

Sy = Q. stellata F. von Wangenheim var. mississippiensis (W. Ashe) E. Little
Sy = Q. stellata var. paludosa C. Sargent
Q. sinuata T. Walter var. ***breviloba*** (J. Torrey) C. H. Muller
Sy = Q. durandii S. Buckley var. breviloba (J. Torrey) E. Palmer
Sy = Q. sinuata var. san-sabeana (S. Buckley) S. Buckley
Q. sinuata T. Walter var. ***sinuata***
Sy = Q. durandii S. Buckley var. durandii
Q. stellata F. von Wangenheim var. ***margarettiae*** (W. Ashe) C. Sargent
Sy = Q. drummondii F. Liebmann
Sy = Q. margarettiae W. Ashe *ex* J. K. Small
Sy = Q. stellata var. araniosa C. Sargent
Q. stellata F. von Wangenheim var. ***stellata***
Sy = Q. stellata var. attenuata C. Sargent
Sy = Q. stellata var. parviloba C. Sargent
Q. suber C. Linnaeus [**cultivated**]
* ***Q. tardifolia*** C. H. Muller [**TOES: V**]
Q. texana S. Buckley
Sy = Q. nuttallii E. Palmer
Sy = Q. shumardii S. Buckley var. microcarpa (J. Torrey) L. Shinners
Sy = Q. shumardii var. texana (S. Buckley) W. Ashe
Q. toumeyi C. Sargent
Q. turbinella E. Greene
Sy = Q. turbinella var. ajoënsis (C. H. Muller) E. Little
Q. velutina J. de Lamarck
Sy = Q. velutina var. missouriensis C. Sargent
Q. virginiana P. Miller var. ***fusiformis*** (J. K. Small) C. Sargent
Sy = Q. oleoides var. quaterna C. H. Muller
Sy = Q. virginiana var. macrophylla C. Sargent
Q. virginiana P. Miller var. ***virginiana***
Q. × ***arkansana*** C. Sargent × ***Q. nigra*** C. Linnaeus
Q. × ***atlantica*** W. Ashe [***incana*** × ***laurifolia***]
Q. × ***beadlei*** W. Trelease *ex* E. Palmer [***alba*** × ***michauxii***]
Q. × ***beaumontiana*** C. Sargent [***falcata*** × ***laurifolia***]
Q. × ***bebbiana*** C. Schneider [***alba*** × ***macrocarpa***]
Q. × ***burnetensis*** E. Little [***macrocarpa*** × ***virginiana***]
Q. × ***bushii*** C. Sargent [***marilandica*** × ***velutina***]
Q. × ***byarsii*** G. Sudworth [***macrocarpa*** × ***michauxii***]
Q. × ***caduca*** W. Trelease [***incana*** × ***nigra***]
Q. × ***capesii*** W. Wolf [***nigra*** × ***phellos***]
Q. × ***cocksii*** C. Sargent [***laurifolia*** × ***velutina***]
Q. × ***comptoniae*** C. Sargent [***lyrata*** × ***virginiana***]
Q. × ***cravenensis*** E. Little [***incana*** × ***marilandica***]
Sy = Q. × carolinensis W. Trelease
Q. × ***deamii*** W. Trelease [***macrocarpa*** × ***muhlenbergii***]
Q. × ***demareei*** W. Ashe [***nigra*** × ***velutina***]
Q. × ***discreta*** K. Laughlin [***shumardii*** × ***velutina***]
Q. × ***diversiloba*** B. Tharp *ex* A. Camus [***laurifolia*** × ***marilandica***]
Q. × ***fernowii*** W. Trelease [***alba*** × ***stellata***]
Q. × ***filialis*** E. Little [***phellos*** × ***velutina***]
Q. × ***garlandensis*** E. Palmer [***falcata*** × ***nigra***]
Q. × ***guadalupensis*** C. Sargent [***macrocarpa*** × ***stellata***]
Q. × ***harbisonii*** C. Sargent [***stellata*** × ***virginiana***]
Q. × ***hastingsii*** C. Sargent [***marilandica*** × ***shumardii***]
Q. × ***incomita*** E. Palmer [***falcata*** × ***marilandica***]
Q. × ***inconstans*** E. Palmer [***gravesii*** × ***hypoleucoides***]
Sy = Q. livermorensis C. H. Muller
Q. × ***joorii*** W. Trelease [***falcata*** × ***shumardii***]
Q. laurifolia A. Michaux × ***Q. velutina*** J. de Lamarck
Q. × ***ludoviciana*** C. Sargent [***falcata*** var. ***pagodifolia*** × ***phellos***]
Q. × ***macnabiana*** G. Sudworth [***sinuata*** × ***stellata***]
Q. × ***megaleia*** K. Laughlin [***lyrata*** × ***macrocarpa***]
Q. × ***moultonensis*** W. Ashe [***phellos*** × ***shumardii***]
Q. × ***neopalmeri*** G. Sudworth *ex* E. Palmer [***nigra*** × ***shumardii***]
Q. × ***neotharpii*** A. Camus [***minima*** × ***stellata***]
Q. × ***organensis*** W. Trelease [***arizonica*** × ***grisea***]
Q. × ***pauciloba*** P. Rydberg [***gambelii*** × ***turbinella***]
Sy = Q. fendleri F. Liebmann
Sy = Q. venustula E. Greene
Sy = Q. × andrewsii C. Sargent
Q. × ***podophylla*** W. Trelease [***incana*** × ***velutina***]
Q. × ***pseudomargarettiae*** W. Trelease [***stellata*** var. ***margarettiae*** × ***stellata*** var. ***stellata***]
* ***Q.*** × ***robusta*** C. H. Muller [***emoryi*** × ***gravesii***] [**TOES: V**]
Q. × ***rudkinii*** N. Britton [***marilandica*** × ***phellos***]
Q. × ***sterilis*** W. Trelease *ex* E. Palmer [***marilandica*** × ***nigra***]
Q. × ***sterretii*** W. Trelease [***lyrata*** × ***stellata***]
Q. × ***subfalcata*** W. Trelease [***falcata*** × ***phellos***]
Q. × ***subintegra*** (G. Engelmann) W. Trelease [***falcata*** × ***incana***]
Q. × ***tharpii*** C. H. Muller [***emoryi*** × ***graciliformis***]

Q. × ***tottenii*** L. Melvin [***lyrata*** × ***michauxii***]
Q. × ***willdenowiana*** (L. Dippel) L. Beissner, E. Schelle, & H. Zabel [***falcata*** × ***velutina***]
Sy = Q. × pinetorum H. Moldenke

FLACOURTIACEAE, *nomen conservandum*

DORYALIS E. Meyer *ex* G. Arnott
Sy = Dovyalis (typographic error)
D. caffra T. R. Sim [**cultivated**]

XYLOSMA J. G. Forster, *nomen conservandum*
X. flexuosa (K. Kunth) W. Hemsley
X. japonicum (W. Walpers) A. Gray [**cultivated**]
Sy = Apactis japonica C. Thunberg
Sy = Flacourtia japonica W. Walpers
Sy = Xylosma congestum E. Merrill

FOUQUIERIACEAE, *nomen conservandum*

FOUQUIERIA K. Kunth
F. splendens G. Engelmann subsp. ***splendens***

FRANKENIACEAE, *nomen conservandum*

FRANKENIA C. Linnaeus
F. jamesii J. Torrey *ex* A. Gray
* ***F. johnstonii*** D. Correll [**TOES: I**]

FUMARIACEAE, *nomen conservandum*

CORYDALIS A. P. de Candolle, *nomen conservandum*
C. aurea C. von Willdenow subsp. ***aurea***
Sy = Capnoides aureum (C. von Willdenow) K. E. O. Kuntze
Sy = C. euchlamydeum E. Wooton & P. Standley
Sy = Corydalis aurea var. aurea
C. aurea C. von Willdenow subsp. ***occidentalis*** (G. Engelmann *ex* A. Gray) G. Ownbey
Sy = Capnoides montanum (G. Engelmann) N. Britton
Sy = Corydalis aurea var. occidentalis G. Engelmann *ex* A. Gray
Sy = C. curvisiliqua G. Engelmann subsp. occidentalis (G. Engelmann *ex* A. Gray) W. A. Weber
Sy = C. montana G. Engelmann
C. crystallina G. Engelmann
Sy = Capnoides crystallinum (G. Engelmann) K. E. O. Kuntze
C. curvisiliqua G. Engelmann subsp. ***curvisiliqua***
Sy = C. curvisiliqua var. curvisiliqua
C. curvisiliqua G. Engelmann subsp. ***grandibracteata*** (F. Fedde) G. Ownbey
Sy = C. curvisiliqua var. grandibracteata F. Fedde
C. micrantha (G. Engelmann *ex* A. Gray) A. Gray subsp. ***australis*** (A. Chapman) G. Ownbey
Sy = C. aurea C. von Willdenow var. australis A. Chapman
Sy = C. halei (J. K. Small) M. Fernald & B. Schubert
Sy = C. micrantha var. australis (A. Chapman) L. Shinners
C. micrantha (G. Engelmann *ex* A. Gray) A. Gray subsp. ***micrantha***
Sy = Capnoides micranthum (G. Engelmann *ex* A. Gray) N. Britton
Sy = Corydalis aurea C. von Willdenow var. micrantha G. Engelmann *ex* A. Gray
Sy = C. micrantha var. micrantha
C. micrantha (G. Engelmann *ex* A. Gray) A. Gray subsp. ***texensis*** G. Ownbey
Sy = C. micrantha var. texensis (G. Ownbey) L. Shinners

FUMARIA C. Linnaeus
F. officinalis C. Linnaeus
F. parviflora J. de Lamarck

GARRYACEAE, *nomen conservandum*

GARRYA D. Douglas *ex* J. Lindley
G. ovata G. Bentham subsp. ***goldmanii*** (E. Wooton & P. Standley) G. Dahling
Sy = G. goldmanii E. Wooton & P. Standley
G. ovata G. Bentham subsp. ***lindheimeri*** (J. Torrey) G. Dahling
Sy = G. lindheimeri J. Torrey
G. wrightii J. Torrey

GENTIANACEAE, *nomen conservandum*

BARTONIA G. H. Muhlenberg *ex* C. von Willdenow, *nomen conservandum*
B. paniculata (A. Michaux) G. H. Muhlenberg subsp. ***paniculata***
Sy = B. lanceolata J. K. Small
Sy = B. virginica (C. Linnaeus) N. Britton, E. Sterns, J. Poggenburg var. paniculata (A. Michaux) J. Boivin
Sy = Centaurella paniculata A. Michaux

* ***B. texana*** D. Correll [**TOES**: V]
B. verna (A. Michaux) C. Rafinesque-Schmaltz *ex* W. Barton
Sy = Centaurella verna A. Michaux

CENTAURIUM J. Hill
C. arizonicum (A. Gray) A. A. Heller
Sy = C. calycosum (S. Buckley) M. Fernald var. arizonicum (A. Gray) I. Tidestrom
Sy = Erythraea calycosa S. Buckley var. arizonica A. Gray
C. beyrichii (J. Torrey & A. Gray) B. Robinson
Sy = Erythraea beyrichii J. Torrey & A. Gray
C. breviflorum (L. Shinners) B. L. Turner
Sy = C. calycosum (S. Buckley) M. Fernald var. breviflorum L. Shinners
C. calycosum (S. Buckley) M. Fernald
Sy = Erythraea calycosa S. Buckley
C. glanduliferum (D. Correll) B. L. Turner
Sy = C. beyrichii (J. Torrey & A. Gray) B. Robinson var. glanduliferum D. Correll
C. maryannum B. L. Turner
C. nudicaule (G. Engelmann) B. Robinson
Sy = Erythraea nudicaulis G. Engelmann
C. pulchellum (O. Swartz) G. Druce
C. texense (A. Grisebach *ex* W. Hooker) M. Fernald
Sy = Erythraea texensis A. Grisebach

EUSTOMA R. A. Salisbury *ex* G. Don
E. exaltatum (C. Linnaeus) A. Salisbury *ex* G. Don
Sy = E. barkleyi P. Standley *ex* L. Shinners
Sy = E. silenifolium A. Salisbury
Sy = Gentiana exaltata C. Linnaeus
E. russellianum (W. Hooker) G. Don
Sy = E. grandiflorum (C. Rafinesque-Schmaltz) L. Shinners

FRASERA T. Walter
F. speciosa D. Douglas *ex* A. Grisebach
Sy = F. macrophylla E. Greene
Sy = F. speciosa var. scabra M. E. Jones
Sy = Swertia radiata (A. Kellogg) K. E. O. Kuntze var. radiata
Sy = S. radiata var. macrophylla (E. Greene) H. St. John
Sy = Tesseranthium macrophyllum (E. Greene) P. Rydberg
Sy = T. radiatum A. Kellogg

GENTIANA C. Linnaeus
G. affinis A. Grisebach
Sy = Dasystephana affinis (A. Grisebach) P. Rydberg
Sy = Pneumonanthe affinis E. Greene
G. saponaria C. Linnaeus
Sy = Dasystephana saponaria (C. Linnaeus) J. K. Small

OBOLARIA C. Linnaeus
O. virginica C. Linnaeus

SABATIA M. Adanson
S. angularis (C. Linnaeus) F. Pursh
S. arenicola J. Greenman
Sy = S. carnosa J. K. Small
S. brachiata S. Elliott
S. calycina (J. de Lamarck) A. A. Heller
Sy = S. cubensis (A. Grisebach) I. Urban
* ***S. campanulata*** (C. Linnaeus) J. Torrey [**TOES**: III]
Sy = S. campanulata var. gracilis (A. Michaux) M. Fernald
S. campestris T. Nuttall
S. dodecandra (C. Linnaeus) N. Britton, E. Sterns, & J. Poggenburg var. ***foliosa*** (M. Fernald) R. Wilbur
Sy = S. foliosa M. Fernald
Sy = S. harperi J. K. Small
Sy = S. obtusata S. F. Blake
S. formosa S. Buckley
S. gentianoides S. Elliott
Sy = Lapithea gentianoides (S. Elliott) A. Grisebach

GERANIACEAE, *nomen conservandum*

ERODIUM L'Héritier de Brutelle *ex* W. Aiton
E. botrys (A. Cavanilles) A. Bertoloni
E. cicutarium (C. Linnaeus) L'Héritier de Brutelle *ex* W. Aiton
Sy = Geranium cicutarium C. Linnaeus
E. texanum A. Gray

GERANIUM C. Linnaeus
G. atropurpureum A. A. Heller var. ***atropurpureum***
Sy = G. caespitosum E. James subsp. atropurpureum (A. A. Heller) W. A. Weber
Sy = G. caespitosum var. atropurpureum A. A. Heller
G. carolinianum C. Linnaeus var. ***carolinianum***
G. dissectum C. Linnaeus
Sy = G. laxum L. Hanks
G. lentum E. Wooton & P. Standley

G. texanum (W. Trelease) A. A. Heller
Sy = G. carolinianum C. Linnaeus var. texanum W. Trelease
G. wislizenii S. Watson

PELARGONIUM L'Héritier de Brutelle *ex* W. Aiton
P. peltatum (C. Linnaeus) L'Héritier de Brutelle *ex* W. Aiton **[cultivated]**
P. × hortorum L. Bailey [***inquinans × zonale***] **[cultivated]**

GOODENIACEAE, *nomen conservandum*

SCAEVOLA C. Linnaeus, *nomen conservandum*
S. plumieri (C. Linnaeus) M. H. Vahl
Sy = Lobelia plumieri C. Linnaeus
Sy = Scaevola lobelia J. Murray

GROSSULARIACEAE, *nomen conservandum*

ESCALLONIA J. Mutis *ex* C. von Linné
E. rubra (H. Ruiz López & J. Pavón) C. Persoon var. ***macrantha*** (W. Hooker & G. Arnott) K. Reiche **[cultivated]**
Sy = E. macrantha W. Hooker & G. Arnott

ITEA C. Linnaeus
I. virginica C. Linnaeus

RIBES C. Linnaeus
R. aureum F. Pursh var. ***aureum***
Sy = Chrysobotrya aurea (F. Pursh) P. Rydberg
R. aureum F. Pursh var. ***villosum*** A. P. de Candolle
Sy = R. odoratum S. Endlicher
R. curvatum J. K. Small
Sy = Grossularia campestris J. K. Small
Sy = G. texensis F. Coville & A. Berger
R. leptanthum A. Gray
Sy = Grossularia leptantha (A. Gray) F. Coville & N. Britton
R. mescalerium F. Coville

HALORAGACEAE, *nomen conservandum*

MYRIOPHYLLUM C. Linnaeus
M. aquaticum (J. M. da Conceicão Vellozo) B. Verdcourt
Sy = Enydria aquatica J. M. da Conceicão Vellozo
Sy = Myriophyllum brasiliense J. Cambessèdes
Sy = M. proserpinacoides J. Gillies *ex* W. Hooker & G. Arnott
M. heterophyllum A. Michaux
M. hippuroides T. Nuttall *ex* J. Torrey & A. Gray
M. pinnatum (T. Walter) N. Britton, E. Sterns, & J. Poggenburg
Sy = M. scabratum A. Michaux
Sy = Potamogeton pinnatum T. Walter
M. spicatum C. Linnaeus var. ***exalbescens*** (M. Fernald) W. Jepson
Sy = M. exalbescens M. Fernald
M. verticillatum C. Linnaeus
Sy = M. verticillatum var. cheneyi N. Fassett
Sy = M. verticillatum var. intermedium W. Koch
Sy = M. verticillatum var. pectinatum C. Wallroth
Sy = M. verticillatum var. pinnatifidum C. Wallroth

PROSERPINACA C. Linnaeus
P. palustris C. Linnaeus var. ***amblyogona*** M. Fernald
Sy = P. amblyogona (M. Fernald) J. K. Small
P. palustris C. Linnaeus var. ***crebra*** M. Fernald & L. Griscom
P. palustris C. Linnaeus var. ***palustris***
Sy = P. palustris var. latifolia A. Schindler
Sy = P. platycarpa J. K. Small
P. pectinata J. de Lamarck

HAMAMELIDACEAE, *nomen conservandum*

HAMAMELIS C. Linnaeus
H. virginiana C. Linnaeus
Sy = H. macrophylla F. Pursh
Sy = H. vernalis C. Sargent var. tomentella (A. Rehder) E. Palmer
Sy = H. vernalis var. venalis
Sy = H. virginiana var. macrophylla (F. Pursh) T. Nuttall
Sy = H. virginiana var. parvifolia T. Nuttall

LIQUIDAMBAR C. Linnaeus
L. styraciflua C. Linnaeus

LOROPETALUM R. Brown
L. chinense (R. Brown) D. Oliver **[cultivated]**

HIPPOCASTANACEAE, *nomen conservandum*

AESCULUS C. Linnaeus
A. flava W. Aiton **[cultivated]**
Sy = A. octandra H. Marshall

A. glabra C. von Willdenow var. ***arguta*** (S. Buckley) B. Robinson
Sy = A. arguta S. Buckley
Sy = A. arguta var. buckleyi C. Sargent
A. glabra C. von Willdenow var. ***glabra***
Sy = A. arguta S. Buckley var. leucodermis C. Sargent
Sy = A. arguta var. micrantha C. Sargent
Sy = A. arguta var. monticola C. Sargent
Sy = A. arguta var. pallida (C. von Willdenow) E. von Kirchner
Sy = A. arguta var. sargentii A. Rehder
A. pavia C. Linnaeus var. ***flavescens*** (C. Sargent) D. Correll
A. pavia C. Linnaeus var. ***pavia***
Sy = A. pavia var. discolor (F. Pursh) A. Gray

HYDRANGEACEAE, *nomen conservandum*

DEUTZIA C. Thunberg
D. gracilis P. von Siebold & J. Zuccarini [**cultivated**]
D. scabra C. Thunberg [**cultivated**]
Sy = D. scabra var. candidissima (K. Fröbel) A. Rehder

FENDLERA G. Engelmann & A. Gray
F. rigida I. M. Johnston
Ma = F. linearis *auct.*, non A. Rehder
F. rupicola A. Gray var. ***falcata*** (J. Thornber) A. Rehder
F. rupicola A. Gray var. ***rupicola***
F. utahensis (S. Watson) A. A. Heller var. ***cymosa*** (E. Greene *ex* E. Wooton & P. Standley) T. Kearney & R. Peebles
F. wrightii (A. Gray) A. A. Heller
Sy = F. rupicola A. Gray var. wrightii A. Gray
Sy = F. tomentella J. Thornber

HYDRANGEA C. Linnaeus
H. macrophylla (C. Thunberg) N. Séringe [**cultivated**]
Sy = H. otaksa P. von Siebold & J. Zuccarini
H. paniculata P. von Siebold [**cultivated**]
H. quercifolia W. Bartram [**cultivated**]

PHILADELPHUS C. Linnaeus
P. argenteus P. Rydberg
Sy = P. argyrocalyx E. Wooton var. argenteus (P. Rydberg) A. Engler
Sy = P. microphyllus A. Gray subsp. argenteus (P. Rydberg) C. Hitchcock
Sy = P. microphyllus var. argenteus (P. Rydberg) T. Kearney & R. Peebles
P. argyrocalyx E. Wooton
Sy = P. ellipticus P. Rydberg
Sy = P. microphyllus A. Gray subsp. argyrocalyx (E. Wooton) C. Hitchcock
Sy = P. serpyllifolius A. Gray subsp. argyrocalyx (E. Wooton) C. Hitchcock
Sy = P. serpyllifolius var. argyrocalyx (E. Wooton) M. E. Jones
P. coronarius C. Linnaeus [**cultivated**]
* ***P. crinitus*** (C. Hitchcock) Shiu-ying Hu [**TOES: V**]
Sy = P. microphyllus A. Gray subsp. crinitus C. Hitchcock
* ***P. ernestii*** Shiu-ying Hu [**TOES: V**]
P. hitchcockianus Shiu-ying Hu
P. mearnsii W. Evans *ex* B. Köhne
P. microphyllus A. Gray var. ***microphyllus***
P. occidentalis A. Nelson var. ***occidentalis***
Sy = P. microphyllus A. Gray subsp. occidentalis (A. Nelson) C. Hitchcock
Sy = P. microphyllus var. occidentalis (A. Nelson) R. Dorn
P. palmeri P. Rydberg
P. pubescens J. Loiseleur-Deslongchamps
Sy = P. intectus C. Beadle
Sy = P. intectus var. pubigerus Shiu-ying Hu
Sy = P. latifolius H. A. Schrader *ex* A. P. de Candolle
Sy = P. pubescens var. intectus (C. Beadle) A. Moore
Sy = P. pubescens var. verrucosus (H. A. Schrader) Shiu-ying Hu
P. serpyllifolius A. Gray
P. texensis Shiu-ying Hu var. ***coryanus*** Shiu-ying Hu
* ***P. texensis*** Shiu-ying Hu var. ***texensis*** [**TOES: V**]

HYDROPHYLLACEAE, *nomen conservandum*

ELLISIA C. Linnaeus
E. nyctelea C. Linnaeus
Sy = Ipomoea nyctelea C. Linnaeus
Sy = Macrocalyx nyctelea (C. Linnaeus) K. E. O. Kuntze
Sy = Nyctelea nyctelea (C. Linnaeus) N. Britton

EUCRYPTA T. Nuttall
E. micrantha (J. Torrey) A. A. Heller
Sy = Ellisia micrantha (J. Torrey) A. Brand
Sy = Phacelia micrantha J. Torrey

HYDROLEA C. Linnaeus, *nomen conservandum*
H. ovata T. Nuttall *ex* J. Choisy
Sy = Nama ovata (T. Nuttall *ex* J. Choisy) N. Britton
H. spinosa C. Linnaeus
H. uniflora C. Rafinesque-Schmaltz
Sy = H. affinis A. Gray
Sy = Nama affinis (A. Gray) K. E. O. Kuntze

NAMA C. Linnaeus, *nomen conservandum*
N. carnosum (E. Wooton) C. Hitchcock
Sy = Andropus carnosus (E. Wooton) A. Brand
Sy = Conanthus carnosus E. Wooton
Sy = Nama stenophyllum A. Gray *ex* W. Hemsley var. egenum J. Macbride
N. dichotomum (H. Ruiz López & J. Pavón) J. Choisy
Sy = Conanthus angustifolius (A. Gray) A. A. Heller
Sy = Marilaunidium angustifolium (A. Gray) K. E. O. Kuntze
Sy = Nama angustifolium (A. Gray) A. Nelson
Sy = N. dichotomum var. angustifolium A. Gray
Sy = N. serpylloides W. Hemsley
N. havardii A. Gray
N. hispidum A. Gray
Sy = Conanthus hispidus (A. Gray) A. A. Heller
Sy = Marilaunidium hispidum (A. Gray) K. E. O. Kuntze
Sy = Nama foliosum (W. Wooton & P. Standley) I. Tidestrom
Sy = N. hispida A. Gray
Sy = N. hispidum var. mentzelii A. Brand
Sy = N. hispidum var. revolutum W. Jepson
Sy = N. hispidum var. spathulatum (J. Torrey) C. Hitchcock
Sy = N. tenue (E. Wooton & P. Standley) I. Tidestrom
N. jamaicense C. Linnaeus
Sy = Marilaunidium jamaicense (C. Linnaeus) K. E. O. Kuntze
N. parvifolium (J. Torrey) J. Greenman
N. stenocarpum A. Gray
Sy = Conanthus stenocarpus A. A. Heller
Sy = Marilaunidium stenocarpum K. E. O. Kuntze
Sy = Nama humifusum A. Brand
N. stevensii C. Hitchcock
N. torynophyllum J. Greenman
N. undulatum K. Kunth
N. xylopodum (E. Wooton & P. Standley) C. Hitchcock
Sy = Marilaunidium xylopodum E. Wooton & P. Standley

NEMOPHILA T. Nuttall, *nomen conservandum*
N. aphylla (C. Linnaeus) R. Brummitt
Sy = N. microcalyx (T. Nuttall) F. von Fischer & C. von Meyer
Sy = N. triloba (C. Rafinesque-Schmaltz) J. Thieret
N. phacelioides T. Nuttall

PHACELIA A. L. de Jussieu
P. coerulea E. Greene
Sy = P. invenusta A. Gray
P. congesta W. Hooker
Sy = P. congesta var. dissecta A. Gray
P. crenulata J. Torrey *ex* S. Watson var. ***corrugata*** (A. Nelson) A. Brand
Sy = P. corrugata A. Nelson
P. crenulata J. Torrey *ex* S. Watson var. ***crenulata***
Sy = P. ambigua M. E. Jones
Sy = P. crenulata var. ambigua J. Macbride
Sy = P. crenulata var. funera J. Voss *ex* P. Munz
Sy = P. crenulata var. vulgaris A. Brand
Ma = P. intermedia *auct.*, non E. Wooton
P. glabra T. Nuttall
P. hirsuta T. Nuttall
P. infundibuliformis J. Torrey
P. integrifolia J. Torrey var. ***integrifolia***
P. integrifolia J. Torrey var. ***texana*** (J. Voss) N. Atwood
Sy = P. texana J. Voss
P. laxa J. K. Small
* ***P. pallida*** I. M. Johnston **[TOES: V]**
P. patuliflora (G. Engelmann & A. Gray) A. Gray var. ***austrotexana*** J. A. Moyer
P. patuliflora (G. Engelmann & A. Gray) A. Gray var. ***patuliflora***
Sy = Eucota patuliflora G. Engelmann & A. Gray
Sy = Phacelia hispida S. Buckley
P. patuliflora (G. Engelmann & A. Gray) A. Gray var. ***teucriifolia*** (I. M. Johnston) L. Constance
Sy = P. teucriifolia I. M. Johnston
P. popei J. Torrey & A. Gray var. ***popei***
Sy = P. depauperata E. Wooton & P. Standley
P. popei J. Torrey & A. Gray var. ***similis*** (E. Wooton & P. Standley) J. Voss
P. robusta (J. Macbride) I. M. Johnston
Sy = P. integrifolia var. robusta J. Macbride
P. rupestris E. Greene

Sy = P. congesta W. Hooker var. rupestris J. Macbride
P. strictiflora (G. Engelmann & A. Gray) A. Gray var. ***connexa*** L. Constance
P. strictiflora (G. Engelmann & A. Gray) A. Gray var. ***lundelliana*** L. Constance
P. strictiflora (G. Engelmann & A. Gray) A. Gray var. ***robbinsii*** L. Constance
P. strictiflora (G. Engelmann & A. Gray) A. Gray var. ***strictiflora***

ILLICIACEAE, *nomen conservandum*

ILLICIUM C. Linnaeus
I. floridanum J. Ellis [**cultivated**]

JUGLANDACEAE, *nomen conservandum*

CARYA T. Nuttall, *nomen conservandum*
C. alba (C. Linnaeus) T. Nuttall *ex* S. Elliott
Sy = C. tomentosa (J. Poiret) T. Nuttall
Sy = C. tomentosa var. subcoriacea (C. Sargent) E. Palmer & J. Steyermark
Sy = Hicoria tomentosa (J. de Lamarck *ex* J. Poiret) C. Rafinesque-Schmaltz
C. aquatica (F. Michaux) T. Nuttall
Sy = C. aquatica var. australis C. Sargent
Sy = Hicoria aquatica (F. Michaux) N. Britton
C. cordiformis (F. von Wangenheim) K. Koch
Sy = C. cordiformis var. latifolia C. Sargent
Sy = Hicoria cordiformis (F. von Wangenheim) N. Britton
C. glabra (P. Miller) R. Sweet var. ***hirsuta*** (W. Ashe) W. Ashe
Sy = C. leiodermis C. Sargent
Sy = C. ovalis (F. von Wangenheim) C. Sargent var. hirsuta (W. Ashe) C. Sargent
Sy = Hicoria glabra (P. Miller) N. Britton var. hirsuta W. Ashe
C. illinoinensis (F. von Wangenheim) K. Koch [**state tree**]
Sy = C. oliviformis (F. Michaux) T. Nuttall
Sy = C. pecan (H. Marshall) G. Engelmann & K. Graebner
Sy = Hicoria pecan (H. Marshall) N. Britton
C. laciniosa (F. Michaux) G. Don
Sy = Hicoria laciniosa (F. Michaux) C. Sargent
C. myristiciformis (F. Michaux) T. Nuttall
Sy = Hicoria myristiciformis (F. Michaux) N. Britton
C. ovata (P. Miller) K. Koch
Sy = C. ovata var. fraxinifolia C. Sargent
Sy = C. ovata var. nuttallii C. Sargent
Sy = C. ovata var. pubescens C. Sargent
Sy = Hicoria borealis W. Ashe
Sy = H. ovata (P. Miller) N. Britton
C. texana S. Buckley
Sy = C. buckleyi M. Durieu de Maisonneuve
Sy = C. buckleyi var. arkansana (C. Sargent) C. Sargent
Sy = C. texana var. arkansana (C. Sargent) E. Little
Sy = C. texana var. villosa (C. Sargent) E. Little
Sy = Hicoria villosa (C. Sargent) W. Ashe
Ma = H. texana *auct.*, non J. Le Conte

JUGLANS C. Linnaeus
J. major (J. Torrey) A. A. Heller
Sy = J. elaeopyren L.-A. Dode
Sy = J. microcarpa J. Berlandier var. major (J. Torrey) L. Benson
Sy = J. rupestris G. Engelmann *ex* J. Torrey var. major J. Torrey
J. microcarpa J. Berlandier var. ***microcarpa***
Sy = J. rupestris G. Engelmann *ex* J. Torrey
J. microcarpa J. Berlandier var. ***stewartii*** (I. M. Johnston) W. Manning
J. nigra C. Linnaeus
Sy = Wallia nigra (C. Linnaeus) F. Alefeld
J. regia C. Linnaeus

PTEROCARYA K. Kunth
P. stenoptera A. P. de Candolle [**cultivated**]

KRAMERIACEAE, *nomen conservandum*

KRAMERIA P. Köfling
K. erecta C. von Willdenow *ex* J. A. Schultes
Sy = K. glandulosa J. Rose & J. Painter
Sy = K. imparata (J. Macbride) N. Britton
Sy = K. interior J. Rose & J. Painter
Sy = K. navae J. Rzedowski
Sy = K. palmeri J. Rose
Sy = K. parviflora G. Bentham
Sy = K. parviflora var. glandulosa (J. Rose & J. Painter) J. Macbride
Sy = K. parviflora var. imparata J. Macbride
Sy = K. pauciflora G. Bentham
Sy = K. rosmarinifolia J. Pavón
K. grayi J. Rose & J. Painter
Sy = K. bicolor S. Watson
Sy = K. canescens A. Gray

K. lanceolata J. Torrey
Sy = Dimenops lanceolata (J. Torrey) C. Rafinesque-Schmaltz
Sy = Krameria beyrichii F. Sporleder *ex* O. Berg
Sy = K. spathulata J. K. Small
K. ramosissima (A. Gray) S. Watson
Sy = K. parvifolia G. Bentham var. ramosissima A. Gray

LAMIACEAE, *nomen conservandum*

AGASTACHE J. Clayton *ex* J. Gronovius
A. breviflora (A. Gray) C. Epling
Sy = Cedronella breviflora A. Gray
A. cana (W. Hooker) E. Wooton & P. Standley
Sy = Cedronella cana W. Hooker
A. micrantha (A. Gray) E. Wooton & P. Standley var. ***micrantha***
Sy = Cedronella micrantha A. Gray
A. pallidiflora (A. A. Heller) P. Rydberg subsp. ***neomexicana*** (J. Briquet) H. Lint & C. Epling
Sy = A. breviflora (A. Gray) C. Epling var. havardii (A. Gray) L. Shinners
Sy = A. neomexicana (J. Briquet) P. Standley
Sy = A. pallidiflora subsp. havardii (A. Gray) H. Lint & C. Epling
Sy = A. pallidiflora var. neomexicana (J. Briquet) R. Sanders

AJUGA C. Linnaeus
A. reptans C. Linnaeus [**cultivated**]

BRAZORIA G. Engelmann & A. Gray
B. arenaria C. Lundell
* ***B. pulcherrima*** C. Lundell [**TOES: V**]
B. scutellarioides G. Engelmann & A. Gray
B. truncata (G. Bentham) G. Engelmann & A. Gray

CALAMINTHA P. Miller
C. arkansana (T. Nuttall) L. Shinners
Sy = C. glabella (A. Michaux) G. Bentham var. angustifolia (J. Torrey) G. DeWolf
Sy = Clinopodium arkansanum (T. Nuttall) H. House
Sy = Hedeoma arkansana T. Nuttall
Sy = Satureja arkansana (T. Nuttall) J. Briquet

COLEUS J. de Loureiro
C. scutellarioides (C. Linnaeus) G. Bentham [**cultivated**]
Sy = C. blumei G. Bentham
Sy = Plectranthus blumei (G. Bentham) G. Launert
Sy = P. scutellarioides (C. Linnaeus) R. Brown

CUNILA D. Royen *ex* C. Linnaeus, *nomen conservandum*
C. origanoides (C. Linnaeus) N. Britton
Sy = C. marina C. Linnaeus
Sy = Mappia origanoides (C. Linnaeus) H. House

GLECHOMA C. Linnaeus, orthographic *conservandum*
G. hederacea C. Linnaeus [**cultivated**]
Sy = G. hederacea var. micrantha M. Moricand
Sy = G. hederacea var. parviflora (G. Bentham) H. House
Sy = Neptea glechoma G. Bentham
Sy = N. glechoma var. parviflora G. Bentham
Sy = N. hederacea V. Trevisan de Saint-Léon var. hederacea
Sy = N. hederacea var. parviflora G. Druce

HEDEOMA C. Persoon
H. acinoides G. Scheele
* ***H. apiculata*** W. Stewart [**TOES: II**]
H. costata A. Gray var. ***pulchella*** (E. Greene) R. Irving
Sy = H. convisae A. Nelson
Sy = H. pulchella E. Greene
H. drummondii G. Bentham
Sy = H. ciliata T. Nuttall
Sy = H. compora P. Rydberg
Sy = H. drummondii var. crenulata R. Irving
Sy = H. longiflora P. Rydberg
Sy = H. ovata A. Nelson
H. hispida F. Pursh
H. molle J. Torrey
Sy = Poliomintha mollis (J. Torrey) A. Gray
H. nana (J. Torrey) J. Briquet subsp. ***nana***
Sy = H. dentata J. Torrey var. nana J. Torrey
Sy = H. thymoides A. Gray
* ***H. pilosa*** R. Irving [**TOES: III**]
H. plicata J. Torrey
H. reverchonii (A. Gray) A. Gray var. ***reverchonii***
Sy = H. drummondii G. Bentham var. reverchonii A. Gray
Sy = H. lata J. K. Small
H. reverchonii (A. Gray) A. Gray var. ***serpyllifolia*** (J. K. Small) R. Irving
Sy = H. drummondii G. Bentham var. serpyllifolium (J. K. Small) R. Irving
Sy = H. sancta J. K. Small
Sy = H. serpyllifolia J. K. Small

HYPTIS N. von Jacquin, *nomen conservandum*
H. alata (C. Rafinesque-Schmaltz) L. Shinners
Sy = H. alata var. stenophylla L. Shinners
Sy = H. radiata C. von Willdenow

LAMIUM C. Linnaeus
L. amplexicaule C. Linnaeus var. **amplexicaule**
Sy = L. amplexicaule var. calyciflorum M. Tenore
Sy = L. amplexicaule var. cryptantha A. Cariot
Sy = L. amplexicaule var. fallax E. Junger *ex* E. Fiek
Sy = L. amplexicaule var. genuinum N. Stojanov
Sy = L. amplexicaule var. lassithicum (P. Cousturier & M. Gandoger *ex* M. Gandoger) A. von Hayek
Sy = L. amplexicaule var. nanum H. Gams
Sy = L. amplexicaule var. parviflorum P. Schur
Sy = L. amplexicaule var. tel-avivensis A. Eig
Sy = L. amplexicaule var. thracicum J. Velenovský
Sy = L. mauritanicum M. Gandoger *ex* J. Battandier & L. Trabut
Sy = L. stepposum I. Kossko
L. maculatum C. Linnaeus [**cultivated**]
L. purpureum C. Linnaeus var. **incisum** (C. von Willdenow) C. Persoon
Sy = L. aeolicum M. Lojacono
Sy = L. amplexicaule-purpureum G. Meyer
Sy = L. bifidum D. Cirillo subsp. albimontanum K. H. Rechinger
Sy = L. confusum J. de Martrin-Donos
Sy = L. dissectum W. Withering
Sy = L. incisum C. von Willdenow var. obscurum B. Dumortier
Sy = L. purpureum var. decipiens O. Sonder *ex* J. F. Koch
Sy = L. purpureum var. exanulatum V. Loret
Sy = L. urticifolium C. Weihe
Sy = L. westphalicum C. Weihe
L. purpureum C. Linnaeus var. **purpureum**
Sy = L. purpureum var. albiflorum B. Dumortier
Sy = L. purpureum var. aznavouri M. Gandoger *ex* G. Aznavour
Sy = L. purpureum var. genuinum N. Stojanov
Sy = L. purpureum var. grandifolium W. Petermann
Sy = L. purpureum var. integrum A. Gray
Sy = L. purpureum var. longidens J. Podpěra
Sy = L. purpureum var. lumbii G. Druce
Sy = L. purpureum var. microdonton W. Petermann
Sy = L. purpureum var. niveum P. Schur

LAVANDULA C. Linnaeus
L. angustifolia P. Miller [**cultivated**]
Sy = L. spica C. Linnaeus
L. burmanii G. Bentham [**cultivated**]
Sy = L. multifida N. Burman

LEONOTIS (C. Persoon) W. T. Aiton
L. nepetifolia (C. Linnaeus) W. T. Aiton [**cultivated**]
Sy = Phlomis nepetifolia C. Linnaeus

LEONURUS C. Linnaeus
L. sibiricus C. Linnaeus
Ma = L. artemisia *auct.*, non J. de Loureiro

LYCOPUS C. Linnaeus
L. americanus G. H. Muhlenberg *ex* W. Barton
Sy = L. americanus var. longii W. Benner
Sy = L. americanus var. scabrifolius M. Fernald
Sy = L. sinuatus S. Elliott
Sy = Phytosalpinx americana J. Lunell
L. asper E. Greene
Sy = L. lucidus P. Turczaninow *ex* G. Bentham subsp. americanus (A. Gray) O. Hultén
Sy = L. lucidus var. americanus A. Gray
L. rubellus C. Moench
Sy = L. rubellus var. arkansanus (J. Fresenius) W. Benner
Sy = L. rubellus var. lanceolatus W. Benner
Sy = L. velutinus P. Rydberg
L. virginicus C. Linnaeus

MARRUBIUM C. Linnaeus
M. vulgare C. Linnaeus

MELISSA C. Linnaeus
M. officinalis C. Linnaeus [**cultivated**]

MENTHA C. Linnaeus
M. arvensis C. Linnaeus
Sy = M. arvensis var. lanata C. Piper
Sy = M. arvensis var. penardii J. Briquet
Sy = M. borealis A. Michaux
Sy = M. canadensis C. Linnaeus
Sy = M. gentilis C. Linnaeus
M. spicata C. Linnaeus
Sy = M. spicata var. viridis C. Linnaeus
Sy = M. viridis C. Linnaeus
M. × piperita C. Linnaeus [***aquatica* × *spicata***]
Sy = M. aquatica C. Linnaeus var. crispa (C. Linnaeus) G. Bentham

Sy = M. crispa C. Linnaeus
Sy = M. dumetorum J. A. Schultes
M.* × *rotundifolia (C. Linnaeus) W. Hudson [***longifolia* × *suaveolens***]
Sy = M. spicata C. Linnaeus var. rotundifolia C. Linnaeus

MICROMERIA G. Bentham, *nomen conservandum*
M. brownei (O. Swartz) G. Bentham var. ***pilosiuscula*** A. Gray
Sy = M. pilosiuscula (A. Gray) J. K. Small
Sy = Satureja brownei (O. Swartz) J. Briquet var. pilosiuscula (A. Gray) J. Briquet

MOLLUCELLA C. Linnaeus
M. laevis C. Linnaeus

MONARDA C. Linnaeus
M. citriodora V. de Cervantes *ex* M. Lagasca y Segura var. ***citriodora***
Sy = M. aristata T. Nuttall
Sy = M. citriodora subsp. citriodora
Sy = M. citriodora var. attenuata R. Scora
Sy = M. dispersa J. K. Small
Sy = M. tenuiaristata A. Gray *ex* J. K. Small
M. citriodora V. de Cervantes *ex* M. Lagasca y Segura var. ***parva*** R. Scora
M. clinopodioides A. Gray
Sy = M. aristata W. Hooker
M. didyma C. Linnaeus [**cultivated**]
M. fistulosa C. Linnaeus var. ***menthifolia*** (R. Graham) M. Fernald
Sy = M. comata P. Rydberg
Sy = M. fistulosa subsp. menthifolia (R. Graham) L. S. Gill
Sy = M. menthifolia R. Graham
Sy = M. stricta E. Wooton
M. fistulosa C. Linnaeus var. ***mollis*** G. Bentham
Sy = M. mollis C. Linnaeus
M. fruticulosa C. Epling
Sy = M. punctata C. Linnaeus subsp. fruticulosa (C. Epling) R. Scora
Sy = M. punctata var. fruticulosa (C. Epling) R. Scora
M. lindheimeri G. Engelmann & A. Gray *ex* A. Gray
Sy = M. hirsutissima J. K. Small
M. maritima (V. Cory) D. Correll
Sy = M. punctata C. Linnaeus subsp. maritima (V. Cory) R. Scora
Sy = M. punctata var. maritima V. Cory
M. pectinata T. Nuttall
M. punctata C. Linnaeus var. ***correllii*** B. L. Turner
M. punctata C. Linnaeus var. ***intermedia*** (E. McClintock & C. Epling) U. Waterfall
Sy = M. punctata subsp. intermedia E. McClintock & C. Epling
M. punctata C. Linnaeus var. ***lasiodonta*** A. Gray
Sy = M. lasiodonta (A. Gray) J. K. Small
Sy = M. punctata subsp. coryi E. McClintock & C. Epling
Sy = M. punctata subsp. immaculata F. Pennell
Sy = M. punctata var. coryi (E. McClintock & C. Epling) V. Cory *ex* L. Shinners
Sy = M. punctata var. immaculata (F. Pennell) R. Scora
M. punctata C. Linnaeus var. ***occidentalis*** (C. Epling) E. Palmer & J. Steyermark
Sy = M. punctata subsp. occidentalis C. Epling
M. punctata C. Linnaeus var. ***punctata***
Sy = M. punctata subsp. punctata
Ma = M. punctata var. arkansana E. McClintock & C. Epling. NOTE: B. L. Turner (pers. comm.) believes all of McClintock & Epling's reports of this taxon for Texas are M. punctata var. punctata.
M. russeliana T. Nuttall *ex* J. Sims
Sy = M. virgata C. Rafinesque-Schmaltz
M. stanfieldii J. K. Small
Sy = M. punctata C. Linnaeus subsp. stanfieldii (J. K. Small) C. Epling
Sy = M. punctata var. stanfieldii (J. K. Small) V. Cory
M. viridissima D. Correll

NEPETA C. Linnaeus
N. cataria C. Linnaeus [**cultivated**]

OCIMUM C. Linnaeus
O. basilicum C. Linnaeus [**cultivated**]

ORIGANUM C. Linnaeus
O. hirtum J. Link [**cultivated**]
Sy = O. heracleoticum G. Bentham
O. majorana C. Linnaeus [**cultivated**]
Sy = Majorana hortensis C. Moench
Sy = M. majorana (C. Linnaeus) G. Karsten
O. vulgare C. Linnaeus [**cultivated**]

PERILLA C. Linnaeus
P. frutescens (C. Linnaeus) N. Britton var. ***frutescens***

PHLOMIS C. Linnaeus
P. fruticosa C. Linnaeus [**cultivated**]

PHYSOSTEGIA G. Bentham
P. angustifolia M. Fernald
Sy = P. edwardsiana L. Shinners
* **P. correllii** (C. Lundell) L. Shinners **[TOES: V]**
Sy = Dracocephalum correllii C. Lundell
P. digitalis J. K. Small
P. intermedia (T. Nuttall) G. Engelmann & A. Gray
Sy = Dracocephalum intermedium T. Nuttall
Sy = Physostegia micrantha C. Lundell
P. longisepala P. Cantino
P. pulchella C. Lundell
P. virginiana (C. Linnaeus) G. Bentham subsp. **praemorsa** (L. Shinners) P. Cantino
Sy = P. praemorsa L. Shinners
Sy = P. serotina L. Shinners
Sy = P. virginiana var. arenaria B. Shimek
Sy = P. virginiana var. reducta J. Boivin
P. virginiana (C. Linnaeus) G. Bentham subsp. **virginiana**
Sy = Dracocephalum virginianum C. Linnaeus
Sy = Physostegia virginiana var. elongata J. Boivin
Sy = P. virginiana var. formosior (C. Lundell) J. Boivin
Sy = P. virginiana var. granulosa (N. Fassett) M. Fernald
Sy = P. virginiana var. speciosa (R. Sweet) A. Gray

POLIOMINTHA A. Gray
P. glabrescens A. Gray
Sy = Hedeoma glabrescens (A. Gray) J. Briquet
P. incana (J. Torrey) A. Gray
Sy = Hedeoma incana J. Torrey
P. longiflora A. Gray **[cultivated]**
Sy = Hedeoma longiflora (A. Gray) J. Briquet

PRUNELLA C. Linnaeus
P. vulgaris C. Linnaeus subsp. **lanceolata** (W. Barton) Shiu-ying Hu
Sy = P. vulgaris var. elongata G. Bentham
Sy = P. vulgaris var. lanceolata (W. Barton) M. Fernald
P. vulgaris C. Linnaeus subsp. **vulgaris**
Sy = P. vulgaris var. atropurpurea M. Fernald
Sy = P. vulgaris var. calvescens M. Fernald
Sy = P. vulgaris var. hispida G. Bentham
Sy = P. vulgaris var. minor J. G. Smith
Sy = P. vulgaris var. nana W. Clute
Sy = P. vulgaris var. parviflora (J. Poiret) A. P. de Candolle
Sy = P. vulgaris var. rouleauiana J. Marie-Victorin

PYCNANTHEMUM A. Michaux, *nomen conservandum*
P. albescens J. Torrey & A. Gray
Sy = Koellia albescens (J. Torrey & A. Gray) K. E. O. Kuntze
P. clinopodioides J. Torrey & A. Gray
Sy = Koellia clinopodioides (J. Torrey & A. Gray) K. E. O. Kuntze
P. muticum (A. Michaux) C. Persoon
Sy = Brachystemum muticum A. Michaux
Sy = Koellia mutica (A. Michaux) K. E. O. Kuntze
P. tenuifolium H. A. Schrader
Ma = Koellia flexuosa *auct.*, non (T. Walter) J. Macbride
Ma = Pycnanthemum flexuosum *auct.*, non (T. Walter) N. Britton, E. Sterns, & J. Poggenburg

RHODODON C. Epling
R. angulatus (B. Tharp) B. L. Turner
Sy = Stachydeoma angulatus B. Tharp
R. ciliatus (G. Bentham) C. Epling
Sy = Hedeoma ciliata G. Bentham
Sy = H. texanum V. Cory

ROSMARINUS C. Linnaeus
R. officinalis C. Linnaeus **[cultivated]**

SALAZARIA J. Torrey
S. mexicana J. Torrey

SALVIA C. Linnaeus
S. arizonica A. Gray
S. azurea A. Michaux *ex* J. de Lamarck var. **azurea**
S. azurea A. Michaux *ex* J. de Lamarck var. **grandiflora** G. Bentham
Sy = S. azurea subsp. intermedia C. Epling
Sy = S. azurea subsp. pitcheri (J. Torrey *ex* G. Bentham) C. Epling
Sy = S. azurea var. pitcheri (J. Torrey *ex* G. Bentham) E. Sheldon
Sy = S. pitcheri J. Torrey *ex* G. Bentham
S. ballotiflora G. Bentham
Sy = S. laxa G. Bentham
S. blepharophylla T. Brandegee *ex* C. Epling **[cultivated]**
S. buchananii R. Hedge **[cultivated]**
S. chamaedroides A. Cavanilles **[cultivated]**
S. coccinea P. Buc'hoz *ex* A. Etlinger
Sy = S. ciliata G. Bentham

Sy = S. coccinea var. pseudococcinea (N. von Jacquin) K. E. O. Kuntze
Sy = S. galeottii M. Martens
Sy = S. pseudococcinea N. von Jacquin
S. dolichantha (V. Cory) E. Whitehouse
Sy = Salviastrum dolichanthum V. Cory
S. elegans M. H. Vahl **[cultivated]**
S. engelmannii A. Gray
S. farinacea G. Bentham var. ***farinacea***
Sy = S. earlei E. Wooton & P. Standley
S. farinacea G. Bentham var. ***latifolia*** L. Shinners
S. greggii A. Gray
S. guaranitica A. Saint-Hilaire *ex* G. Bentham **[cultivated]**
S. henryi A. Gray
S. hispanica C. Linnaeus
S. involucrata A. Cavanilles **[cultivated]**
Sy = S. laevigata K. Kunth
S. leptophylla G. Bentham
S. leucantha A. Cavanilles **[cultivated]**
Sy = S. bicolor M. Sessé y Lacasta & J. Mociño
S. lycioides A. Gray
Sy = S. ramosissima M. Fernald
S. lyrata C. Linnaeus
S. microphylla K. Kunth **[cultivated]**
Sy = S. grahamii G. Bentham
Sy = S. microphylla var. canescens A. Gray
Sy = S. microphylla var. wislizeni A. Gray
S. miniata M. Fernald **[cultivated]**
S. officinalis C. Linnaeus **[cultivated]**
* ***S. penstemonoides*** K. Kunth & P. C. Bouché **[TOES: V]**
S. pinguifolia (M. Fernald) E. Wooton & P. Standley
Sy = S. ballotiflora G. Bentham var. pinguifolia M. Fernald
S. pratensis C. Linnaeus **[cultivated]**
Sy = S. haematodes C. Linnaeus
S. reflexa J. Hornemann
Sy = S. lanceolata P. Broussonet
Sy = S. lancifolia J. Poiret
S. regla A. Cavanilles
Sy = S. crenata M. Martens & H. Galeottii
Sy = S. deltoidea C. Persoon
S. roemeriana G. Scheele
S. sclarea C. Linnaeus **[cultivated]**
S. sinaloënsis M. Fernald **[cultivated]**
S. splendens F. Sellow *ex* J. J. Römer & J. A. Schultes **[cultivated]**
S. subincisa G. Bentham
* ***S. summa*** A. Nelson **[TOES: V]**
S. texana (G. Scheele) J. Torrey
Sy = Salviastrum texanum G. Scheele
S. uligunosa G. Bentham **[cultivated]**
S. vinacea E. Wooton & P. Standley

SATUREJA C. Linnaeus
S. hortensis C. Linnaeus **[cultivated]**
S. montana C. Linnaeus **[cultivated]**

SCUTELLARIA C. Linnaeus
S. cardiophylla G. Engelmann & A. Gray
S. drummondii G. Bentham var. ***drummondii***
S. drummondii G. Bentham var. ***edwardsiana*** B. L. Turner
S. drummondii G. Bentham var. ***runyonii*** B. L. Turner
S. elliptica G. H. Muhlenberg *ex* K. Sprengel var. ***elliptica***
Sy = S. ovalifolia C. Persoon
Sy = S. pilosa A. Michaux
S. elliptica G. H. Muhlenberg *ex* K. Sprengel var. ***hirsuta*** (C. Short & R. Peter) M. Fernald
Sy = S. ovalifolia C. Persoon subsp. hirsuta (C. Short & R. Peter) M. Fernald
S. galericulata C. Linnaeus
Sy = S. galericulata subsp. pubescens (G. Bentham) A. Löve & D. Löve
Sy = S. galericulata var. epilobifolia (A. Hamilton) L. Jordal
Sy = S. galericulata var. pubescens G. Bentham
S. integrifolia C. Linnaeus
Sy = S. integrifolia var. hispida G. Bentham
* ***S. laevis*** L. Shinners **[TOES: V]**
S. lateriflora C. Linnaeus var. ***lateriflora***
S. muriculata C. Epling
S. ovata J. Hill subsp. ***bracteata*** (G. Bentham) C. Epling
Sy = S. ovata var. bracteata (G. Bentham) S. F. Blake
Sy = S. versicolor T. Nuttall var. bracteata G. Bentham
S. ovata J. Hill subsp. ***mexicana*** C. Epling
S. ovata J. Hill subsp. ***ovata***
Sy = S. ovata subsp. mississippiensis (K. von Martius) C. Epling
Sy = S. ovata subsp. versicolor (T. Nuttall) C. Epling
Sy = S. ovata var. calcarea (C. Epling) H. Gleason
Sy = S. ovata var. versicolor (T. Nuttall) M. Fernald
S. parvula A. Michaux var. ***australis*** N. Fassett
Sy = S. australis (N. Fassett) C. Epling

S. parvula A. Michaux var. ***leonardii*** (C. Epling) M. Fernald
Sy = S. ambigua T. Nuttall
Sy = S. leonardii C. Epling
S. parvula A. Michaux var. ***parvula***
S. potosina T. Brandegee subsp. ***parviflora*** C. Epling
S. potosina T. Brandegee subsp. ***potosina*** var. ***davisiana*** B. L. Turner
Ma = S. potosina subsp. platyphylla C. Epling: Texas authors
S. potosina T. Brandegee subsp. ***potosina*** var. ***tessellata*** (C. Epling) B. L. Turner
S. racemosa C. Persoon
Ma = S. minor *auct.*, non W. Hudson: Texas authors
S. resinosa J. Torrey
S. texana B. L. Turner
S. wrightii A. Gray
Sy = S. brevifolia (A. Gray) A. Gray
Sy = S. integrifolia C. Linnaeus var. brevifolia A. Gray
Sy = S. resinosa J. Torrey var. brevifolia (A. Gray) C. Penland

SIDERITIS C. Linnaeus
S. lanata C. Linnaeus

STACHYS C. Linnaeus
S. bigelovii A. Gray
S. byzantina K. Koch *ex* G. Scheele [**cultivated**]
Sy = S. olympica J. Poiret
S. coccinea C. Ortega
Sy = S. cardinalis K. E. O. Kuntze
Sy = S. oxacana M. Fernald
S. crenata C. Rafinesque-Schmaltz
Ma = S. agraria *auct.*, non A. von Chamisso & D. von Schlechtendal
S. drummondii G. Bentham
S. floridana R. Shuttleworth *ex* G. Bentham
Sy = S. sieboldii F. Miquel
S. tenuifolia C. von Willdenow subsp. ***tenuifolia***
Sy = S. tenuifolia var. aspera (A. Michaux) M. Fernald
Sy = S. tenuifolia var. hispida (F. Pursh) M. Fernald
Sy = S. tenuifolia var. platyphylla M. Fernald

TEUCRIUM C. Linnaeus
T. canadense C. Linnaeus var. ***canadense***
Sy = T. canadense var. angustatum A. Gray
Sy = T. canadense var. littorale (E. Bicknell) M. Fernald
T. canadense C. Linnaeus var. ***occidentale*** (A. Gray) E. McClintock & C. Epling
Sy = T. canadense subsp. boreale (E. Bicknell) L. Shinners
Sy = T. canadense subsp. occidentale (A. Gray) W. A. Weber
Sy = T. canadense subsp. viscidum R. L. Taylor & B. MacBryde
Sy = T. canadense var. boreale (E. Bicknell) L. Shinners
Sy = T. occidentale A. Gray
T. cubense N. von Jacquin var. ***cubense***
Sy = T. cubense subsp. chamaedrifolium (P. Miller) C. Epling
Sy = T. cubense subsp. cubense
T. cubense N. von Jacquin var. ***densum*** W. Jepson
Sy = T. cubense subsp. depressum (J. K. Small) E. McClintock & C. Epling
Sy = T. depressum (J. K. Small) E. McClintock & C. Epling
T. cubense N. von Jacquin var. ***laevigatum*** (M. H. Vahl) L. Shinners
Sy = Melosma laevigatum (M. H. Vahl) J. K. Small
Sy = Melosmon laciniatum (J. Torrey) J. K. Small
Sy = Teucrinum cubense subsp. laevigatum (M. H. Vahl) E. McClintock & C. Epling
Sy = T. laciniatum J. Torrey
Sy = T. laevigatum M. H. Vahl

THYMUS C. Linnaeus
T. vulgaris C. Linnaeus [**cultivated**]

TRICHOSTEMA C. Linnaeus
T. arizonicum A. Gray
T. brachiatum C. Linnaeus
Sy = Isanthus brachiatus (C. Linnaeus) N. Britton, E. Sterns, & J. Poggenburg
Sy = I. brachiatus var. linearis N. Fassett
Sy = Tetraclea viscida C. Lundell
T. dichotomum C. Linnaeus
Sy = T. dichotomum var. puberulum M. Fernald & L. Griscom
T. setaceum M. Houttuyn
Sy = T. dichotomum C. Linnaeus var. lineare (T. Walter) F. Pursh

LAURACEAE, *nomen conservandum*

CASSYTHA C. Linnaeus
C. filiformis C. Linnaeus

CINNAMOMUM J. C. von Schaeffer, *nomen conservandum*
C. camphora (C. Linnaeus) J. Presl [**cultivated**]
Sy = C. camphora (C. Linnaeus) G. Karsten

LAURUS C. Linnaeus
L. nobilis C. Linnaeus [**cultivated**]

LINDERA C. Thunberg, *nomen conservandum*
L. benzoin (C. Linnaeus) K. von Blume var. ***pubescens*** (E. Palmer & J. Steyermark) A. Rehder
Sy = Benzoin aestivale (C. Linnaeus) C. Nees von Esenbeck var. pubescens E. Palmer & J. Steyermark

PERSEA P. Miller, *nomen conservandum*
P. americana (C. Linnaeus) P. Miller var. ***americana*** [**cultivated**]
Sy = P. persea (C. Linnaeus) T. Cockerell
P. borbonia (C. Linnaeus) K. Sprengel
Sy = P. littoralis J. K. Small
Sy = Tamala borbonia (C. Linnaeus) C. Rafinesque-Schmaltz
P. humilis G. Nash [**cultivated**]
Sy = P. borbonia (C. Linnaeus) K. Sprengel var. humilis (G. Nash) J. Kopp
P. palustris (C. Rafinesque-Schmaltz) C. Sargent
Sy = P. borbonia (C. Linnaeus) K. Sprengel var. pubescens (F. Pursh) E. Little
Sy = P. pubescens (F. Pursh) C. Sargent
Sy = Tamala pubescens (F. Pursh) J. K. Small

SASSAFRAS C. Nees von Esenbeck & J. Ebermaier
S. albidum (T. Nuttall) C. Nees von Esenbeck
Sy = S. albidum var. molle (C. Rafinesque-Schmaltz) M. Fernald
Sy = S. sassafras (C. Linnaeus) G. Karsten

LEITNERIACEAE, *nomen conservandum*

LEITNERIA A. Chapman
L. floridana A. Chapman

LENTIBULARIACEAE, *nomen conservandum*

PINGUICULA C. Linnaeus
P. pumila A. Michaux
Sy = P. pumila var. buswellii H. Moldenke

UTRICULARIA C. Linnaeus
U. cornuta A. Michaux
Sy = Stomoisia cornuta (A. Michaux) C. Rafinesque-Schmaltz
U. foliosa C. Linnaeus
Sy = U. mixta J. Barnhart
Sy = U. olygosperma A. de Saint-Hilaire
U. gibba C. Linnaeus
Sy = U. alba J. von Hoffmannsegg *ex* J. Link
Sy = U. anomala A. de Saint-Hilaire & F. de Girard
Sy = U. aphylla H. Ruiz López & J. Pavón
Sy = U. biflora J. de Lamarck
Sy = U. bipartita S. Elliott
Sy = U. crenata M. H. Vahl
Sy = U. diantha W. Roxburgh *ex* J. J. Römer & J. A. Schultes
Sy = U. diflora W. Roxburgh
Sy = U. exoleta R. Brown
Sy = U. fibrosa T. Walter
Sy = U. furcata C. Persoon
Sy = U. gibbosa J. Hill
Sy = U. integra J. Le Conte *ex* S. Elliott
Sy = U. obtusa O. Swartz
Sy = U. pumila T. Walter
Sy = U. roxburghii K. Sprengel
Sy = U. tenuis A. Cavanilles
Sy = Vesiculina gibba (C. Linnaeus) C. Rafinesque-Schmaltz
U. inflata T. Walter
Sy = Plectoma inflata (T. Walter) C. Rafinesque-Schmaltz
Sy = P. stellata C. Rafinesque-Schmaltz
Sy = Utricularia ceratophylla A. Michaux
U. juncea M. H. Vahl
Sy = Personula grandiflora C. Rafinesque-Schmaltz
Sy = Stomoisia juncea (M. H. Vahl) J. Barnhart
Sy = Utricularia angulosa J. Poiret
Sy = U. cornuta A. Michaux var. michauxii Gómez
Sy = U. personata J. Le Conte *ex* S. Elliott
Sy = U. sclerocarpa J. Wright *ex* F. Sauvalle
Sy = U. virgatula J. Barnhart
U. macrorhiza J. Le Conte
Sy = Lentibularia vulgaris (C. Linnaeus) C. Moench var. americana (A. Gray) J. Nieuwland & J. Lunell
Sy = Megozipa macrorhiza (J. Le Conte) C. Rafinesque-Schmaltz
Sy = Utricularia grandiflora M. Martens
Sy = U. robbinsii (A. Wood) A. Wood

Sy = U. vulgaris J. Bigelow
Sy = U. vulgaris C. Linnaeus var. americana A. Gray
Sy = U. vulgaris subsp. macrorhiza (J. Le Conte) R. Clausen
Ma = U. vulgaris *auct.*, non C. Linnaeus: Texas authors

* ***U. purpurea*** T. Walter [**TOES: IV**]
Sy = U. violacea W. Barton, non R. Brown
Sy = Vesiculina purpurea (T. Walter) C. Rafinesque-Schmaltz
Sy = V. saccata (S. Elliott) C. Rafinesque-Schmaltz

U. radiata J. K. Small
Sy = U. inflata T. Walter var. minor A. Chapman
Sy = U. inflata var. radiata (J. K. Small) W. Stone

U. striata J. Le Conte *ex* J. Torrey
Sy = Trilobulina striata (J. Le Conte) C. Rafinesque-Schmaltz
Sy = Utricularia fibrosa N. Britton, non T. Walter

U. subulata C. Linnaeus
Sy = Setiscapella subulata (C. Linnaeus) J. Barnhart
Sy = Utricularia bradei F. Markgraf
Sy = U. capillaris J. von Hoffmannsegg *ex* J. J. Römer & J. A. Schultes
Sy = U. cleistogama (A. Gray) N. Britton
Sy = U. filiformis J. J. Römer & J. A. Schultes
Sy = U. media P. Salzmann *ex* A. de Saint-Hilaire *ex* F. de Girard
Sy = U. rendlei C. G. Lloyd
Sy = U. setacea A. Michaux
Sy = U. triloba R. Good

LINACEAE, *nomen conservandum*

LINUM C. Linnaeus

L. alatum (J. K. Small) H. Winkler
Sy = Cathartolinum alatum J. K. Small
Sy = Mesynium alatum (J. K. Small) W. A. Weber

L. aristatum G. Engelmann
Sy = Cathartolinum aristatum (G. Engelmann) J. K. Small
Sy = Mesynium aristatum (G. Engelmann) W. A. Weber

L. australe A. A. Heller var. ***australe***
Sy = Cathartolinum australe (A. A. Heller) J. K. Small
Sy = Linum aristatum G. Engelmann var. australe (A. A. Heller) T. Kearney & R. Peebles
Sy = Mesynium australe (A. A. Heller) W. A. Weber

L. australe A. A. Heller var. ***glandulosum*** C. Rogers
Sy = Mesynium australe (A. A. Heller) J. K. Small var. glandulosum (C. Rogers) W. A. Weber

L. berlandieri W. Hooker var. ***berlandieri***
Sy = Cathartolinum berlandieri (W. Hooker) J. K. Small
Sy = Linum rigidum F. Pursh var. berlandieri (W. Hooker) J. Torrey & A. Gray

L. berlandieri W. Hooker var. ***filifolium*** (L. Shinners) C. Rogers
Sy = L. rigidum F. Pursh var. filifolium L. Shinners

L. compactum A. Nelson
Sy = Cathartolinum compactum (A. Nelson) J. K. Small
Sy = Linum rigidum F. Pursh var. compactum (A. Nelson) C. Rogers

L. elongatum (J. K. Small) H. Winkler
Sy = Cathartolinum elongatum J. K. Small

L. floridanum (J. Planchon) W. Trelease var. ***floridanum***
Sy = Cathartolinum floridanum (J. Planchon) J. K. Small
Sy = Linum virginianum C. Linnaeus var. floridanum J. Planchon

L. hudsonioides J. Planchon
Sy = Cerastium clawsonii D. Correll
Sy = Mesynium hudsonioides (J. Planchon) W. A. Weber

L. imbricatum (C. Rafinesque-Schmaltz) L. Shinners
Sy = Mesynium imbricatum (C. Rafinesque-Schmaltz) W. A. Weber
Sy = Nezera imbricata C. Rafinesque-Schmaltz

L. lewisii F. Pursh var. ***lewisii***
Sy = Adenolinum lewisii (F. Pursh) A. Löve & D. Löve
Sy = Linum perenne C. Linnaeus subsp. lewisii (F. Pursh) O. Hultén
Sy = L. perenne var. lewisii (F. Pursh) A. Eaton & J. Wright

L. lundellii C. Rogers

L. medium (J. Planchon) N. Britton var. ***texanum*** (J. Planchon) M. Fernald
Sy = L. striatum T. Walter var. texanum (J. Planchon) J. Boivin

L. pratense (J. Norton) J. K. Small
Sy = L. lewisii F. Pursh var. pratense J. Norton

L. puberulum (G. Engelmann) A. A. Heller
Sy = Cathartolinum puberulum (G. Engelmann) J. K. Small
Sy = Linum rigidulum F. Pursh var. puberulum G. Engelmann

Sy = L. vestitum E. Wooton & P. Standley
Sy = Mesynium puberulum (G. Engelmann) W. A. Weber
L. rigidum F. Pursh var. ***rigidulum***
Sy = Cathartolinum rigidum (F. Pursh) J. K. Small
Sy = Mesynium rigidum (F. Pursh) A. Löve & D. Löve
L. rupestre (A. Gray) G. Engelmann *ex* A. Gray
Sy = L. boottii J. Planchon var. rupestre A. Gray
L. schiedeanum D. von Schlechtendal & A. von Chamisso
Sy = L. greggii G. Engelmann
L. striatum T. Walter
Sy = Cathartolinum striatum (T. Walter) J. K. Small
Sy = Linum striatum var. multijugum M. Fernald
Sy = Nezera striata (T. Walter) J. Nieuwland
L. sulcatum J. Riddell var. ***sulcatum***
Sy = Cathartolinum sulcatum (J. Riddell) J. K. Small
L. usitatissimum C. Linnaeus
Sy = L. humile P. Miller
Sy = L. usitatissimum var. humile (P. Miller) C. Persoon
L. vernale E. Wooton
Sy = Cathartolinum vernale (E. Wooton) J. K. Small
Sy = C. virginianum (C. Linnaeus) H. G. Reichenbach
Sy = Nezera virginiana (C. Linnaeus) J. Nieuwland

LOASACEAE, *nomen conservandum*

CEVALLIA M. Lagasca y Segura
C. sinuata M. Lagasca y Segura

EUCNIDE J. Zuccarini, *nomen conservandum*
E. bartonioides J. Zuccarini

MENTZELIA C. Linnaeus
M. albescens (J. Gillies & G. Arnott) A. Grisebach
Sy = M. wrightii A. Gray
M. albicaulis (D. Douglas *ex* W. Hooker) D. Douglas *ex* J. Torrey & A. Gray
Sy = Acrolasia albicaulis (D. Douglas *ex* W. Hooker) P. Rydber
Sy = Bartonia albicaulis D. Douglas *ex* W. Hooker
Sy = Mentzelia albicaulis var. ctenophora (P. Rydber) H. St. John
Sy = M. albicaulis var. gracilis J. Darlington
Sy = M. albicaulis var. tenerrima (P. Rydberg) H. St. John
Sy = M. parviflora A. A. Heller
M. asperula E. Wooton & P. Standley
M. decapetala (J. Sims) I. Urban & E. Gilg *ex* E. Gilg
Sy = Bartonia decapetala F. Pursh *ex* J. Sims
Sy = Mentzelia ornata J. Torrey & A. Gray
Sy = Nuttallia decapetala (F. Pursh *ex* J. Sims) E. Greene
M. humilis (A. Gray) J. Darlington
Sy = M. multiflora (T. Nuttall) A. Gray var. humilis A. Gray
Sy = Nuttallia gypsea E. Wooton & P. Standley
Sy = N. humilis (A. Gray) P. Rydberg
M. incisa I. Urban & E. Gilg
M. lindheimeri I. Urban & E. Gilg
M. montana (A. Davidson) A. Davidson
Sy = Acrolasia montana A. Davidson
M. multiflora (T. Nuttall) A. Gray var. ***multiflora***
Sy = M. pumila T. Nuttall *ex* J. Torrey & A. Gray var. multiflora (T. Nuttall) I. Urban & E. Gilg
Sy = Nuttallia multiflora (T. Nuttall) E. Greene
M. nuda (F. Pursh) J. Torrey & A. Gray var. ***stricta*** (G. Osterhout) H. Harrington
Sy = M. stricta (G. Osterhout) R. Jeffs & E. Little
Sy = Nuttallia stricta (G. Osterhout) E. Greene
M. oligosperma T. Nuttall *ex* J. Sims
Sy = M. monosperma E. Wooton & P. Standley
M. pachyrhiza I. M. Johnston
M. reverchonii (I. Urban & E. Gilg) H. Thompson & J. Zavortink
Sy = Nuttallia reverchonii (I. Urban & E. Gilg) W. A. Weber
M. saxicola H. Thompson & J. Zavortink
M. strictissima (E. Wooton & P. Standley) J. Darlington
Sy = Nuttallia strictissima (E. Wooton & P. Standley) J. Darlington
M. texana I. Urban & E. Gilg

LOGANIACEAE, *nomen conservandum*

GELSEMIUM A. L. de Jussieu
G. sempervirens A. de Saint-Hilaire
Sy = Bignonia sempervirens C. Linnaeus

MITREOLA C. Linnaeus
M. petiolata (T. Walter) J. Torrey & A. Gray
Sy = Cynoctonum mitreola (C. Linnaeus) N. Britton
Sy = C. succulentum R. Long
M. sessilifolia (J. F. Gmelin) G. Don
Sy = Cynoctonum sessilifolium (T. Walter) J. F. Gmelin

Sy = C. sessilifolium var. angustifolium J. Torrey & A. Gray
Sy = C. sessilifolium var. microphyllum R. Long

SPIGELIA C. Linnaeus
S. lindheimeri A. Gray
S. marilandica (C. Linnaeus) C. Linnaeus
S. texana (J. Torrey & A. Gray) A. L. de Candolle
Sy = Coelostylis texana J. Torrey & A. Gray

LYTHRACEAE, *nomen conservandum*

AMMANNIA C. Linnaeus
A. auriculata C. von Willdenow
Sy = A. auriculata var. arenaria (K. Kunth) B. Köhne
A. coccinea C. Rottbøll
Sy = A. coccinea subsp. purpurea (J. de Lamarck) B. Köhne
Sy = A. teres C. Rafinesque-Schmaltz
Sy = A. texana G. Scheele
A. latifolia C. Linnaeus
Sy = A. koehnei N. Britton
Sy = A. teres C. Rafinesque-Schmaltz var. exauriculata (M. Fernald) M. Fernald
A. robusta O. von Heer & E. von Regel
Sy = A. coccinea C. Rottbøll subsp. robusta (O. von Heer & E. von Regel) B. Köhne

CUPHEA P. Browne
C. carthagenensis (N. von Jacquin) J. Macbride
Sy = Parsonsia balsamona (A. von Chamisso & D. von Schlechtendal) P. Standley
C. glutinosa A. von Chamisso & D. von Schlechtendal
C. hyssopifolia K. Kunth [**cultivated**]
C. ignea A. P. de Candolle [**cultivated**]
C. micropetala K. Kunth [**cultivated**]

DECODON J. F. Gmelin
D. verticillatus (C. Linnaeus) S. Elliott
Sy = D. verticillatus var. nematopetala R. C. Bacigalupo

DIDIPLIS C. Rafinesque-Schmaltz
D. diandra (T. Nuttall *ex* A. P. de Candolle) A. Wood
Sy = Peplis diandra T. Nuttall *ex* A. P. de Candolle

HEIMIA J. Link
H. salicifolia (K. Kunth) J. Link
Sy = Nesaea salicifolia K. Kunth

LAGERSTROEMIA C. Linnaeus
L. fauriei B. Köhne [**cultivated**]
L. indica C. Linnaeus [**cultivated**]
L. fauriei B. Köhne × ***L. indica*** C. Linnaeus [**cultivated**]

LAWSONIA C. Linnaeus
L. inermis C. Linnaeus [**cultivated**]

LYTHRUM C. Linnaeus
L. alatum F. Pursh var. ***lanceolatum*** (S. Elliott) J. Torrey & A. Gray *ex* J. Rothrock
Sy = L. lanceolatum S. Elliott
L. californicum J. Torrey & A. Gray
L. ovalifolium B. Köhne
L. salicaria C. Linnaeus [**cultivated**]
Sy = L. salicaria var. gracilior P. Turczaninow
Sy = L. salicaria var. tomentosum (P. Miller) A. P. de Candolle

NESAEA P. Commerson *ex* K. Kunth, *nomen conservandum*
N. longipes A. Gray
Sy = Heimia longipes (A. Gray) V. Cory

ROTALA C. Linnaeus
R. ramosior (C. Linnaeus) B. Köhne
Sy = R. ramosior var. interior M. Fernald & L. Griscom

MAGNOLIACEAE, *nomen conservandum*

LIRIODENDRON C. Linnaeus
L. tulipifera C. Linnaeus [**cultivated**]

MAGNOLIA C. Linnaeus
M. ashei C. Weatherby [**cultivated**]
Sy = M. macrophylla A. Michaux subsp. ashei (C. Weatherby) S. Spongberg
M. grandiflora C. Linnaeus
M. pyramidata W. Bartram
Ma = M. fraseri *auct.*, non T. Walter: Texas authors
M. stellata (P. von Siebold & J. Zuccarini) C. Maximowicz [**cultivated**]
Sy = Buergeria stellata P. von Siebold & J. Zuccarini
M. virginiana C. Linnaeus
Sy = M. virginiana var. australis C. Sargent
Sy = M. virginiana var. parva W. Ashe
M. × ***soulangiana*** É. Soulange-Bodin [***denudata*** × ***liliflora***] [**cultivated**]

MICHELIA C. Linnaeus
- ***M. champaca*** C. Linnaeus [**cultivated**]
- ***M. figo*** K. Sprengel [**cultivated**]

MALPIGHIACEAE, *nomen conservandum*

ASPICARPA L. C. Richard
- ***A. hyssopifolia*** A. Gray
- ***A. longipes*** A. Gray
  - Sy = A. humilis (G. Bentham) A. L. de Jussieu

GALPHIMIA A. Cavanilles
- ***G. angustifolia*** G. Bentham
  - Sy = G. linifolia A. Gray
  - Sy = Thryallis angustifolia (G. Bentham) K. E. O. Kuntze

HETEROPTERIS K. Kunth
- ***H. glabra*** W. Hooker & G. Arnott [**cultivated**]

HIRAEA N. von Jacquin
- ***H. septentrionalis*** A. L. de Jussieu [**cultivated**]

JANUSIA A. L. de Jussieu
- ***J. gracilis*** A. Gray

MALPIGHIA C. Linnaeus
- ***M. glabra*** C. Linnaeus

STIGMAPHYLLON A. L. de Jussieu
- ***S. ciliatum*** (J. de Lamarck) A. L. de Jussieu [**cultivated**]
  - Sy = Banisteria ciliata J. de Lamarck

MALVACEAE, *nomen conservandum*

ABELMOSCHUS F. Mekicus
- ***A. esculentus*** (C. Linnaeus) C. Moench [**cultivated**]
  - Sy = Hibiscus esculentus C. Linnaeus

ABUTILON P. Miller
- ***A. abutiloides*** (N. von Jacquin) C. Garcke *ex* N. Britton & Percy Wilson
  - Sy = A. americanum (C. Linnaeus) R. Sweet
  - Sy = A. dentatum J. Rose
  - Sy = A. domingense P. Turczaninow
  - Sy = A. lignosum (A. Cavanilles) G. Don
  - Sy = A. scabrum S. Watson
  - Sy = Lavatera americana C. Linnaeus
  - Sy = Sida abutiloides N. von Jacquin
  - Sy = S. crassifolia L'Héritier de Brutelle
  - Sy = S. lignosa A. Cavanilles
  - Sy = S. tomentosa A. Cavanilles
  - Sy = S. tricuspidata A. Cavanilles
- ***A. berlandieri*** A. Gray *ex* S. Watson
- ***A. fruticosum*** J. Guillemin & G. Perrottet
  - Sy = A. fruticosum var. microphyllum (A. Richard) S. Abedin
  - Sy = A. microphyllum A. Richard
  - Sy = A. nuttallii J. Torrey & A. Gray
  - Sy = A. texense J. Torrey & A. Gray
  - Sy = Sida perrottetiana D. Dietrich
  - Ma = Abutilon incanum *auct.*, non (J. Link) R. Sweet: Texas authors
- ***A. hulseanum*** (J. Torrey & A. Gray) J. Torrey *ex* A. Gray
  - Sy = A. leucophaeum B. P. G. Hochreutiner
  - Sy = Sida hulseana J. Torrey & A. Gray
- ***A. hypoleucum*** A. Gray
  - Sy = A. selerianum O. Ulbrich
  - Sy = A. subsagittatum B. P. G. Hochreutiner
- ***A. malacum*** S. Watson
- ***A. mollicomum*** (C. von Willdenow) R. Sweet
  - Sy = A. sonorae A. Gray
  - Sy = Sida mollicoma C. von Willdenow
  - Sy = S. sericea A. Cavanilles, non P. Miller
- ***A. parvulum*** A. Gray
- ***A. theophrasti*** F. Medikus
  - Sy = A. abutilon (C. Linnaeus) H. Rusby
  - Sy = A. avicennae J. Gaertner
  - Sy = Sida abutilon C. Linnaeus
- ***A. trisulcatum*** (N. von Jacquin) I. Urban
  - Sy = A. nealleyi J. Coulter
  - Sy = A. ramosissimum K. Presl
  - Sy = A. triquetrum (C. Linnaeus) R. Sweet
  - Sy = Bastardia triquetra (C. Linnaeus) S. A. de Morales
  - Sy = Sida trisulcata N. von Jacquin
- ***A. umbellatum*** (C. Linnaeus) R. Sweet
  - Sy = Sida umbellata C. Linnaeus
- ***A. wrightii*** A. Gray

ALCEA C. Linnaeus
- ***A. rosea*** C. Linnaeus [**cultivated**]
  - Sy = Althaea mexicana G. Kunze
  - Sy = A. rosea (C. Linnaeus) A. Cavanilles
  - Sy = A. sinensis A. Cavanilles

ALLOWISSADULA D. Bates
- ***A. holosericea*** (G. Scheele) D. Bates
  - Sy = Abutilon holosericeum G. Scheele
  - Sy = A. marshii P. Standley

Sy = A. velutinum A. Gray, non G. Don
Sy = Wissadula holosericea (G. Scheele) C. Garcke
Sy = W. insignis R. Fries
A. lozanii (J. Rose) D. Bates
Sy = Pseudoabutilon lozanii (J. Rose) R. Fries
Sy = Wissadula lozanii J. Rose

ALYOGYNE F. Alefeld
A. huegelii (S. Endlicher) P. Fryxell [**cultivated**]

ANODA A. Cavanilles
A. crenatiflora C. Ortega
Sy = A. crenatiflora var. glabrata J. Rose
Sy = A. ortegae K. Sprengel
Sy = A. parviflora A. Cavanilles
Sy = Sida crenatiflora (C. Ortega) C. Persoon
Sy = S. ortegae E. von Steudel
A. cristata (C. Linnaeus) D. von Schlechtendal
Sy = A. arizonica A. Gray
Sy = A. arizonica var. digitata A. Gray
Sy = A. brachyantha G. H. L. Reichenbach
Sy = A. cristata var. albiflora B. P. G. Hochreutiner
Sy = A. cristata var. brachyantha (G. H. L. Reichenbach) B. P. G. Hochreutiner
Sy = A. cristata var. digitata (A. Gray) B. P. G. Hochreutiner
Sy = A. hastata A. Cavanilles
Sy = A. lavateroides F. Medikus
Sy = A. populifolia R. Philippi
Sy = A. triloba A. Cavanilles
Sy = Sida cristata C. Linnaeus
A. lanceolata W. Hooker & G. Arnott
Sy = A. wrightii A. Gray
Sy = Sida unidentata D. Dietrich
A. pentaschista A. Gray
Sy = A. extrema B. P. G. Hochreutiner
Sy = A. pentaschista var. obtusior B. Robinson
Sy = Sida integrifolia M. Sessé y Lacasta & J. Mociño
Sy = S. palmeri J. G. Smith
Sy = Sidanoda pentaschista (A. Gray) E. Wooton & P. Standley

BASTARDIA K. Kunth
B. viscosa (C. Linnaeus) K. Kunth
Sy = B. guayaquilensis P. Turczaninow
Sy = B. parvifolia K. Kunth
Sy = Sida circinnata C. von Willdenow *ex* K. Sprengel
Sy = S. foetida A. Cavanilles
Sy = S. magdalenae A. P. de Candolle
Sy = S. pannosa P. Turczaninow
Sy = S. viscosa C. Linnaeus

BATESIMALVA P. Fryxell
* ***B. violacea*** (J. Rose) P. Fryxell [**TOES: V**]
Sy = Gaya violacea J. Rose

BILLIETURNERA P. Fryxell
B. helleri (J. Rose *ex* A. A. Heller) P. Fryxell
Sy = Disella cuneifolia (A. Gray) E. Greene
Sy = Sida cuneifolia A. Gray, non W. Roxburgh
Sy = S. grayana I. Clement
Sy = S. helleri J. Rose *ex* A. A. Heller

CALLIRHOË T. Nuttall
C. alcaeoides (A. Michaux) A. Gray
Sy = Sida alcaeoides A. Michaux
C. involucrata (T. Nuttall *ex* J. Torrey & A. Gray) A. Gray var. ***involucrata***
Sy = C. involucrata var. novomexicana E. Baker
Sy = Malva involucrata J. Torrey & A. Gray
C. involucrata (T. Nuttall *ex* J. Torrey & A. Gray) A. Gray var. ***lineariloba*** (J. Torrey & A. Gray) A. Gray *ex* S. Watson
Sy = C. geranioides J. K. Small
C. leiocarpa R. Martin
C. papaver (A. Cavanilles) A. Gray
C. pedata (T. Nuttall *ex* W. Hooker) A. Gray
Sy = C. digitata T. Nuttall var. stipulata U. Waterfall
* ***C. scabriuscula*** B. Robinson [**TOES: I**]

CIENFUEGOSIA A. Cavanilles
C. drummondii (A. Gray) F. Lewton
Sy = C. sulphurea (K. Kunth) E. Hassler var. glabra C. Garcke
Sy = Fugosia drummondii A. Gray

FRYXELLIA D. Bates
* ***F. pygmaea*** (D. Correll) D. Bates [**TOES: V**]
Sy = Anoda pygmaea D. Correll

GOSSYPIUM C. Linnaeus
G. hirsutum C. Linnaeus [**cultivated**]
Sy = G. hopi F. Lewton
Sy = G. latifolium J. Murray
Sy = G. mexicanum A. Todaro
Sy = G. taitense F. Parlatore

HERISSANTIA F. Medikus
H. crispa (C. Linnaeus) G. Brizicky
Sy = Abutilon crispum (C. Linnaeus) F. Medikus
Sy = A. sessilifolium K. Presl

Sy = Beloëre crispa (C. Linnaeus) R. Shuttleworth *ex* A. Gray
Sy = Bogenhardia crispa (C. Linnaeus) T. Kearney
Sy = Gayoides crispum (C. Linnaeus) J. K. Small
Sy = Pseudobastardia crispa (C. Linnaeus) E. Hassler
Sy = Sida amplexicaulis J. de Lamarck
Sy = S. crispa C. Linnaeus
Sy = S. imberbis A. P. de Candolle
Sy = S. retrofacta A. P. de Candolle
Sy = S. sessilis J. Vellozo

HIBISCUS C. Linnaeus, *nomen conservandum*
H. aculeatus T. Walter
Sy = H. scaber A. Michaux, non J. de Lamarck
H. cannabinus C. Linnaeus [**cultivated**]
H. coccineus T. Walter [**cultivated**]
Sy = H. semilobatus A. Chapman
H. coulteri W. Harvey *ex* A. Gray
Sy = H. coulteri var. brevipedunculatus M. E. Jones
* ***H. dasycalyx*** S. F. Blake & I. Shiller [**TOES: III**]
H. denudatus G. Bentham
Sy = H. denudatus var. involucellatus A. Gray
Sy = H. involucellatus (A. Gray) E. Wooton & P. Standley
H. grandiflorus A. Michaux
H. laevis C. Allioni
Sy = H. militaris A. Cavanilles
H. martianus J. Zuccarini
Sy = H. cardiophyllus A. Gray
H. moscheutos C. Linnaeus subsp. ***lasiocarpos*** (A. Cavanilles) O. Blanchard [***ined.***]
Sy = H. californicus A. Kellogg
Sy = H. langloisii E. Greene
Sy = H. lasiocarpos A. Cavanilles
Sy = H. leucophyllus I. Shiller
Sy = H. platanoides E. Greene
H. mutabilis C. Linnaeus [**cultivated**]
Sy = Abelmoschus mutabilis (C. Linnaeus) C. Moench
Sy = Hibiscus immutabilis F. Dehnhardt
Sy = H. sinensis P. Miller
H. rosa-sinensis C. Linnaeus var. ***rosa-sinensis*** [**cultivated**]
Sy = H. festalis R. A. Salisbury
H. rosa-sinensis C. Linnaeus var. ***schizopetalus*** W. T. Dyer [**cultivated**]
Sy = H. schizopetalus (W. T. Dyer) J. Hooker
H. striatus A. Cavanilles subsp. ***lambertianus*** (K. Kunth) O. Blanchard *ex* G. Proctor
Sy = H. cubensis A. Richard
Sy = H. lambertianus K. Kunth
Sy = H. sagraeanus E. Mercier
Sy = H. salviifolius A. de Saint-Hilaire
H. syriacus C. Linnaeus [**cultivated**]
Sy = H. acerifolius R. A. Salisbury, non A. P. de Candolle
Sy = H. floridus R. A. Salisbury
Sy = H. rhombifolius A. Cavanilles
Sy = Ketmia syriaca (C. Linnaeus) J. Scopoli
H. tiliaceus C. Linnaeus [**cultivated**]
Sy = Pariti grande N. Britton *ex* J. K. Small
Sy = P. tiliaceum (C. Linnaeus) A. L. de Jussieu *ex* N. Britton & C. Millspaugh
H. trionum C. Linnaeus
Sy = Trionum trionum (C. Linnaeus) E. Wooton & P. Standley

KOSTELETZKYA K. Presl, *nomen conservandum*
K. depressa (C. Linnaeus) O. Blanchard, P. Fryxell, & D. Bates
Sy = Hibiscus pentaspermus C. Bertero *ex* A. P. de Candolle, non T. Nuttall
Sy = H. tampicensis M. Moricand *ex* N. Séringe
Sy = Kosteletzkya asterocarpa P. Turczaninow
Sy = K. cordata K. Presl
Sy = K. hastata K. Presl
Sy = K. hispida K. Presl
Sy = K. sagittata K. Presl
Sy = K. stellata M. Fernald
Sy = K. violacea J. Rose
Sy = Melochia depressa C. Linnaeus
Sy = M. diffusa C. Bertero *ex* L. Colla
Sy = Riedleia depressa (C. Linnaeus) A. P. de Candolle
Sy = Sida carnea A. P. de Candolle
Sy = Visenia depressa (C. Linnaeus) K. Sprengel
K. virginica (C. Linnaeus) K. Presl *ex* A. Gray
Sy = K. althaeifolia (A. Chapman) A. Gray
Sy = K. virginica var. althaeifolia A. Chapman
Sy = K. virginica var. aquilonia M. Fernald

KRAPOVICKASIA P. Fryxell
K. physaloides (K. Presl) P. Fryxell
Sy = Sida physaloides K. Presl
Sy = S. standleyi I. Clement

LAVATERA C. Linnaeus
L. trimestris C. Linnaeus [**cultivated**]
Sy = Stegia lavatera A. P. de Candolle

MALACHRA C. Linnaeus
M. capitata (C. Linnaeus) C. Linnaeus
Sy = M. mexicana H. A. Schrader
Sy = M. palmata C. Moench

Sy = Sida capitata C. Linnaeus
Sy = Urena capitata (C. Linnaeus) M. Gómez

MALVA C. Linnaeus
M. neglecta C. Wallroth
M. parviflora C. Linnaeus
M. rotundifolia C. Linnaeus
Sy = M. borealis C. Wallroth
M. sylvestris C. Linnaeus
Sy = M. sylvestris subsp. mauritiana (C. Linnaeus) A. Thellung
Sy = M. sylvestris var. mauritiana (C. Linnaeus) P. Boissier

MALVASTRUM A. Gray, *nomen conservandum*
M. americanum (C. Linnaeus) J. Torrey
Sy = M. americana C. Linnaeus
Sy = Malveopsis americana (C. Linnaeus) K. E. O. Kuntze
Sy = Sphaeralcea americana (C. Linnaeus) M. C. Metz
M. aurantiacum (G. Scheele) W. Walpers
Sy = Malva aurantiaca G. Scheele
Sy = Malvastrum wrightii A. Gray
M. coromandelianum (C. Linnaeus) C. Garcke
Sy = Malva coromandeliana C. Linnaeus
Sy = M. lindheimeriana G. Scheele
Sy = M. tricuspidata R. Brown *ex* W. T. Aiton
Sy = Malvastrum tricuspidatum (R. Brown *ex* W. T. Aiton) A. Gray
Sy = Malveopsis coromandeliana (C. Linnaeus) T. Morong

MALVAVISCUS P. Fabricius
M. drummondii J. Torrey & A. Gray
Sy = M. arboreus J. Dillenius *ex* A. Cavanilles var. drummondii (J. Torrey & A. Gray) R. Schery
Ma = M. arboreus *auct.*, non J. Dillenius *ex* A. Cavanilles: Texas authors
Ma = M. arboreus var. mexicanus *auct.*, non D. von Schlechtendal: Texas authors
M. penduliflorus A. P. de Candolle [**cultivated**]
Sy = M. arboreus J. Dillenius *ex* A. Cavanilles subsp. penduliflorus (A. P. de Candolle) E. Hadac
Sy = M. arboreus var. longifolius (C. Garcke) R. Schery
Sy = M. arboreus var. penduliflorus (A. P. de Candolle) R. Schery
Sy = M. longifolius C. Garcke

MALVELLA H. Jaubert & E. Spach
M. lepidota (A. Gray) P. Fryxell
Sy = Disella lepidota (A. Gray) E. Greene
Sy = Sida lepidota A. Gray
Sy = S. lepidota var. depauperata A. Gray
Sy = S. leprosa (C. Ortega) K. Schumann var. depauperata (A. Gray) I. Clement
M. leprosa (C. Ortega) A. Krapovickas
Sy = Malva leprosa C. Ortega
Sy = Sida hederacea (D. Douglas *ex* W. Hooker) J. Torrey *ex* A. Gray
Sy = S. leprosa (C. Ortega) K. Schumann
M. sagittifolia (A. Gray) P. Fryxell
Sy = Sida lepidota A. Gray var. sagittifolia A. Gray
Sy = S. sagittifolia (A. Gray) P. Rydberg

MEXIMALVA P. Fryxell
M. filipes (A. Gray) P. Fryxell
Sy = Sida filipes A. Gray

MODIOLA C. Moench
M. caroliniana (C. Linnaeus) G. Don
Sy = Abutilodes carolinianum (C. Linnaeus) K. E. O. Kuntze
Sy = Malva caroliniana C. Linnaeus
Sy = Modanthos caroliniana (C. Linnaeus) F. Alefeld
Sy = Modiola multifida C. Moench

PAVONIA A. Cavanilles, *nomen conservandum*
P. lasiopetala G. Scheele
Sy = Malache lasiopetala (G. Scheele) K. E. O. Kuntze
Sy = M. wrightii (A. Gray) K. E. O. Kuntze
Sy = Pavonia wrightii A. Gray

RHYNCHOSIDA P. Fryxell
R. physocalyx (A. Gray) P. Fryxell
Sy = Physaliastrum physocalyx (A. Gray) H. C. Monteiro
Sy = Sida hastata A. de Saint-Hilaire
Sy = S. inflata D. Larrañaga
Sy = S. physocalyx A. Gray

SIDA C. Linnaeus
S. abutifolia P. Miller
Sy = S. diffusa F. Humboldt, A. Bonpland, & K. Kunth
Sy = S. editorum M. Gandoger
Sy = S. filicaulis J. Torrey & A. Gray
Sy = S. filicaulis var. setosa A. Gray
Sy = S. filiformis M. Moricand *ex* N. Séringe

Sy = S. procumbens O. Swartz
Sy = S. supina L'Héritier de Brutelle

S. ciliaris C. Linnaeus
Sy = Malvastrum linearifolium S. Buckley
Sy = Pseudomalachra ciliaris (C. Linnaeus) H. C. Monteiro
Sy = Sida anomala A. de Saint-Hilaire var. mexicana (M. Moricand) L. Shinners
Sy = S. fasciculata J. Torrey & A. Gray, non C. von Willdenow
Sy = S. muricata A. Cavanilles

S. cordifolia C. Linnaeus
Sy = S. althaeifolia O. Swartz
Sy = S. conferta J. Link, non P. Salzmann *ex* J. Triana & J. Planchon
Sy = S. cordifolia var. althaeifolia (O. Swartz) A. Grisebach
Sy = S. decagyna C. Schumacher & P. Thonning *ex* C. Schumacher
Sy = S. holosericea C. von Willdenow *ex* K. Sprengel
Sy = S. micans A. Cavanilles
Sy = S. pellita K. Kunth
Sy = S. pungens K. Kunth

S. elliottii J. Torrey & A. Gray
Sy = S. gracilis S. Elliott, non A. Richard
Sy = S. leptophylla J. K. Small

S. lindheimeri G. Engelmann & A. Gray
Sy = S. elliottii J. Torrey & A. Gray var. texana J. Torrey & A. Gray

S. longipes A. Gray

S. neomexicana A. Gray

S. rhombifolia C. Linnaeus
Sy = Malva rhombifolia (C. Linnaeus) J. W. Krause
Sy = Napaea rhombifolia (C. Linnaeus) C. Moench

S. spinosa C. Linnaeus
Sy = Malvinda spinosa (C. Linnaeus) C. Moench
Sy = Sida heterocarpa G. Engelmann *ex* A. Gray

S. tragiifolia A. Gray

SIDASTRUM E. Baker

S. paniculatum (C. Linnaeus) P. Fryxell
Sy = Sida capillaris A. Cavanilles
Sy = S. floribunda K. Kunth
Sy = S. paniculata C. Linnaeus

SPHAERALCEA A. de Saint-Hilaire

S. angustifolia (A. Cavanilles) G. Don subsp. ***angustifolia***
Sy = Malva angustifolia A. Cavanilles
Sy = Sphaeroma angustifolia (A. Cavanilles) G. Don
Sy = S. angustifolium (A. Cavanilles) D. von Schlechtendal
Sy = S. stellata J. Torrey

S. angustifolia (A. Cavanilles) G. Don subsp. ***cuspidata*** (A. Gray) T. Kearney
Sy = S. angustifolia var. cuspidata A. Gray

S. angustifolia (A. Cavanilles) G. Don subsp. ***lobata*** (E. Wooton) T. Kearney
Sy = S. angustifolia var. lobata (E. Wooton) T. Kearney
Sy = S. angustifolia var. oblongifolia (A. Gray) L. Shinners
Sy = S. lobata E. Wooton

S. coccinea (T. Nuttall) P. Rydberg
Sy = Malva coccinea T. Nuttall
Sy = Malvastrum coccineum (T. Nuttall) A. Gray
Sy = Sphaeralcea coccinea subsp. coccinea
Sy = S. coccinea subsp. elata (E. Baker) T. Kearney
Sy = S. coccinea var. dissecta (T. Nuttall *ex* J. Torrey) T. Kearney
Sy = S. coccinea var. elata (E. Baker) T. Kearney

S. digitata (E. Greene) P. Rydberg var. ***angustiloba*** (A. Gray) L. Shinners
Sy = S. digitata subsp. tenuipes (E. Wooton & P. Standley) T. Kearney
Sy = S. pedata var. angustiloba A. Gray
Sy = S. tenuipes E. Wooton & P. Standley

S. fendleri A. Gray var. ***fendleri***
Sy = S. leiocarpa E. Wooton & P. Standley

S. hastulata A. Gray
Sy = S. arenaria E. Wooton & P. Standley
Sy = S. simulans E. Wooton & P. Standley
Sy = S. subhastata J. Coulter

S. incana J. Torrey *ex* A. Gray var. ***incana***

S. laxa E. Wooton & P. Standley
Sy = S. ribifolia E. Wooton & P. Standley

S. leptophylla (A. Gray) P. Rydberg
Sy = Malvastrum leptophyllum A. Gray

S. lindheimeri A. Gray

S. pedatifida (A. Gray) A. Gray
Sy = Malvastrum pedatifidum A. Gray
Sy = Sidalcea atacosa S. Buckley

S. polychroma J. La Duke

S. wrightii A. Gray

THESPESIA D. Solander *ex* J. F. Corrêa da Serra, *nomen conservandum*

T. populnea (C. Linnaeus) D. Solander *ex* J. F. Corrêa da Serra [**cultivated**]

Sy = Hibiscus bacciferus K. von Blume
Sy = H. blumei K. E. O. Kuntze
Sy = H. populifolius R. A. Salisbury
Sy = H. populneus C. Linnaeus
Sy = Malvaviscus populneus (C. Linnaeus) J. Gaertner
Sy = Thespesia macrophylla K. von Blume

WISSADULA F. Medikus
W. amplissima (C. Linnaeus) R. Fries
Sy = Abutilon amplissimum (C. Linnaeus) K. E. O. Kuntze
Sy = A. mucronulatum (A. Gray *ex* J. Torrey) A. Gray
Sy = Sida amplissima C. Linnaeus
Sy = Wissadula mucronulata A. Gray *ex* J. Torrey
W. periplocifolia (C. Linnaeus) K. Presl *ex* G. Thwaites
Sy = Abutilon periplocifolium (C. Linnaeus) R. Sweet
Sy = Sida periplocifolia C. Linnaeus
Sy = Wissadula periplocifolia var. gracillima R. Fries

MELASTOMATACEAE, *nomen conservandum*

RHEXIA C. Linnaeus
R. alifanus T. Walter
Sy = R. glabella A. Michaux
R. lutea T. Walter
R. mariana C. Linnaeus var. ***interior*** (F. Pennell) R. Kral & P. Bostick
Sy = R. interior F. Pennell
R. mariana C. Linnaeus var. ***mariana***
Sy = R. mariana var. exalbida A. Michaux
Sy = R. mariana var. leiosperma M. Fernald & L. Griscom
R. petiolata T. Walter
Sy = R. ciliosa A. Michaux
R. virginica C. Linnaeus
Sy = R. virginica var. purshii (K. Sprengel) C. James
Sy = R. virginica var. septemnervia (T. Walter) F. Pursh

TIBOUCHINA J. Aublet
T. urvilleana (A. P. de Candolle) C. Cogniaux [**cultivated**]
Ma = T. semidecandra *auct.*, non (A. P. de Candolle) C. Cogniaux

MELIACEAE, *nomen conservandum*

MELIA C. Linnaeus
M. azedarach C. Linnaeus

MENISPERMACEAE, *nomen conservandum*

CALYCOCARPUM (T. Nuttall) E. Spach
C. lyonii (F. Pursh) A. Gray

COCCULUS A. P. de Candolle, *nomen conservandum*
C. carolinus (C. Linnaeus) A. P. de Candolle
Sy = Cebatha carolina (C. Linnaeus) N. Britton
Sy = Epibaterium carolinum (C. Linnaeus) N. Britton
Sy = Menispermum carolinum C. Linnaeus
C. diversifolius A. P. de Candolle
Sy = Cebatha diversifolia (A. P. de Candolle) K. E. O. Kuntze
Sy = Cocculus oblongifolius A. P. de Candolle

MENISPERMUM C. Linnaeus
M. canadense C. Linnaeus
Sy = M. mexicanum J. Rose

MENYANTHACEAE, *nomen conservandum*

NYMPHOIDES J. Hill
N. aquatica (J. F. Gmelin) K. E. O. Kuntze
N. peltata (J. G. Gmelin) K. E. O. Kuntze
Sy = Limnanthemum peltatum J. G. Gmelin
Sy = Nymphoides nymphaeoides (C. Linnaeus) N. Britton

MOLLUGINACEAE, *nomen conservandum*

GLINUS C. Linnaeus
G. lotoides C. Linnaeus
Sy = Mollugo lotoides (C. Linnaeus) C. B. Clarke
G. radiatus (H. Ruiz López & J. Pavón) P. Rohrbach
Sy = G. cambessidesii E. Fenzl
Sy = Mollugo radiata H. Ruiz López & J. Pavón
Sy = M. glinoides J. Cambessèdes

MOLLUGO C. Linnaeus
M. cerviana (C. Linnaeus) N. Séringe
Sy = Pharnaceum cerviana C. Linnaeus
M. verticillata C. Linnaeus
Sy = M. berteriana N. Séringe

MONOTROPACEAE, *nomen conservandum*

MONOTROPA C. Linnaeus
M. *hypopithys* C. Linnaeus
Sy = Hypopithys latisquama P. Rydberg
Sy = Monotropa hypopithys subsp. lanuginosa (A. Michaux) H. Hara
Sy = M. hypopithys var. americana (A. P. de Candolle) K. Domin
Sy = M. hypopithys var. latisquama (P. Rydberg) T. Kearney & R. Peebles
Sy = M. hypopithys var. rubra (J. Torrey) O. Farwell
Sy = M. latisquama (P. Rydberg) Shiu-ying Hu
M. *uniflora* C. Linnaeus
Sy = M. brittonii J. K. Small

PTEROSPORA T. Nuttall
P. andromedea T. Nuttall

MORACEAE, *nomen conservandum*

BROUSSONETIA L'Héritier de Brutelle *ex* E. Ventenat, *nomen conservandum*
B. papyrifera (C. Linnaeus) L'Héritier de Brutelle *ex* E. Ventenat [**cultivated**]
Sy = Morus papyrifera C. Linnaeus
Sy = Papyrius papyriferus (C. Linnaeus) K. E. O. Kuntze

FATOUA J. Gaudin
F. villosa (C. Thunberg) T. Nakai

FICUS C. Linnaeus
F. carica C. Linnaeus [**cultivated**]
F. pumila C. Linnaeus [**cultivated**]

MACLURA T. Nuttall, *nomen conservandum*
M. pomifera (C. Rafinesque-Schmaltz) C. Schneider
Sy = Ioxylon pomiferum C. Rafinesque-Schmaltz
Sy = Toxylon pomiferum C. Rafinesque-Schmaltz *ex* C. Sargent

MORUS C. Linnaeus
M. alba C. Linnaeus
Sy = M. alba var. tatarica (C. Linnaeus) N. Séringe
M. microphylla S. Buckley
Sy = M. confinis E. Greene
Sy = M. crataegifolia E. Greene
Sy = M. grisea E. Greene
Sy = M. microphylyra E. Greene
Sy = M. radulina E. Greene
M. nigra C. Linnaeus [**cultivated**]
M. rubra C. Linnaeus var. ***rubra***
M. rubra C. Linnaeus var. ***tomentosa*** (C. Rafinesque-Schmaltz) E. Bureau

MYRICACEAE, *nomen conservandum*

MORELLA J. de Loureiro
M. caroliniensis (P. Miller) J. K. Small
Sy = Cerothamnus caroliniensis (P. Miller) I. Tidestrom
Sy = Myrica caroliniensis P. Miller
Sy = M. heterophylla C. Rafinesque-Schmaltz
Sy = M. heterophylla var. curtissii (E. Chevalier) M. Fernald
M. cerifera (C. Linnaeus) J. K. Small
Sy = Cerothamnus ceriferus (C. Linnaeus) J. K. Small
Sy = C. pumilus (A. Michaux) J. K. Small
Sy = Myrica cerifera C. Linnaeus
Sy = M. cerifera var. pumila A. Michaux
Sy = M. pusilla C. Rafinesque-Schmaltz

MYRSINACEAE, *nomen conservandum*

ARDISIA O. Swartz, *nomen conservandum*
A. crispa (C. Thunberg) A. P. de Candolle [**cultivated**]
A. japonica (C. Thunberg) K. von Blume [**cultivated**]

MYRSINE C. Linnaeus
M. africana C. Linnaeus [**cultivated**]

MYRTACEAE, *nomen conservandum*

ACCA O. Berg
A. sellowiana (O. Berg) K. Burret [**cultivated**]
Sy = Feijoa sellowiana O. Berg

CALLISTEMON R. Brown
C. citrinus (M. Curtis) O. Stapf [**cultivated**]
Sy = C. lanceolatus (O. Swartz) A. P. de Candolle

EUCALYPTUS L'Héritier de Brutelle
E. camaldulensis F. Dehnhardt [**cultivated**]
Sy = E. rostrata D. von Schlechtendal
E. cinerea F. von Mueller *ex* F. Miquel [**cultivated**]
E. gunnii J. Hooker [**cultivated**]

E. leucoxylon F. von Mueller **[cultivated]**
E. pulverulenta J. Sims **[cultivated]**
E. tereticornis J. E. Smith **[cultivated]**
Sy = E. umbellatus J. Gaertner
E. viminalis J. Houtton de Labillardière **[cultivated]**
E. × ***algeriensis*** P. Taubert [***rostrata*** × ***viminalis***] **[cultivated]**

EUGENIA C. Linnaeus
E. myrtifolia J. Cambessèdes **[cultivated]**

MELALEUCA C. Linnaeus, *nomen conservandum*
M. armillaris (D. Solander *ex* J. Gaertner) J. E. Smith **[cultivated]**
M. elliptica J. Houtton de Labillardière **[cultivated]**
M. quinquenervia (A. Cavanilles) S. T. Blake **[cultivated] [federal noxious weed]**
Ma = M. leucadendron *auct.*, non (C. Linnaeus) C. Linnaeus: American authors

MYRTUS J. de Tournefort *ex* C. Linnaeus
M. communis C. Linnaeus **[cultivated]**

PSIDIUM C. Linnaeus
P. longipes (O. Berg) R. McVaugh **[cultivated]**
Sy = Eugenia longipes O. Berg
Sy = Morsiera longipes (O. Berg) J. K. Small
Sy = Psidium littorale G. Raddi var. longipes (O. Berg) F. Fosberg

NELUMBONACEAE, *nomen conservandum*

NELUMBO M. Adanson
N. lutea C. von Willdenow
Sy = Nelumbium luteum C. von Willdenow

NYCTAGINACEAE, *nomen conservandum*

ABRONIA A. L. de Jussieu
A. ameliae C. Lundell
A. angustifolia E. Greene
Sy = A. angustifolia var. arizonica (P. Standley) T. Kearney & R. Peebles
Sy = A. torreyi P. Standley
A. carletoni J. Coulter & E. Fisher
A. fragrans T. Nuttall *ex* W. Hooker
Sy = A. fragrans var. glaucescens A. Nelson
* ***A. macrocarpa*** L. Galloway **[TOES: I]**

ACLEISANTHES A. Gray
A. acutifolia P. Standley
A. anisophylla A. Gray
* ***A. crassifolia*** A. Gray **[TOES: V]**
A. longiflora A. Gray
A. obtusa (J. Choisy) P. Standley
Sy = A. berlandieri A. Gray
Sy = A. greggii P. Standley
* ***A. wrightii*** (A. Gray) G. Bentham & J. Hooker *ex* W. Hemsley **[TOES: V]**
Sy = Pentacrophys wrightii A. Gray

ALLIONIA C. Linnaeus, *nomen conservandum*
A. choisyi P. Standley
Sy = A. glabra (J. Choisy) P. Standley, non (S. Watson) K. E. O. Kuntze
Sy = A. incarnata C. Linnaeus var. glabra J. Choisy
Sy = Wedelia glabra P. Standley
Sy = Wedeliella glabra (P. Standley) T. Cockerell
A. incarnata C. Linnaeus var. ***incarnata***
Sy = Wedelia cristata P. Standley
Sy = W. incarnata C. Linnaeus
Sy = W. incarnata (C. Linnaeus) K. E. O. Kuntze subsp. anodonta P. Standley

AMMOCODON P. Standley
A. chenopodioides (A. Gray) P. Standley
Sy = Selinocarpus cheopodioides A. Gray

ANULOCAULIS P. Standley
A. eriosolenus (A. Gray) P. Standley
Sy = Boerhavia eriosolena A. Gray
A. gypsogenus U. Waterfall
* ***A. leiosolenus*** (J. Torrey) P. Standley var. ***lasianthus*** I. M. Johnston **[TOES: V]**
A. leiosolenus (J. Torrey) P. Standley var. ***leiosolenus***
Sy = Boerhavia leiosolena J. Torrey
A. reflexus I. M. Johnston

BOERHAVIA C. Linnaeus
B. anisophylla J. Torrey
B. coccinea P. Miller
Sy = B. ixodes P. Standley
Sy = B. ramulosa M. E. Jones
Sy = B. viscosa M. Lagasca y Segura & J. D. Rodríguez var. oligadena A. Heimerl
B. diffusa C. Linnaeus
Sy = B. caribaea N. von Jacquin
B. erecta C. Linnaeus
Sy = B. erecta var. thornberi (M. E. Jones) P. Standley
Sy = B. thornberi M. E. Jones
B. gracillima A. Heimerl
Sy = B. organensis P. Standley

B. intermedia M. E. Jones
Sy = B. erecta C. Linnaeus var. intermedia (M. E. Jones) T. Kearney & R. Peebles
Sy = B. lateriflora P. Standley
Sy = B. universitatis P. Standley
B. linearifolia A. Gray
Sy = B. lindheimeri P. Standley
Sy = B. tenuifolia A. Gray
* ***B. mathesiana*** S. B. Jones [**TOES: III**]
B. purpurascens A. Gray
B. scandens C. Linnaeus
Sy = Commicarpus scandens (C. Linnaeus) P. Standley
B. spicata J. Choisy
Sy = B. palmeri S. Watson
Sy = B. torreyana (S. Watson) P. Standley
B. wrightii A. Gray
Sy = B. bracteosa S. Watson

BOUGAINVILLEA P. Commerson *ex* A. L. de Jussieu, *nomen conservandum*
B. glabra J. Choisy [**cultivated**]
B. spectabilis C. von Willdenow [**cultivated**]
Sy = B. bracteata C. Persoon
Sy = B. virescens J. Choisy
Sy = Tricycla spectabilis J. Poiret
B. × buttiana G. Holt & P. Standley [***glabra*** × ***peruviana*** F. von Humboldt & A. Bonpland] [**cultivated**]

CYPHOMERIS P. Standley
C. crassifolia (P. Standley) P. Standley
Sy = Senkenbergia crassifolia P. Standley
C. gypsophiloides (M. Martens & H. Galeotti) P. Standley
Sy = Boerhavia gibbosa J. Pavón *ex* J. Choisy
Sy = B. gypsophiloides (M. Martens & H. Galeotti) J. Coulter
Sy = Lindenia gypsophiloides M. Martens & H. Galeotti
Sy = Senkenbergia annulata J. Schauer
Sy = S. gypsophiloides (M. Martens & H. Galeotti) G. Bentham & J. Hooker
Sy = Tinantina gypsophiloides (M. Martens & H. Galeotti) M. Martens & H. Galeotti

MIRABILIS C. Linnaeus
M. albida (T. Walter) A. Heimerl
Sy = Allionia albida T. Walter
Sy = A. bracteata P. Rydberg
Sy = Mirabilis albida var. lata L. Shinners
Sy = M. albida var. uniflora A. Heimerl
Sy = M. coahuilensis (P. Standley) P. Standley
Sy = M. dumetorum L. Shinners
Sy = M. eutricha L. Shinners
Sy = M. grayana (P. Standley) P. Standley
Sy = M. oblongifolia (A. Gray) A. Heimerl
Sy = M. pauciflorus (S. Buckley) P. Standley
Sy = M. pseudaggregata A. Heimerl
Sy = M. rotata (P. Standley) I. M. Johnston
Sy = Oxybaphus albidus (T. Walter) R. Sweet
M. austrotexana B. L. Turner
M. comata (J. K. Small) P. Standley
Sy = Allionia comata J. K. Small
Sy = Oxybaphus comatus (J. K. Small) C. Weatherby
M. gigantea (P. Standley) L. Shinners
Sy = Allionia gigantea P. Standley
Sy = Oxybaphus giganteus (P. Standley) C. Weatherby
M. glabra (S. Watson) P. Standley
Sy = Allionia glabra (S. Watson) K. E. O. Kuntze
Sy = Mirabilis carletonii (P. Standley) P. Standley
Sy = M. ciliata (P. Standley) P. Standley
Sy = M. exaltata (P. Standley) P. Standley
Sy = Oxybaphus glabrus S. Watson
M. hirsuta (F. Pursh) C. MacMillan
Sy = Allionia hirsuta F. Pursh
Sy = Oxybaphus hirsutus (F. Pursh) R. Sweet
M. jalapa C. Linnaeus [**cultivated**]
Sy = M. lindheimeri (P. Standley) L. Shinners
M. linearis (F. Pursh) A. Heimerl
Sy = Allionia boldinii (J. Holzinger) A. Heimerl
Sy = A. gausapoides P. Standley
Sy = A. linearis F. Pursh
Sy = A. pinetorum P. Standley
Sy = Mirabilis decumbens (T. Nuttall) F. Daniels
Sy = M. diffusa (A. A. Heller) C. F. Reed
Sy = M. gausapoides (P. Standley) P. Standley
Sy = M. hirsuta (F. Pursh) C. MacMillan var. linearis (F. Pursh) J. Boivin
Sy = M. lanceolatus (P. Rydberg) B. Robinson
Sy = M. linearis (F. Pursh) B. Robinson
Sy = M. linearis var. subhispida A. Heimerl
Sy = Oxybaphus linearis (F. Pursh) B. Robinson
Sy = O. linearis var. subhispidus (A. Heimerl) W. A. Dayton
M. longiflora C. Linnaeus var. ***wrightiana*** (N. Britton & T. Kearney) T. Kearney & R. Peebles
Sy = M. suaveolens K. Kunth
Sy = M. tubiflora E. Fries *ex* A. Heimerl
Sy = M. wrightiana A. Gray *ex* N. Britton & T. Kearney

M. multiflora (J. Torrey) A. Gray
Sy = Oxybaphus multiflora J. Torrey
Sy = Quamoclidion cordifolium G. Osterhout
M. nyctaginea (A. Michaux) C. MacMillan
Sy = Allionia nyctaginea A. Michaux
* Sy = Mirabilis collina L. Shinners [**TOES: V**]
Sy = Oxybaphus nyctagineus (A. Michaux) R. Sweet
M. oxybaphoides (A. Gray) A. Gray
Sy = Allioniella oxybaphoides (A. Gray) P. Rydberg
M. texensis (J. Coulter) B. L. Turner

NYCTAGINIA J. Choisy
N. capitata J. Choisy
Sy = N. cockerellae A. Nelson

PISONIA C. Linnaeus
P. aculeata C. Linnaeus
Sy = P. aculeata var. macranthocarpa J. D. Smith

SELINOCARPUS A. Gray
S. angustifolius J. Torrey
S. diffusus A. Gray
S. lanceolatus E. Wooton var. ***lanceolatus***
S. maloneanus B. L. Turner
S. parvifolius (J. Torrey) P. Standley

TRIPTEROCALYX W. Hooker *ex* P. Standley
T. carnea (E. Greene) L. Galeotti var. ***carnea***
Sy = Abronia carnea E. Greene

NYMPHAEACEAE, *nomen conservandum*

NUPHAR J. E. Smith, *nomen conservandum*
N. lutea (C. Linnaeus) J. E. Smith subsp. ***advena*** (W. Aiton) J. Kartesz & K. Gandhi
Sy = N. advena (W. Aiton) W. T. Aiton var. advena
Sy = N. advena var. tomentosa J. Torrey & A. Gray
Sy = N. fluviatilis (R. Harper) P. Standley
Sy = N. lutea subsp. macrophyllum (J. K. Small) E. Beal
Sy = N. lutea subsp. ozarkana (P. Miller & P. Standley) E. Beal
Sy = N. microcarpa (P. Miller & P. Standley) P. Standley
Sy = N. ovata (P. Miller & P. Standley) P. Standley
Sy = N. puteora M. Fernald
Sy = Nymphaea advena W. Aiton
Sy = N. chartacea P. Miller & P. Standley
Sy = N. fluviatilis R. Harper
Sy = N. macrophylla J. K. Small
Sy = Nymphozanthus advena (W. Aiton) M. Fernald
Sy = N. ozarkanus (P. Miller & P. Standley) E. Palmer & J. Steyermark

NYMPHAEA C. Linnaeus, *nomen conservandum*
N. ampla (R. A. Salisbury) A. P. de Candolle
N. elegans W. Hooker
Sy = Castalia elegans (W. Hooker) E. Greene
N. mexicana J. Zuccarini
Sy = Castalia flava (E. Leitner) E. Greene
N. odorata W. Aiton var. ***odorata***
Sy = Castalia lekophylla J. K. Small
Sy = C. minor (J. Sims) A. P. de Candolle
Sy = C. odorata (W. Aiton) A. Wood
Sy = C. reniformis A. P. de Candolle
Sy = C. tuberosa (J. Paine) E. Greene
Sy = Nymphaea minor (J. Sims) A. P. de Candolle
Sy = N. odorata var. gigantea C. Tricker
Sy = N. odorata var. godfreyi G. Ward
Sy = N. odorata var. maxima (S. Conrad) J. Boivin
Sy = N. odorata var. minor J. Sims
Sy = N. odorata var. rosea F. Pursh
Sy = N. odorata var. stenopetala M. Fernald
Sy = N. odorata var. villosa J. Caspary
Sy = N. tuberosa J. Paine

NYSSACEAE, *nomen conservandum*

NYSSA C. Linnaeus
N. aquatica C. Linnaeus
Sy = N. uniflora F. von Wangenheim
N. biflora T. Walter
Sy = N. sylvatica H. Marshall var. biflora (T. Walter) C. Sargent
Sy = N. ursina J. K. Small
N. sylvatica H. Marshall
Sy = N. sylvatica var. caroliniana (J. Poiret) M. Fernald
Sy = N. sylvatica var. dilatata M. Fernald

OLEACEAE, *nomen conservandum*

CHIONANTHUS C. Linnaeus
C. retusus J. Lindley & J. Paxton [**cultivated**]
Sy = C. chinensis C. Maximowicz
Sy = C. serrulatus B. Hayata

C. virginicus C. Linnaeus
Sy = C. virginicus var. maritimus F. Pursh

FORESTIERA J. Poiret, *nomen conservandum*
F. acuminata (A. Michaux) J. Poiret
Sy = Adelina acuminata A. Michaux
Sy = Forestiera acuminata var. vestita E. Palmer
F. angustifolia J. Torrey
Sy = F. puberula A. Eastwood
Sy = F. texana V. Cory var. palmeri V. Cory
Sy = F. texana var. texana
F. ligustrina (A. Michaux) J. Poiret
Sy = Adelia ligustrina A. Michaux
Sy = Forestiera autumnalis (A. Michaux) J. Poiret
F. pubescens T. Nuttall var. ***glabrifolia*** L. Shinners
F. pubescens T. Nuttall var. ***pubescens***
Sy = F. neomexicana A. Gray
Sy = F. pubescens subsp. neomexicana (A. Gray) E. Murray
Sy = F. pubescens var. neomexicana (A. Gray) E. Murray
Sy = F. sphaerocarpa J. Torrey
F. reticulata J. Torrey
Sy = F. racemosa S. Watson
Sy = Gymnanthes texana P. Standley

FORSYTHIA M. H. Vahl, *nomen conservandum*
F. viridissima J. Lindley [**cultivated**]
F. × intermedia H. Zabel [***suspensa × viridissima***] [**cultivated**]

FRAXINUS C. Linnaeus
F. americana C. Linnaeus
Sy = F. americana var. biltmoreana (C. Beadle) J. Wright *ex* M. Fernald
Sy = F. americana var. crassifolia C. Sargent
Sy = F. americana var. curtissii (G. Vasey) J. K. Small
Sy = F. americana var. juglandifolia (J. de Lamarck) A. Rehder
Sy = F. americana var. microcarpa A. Gray
F. berlandieriana A. P. de Candolle
F. caroliniana P. Miller
Sy = F. caroliniana var. cubensis (A. Grisebach) A. Lingelsheim
Sy = F. caroliniana var. oblanceolata (M. Curtis) M. Fernald & B. Schubert
F. cuspidata J. Torrey
Sy = F. cuspidata subsp. macropetala (A. Eastwood) E. Murray
Sy = F. cuspidata var. macropetala (A. Eastwood) A. Rehder
Sy = F. cuspidata var. serrata A. Rehder
Sy = F. macropetala A. Eastwood
Sy = Ornus cuspidata J. Nieuwland
F. greggii A. Gray
Sy = F. schiedeana D. von Schlechtendal & A. von Chamisso var. parvifolia J. Torrey
F. oxycarpa C. von Willdenow [**cultivated**]
F. papillosa A. Lingelsheim
F. pennsylvanica H. Marshall
Sy = F. pennsylvanica var. austinii M. Fernald
Sy = F. pennsylvanica var. integerrima (M. H. Vahl) M. Fernald
Sy = F. pennsylvanica var. lanceolata (M. Borkhausen) C. Sargent
Sy = F. pennsylvanica var. subintegerrima (M. H. Vahl) M. Fernald
F. texensis (A. Gray) C. Sargent
Sy = F. americana C. Linnaeus subsp. texensis (A. Gray) G. Miller
Sy = F. americana var. texensis A. Gray
F. velutina J. Torrey
Sy = F. attenuata M. E. Jones
Sy = F. coriacea S. Watson
Sy = F. pennsylvanica H. Marshal subsp. velutina (J. Torrey) G. Miller
Sy = F. toumeyi N. Britton
Sy = F. velutina var. coriacea (S. Watson) A. Rehder
Sy = F. velutina var. glabra (J. Thornber) A. Rehder
Sy = F. velutina var. toumeyi (N. Briton) A. Rehder

JASMINUM C. Linnaeus
J. floridum A. von Bunge [**cultivated**]
J. humile C. Linnaeus [**cultivated**]
J. mesnyi H. Hance [**cultivated**]
J. nitidum S. Skan [**cultivated**]
J. nudiflorum J. Lindley [**cultivated**]
J. sambac (C. Linnaeus) W. Aiton [**cultivated**]

LIGUSTRUM C. Linnaeus
L. amurense É. Carrière [**cultivated**]
L. japonicum C. Thunberg [**cultivated**]
L. lucidum W. Aiton [**cultivated**]
L. ovalifolium J. Hasskarl [**cultivated**]
L. quibhoui É. Carrière [**cultivated**]
L. sinense J. de Loureiro
Sy = L. villosum May

MENODORA A. Bonpland
M. decemfida (L. Gill *ex* W. Hooker & G. Arnott) A. Gray var. ***longifolia*** J. Steyermark

M. heterophylla M. Moricand *ex* A. P. de Candolle
Sy = Bolivaria grisebachii G. Scheele
M. longiflora A. Gray
Sy = M. hispida E. Palmer
Sy = Menodoropsis longiflora (A. Gray) J. K. Small
M. scabra A. Gray
Sy = Bolivaria scabra G. Engelmann *ex* A. Gray
Sy = Menodora laevis E. Wooton & P. Standley
Sy = M. scabra A. Gray var. glabrescens A. Gray
Sy = M. scabra var. laevis (E. Wooton & P. Standley) J. Steyermark
Sy = M. scabra var. ramosissima J. Steyermark
Sy = M. scoparia G. Engelmann *ex* A. Gray

OLEA C. Linnaeus
O. europaea C. Linnaeus [**cultivated**]

OSMANTHUS J. de Loureiro
O. americanus (C. Linnaeus) G. Bentham & J. Hooker *ex* A. Gray var. ***americanus*** [**cultivated**]
Sy = Amarolea americana (C. Linnaeus) J. K. Small
Sy = Olea americana C. Linnaeus
Sy = Osmanthus floridanus A. Chapman
O. fragrans J. de Loureiro [**cultivated**]
O. heterophyllus (G. Don) P. S. Green [**cultivated**]

SYRINGA C. Linnaeus
S. laciniata P. Miller [**cultivated**]
S. vulgaris C. Linnaeus [**cultivated**]
S. × chinensis C. von Willdenow [× ***persica*** × ***vulgaris***] [**cultivated**]
S. × persica C. Linnaeus [***afghanica*** × ***laciniata***] [**cultivated**]

ONAGRACEAE, *nomen conservandum*

CALYLOPHUS E. Spach
C. berlandieri E. Spach subsp. ***berlandieri***
Sy = C. drummondianus E. Spach subsp. berlandieri (E. Spach) H. Towner & P. Raven
Sy = Oenothera serrulata T. Nuttall subsp. drummondii (J. Torrey & A. Gray) P. Munz
C. berlandieri E. Spach subsp. ***pinifolius*** (G. Engelmann *ex* A. Gray) H. Towner
Sy = C. drummondianus E. Spach
Sy = C. serrulatus (T. Nuttall) P. Raven var. spinulosus (J. Torrey & A. Gray) L. Shinners
Sy = Oenothera serrulata T. Nuttall subsp. pinifolia (G. Engelmann *ex* A. Gray) P. Munz
C. hartwegii (G. Bentham) P. Raven subsp. ***fendleri*** (A. Gray) H. Towner & P. Raven
Sy = Oenothera fendleri A. Gray
Sy = O. hartwegii G. Bentham var. fendleri A. Gray
C. hartwegii (G. Bentham) P. Raven subsp. ***filifolius*** (A. Eastwood) H. Towner & P. Raven
Sy = C. hartwegii var. filifolius (A. Eastwood) L. Shinners
Sy = Oenothera filifolia (A. Eastwood) P. Munz
Sy = O. hartwegii var. filifolia (A. Eastwood) P. Munz
C. hartwegii (G. Bentham) P. Raven subsp. ***hartwegii***
Sy = Galpinsia hartwegii (G. Bentham) N. Britton
Sy = Oenothera hartwegii G. Bentham
Ma = O. greggii *auct.*, non A. Gray var. pringlei: *sensu* P. Munz
C. hartwegii (G. Bentham) P. Raven subsp. ***maccartii*** (L. Shinners) H. Towner & P. Raven
Sy = C. hartwegii var. maccartii L. Shinners
Sy = Oenothera greggii A. Gray var. pringlei P. Munz
C. hartwegii (G. Bentham) P. Raven subsp. ***pubescens*** (A. Gray) H. Towner & P. Raven
Sy = C. hartwegii var. pubescens (A. Gray) L. Shinners
Sy = Oenothera greggii A. Gray var. lampasana (S. Buckley) P. Munz
Sy = O. greggii A. Gray var. pubescens A. Gray
C. lavandulifolius (J. Torrey & A. Gray) P. Raven
Sy = C. hartwegii (G. Bentham) P. Raven subsp. lavandulifolius (J. Torrey & A. Gray) H. Towner & P. Raven
Sy = Galpinsia lavandulifolia (J. Torrey & A. Gray) J. K. Small
Sy = Oenothera lavandulifolia J. Torrey & A. Gray var. glandulosa P. Munz
C. serrulatus (T. Nuttall) P. Raven
Sy = C. australis H. Towner & P. Raven
Sy = Meriolix serrulata (T. Nuttall) W. Walpers
Sy = Oenothera serrulata T. Nuttall
C. tubicula (A. Gray) P. Raven
Sy = Galpinsia tubicula (A. Gray) J. K. Small
Sy = Oenothera tubicula A. Gray

CAMISSONIA J. Link
C. chamaenerioides (A. Gray) P. Raven
Sy = Oenothera chamaenerioides A. Gray
Sy = Sphaerostigma chamaenerioides (A. Gray) J. K. Small

EPILOBIUM C. Linnaeus
E. ciliatum C. Rafinesque-Schmaltz subsp. ***ciliatum***
Sy = E. adenocaulon H. Haussknecht
Sy = E. adenocaulon var. perplexans W. Trelease
Sy = E. ciliatum var. ecomosum (N. Fassett) J. Boivin
Sy = E. fendleri H. Haussknecht
Sy = E. novomexicanum H. Haussknecht
E. coloratum J. Biehler

GAURA C. Linnaeus
* ***G. boquillensis*** P. Raven & D. Gregory [**TOES**: **V**]
G. brachycarpa J. K. Small
G. calcicola P. Raven & D. Gregory
G. coccinea T. Nuttall *ex* F. Pursh
Sy = G. coccinea var. arizonica P. Munz
Sy = G. coccinea var. epilobioides (K. Kunth) P. Munz
Sy = G. coccinea var. glabra (J. Lehmann) J. Torrey & A. Gray
Sy = G. coccinea var. parviflora (J. Torrey) J. Torrey & A. Gray
Sy = G. odorata M. Sessé y Lacasta *ex* M. Lagasca y Segura
G. demareei P. Raven & D. Gregory
G. drummondii (E. Spach) J. Torrey & A. Gray
Sy = G. odorata M. Lagasca y Segura
G. hexandra C. Ortega subsp. ***gracilis*** (E. Wooton & P. Standley) P. Raven & D. Gregory
Sy = G. brassicacea E. Wooton & P. Standley
Sy = G. glandulosa E. Wooton & P. Standley
Sy = G. gracilis E. Wooton & P. Standley
Sy = G. podocarpa E. Wooton & P. Standley
Sy = G. strigillosa E. Wooton & P. Standley
G. lindheimeri G. Engelmann & A. Gray
Sy = G. filiformis J. K. Small var. munzii V. Cory
G. longiflora E. Spach
Sy = G. filiformis J. K. Small
G. macrocarpa J. Rothrock
G. mckelveyae (P. Munz) P. Raven & D. Gregory
Sy = G. villosa J. Torrey var. mckelveyae P. Munz
G. parviflora D. Douglas *ex* J. Lehmann
Sy = G. australis A. Grisebach
Sy = G. parviflora var. lachnocarpa C. Weatherby
G. sinuata T. Nuttall *ex* N. Séringe
G. suffulta A. Gray subsp. ***nealleyi*** (J. Coulter) P. Raven & D. Gregory
Sy = G. nealleyi J. Coulter
G. suffulta A. Gray subsp. ***suffulta***
G. triangulata S. Buckley
Sy = G. tripetala A. Cavanilles var. triangulata (S. Buckley) P. Munz
G. tripetala A. Cavanilles var. ***coryi*** P. Munz
G. villosa J. Torrey subsp. ***parksii*** (P. Munz) P. Raven & D. Gregory
Sy = G. villosa var. parksii P. Munz
G. villosa J. Torrey subsp. ***villosa***
Sy = G. cinerea E. Wooton & P. Standley
Sy = G. villosa var. arenicola P. Munz
Sy = G. villosa var. villosa

LUDWIGIA C. Linnaeus
L. alternifolia C. Linnaeus
Sy = L. alternifolia var. linearifolia N. Britton
Sy = L. alternifolia var. pubescens E. Palmer & J. Steyermark
L. decurrens T. Walter
Sy = Jussiaea decurrens (T. Walter) A. P. de Candolle
L. glandulosa T. Walter
Sy = L. cylindrica S. Elliott var. brachycarpa J. Torrey & A. Gray
Sy = L. glandulosa subsp. brachycarpa (J. Torrey & A. Gray) Peng
Sy = L. glandulosa var. torreyi P. Munz
L. hirtella C. Rafinesque-Schmaltz
L. leptocarpa (T. Nuttall) H. Hara
Sy = Jussiaea leptocarpa T. Nuttall
Sy = J. suffruticosa C. Linnaeus
Sy = J. suffruticosa var. ligustrifolia (K. Kunth) A. Grisebach
Sy = J. suffruticosa var. octofila (A. P. de Candolle) P. Munz
Sy = Ludwigia leptocarpa var. meyeriana (K. E. O. Kuntze) Brother Alain
L. microcarpa A. Michaux
L. octovalvis (N. von Jacquin) P. Raven subsp. ***octovalvis***
Sy = Jussiaea angustifolia J. de Lamarck
Sy = J. clavata (K. Presl) M. E. Jones
Sy = J. octofila A. P. de Candolle
Sy = Ludwigia octovalvis var. ligustrifolia (K. Kunth) Brother Alain
Sy = L. octovalvis var. macropoda (K. Presl) L. Shinners
Sy = L. octovalvis var. octofila (A. P. de Candolle) Brother Alain
L. peploides (K. Kunth) P. Raven subsp. ***glabrescens*** (K. E. O. Kuntze) P. Raven
Sy = Jussiaea repens C. Linnaeus
Sy = J. repens var. glabrescens K. E. O. Kuntze
Sy = Ludwigia peploides var. glabrescens (K. E. O. Kuntze) L. Shinners
L. pilosa T. Walter
L. repens J. Forster
Sy = Isnardia repens (J. Forster) A. P. de Candolle

Sy = Ludwigia natans S. Elliott
Sy = L. repens var. rotundata (A. Grisebach) M. Fernald & L. Griscom
Sy = L. repens var. stipitata (M. Fernald & L. Griscom) P. Munz

L. sphaerocarpa S. Elliott
Sy = L. sphaerocarpa var. deamii M. Fernald & L. Griscom
Sy = L. sphaerocarpa var. jungens M. Fernald & L. Griscom
Sy = L. sphaerocarpa var. macrocarpa M. Fernald & L. Griscom

L. uruguayensis (J. Cambessèdes) H. Hara
Sy = Jussiaea uruguayensis J. Cambessèdes

OENOTHERA C. Linnaeus

O. albicaulis F. Pursh
Sy = Anogra albicaulis (F. Pursh) N. Britton
Sy = A. ctenophylla E. Wooton & P. Standley
Sy = Oenothera ctenophylla (E. Wooton & P. Standley) I. Tidestrom

O. biennis C. Linnaeus
Sy = O. biennis subsp. caeciarum P. Munz
Sy = O. biennis subsp. centralis P. Munz
Sy = O. muricata C. Linnaeus
Sy = O. pratincola H. Bartlett
Sy = O. pratincola var. pycnocarpa (G. Atkinson & H. Bartlett) K. Wiegand
Sy = O. pycnocarpa G. Atkinson & H. Bartlett

O. brachycarpa A. Gray
Sy = Lavauxia brachycarpa (A. Gray) N. Britton
Sy = L. wrightii (A. Gray) J. K. Small
Sy = Megapterium brachycarpum (A. Gray) A. A. Léveillé
Sy = Oenothera brachycarpa var. wrightii (A. Gray) A. A. Léveillé
Sy = O. wrightii A. Gray

O. caespitosa T. Nuttall subsp. ***marginata*** (T. Nuttall *ex* W. Hooker & G. Arnott) P. Munz
Sy = O. caespitosa subsp. eximia (A. Gray) P. Munz
Sy = O. caespitosa var. eximia (A. Gray) P. Munz
Sy = O. caespitosa var. marginata (T. Nuttall) *ex* W. Hooker & G. Arnott) P. Munz
Sy = Pachylophus marginatus (T. Nuttall) P. Rydberg

O. canescens J. Torrey & J. Frémont
Sy = Gaurella canescens (J. Torrey & J. Frémont) A. Nelson
Sy = G. guttulata J. K. Small

O. cordata J. Loudon

O. coronopifolia J. Torrey & A. Gray
Sy = Anogra coronopifolia (J. Torrey & A. Gray) N. Britton

O. coryi W. Wagner

O. drummondii W. Hooker subsp. ***drummondii***
Sy = Raimannia drummondii (W. Hooker) J. Rose *ex* T. A. Sprague & J. Riley

O. elata K. Kunth subsp. ***hirsutissima*** (A. Gray *ex* S. Watson) W. Dietrich
Sy = O. biennis C. Linnaeus subsp. hirsutissima (A. Gray *ex* S. Watson) P. Munz
Sy = O. hookeri J. Torrey & A. Gray subsp. hirsutissima (A. Gray *ex* S. Watson) P. Munz
Sy = O. hookeri subsp. hewettii T. Cockerell

O. elata K. Kunth subsp. ***hookeri*** (J. Torrey & A. Gray) W. Dietrich & W. Wagner
Sy = O. biennis C. Linnaeus var. hookeri (J. Torrey & A. Gray) J. Boivin
Sy = O. hookeri J. Torrey & A. Gray
Sy = O. hookeri subsp. montereyensis P. Munz

O. elata K. Kunth subsp. ***texensis*** W. Dietrich & W. Wagner

O. engelmannii (J. K. Small) P. Munz
Sy = Anogra engelmannii (J. K. Small) E. Wooton & P. Standley

O. falfurriae W. Dietrich & W. Wagner

O. grandis (N. Britton) B. Smyth
Sy = O. laciniata J. Hill var. grandiflora (S. Watson) B. Robinson
Sy = Raimannia grandis (N. Britton) J. Rose

O. havardii S. Watson

O. heterophylla E. Spach subsp. ***heterophylla***

O. jamesii J. Torrey & A. Gray
Sy = Onagra jamesii (J. Torrey & A. Gray) J. K. Small

O. kunthiana (E. Spach) P. Munz
Sy = Hartmannia kunthiana E. Spach

O. laciniata J. Hill
Sy = O. rosea L'Héritier de Brutelle *ex* W. Aiton
Sy = Raimannia laciniata (J. Hill) J. Rose

O. linifolia T. Nuttall
Sy = Kneiffia linifolia (T. Nuttall) E. Spach
Sy = Oenothera linifolia var. glandulosa P. Munz
Sy = Peniophyllum linifolium (T. Nuttall) F. Pennell

O. macrocarpa T. Nuttall subsp. ***incana*** (A. Gray) W. Wagner
Sy = O. macrocarpa var. incana (A. Gray) J. Reveal
Sy = O. missouriensis J. Sims var. incana A. Gray

O. macrocarpa T. Nuttall subsp. ***macrocarpa***
Sy = O. missouriensis J. Sims

O. macrocarpa T. Nuttall subsp. ***oklahomensis***

(J. Norton) W. Wagner
Sy = Megapterium oklahomense J. Norton
Sy = Oenothera macrocarpa T. Nuttall var. oklahomensis (J. Norton) J. Reveal
Sy = O. missouriensis var. oklahomensis (J. Norton) P. Munz
O. mexicana E. Spach
O. neomexicana (J. K. Small) P. Munz
Sy = Anogra neomexicana J. K. Small
O. pallida J. Lindley subsp. ***runcinata*** (G. Engelmann) P. Munz & W. Klein
Sy = Anogra albicaulis (F. Pursh) N. Britton var. runcinata G. Engelmann
Sy = Oenothera gypsophila (A. Eastwood) A. A. Heller
Sy = O. pallida var. runcinata (G. Engelmann) A. Cronquist
Sy = O. runcinata (G. Engelmann) P. Munz
Sy = O. runcinata var. brevifolia (G. Engelmann) P. Munz
Sy = O. runcinata var. leucotricha (E. Wooton & P. Standley) P. Munz
* ***O. pilosella*** C. Rafinesque-Schmaltz subsp. ***sessilis*** (F. Pennell) G. Straley **[TOES: V]**
Sy = O. sessilis (F. Pennell) P. Munz
O. primiveris A. Gray subsp. ***primiveris***
Sy = Lavauxia primiveris (A. Gray) J. K. Small
Sy = Oenothera primiveris subsp. caulescens (P. Munz) P. Munz
Sy = O. primiveris var. caulescens P. Munz
O. pubescens C. von Willdenow *ex* K. Sprengel
Sy = O. amplexicaulis (E. Wooton & P. Standley) I. Tidestrom
Sy = O. laciniata J. Hill subsp. pubescens (C. von Willdenow *ex* K. Sprengel) P. Munz
Sy = O. laciniata var. pubescens (C. von Willdenow *ex* K. Sprengel) P. Munz
O. rhombipetala T. Nuttall *ex* J. Torrey & A. Gray
O. spachiana J. Torrey & A. Gray
O. speciosa T. Nuttall
Sy = Hartmannia speciosa (T. Nuttall) J. K. Small
Sy = Oenothera speciosa var. childsii (L. Bailey) P. Munz
O. tetraptera A. Cavanilles
O. texensis P. Raven & D. Parnell
O. triloba T. Nuttall
Sy = Lavauxia trilobata (T. Nuttall) E. Spach

STENOSIPHON E. Spach
S. linifolius (T. Nuttall *ex* E. James) G. Heynhold
Sy = Gaura linifolia T. Nuttall
Sy = Stenosiphon virgatus E. Spach

OROBANCHACEAE, *nomen conservandum*

CONOPHOLIS C. Wallroth
C. alpina F. Liebmann var. ***mexicana*** (A. Gray *ex* S. Watson) R. Haynes
Sy = C. mexicana A. Gray *ex* S. Watson

EPIFAGUS T. Nuttall, *nomen conservandum*
E. virginiana (C. Linnaeus) W. Barton
Sy = Leptamnium virginianum (C. Linnaeus) C. Rafinesque-Schmaltz

OROBANCHE C. Linnaeus
O. cooperi (A. Gray) A. A. Heller subsp. ***cooperi***
Sy = Aphyllon cooperi A. Gray
Sy = A. ludovicianum A. Gray var. cooperi A. Gray
Sy = Myzorrhiza cooperi (A. Gray) P. Rydberg
Sy = Orobanche ludoviciana T. Nuttall var. cooperi (A. Gray) G. Beck
O. fasciculata T. Nuttall
Sy = Anoplanthus fasciculatus (T. Nuttall) W. Walpers
Sy = A. luteus P. Rydberg
Sy = Aphyllon fasciculatum J. Torrey & A. Gray var. luteum A. Gray
Sy = Loxanthes fasciculata C. Rafinesque-Schmaltz
Sy = Orobanche fasciculata var. franciscana D. Achey
Sy = O. fasciculata var. lutea (C. Parry) D. Achey
Sy = O. fasciculata var. subulata G. Goodman
Sy = Phelipaea lutea C. Parry
Sy = Thalesia fasciculata (T. Nuttall) N. Britton
Sy = T. fasciculata var. lutea N. Britton
O. ludoviciana T. Nuttall subsp. ***ludoviciana***
Sy = Aphyllon ludovicianum A. Gray
Sy = Conopholis ludoviciana (T. Nuttall) A. Wood
Sy = Myzorrhiza ludoviciana (T. Nuttall) P. Rydberg
Sy = Orobanche ludoviciana var. arenosa (W. Suksdorf) A. Cronquist
Sy = O. ludoviciana var. genuina G. Beck
Sy = Phelipaea ludoviciana G. Don
O. ludoviciana T. Nuttall subsp. ***multiflora*** (T. Nuttall) F. S. Collins
Sy = Myzorrhiza multiflora (T. Nuttall) P. Rydberg
Sy = Orobanche ludoviciana var. multiflora (T. Nuttall) G. Beck
Sy = O. multiflora T. Nuttall

O. ramosa C. Linnaeus [**federal noxious weed**]
O. uniflora C. Linnaeus
Sy = Aphyllon inundatum W. Suksdorf
Sy = A. minutum W. Suksdorf
Sy = A. sedii W. Suksdorf
Sy = A. uniflorum A. Gray var. occidentale E. Greene
Sy = Orobanche uniflora subsp. occidentalis (E. Greene) L. Abrams *ex* R. J. Ferris
Sy = O. uniflora var. minuta (W. Suksdorf) G. Beck
Sy = O. uniflora var. occidentalis (E. Greene) R. L. Taylor & B. MacBryde
Sy = O. uniflora var. purpurea (A. A. Heller) D. Achey
Sy = O. uniflora var. sedii (W. Suksdorf) D. Achey
Sy = O. uniflora var. terrae-novae (M. Fernald) P. Munz
Sy = Thalesia purpurea A. A. Heller
Sy = T. uniflora (C. Linnaeus) N. Britton

OXALIDACEAE, *nomen conservandum*

OXALIS C. Linnaeus
O. albicans K. Kunth subsp. ***albicans***
Sy = O. corniculata C. Linnaeus subsp. albicans (K. Kunth) A. Lourteig
Sy = O. wrightii A. Gray
Sy = Xanthoxalis albicans (K. Kunth) J. K. Small
Sy = X. wrightii (A. Gray) L. Abrams
O. albicans K. Kunth subsp. ***pilosa*** (T. Nuttall) G. Eiten
Sy = O. corniculata C. Linnaeus subsp. pilosa (T. Nuttall) A. Lourteig
Sy = O. pilosa T. Nuttall
Sy = O. wrightii (A. Gray) L. Abrams var. pilosa (T. Nuttall) K. Wiegand
Sy = Xanthoxalis pilosa (T. Nuttall) J. K. Small
O. alpina (J. Rose) J. Rose *ex* R. Knuth
O. articulata M. de Savigny subsp. ***rubra*** (A. de Saint-Hilaire) A. Lourteig
Sy = O. rubra A. de Saint-Hilaire
O. corniculata C. Linnaeus var. ***wrightii*** (A. Gray) B. L. Turner
Sy = Acetosella corniculata (C. Linnaeus) K. E. O. Kuntze
O. debilis K. Kunth var. ***corymbosa*** (A. P. de Candolle) A. Lourteig
Sy = O. corymbosa A. P. de Candolle
O. dichondrifolia A. Gray
O. dillenii N. von Jacquin subsp. ***dillenii***
Sy = O. corniculata C. Linnaeus var. dillenii (N. von Jacquin) W. Trelease
Sy = O. dilleni var. radicans L. Shinners
Sy = Xanthoxalis dillenii (N. von Jacquin) J. Holub
O. dillenii N. von Jacquin subsp. ***filipes*** (J. K. Small) G. Eiten
Sy = O. filipes J. K. Small
Sy = O. florida R. A. Salisbury
Sy = O. florida subsp. prostrata (A. Haworth) A. Lourteig
Sy = Xanthoxalis filipes (J. K. Small) J. K. Small
Sy = X. florida (R. A. Salisbury) H. Moldenke
O. drummondii A. Gray
Ma = O. amplifolia *auct.*, non (W. Trelease) R. Knuth
O. frutescens C. Linnaeus subsp. ***angustifolia*** (K. Kunth) A. Lourteig
Sy = O. angustifolia K. Kunth
Sy = O. berlandieri J. Torrey
O. latifolia K. Kunth [**cultivated**]
O. lyonii F. Pursh
Sy = O. priceae J. K. Small subsp. texana (J. K. Small) G. Eiten
Sy = O. recurva S. Elliott var. texana (J. K. Small) K. Wiegand
Sy = O. texana (J. K. Small) F. Fedde
Sy = Xanthoxalis texana J. K. Small
O. regnellii F. Miquel [**cultivated**]
O. stricta C. Linnaeus
Sy = Ionoxalis stricta (C. Linnaeus) J. K. Small
Sy = Oxalis corniculata C. Linnaeus var. stricta P. A. L. Savatier
Sy = O. europaea A. Jordan
Sy = O. stricta var. piletocarpa K. Wiegand
Sy = O. stricta var. rufa (J. K. Small) O. Farwell
Sy = O. stricta var. villicaulis (K. Wiegand) O. Farwell
Sy = Xanthoxalis stricta (C. Linnaeus) J. K. Small
O. violacea C. Linnaeus
Sy = Ionoxalis violacea (C. Linnaeus) J. K. Small
Sy = Oxalis violacea var. trichophora N. Fassett

PAPAVERACEAE, *nomen conservandum*

ARGEMONE C. Linnaeus
A. aenea G. Ownbey
A. albiflora J. Hornemann subsp. ***texana*** G. Ownbey
Sy = A. albiflora var. texana (G. Ownbey) L. Shinners

A. aurantiaca G. Ownbey
A. chisosensis G. Ownbey
A. mexicana C. Linnaeus
Sy = A. mexicana var. ochroleuca (R. Sweet) J. Lindley
Sy = A. ochroleuca R. Sweet
Sy = A. ochroleuca var. stenophylla (D. Prain) L. Shinners
A. polyanthemos (F. Fedde) G. Ownbey
Sy = A. intermedia R. Sweet var. polyanthemos F. Fedde
Ma = A. platyceras *auct.*, non J. Link & C. Otto
A. sanguinea E. Greene
Sy = A. platyceras J. Link & C. Otto var. rosea J. Coulter
A. squarrosa E. Greene subsp. ***glabrata*** G. Ownbey
Sy = A. squarrosa var. glabrata (G. Ownbey) L. Shinners

ESCHSCHOLTZIA A. von Chamisso
E. californica A. von Chamisso subsp. ***mexicana*** (E. Greene) J. C. Clark
Sy = E. mexicana E. Greene

GLAUCIUM P. Miller
G. corniculatum (C. Linnaeus) J. Rudolph
Sy = Chelidonium corniculatum C. Linnaeus

PAPAVER C. Linnaeus
P. nudicaule C. Linnaeus [**cultivated**]
P. orientale C. Linnaeus [**cultivated**]
P. rhoeas C. Linnaeus [**cultivated**]
P. somniferum C. Linnaeus [**cultivated**]

SANGUINARIA C. Linnaeus
S. canadensis C. Linnaeus
Sy = S. canadensis var. rotundifolia (E. Greene) F. Fedde

PASSIFLORACEAE, *nomen conservandum*

PASSIFLORA C. Linnaeus
P. affinis G. Engelmann
P. caerulea C. Linnaeus [**cultivated**]
P. foetida C. Linnaeus var. ***gossypifolia*** (N. Desvaux *ex* W. Hamilton) M. T. Masters
Sy = Dysomia gossypifolia M. Roemer
Sy = Passiflora gossypifolia N. Desvaux *ex* W. Hamilton
P. incarnata C. Linnaeus
P. lutea C. Linnaeus
Sy = P. lutea var. glabriflora M. Fernald
P. suberosa C. Linnaeus
Sy = P. pallida C. Linnaeus
P. tenuiloba G. Engelmann
P. vitifolia K. Kunth [**cultivated**]
P. × alatocaerula J. Lindley [***alata × caerulea***] [**cultivated**]
Sy = P. × pfordtii O. Degener

PEDALIACEAE, *nomen conservandum*

PROBOSCIDEA C. Schmidel
P. althaeifolia (G. Bentham) J. Decaisne
Sy = Martynia althaeifolia G. Bentham
Sy = M. arenaria G. Engelmann
Sy = Proboscidea arenaria (G. Engelmann) J. Decaisne
P. louisianica (P. Miller) A. Thellung subsp. ***fragrans*** (J. Lindley) P. Bretting
Sy = Martynia fragrans J. Lindley
Sy = Proboscidea fragrans (J. Lindley) J. Decaisne
P. louisianica (P. Miller) A. Thellung subsp. ***louisianica***
Sy = Martynia louisianica P. Miller
P. parviflora (E. Wooton) E. Wooton & P. Standley subsp. ***parviflora***
Sy = Martynia parviflora E. Wooton
Sy = Proboscidea crassibracteata D. Correll
P. sabulosa D. Correll
* ***P. spicata*** D. Correll [**TOES: V**]

SESAMUM C. Linnaeus
S. orientale C. Linnaeus [**cultivated**]
Sy = S. indicum C. Linnaeus

PHRYMACEAE, *nomen conservandum*

PHRYMA C. Linnaeus
P. leptostachya C. Linnaeus
Sy = P. leptostachya var. confertifolia M. Fernald

PHYTOLACCACEAE, *nomen conservandum*

AGDESTIS J. Mociño & M. Sessé y Lacasta *ex* A. P. de Candolle
A. clematidea J. Mociño & M. Sessé y Lacasta *ex* A. P. de Candolle

PETIVERIA C. Linnaeus
P. alliacea C. Linnaeus
Sy = P. hexandra M. Sessé y Lacasta & J. Mociño
Sy = P. ochroleuca C. Moquin-Tandon
Sy = P. octandra C. Linnaeus

PHYTOLACCA C. Linnaeus
P. americana C. Linnaeus var. **americana**
Sy = P. decandra C. Linnaeus

RIVINA C. Linnaeus
R. humilis C. Linnaeus
Sy = R. laevis C. Linnaeus
Sy = R. portulacoides T. Nuttall
Sy = R. purpurascens H. A. Schrader

PIPERACEAE, *nomen conservandum*

PEPEROMIA H. Ruiz López & J. Pavón
P. pellucida (C. Linnaeus) K. Kunth [**cultivated**]

PITTOSPORACEAE, *nomen conservandum*

PITTOSPORUM J. Banks & D. Solander, *nomen conservandum*
P. tobira (C. Thunberg) W. T. Aiton [**cultivated**]

PLANTAGINACEAE, *nomen conservandum*

PLANTAGO C. Linnaeus
P. aristata A. Michaux
Sy = P. aristata var. nuttallii (D. Rapin) E. Morris
Sy = P. patagonica N. von Jacquin var. aristata (A. Michaux) A. Gray
P. elongata F. Pursh subsp. **elongata**
Sy = P. bigelovii A. Gray
Sy = P. elongata subsp. pentasperma I. Bassett
Sy = P. myosuroides P. Rydberg
P. helleri J. K. Small
P. heterophylla T. Nuttall
Sy = P. hybrida W. Barton
P. hookeriana F. von Fischer & C. von Meyer
P. lanceolata C. Linnaeus
Sy = P. lanceolata var. sphaerostachya F. Mertens & W. Koch
P. major C. Linnaeus var. **major**
Sy = P. asiatica C. Linnaeus
P. ovata P. Forsskål
Sy = P. insularis A. Eastwood
Sy = P. insularis var. fastigiata (E. Morris) W. Jepson
Sy = P. insularis var. scariosa (E. Morris) W. Jepson
P. patagonica N. von Jacquin
Sy = P. ignota J. Morris
Sy = P. lagopus F. Pursh
Sy = P. oblonga J. Morris
Sy = P. patagonica var. breviscapa (L. Shinners) L. Shinners
Sy = P. patagonica var. gnaphalioides (T. Nuttall) A. Gray
Sy = P. patagonica var. oblonga (E. Morris) L. Shinners
Sy = P. patagonica var. spinulosa (J. Decaisne) A. Gray
Sy = P. picta J. Morris
Sy = P. purshii J. J. Römer & J. A. Schultes
Sy = P. purshii var. breviscapa L. Shinners
Sy = P. spinulosa J. Decaisne
P. rhodosperma J. Decaisne
P. rugelii J. Decaisne
Sy = P. rugelii var. asperula O. Farwell
P. virginica C. Linnaeus
Sy = P. caroliniana T. Walter
Sy = P. missouriensis E. von Steudel
Sy = P. purpurascens T. Nuttall *ex* D. Rapin
Sy = P. virginica var. viridescens M. Fernald
P. wrightiana J. Decaisne

PLATANACEAE, *nomen conservandum*

PLATANUS C. Linnaeus
P. occidentalis C. Linnaeus
Sy = P. occidentalis var. glabrata (M. Fernald) C. Sargent

PLUMBAGINACEAE, *nomen conservandum*

CERATOSTIGMA A. von Bunge
C. plumbaginoides A. von Bunge [**cultivated**]
Sy = Plumbago larpentiae J. Lindley

LIMONIUM P. Miller, *nomen conservandum*
L. carolinianum (T. Walter) N. Britton
Sy = L. carolinianum var. angustatum (A. Gray) S. F. Blake
Sy = L. carolinianum var. angustifolium S. F. Blake
Sy = L. carolinianum var. compactum L. Shinners
Sy = L. carolinianum var. nashii (J. K. Small) J. Boivin
Sy = L. carolinianum var. obtusilobum (S. F. Blake) H. Ahles
Sy = L. carolinianum var. trichogonum (S. F. Blake) J. Boivin
Sy = L. nashii J. K. Small var. albiflorum (C. Rafinesque-Schmaltz)

Sy = L. nashii var. angustatum (A. Gray) H. Ahles
Sy = L. nashii var. nashii
Sy = L. nashii var. trichogonum S. F. Blake
L. limbatum J. K. Small
Sy = L. limbatum var. glabrescens D. Correll

PLUMBAGO C. Linnaeus
P. auriculata J. de Lamarck [**cultivated**]
P. capensis C. Thunberg [**cultivated**]
P. scandens C. Linnaeus
Sy = P. mexicana K. Kunth

POLEMONIACEAE, *nomen conservandum*

ERIASTRUM E. Wooton & P. Standley
E. diffusum (A. Gray) H. Mason
Sy = E. diffusum subsp. jonesii H. Mason
Sy = Gilia filifolia T. Nuttall var. diffusa A. Gray
Sy = Hugelia diffusa W. Jespon
Sy = Navarretia filifolia A. Brand var. diffusa A. Brand
Sy = Welwitschia diffusa P. Rydberg

GILIA H. Ruiz López & J. Pavón
G. flavocincta A. Nelson subsp. ***australis*** (A. Grant & V. Grant) A. Day & V. Grant
Sy = G. opthalmoides A. Brand subsp. australis A. Grant & V. Grant
G. incisa G. Bentham
Sy = G. lindheimeriana G. Scheele
Sy = G. perennans L. Shinners
G. insignis (A. Brand) V. Cory & H. Parks
Sy = G. rigidula G. Bentham subsp. insignis A. Brand
G. ludens L. Shinners
G. mexicana A. Grant & V. Grant
G. rigidula G. Bentham subsp. ***acerosa*** (A. Gray) E. Wherry
Sy = G. acerosa (A. Gray) N. Britton
Sy = G. rigidula var. acerosa A. Gray
Sy = Giliastrum rigidula (G. Bentham) P. Rydberg subsp. acerosa (A. Gray) W. A. Weber
G. rigidula G. Bentham subsp. ***rigidula***
G. stewartii I. M. Johnston

IPOMOPSIS A. Michaux
I. aggregata (F. Pursh) V. Grant subsp. ***aggregata***
Sy = Cantua aggregata F. Pursh
Sy = Gilia aggregata (F. Pursh) K. Sprengel
Sy = G. aggregata subsp. euaggregata A. Brand
I. aggregata (F. Pursh) V. Grant subsp. ***formosissima*** (E. Greene) E. Wherry
Sy = Gilia aggregata (F. Pursh) K. Sprengel subsp. formosissima (E. Greene) E. Wherry
Sy = G. aggregata var. maculata M. E. Jones
Sy = G. texana (E. Greene) E. Wooton & P. Standley
Sy = Ipomopsis aggregata subsp. texana (E. Greene) W. Martin & C. Hutchins
Sy = I. arizonica (E. Greene) E. Wherry subsp. texana (E. Greene) E. Wherry
I. arizonica (E. Greene) E. Wherry
Sy = Callisteria arizonica E. Greene
Sy = Gilia aggregata (E. Greene) K. Sprengel var. arizonica F. Fosberg
Sy = G. arizonica (E. Greene) P. Rydberg
Sy = Ipomopsis aggregata (F. Pursh) V. Grant subsp. arizonica (E. Greene) V. Grant & A. Grant
I. havardii (A. Gray) V. Grant
Sy = Gilia havardii A. Gray
Sy = Loeselia havardii A. Gray
Sy = Navarretia havardii (A. Gray) K. E. O. Kuntze
I. longiflora (J. Torrey) V. Grant subsp. ***longiflora***
Sy = Cantua longiflora J. Torrey
Sy = Gilia longiflora (J. Torrey) G. Don
I. polycladon (J. Torrey) V. Grant
Sy = Gilia polycladon J. Torrey
I. pumila (T. Nuttall) V. Grant
Sy = Gilia pumila T. Nuttall
I. rubra (C. Linnaeus) E. Wherry
Sy = Gilia rubra (C. Linnaeus) A. A. Heller
I. thurberi (J. Torrey *ex* A. Gray) V. Grant
Sy = Gilia thurberi (J. Torrey *ex* A. Gray) A. Gray
I. wrightii (A. Gray) F. Gould
Sy = Gilia wrightii A. Gray
Sy = Navarretia wrightii (A. Gray) K. E. O. Kuntze

LINANTHUS G. Bentham
L. bigelovii (A. Gray) E. Greene
Sy = Gilia bigelovii A. Gray

LOESELIA C. Linnaeus
L. greggii S. Watson
Ma = L. scariosa *auct.*, non (M. Martens & H. Galeotti) W. Walpers: American authors

PHLOX C. Linnaeus
P. carolina C. Linnaeus subsp. ***angusta*** E. Wherry
Sy = P. carolina var. angusta (E. Wherry) J. Steyermark
P. cuspidata G. Scheele

Sy = P. cuspidata var. grandiflora E. Whitehouse
Sy = P. cuspidata var. humilis E. Whitehouse
P. divaricata C. Linnaeus subsp. ***laphamii*** (A. Wood) E. Wherry
Sy = P. divaricata var. laphamii A. Wood
P. drummondii W. Hooker subsp. ***drummondii***
Sy = P. drummondii var. peregrina L. Shinners
Sy = P. goldsmithii E. Whitehouse
P. drummondii W. Hooker subsp. ***johnstonii*** (E. Wherry) E. Wherry
Sy = P. johnstonii E. Wherry
P. drummondii W. Hooker subsp. ***mcallisteri*** (E. Whitehouse) E. Wherry
Sy = P. drummondii var. mcallisteri (E. Whitehouse) L. Shinners
P. drummondii W. Hooker subsp. ***tharpii*** (E. Whitehouse) E. Wherry
Sy = P. tharpii E. Whitehouse
P. drummondii W. Hooker subsp. ***wilcoxiana*** (E. Bogusch) E. Wherry
Sy = P. drummondii var. wilcoxiana (E. Bogusch) E. Whitehouse
P. glabriflora (A. Brand) E. Whitehouse subsp. ***glabriflora***
P. glabriflora (A. Brand) E. Whitehouse subsp. ***littoralis*** (V. Cory) E. Wherry
Sy = P. drummondii W. Hooker var. littoralis V. Cory
Sy = P. littoralis (V. Cory) E. Whitehouse
P. mesoleuca E. Greene
P. nana T. Nuttall
* ***P. nivalis*** C. Loddiges *ex* R. Sweet subsp. ***texensis*** C. Lundell [**TOES: I**]
Sy = P. texensis (C. Lundell) C. Lundell
P. oklahomensis E. Wherry
Sy = P. bifida L. Beck var. induta L. Shinners
P. paniculata C. Linnaeus [**cultivated**]
P. pilosa C. Linnaeus subsp. ***detonsa*** (A. Gray) E. Wherry
Sy = P. pilosa var. detonsa A. Gray
P. pilosa C. Linnaeus subsp. ***latisepala*** E. Wherry
Sy = P. pilosa var. aspera (E. Nelson) E. Wherry *ex* F. Gould
P. pilosa C. Linnaeus subsp. ***pilosa***
Sy = P. argillacea W. Clute & R. J. Ferris
Sy = P. pilosa var. virens (A. Michaux) E. Wherry
P. pilosa C. Linnaeus subsp. ***pulcherrima*** C. Lundell
Sy = P. pilosa var. amplexicaulis (C. Rafinesque-Schmaltz) E. Wherry
P. pilosa C. Linnaeus subsp. ***riparia*** E. Wherry
Sy = P. villosissima (A. Gray) E. Whitehouse
P. roemeriana G. Scheele
P. stansburyi (J. Torrey) A. A. Heller subsp. ***stansburyi***
Sy = P. stansburyi subsp. eustansburyi A. Brand
P. subulata C. Linnaeus [**cultivated**]
P. triovulata G. Thurber *ex* J. Torrey
Sy = P. nana T. Nuttall subsp. glabella (A. Gray) A. Brand

POLEMONIUM C. Linnaeus
* ***P. pauciflorum*** S. Watson subsp. ***hinckleyi*** (P. Standley) E. Wherry [**TOES: V**]
Sy = P. hinckleyi P. Standley

POLYGALACEAE, *nomen conservandum*

POLYGALA C. Linnaeus
P. alba T. Nuttall
Sy = P. alba var. suspecta S. Watson
P. balduinii T. Nuttall var. ***balduinii***
Sy = Pylostachya balduinii (T. Nuttall) J. K. Small
P. barbeyana R. Chodat
Sy = P. longa S. F. Blake
Sy = P. racemosa S. F. Blake
Sy = P. reducta S. F. Blake
P. crenata C. James
P. cruciata C. Linnaeus var. ***cruciata***
Sy = P. cruciata var. cuspidata (W. Hooker & G. Arnott) A. Wood
Sy = P. ramosior (G. Nash) J. K. Small
P. glandulosa K. Kunth
Sy = P. greggii S. Watson
P. hemipterocarpa A. Gray
P. hookeri J. Torrey & A. Gray
P. incarnata C. Linnaeus
Sy = Galypola incarnata (C. Linnaeus) N. Nieuwland
P. leptocaulis J. Torrey & A. Gray
P. lindheimeri A. Gray var. ***lindheimeri***
P. lindheimeri A. Gray var. ***parviflora*** W. Wheelock
Sy = P. parvifolia (W. Wheelock) E. Wooton & P. Standley
Sy = P. texensis B. Robinson
Sy = P. tweedyi N. Britton *ex* W. Wheelock
P. macradenia A. Gray var. ***macradenia***
Sy = P. macradenia var. genuina S. F. Blake
* ***P. maravillasensis*** D. Correll [**TOES: V**]
P. mariana P. Miller
Sy = P. harperi J. K. Small
P. nana (A. Michaux) A. P. de Candolle
Sy = Psilostaxis nana (A. Michaux) C. Rafinesque-Schmaltz

Sy = Pylostachya nana (A. Michaux) C. Rafinesque-Schmaltz
P. nitida T. Brandegee var. ***goliadensis*** T. Wendt
P. nitida T. Brandegee var. ***tamaulipana*** T. Wendt
P. nudata T. Brandegee
Ma = P. minutifolia *auct.*, non J. Rose: Texas authors
P. obscura G. Bentham
Sy = P. laeta T. Brandegee
Sy = P. neomexicana E. Wooton & P. Standley
Sy = P. orthotricha S. F. Blake
Sy = P. puberula A. Gray
Sy = P. vagans T. Brandegee
P. ovatifolia A. Gray
P. palmeri S. Watson
P. paniculata C. Linnaeus
P. polygama C. Linnaeus var. ***obtusa*** R. Chodat
P. ramosa S. Elliott
Sy = P. balduinii T. Nuttall var. chlorgena J. Torrey & A. Gray
Sy = Pylostachya ramosa (S. Elliott) J. K. Small
P. rimulicola J. Steyermark var. ***rimulicola***
P. sanguinea C. Linnaeus
Sy = P. viridescens C. Linnaeus
P. scoparioides R. Chodat
Sy = P. scoparioides var. multicaulis A. Gray
Sy = P. wrightii A. Gray
P. verticillata C. Linnaeus var. ***ambigua*** (T. Nuttall) A. Wood
Sy = P. ambigua T. Nuttall
P. verticillata C. Linnaeus var. ***isocycla*** M. Fernald
P. verticillata C. Linnaeus var. ***sphenostachya*** F. Pennell
P. watsonii R. Chodat

POLYGONACEAE, *nomen conservandum*

ANTIGONON S. Endlicher
A. leptopus W. Hooker & G. Arnott [**cultivated**]
Sy = A. cordatum M. Martens & H. Galeotti
Sy = Corculum leptopum (W. Hooker & G. Arnott) S. Stuntz

BRUNNICHIA J. Banks ex D. von Schlechtendal
B. ovata (T. Walter) L. Shinners
Sy = B. cirrhosa J. Gaertner
Sy = Rajania ovata T. Walter

COCCOLOBA C. Linnaeus
C. uvifera (C. Linnaeus) N. von Jacquin [**cultivated**]
Sy = Polygonum uvifera C. Linnaeus

EMEX F. Campderá, *nomen conservandum*
E. spinosa (C. Linnaeus) F. Campderá [**federal noxious weed**]
Sy = Rumex spinosus C. Linnaeus

ERIOGONUM A. Michaux
E. abertianum J. Torrey var. ***abertianum***
Sy = E. abertianum var. gillespiei F. Fosberg
Sy = E. abertianum var. neomexicanum M. Gandoger
Sy = E. abertianum var. villosum F. Fosberg
E. abertianum J. Torrey var. ***cyclosepalum*** (E. Greene) F. Fosberg
Sy = E. cyclosepalum E. Greene
E. alatum J. Torrey var. ***alatum***
Sy = E. alatum subsp. triste (S. Watson) S. Stokes
Sy = E. triste S. Watson
Sy = Pterogonum alatum (J. Torrey) H. Gross
E. alatum J. Torrey var. ***glabriusculum*** J. Torrey
E. annuum T. Nuttall
E. correllii J. Reveal
* ***E. greggii*** J. Torrey & A. Gray [**TOES: V**]
E. havardii S. Watson
Sy = E. leucophyllum E. Wooton & P. Standley
E. hemipterum (J. Torrey) S. Stokes
E. hieracifolium G. Bentham
Sy = E. pannosum E. Wooton & P. Standley
E. jamesii G. Bentham var. ***jamesii***
E. jamesii G. Bentham var. ***undulatum*** (G. Bentham) S. Stokes *ex* M. E. Jones
Sy = E. undulatum G. Bentham
E. lachnogynum J. Torrey *ex* G. Bentham
Sy = E. tetraneuris J. K. Small
E. lonchophyllum J. Torrey & A. Gray var. ***fendlerianum*** (G. Bentham) J. Reveal
Sy = E. ainsliei P. Standley
Sy = E. fendlerianum (G. Bentham) J. K. Small
E. longifolium T. Nuttall var. ***longifolium***
Sy = E. longifolium var. lindheimeri M. Gandoger
Sy = E. longifolium var. plantagineum G. Engelmann & A. Gray
Sy = E. texanum G. Scheele
Sy = E. vespinum L. Shinners
E. multiflorum G. Bentham
E. nealleyi J. Coulter
E. polycladon G. Bentham
Sy = E. densum E. Greene
Sy = E. polycladon var. crispum M. Gandoger
Sy = E. polycladon var. mexicanum M. Gandoger

Sy = E. vimineum D. Douglas *ex* G. Bentham var. densum (E. Greene) S. Stokes
Sy = E. vimineum var. polycladon (G. Bentham) S. Stokes
E. rotundifolium G. Bentham
Sy = E. cernuum T. Nuttall subsp. rotundifolium (G. Bentham) S. Stokes
* ***E. suffruticosum*** S. Watson **[TOES: V]**
E. tenellum J. Torrey var. ***platyphyllum*** (J. Torrey *ex* G. Bentham) J. Torrey
Sy = E. platyphyllum J. Torrey *ex* G. Bentham
E. tenellum J. Torrey var. ***ramosissimum*** G. Bentham
Sy = E. tenellum var. caulescens J. Torrey & A. Gray
E. tenellum J. Torrey var. ***tenellum***
E. wrightii J. Torrey *ex* G. Bentham var. ***wrightii***
Sy = E. trachygonum J. Torrey subsp. wrightii (J. Torrey) S. Stokes
Sy = E. wrightii subsp. glomerulum S. Stokes

POLYGONELLA A. Michaux
P. americana (F. von Fischer & C. von Meyer) J. K. Small
Sy = Gonopyrum americanum F. von Fischer & C. von Meyer
* ***P. parksii*** V. Cory **[TOES: V]**
P. polygama (E. Ventenat) G. Engelmann & A. Gray
Sy = P. polygama var. brachystachya (C. Meisner) R. Wunderlin
Sy = Polygonum polygamum E. Ventenat

POLYGONUM C. Linnaeus, *nomen conservandum*
P. amphibium C. Linnaeus var. ***emersum*** A. Michaux
Sy = Persicaria amphibia (C. Linnaeus) S. Gray var. emersa (A. Michaux) C. Hickman
Sy = P. coccinea (G. H. Muhlenberg *ex* C. von Willdenow) E. Greene
Sy = Polygonum coccineum G. H. Muhlenberg *ex* C. von Willdenow
Sy = P. coccineum G. H. Muhlenberg *ex* C. von Willdenow var. pratincola (E. Greene) E. Stanford
P. argyrocoleon E. von Steudel *ex* G. Kunze
P. aubertii A. C. Henry **[cultivated]**
Sy = Bilderdykia aubertii (A. C. Henry) H. Moldenke
Sy = Fallopia aubertii (A. C. Henry) J. Holub
P. aviculare C. Linnaeus
Sy = P. aviculare var. vegetum C. von Ledebour
Sy = P. heterophyllum J. Lindley
Sy = P. monspeliense C. Persoon
P. buxiforme J. K. Small
Sy = P. aviculare C. Linnaeus var. littorale (J. Link) W. Koch
Sy = P. littorale J. Link
P. caespitosum K. von Blume var. ***longisetum*** (H. de Bruyn) A. Stewart
Sy = P. longisetum H. de Bruyn
P. capitatum F. Buchanan-Hamilton *ex* D. Don **[cultivated]**
P. convolvulus C. Linnaeus var. ***convolvulus***
Sy = Bilderdykia convolvulus (C. Linnaeus) C. Dumortier
Sy = Fallopia convolvulus (C. Linnaeus) A. Löve
Sy = Reynoutria convolvulus (C. Linnaeus) L. Shinners
Sy = Tiniaria convolvulus (C. Linnaeus) P. Webb & C. Moquin-Tandon
P. densiflorum C. Meisner
Sy = Persicaria densiflora (C. Meisner) H. Moldenke
Sy = Polygonum portoricense C. L. Bertero *ex* J. K. Small
P. hydropiper C. Linnaeus
Sy = Persicaria hydropiper (C. Linnaeus) P. M. Opiz
Sy = Polygonum hydropiper var. projectum E. Stanford
P. hydropiperoides A. Michaux
Sy = Persicaria hydropiperoides (A. Michaux) J. K. Small
Sy = Polygonum hydropiperoides var. asperifolium E. Stanford
Sy = P. hydropiperoides var. breviciliatum M. Fernald
Sy = P. hydropiperoides var. bushianum E. Stanford
Sy = P. hydropiperoides var. digitatum M. Fernald
Sy = P. hydropiperoides var. euronotorum M. Fernald
Sy = P. hydropiperoides var. opelousanum (J. Riddell *ex* J. K. Small) J. Riddell *ex* W. Stone
Sy = P. hydropiperoides var. strigosum (J. K. Small) E. Stanford
Sy = P. opelousanum J. Riddell *ex* J. K. Small
P. lapathifolium C. Linnaeus var. ***lapathifolium***
Sy = Persicaria lapathifolia (C. Linnaeus) J. K. Small

Sy = Polygonum incarnatum S. Elliott
Sy = P. lapathifolium subsp. pallidum (W. Withering) E. Fries
Sy = P. lapathifolium var. nodosum (C. Persoon) J. K. Small
Sy = P. lapathifolium var. ovatum A. Braun
Sy = P. lapathifolium var. prostratum C. F. Wimmer

P. orientale C. Linnaeus
Sy = Persicaria orientalis (C. Linnaeus) E. Spach

P. pensylvanicum C. Linnaeus
Sy = Persicaria bicornis (C. Rafinesque-Schmaltz) J. Nieuwland
Sy = P. pensylvanica (C. Linnaeus) G. Maza
Sy = Polygonum bicorne C. Rafinesque-Schmaltz
Sy = P. pensylvanicum var. durum E. Stanford
Sy = P. pensylvanicum var. eglandulosum J. Meyers
Sy = P. pensylvanicum var. genuinum M. Fernald
Sy = P. pensylvanicum var. laevigatum M. Fernald
Sy = P. pensylvanicum var. rosiflorum J. Norton

P. persicaria C. Linnaeus
Sy = Persicaria maculata (C. Rafinesque-Schmaltz) S. Gray
Sy = P. persicaria (C. Linnaeus) J. K. Small
Sy = P. vulgaris P. Webb & C. Moquin-Tandon
Sy = Polygonum persicaria var. angustifolium K. Beckhaus
Sy = P. persicaria var. ruderale (R. A. Salisbury) C. Meisner

P. punctatum S. Elliott var. ***confertiflorum*** (C. Meisner) N. Fassett
Sy = Persicaria punctata (S. Elliott) J. K. Small var. leptostachyum (C. Meisner) J. K. Small
Sy = P. punctatum var. leptostachyum (C. Meisner) J. K. Small
Sy = P. punctatum var. parvum F. Victorin & J. Rousseau

P. ramosissimum A. Michaux var. ***prolificum*** J. K. Small
Sy = P. prolificum (J. K. Small) B. Robinson

P. ramosissimum A. Michaux var. ***ramosissimum***
Sy = P. allocarpum S. F. Blake
Sy = P. autumnale J. Brenckle
Sy = P. interius J. Brenckle
Sy = P. triangulum N. Bicknell

P. sagittatum C. Linnaeus var. ***sagittatum***
Sy = Tracaulon sagittatum (C. Linnaeus) J. K. Small
Sy = Truellum sagittatum (C. Linnaeus) J. Soják

P. scandens C. Linnaeus var. ***cristatum*** (G. Engelmann & A. Gray) H. Gleason
Sy = Bilderdykia cristata (G. Engelmann & A. Gray) E. Greene
Sy = Polygonum cristatum G. Engelmann & A. Gray
Sy = Reynoutria scandens (C. Linnaeus) L. Shinners var. cristata (G. Engelmann & A. Gray) L. Shinners
Sy = Tiniaria cristata (G. Engelmann & A. Gray) J. K. Small

P. scandens C. Linnaeus var. ***scandens***

P. setaceum W. Baldwin var. ***interjectum*** M. Fernald
Sy = Persicaria setacea (W. Baldwin) J. K. Small var. interjecta (M. Fernald) C. F. Reed
Sy = P. setacea var. tonsum (M. Fernald) C. F. Reed
Sy = Polygonum setaceum var. tonsum M. Fernald

P. striatulum B. Robinson
Sy = P. braziliense K. Koch
Sy = P. camporum C. Meisner

P. tenue A. Michaux var. ***tenue***

P. texense M. C. Johnston

P. virginianum C. Linnaeus
Sy = Antenoron virginianum (C. Linnaeus) G. Roberty & S. Vautier
Sy = Persicaria virginiana (C. Linnaeus) K. von Gaertner
Sy = Polygonum virginianum var. glaberrimum (M. Fernald) J. Steyermark
Sy = Tovara virginiana (C. Linnaeus) C. Rafinesque-Schmaltz

RUMEX C. Linnaeus

R. acetosella C. Linnaeus
Sy = Acetosella acetosella (C. Linnaeus) J. K. Small
Sy = Rumex acetosella subsp. angiocarpus (S. Murbeck) S. Murbeck
Sy = R. acetosella var. pyrenaeus (P. Pourret de Figeac) P. Timbal-Lagrave
Sy = R. acetosella var. tenuifolius C. Wallroth

R. altissimus A. Wood
Sy = R. brittanicus C. Meisner, non C. Linnaeus
Sy = R. ellipticus E. Greene

R. chrysocarpus G. Moris
Sy = R. berlandieri C. Meisner

R. conglomeratus J. Murray

R. crispus C. Linnaeus

R. hastatulus W. Baldwin
R. hymenosepalus J. Torrey
Sy = R. hymenosepalus var. euymenosepalus K. H. Rechinger
Sy = R. hymenosepalus var. salinus (A. Nelson) K. H. Rechinger
R. maritimus C. Linnaeus
Sy = R. maritimus subsp. fueginus (R. Philippi) O. Hultén
Sy = R. maritimus var. athrix H. St. John
Sy = R. maritimus var. fueginus (R. Philippi) P. K. Dusén
Sy = R. maritimus var. persicarioides (C. Linnaeus) R. Mitchell
R. obtusifolius C. Linnaeus
Sy = R. obtusifolius subsp. agrestis (E. Fries) B. Danser
Sy = R. obtusifolius subsp. sylvestris (C. Wallroth) K. H. Rechinger
Sy = R. obtusifolius var. sylvestris (C. Wallroth) W. Koch
R. paraguayensis L. Parodi
R. pulcher C. Linnaeus
Sy = R. pulcher subsp. divaricatus (C. Linnaeus) S. Murbeck
R. salicifolius J. Weinmann var. ***mexicanus*** (C. Meisner) C. Hitchcock
Sy = R. mexicanus C. Meisner
Sy = R. salicifolius subsp. triangulivalvis B. Danser
Sy = R. salicifolius var. triangulivalvis (B. Danser) C. Hitchcock
R. spiralis J. K. Small
R. venosus F. Pursh
R. verticillatus C. Linnaeus
Sy = R. fascicularis J. K. Small
Sy = R. floridanus C. Meisner
R. violascens K. H. Rechinger

PORTULACACEAE, *nomen conservandum*

CLAYTONIA C. Linnaeus
C. virginica C. Linnaeus var. ***virginica***

PORTULACA C. Linnaeus
P. grandiflora W. Hooker **[cultivated]**
P. halimoides C. Linnaeus
Sy = P. parvula A. Gray
P. oleracea C. Linnaeus subsp. ***oleracea***
Sy = P. neglecta K. Mackenzie & B. Bush
Sy = P. retusa G. Engelmann
P. pilosa C. Linnaeus
Sy = P. cyanosperma A. Engler
Sy = P. mundula I. M. Johnston
P. suffrutescens G. Engelmann
P. umbraticola K. Kunth subsp. ***coronata*** (J. K. Small) Matthews & Ketron
Sy = P. coronata J. K. Small
P. umbraticola K. Kunth subsp. ***lanceolata*** (G. Engelmann) Matthews & Ketron
Sy = P. lanceolata G. Engelmann

TALINOPSIS A. Gray
T. frutescens A. Gray

TALINUM M. Adanson, *nomen conservandum*
T. aurantiacum G. Engelmann
Sy = Claytonia aurantiaca K. E. O. Kuntze
Sy = Talinum angustissimum (A. Gray) E. Wooton & P. Standley
Sy = T. aurantiacum var. angustissimum A. Gray
T. brevicaule S. Watson
Ma = T. brevifolium *auct.*, non J. Torrey: Texas authors
T. calycinum G. Engelmann
T. chrysanthum J. Rose & P. Standley
T. longipes E. Wooton & P. Standley
T. paniculatum (N. von Jacquin) J. Gaertner var. ***paniculatum***
Sy = Portulaca paniculata J. Jacquin
Sy = Talinum reflexum A. Cavanilles
T. parviflorum T. Nuttall
Sy = T. appalachianum W. Wolf
T. pulchellum E. Wooton & P. Standley
Sy = T. youngiae C. H. Muller
* ***T. rugospermum*** J. Holzinger [**TOES: IV**]

PRIMULACEAE, *nomen conservandum*

ANAGALLIS C. Linnaeus
A. arvensis C. Linnaeus
A. minima (C. Linnaeus) E. Krause
Sy = Centunculus minimus C. Linnaeus

ANDROSACE C. Linnaeus
A. occidentalis F. Pursh
Sy = A. occidentalis var. arizonica (A. Gray) H. St. John
Sy = A. occidentalis var. simplex (P. Rydberg) H. St. John
A. septentrionalis C. Linnaeus subsp. ***glandulosa*** (E. Wooton & P. Standley) G. Robbins

Sy = A. glandulosa E. Wooton & P. Standley
Sy = A. septentrionalis var. glandulosa (E. Wooton & P. Standley) H. St. John

DODECATHEON C. Linnaeus
D. meadia C. Linnaeus subsp. ***brachycarpum*** (J. K. Small) R. Knuth
Sy = D. brachycarpum J. K. Small
Sy = D. meadia var. brachycarpum (J. K. Small) N. Fassett
D. meadia C. Linnaeus subsp. ***meadia***

HOTTONIA C. Linnaeus
H. inflata S. Elliott

LYSIMACHIA C. Linnaeus
L. hybrida A. Michaux
Sy = L. lanceolata T. Walter subsp. hybrida (A. Michaux) J. Ray
Sy = Steironema hybridum (A. Michaux) C. Rafinesque-Schmaltz *ex* B. Jackson
L. lanceolata T. Walter
Sy = L. lanceolata var. angustifolia (J. de Lamarck) A. Gray
Sy = Steironema lanceolatum (T. Walter) A. Gray
L. nummularia C. Linnaeus **[cultivated]**
L. radicans W. Hooker
Sy = Nummularia radicans (W. Hooker) K. E. O. Kuntze
Sy = Steironema radicans (W. Hooker) A. Gray
L. tonsa (A. Wood) A. Engler
Sy = L. tonsa var. simplex (T. Kearney) R. Knuth *ex* A. Engler
Sy = Steironema tonsum (A. Wood) E. Bicknell *ex* N. Britton
Ma = Lysimachia ciliata *auct.*, non C. Linnaeus: Texas authors

SAMOLUS C. Linnaeus
S. ebracteatus K. Kunth subsp. ***alyssoides*** (A. A. Heller) R. Knuth
Sy = S. alyssoides A. A. Heller
Sy = S. ebracteatus K. Kunth var. alyssoides (A. A. Heller) J. Henrickson
S. ebracteatus K. Kunth subsp. ***cuneatus*** (J. K. Small) R. Knuth
Sy = S. cuneatus J. K. Small
Sy = S. ebracteatus K. Kunth var. cuneatus (J. K. Small) J. Henrickson
S. ebracteatus K. Kunth subsp. ***ebracteatus***
Sy = Samodia ebracteata (K. Kunth) F. Baudo
Sy = Samolus ebracteatus K. Kunth var. ebracteatus
S. valerandi C. Linnaeus subsp. ***parviflorus*** (C. Rafinesque-Schmaltz) O. Hultén
Sy = S. floribundus K. Kunth
Sy = S. parviflorus C. Rafinesque-Schmaltz

PROTEACEAE, *nomen conservandum*

GREVILLEA R. Brown *ex* J. Knight, *nomen conservandum*
G. robusta A. Cunningham *ex* R. Brown **[cultivated]**

MACADAMIA F. von Mueller
M. integrifolia J. Maiden & E. Betche **[cultivated]**
Ma = M. ternifolia *auct.*, non F. von Mueller: American authors

STENOCARPUS R. Brown
S. sinuatus (A. Cunningham) S. Endlicher **[cultivated]**

PUNICACEAE, *nomen conservandum*

PUNICA C. Linnaeus
P. granatum C. Linnaeus **[cultivated]**

RAFFLESIACEAE, *nomen conservandum*

PILOSTYLES J. Guillemin
P. thurberi A. Gray

RANUNCULACEAE, *nomen conservandum*

ADONIS C. Linnaeus
A. annua C. Linnaeus **[cultivated]**
Sy = A. autumnalis C. Linnaeus

ANEMONE C. Linnaeus
A. berlandieri G. Pritzel
Sy = A. decapetala var. heterophylla (T. Nuttall) N. Britton
Sy = A. heterophylla T. Nuttall
A. caroliniana T. Walter
A. edwardsiana B. Tharp var. ***edwardsiana***
* ***A. edwardsiana*** B. Tharp var. ***petraea*** D. Correll **[TOES: V]**
A. tuberosa P. Rydberg var. ***texana*** M. Enquist & B. Crozier
Sy = A. okennonii C. Keener & B. Dutton
A. tuberosa P. Rydberg var. ***tuberosa***

AQUILEGIA C. Linnaeus
- ***A. canadensis*** C. Linnaeus
  - Sy = A. canadensis var. australis (J. K. Small) P. Munz
  - Sy = A. canadensis var. coccinea (J. K. Small) P. Munz
  - Sy = A. canadensis var. eminens (E. Greene) J. Boivin
  - Sy = A. canadensis var. hybrida W. Hooker
  - Sy = A. canadensis var. latiuscula (E. Greene) P. Munz
  - Sy = A. phoenicantha V. Cory
- * ***A. chrysantha*** A. Gray var. ***chaplinei*** (P. Standley *ex* E. Payson) H. Lott [**TOES: V**]
  - Sy = A. chaplinei P. Standley *ex* E. Payson
- ***A. chrysantha*** A. Gray var. ***chrysantha***
- * ***A. chrysantha*** A. Gray var. ***hinckleyana*** (P. Munz) H. Lott [**TOES: V**]
  - Sy = A. hinckleyana P. Munz
- * ***A. longissima*** A. Gray [**TOES: V**]

CLEMATIS C. Linnaeus
- ***C. alpina*** (C. Linnaeus) P. Miller
  - Sy = Atragene alpina C. Linnaeus
  - Sy = Clematis columbiana (T. Nuttall) J. Torrey & A. Gray
  - Sy = C. pseudoalpina (G. Kunze) A. Nelson
- ***C. armandi*** A. Franchet [**cultivated**]
- ***C. crispa*** C. Linnaeus
  - Sy = C. crispa var. walteri F. Pursh
  - Sy = Coriflora crispa (C. Linnaeus) W. A. Weber
- ***C. drummondii*** J. Torrey & A. Gray
  - Sy = C. nervata G. Bentham
- ***C. glaucophylla*** J. K. Small
  - Sy = Coriflora glaucophylla (J. K. Small) W. A. Weber
  - Sy = Viorna glaucophylla (J. K. Small) J. K. Small
- ***C. pitcheri*** J. Torrey & A. Gray var. ***dictyota*** (E. Greene) W. Dennis
  - Sy = C. dictyota E. Greene
- ***C. pitcheri*** J. Torrey & A. Gray var. ***pitcheri***
  - Sy = C. filifera G. Bentham
  - Sy = C. pitcheri var. filifera (G. Bentham) B. Robinson
  - Sy = Coriflora pitcheri (J. Torrey & A. Gray) W. A. Weber
  - Sy = Viorna pitcheri (J. Torrey & A. Gray) N. Britton
- ***C. reticulata*** T. Walter
  - Sy = C. subreticulata T. Harbison
  - Sy = Coriflora reticulata (T. Walter) W. A. Weber
  - Sy = Viorna reticulata (T. Walter) J. K. Small
  - Sy = V. subreticulata T. Harbison *ex* J. K. Small
- ***C. ternifolia*** A. P. de Candolle [**cultivated**]
  - Sy = C. dioscoreifolia A. Léveillé & E. Vaniot
  - Sy = C. maximowicziana A. Franchet & P. A. L. Savatier
  - Sy = C. paniculata C. Thunberg, non J. F. Gmelin
- ***C. texensis*** S. Buckley
  - Sy = C. coccinea G. Engelmann
  - Sy = Coriflora texensis (S. Buckley) W. A. Weber
- ***C. versicolor*** J. K. Small *ex* P. Rydberg
  - Sy = Coriflora versicolor (J. K. Small *ex* P. Rydberg) W. A. Weber
  - Sy = Viorna versicolor (J. K. Small *ex* P. Rydberg) J. K. Small
- ***C. viorna*** C. Linnaeus
  - Sy = C. viorna var. flaccida (J. K. Small) Erickson
  - Sy = Coriflora viorna (C. Linnaeus) W. A. Weber
  - Sy = Viorna viorna (C. Linnaeus) J. K. Small
- ***C. virginiana*** C. Linnaeus
  - Sy = C. virginiana var. missouriensis (P. Rydberg) E. Palmer & J. Steyermark

CONSOLIDA S. Gray
- ***C. ajacis*** (C. Linnaeus) P. Schur
  - Sy = C. ambigua (C. Linnaeus) P. Ball & V. Heywood
  - Sy = Delphinium ajacis C. Linnaeus
- ***C. orientalis*** (J. Gay) R. Schrödinger
  - Sy = Delphinium orientale J. Gay

DELPHINIUM C. Linnaeus
- ***D. carolinianum*** T. Walter subsp. ***carolinianum***
  - Sy = D. carolinianum var. crispum L. Perry
  - Sy = D. carolinianum var. nortonianum (K. Mackenzie & B. Bush) L. Perry
- ***D. carolinianum*** T. Walter subsp. ***vimineum*** (D. Don) M. Warnock
  - Sy = D. vimineum D. Don
  - Sy = D. virescens T. Nuttall var. vimineum (D. Don) R. Martin
- ***D. carolinianum*** T. Walter subsp. ***virescens*** (T. Nuttall) C. J. Brooks
  - Sy = D. carolinianum T. Walter subsp. pendardii (E. Huth) M. Warnock
  - Sy = D. pendardii (E. Huth) M. Warnock
  - Sy = D. virescens T. Nuttall var. pendardii (E. Huth) L. Perry
  - Sy = D. virescens var. virescens
- ***D. grandiflorum*** C. Linnaeus [**cultivated**]
- ***D. madrense*** S. Watson
- ***D. wootonii*** P. Rydberg
  - Sy = D. geyeri E. Greene var. wootonii (P. Rydberg) K. Davis

Sy = D. virescens T. Nuttall subsp. wootonii (P. Rydberg) L. Shinners

ISOPYRUM C. Linnaeus

I. biternatum (C. Rafinesque-Schmaltz) J. Torrey & A. Gray

Sy = Enemion biternatum C. Rafinesque-Schmaltz

MYOSURUS C. Linnaeus

M. minimus C. Linnaeus

Sy = M. major E. Greene

Sy = M. shortii C. Rafinesque-Schmaltz

NIGELLA C. Linnaeus

N. damascena C. Linnaeus [**cultivated**]

PULSATILLA P. Miller

P. patens (C. Linnaeus) P. Miller subsp. ***multifida*** (G. Pritzel) A. Zamels

Sy = Anemone patens C. Linnaeus var. multifida G. Pritzel

Sy = A. patens var. nuttalliana (A. P. de Candolle) A. Gray

Sy = A. patens var. wolfgangiana (W. Besser) K. Koch

RANUNCULUS C. Linnaeus

R. abortivus C. Linnaeus

Sy = R. abortivus subsp. acrolasius (M. Fernald) B. Kapoor, A. Löve, & D. Löve

Sy = R. abortivus var. acrolasius M. Fernald

Sy = R. abortivus var. eucyclus M. Fernald

Sy = R. abortivus var. indivisus M. Fernald

R. asiaticus A. von Gueldenstaedt [**cultivated**]

R. cymbalaria F. Pursh var. ***cymbalaria***

Sy = Cyrtorhyncha cymbalaria (F. Pursh) N. Britton

Sy = Ranunculus cymbalaria var. alpinus W. Hooker

R. cymbalaria F. Pursh var. ***saximontanus*** M. Fernald

Sy = Ranunculus cymbalaria subsp. saximontanus (M. Fernald) R. F. Thorne

R. fascicularis G. H. Muhlenberg *ex* J. Bigelow

Sy = R. fascicularis var. apricus (E. Greene) M. Fernald

Sy = R. fascicularis var. cuneiformis (J. K. Small) L. Benson

R. hispidus A. Michaux var. ***nitidus*** (A. Chapman) T. Duncan

Sy = R. carolinianus A. P. de Candolle

Sy = R. septentrionalis J. Poiret var. nitidus A. Chapman

R. laxicaulis (J. Torrey & A. Gray) J. Darby

Sy = R. flammula C. Linnaeus var. laxicaulis J. Torrey & A. Gray

Sy = R. laxicaulis var. mississippiensis (J. K. Small) L. Benson

Sy = R. texensis G. Engelmann

R. longirostris D. Godron

Sy = Batrachium longirostre (D. Godron) F. Schultz

Sy = Ranunculus aquatilis C. Linnaeus var. longirostris (D. Godron) P. Lawson

Sy = R. subrigidus W. Drew

R. macounii N. Britton

Sy = R. macounii var. oreganus (A. Gray) R. Davis

R. macranthus G. Scheele

Sy = R. repens C. Linnaeus var. macranthus A. Gray

R. marginatus J. d'Urville var. ***trachycarpus*** (F. von Fischer & C. von Meyer) G. Arnott

Sy = R. trachycarpus F. von Fischer & C. von Meyer

R. muricatus C. Linnaeus

R. parviflorus C. Linnaeus

R. petiolaris K. Kunth *ex* A. P. de Candolle var. ***arsenei*** (L. Benson) T. Duncan

Sy = R. macranthus G. Scheele var. arsenei L. Benson

R. platensis K. Sprengel

R. pusillus J. Poiret var. ***angustifolius*** (G. Engelmann) L. Benson

Sy = R. pusillus var. trachyspermus G. Engelmann

Sy = R. tener C. Mohr

R. recurvatus J. Poiret

Sy = R. recurvatus var. adpressipilis C. Weatherby

R. repens C. Linnaeus var. ***degeneratus*** P. Schur [**cultivated**]

Sy = R. repens var. pleniflorus M. Fernald

R. repens C. Linnaeus var. ***repens***

Sy = R. repens var. linearilobus A. P. de Candolle

Sy = R. repens var. villosus M. Lamotte

R. sardous H. von Crantz

Sy = R. parvulus C. Linnaeus

R. sceleratus C. Linnaeus var. ***sceleratus***

Sy = Hecatonia scelerata (C. Linnaeus) J. Fourreau

R. trilobus R. Desfontaines

THALICTRUM C. Linnaeus
* ***T. arkansanum*** J. Boivin [**TOES: V**]
T. dasycarpum F. von Fischer & J. Avé-Lallemant
Sy = T. dasycarpum var. hypoglaucum (P. Rydberg) J. Boivin
Sy = T. hypoglaucum P. Rydberg
T. fendleri G. Engelmann *ex* A. Gray var. ***fendleri***
Sy = T. fendleri var. platycarpum W. Trelease
T. fendleri G. Engelmann *ex* A. Gray var. ***wrightii*** (A. Gray) W. Trelease
Sy = T. wrightii A. Gray
* ***T. texanum*** (A. Gray) J. K. Small [**TOES: V**]

XANTHORHIZA H. Marshall
X. simplicissima H. Marshall

RESEDACEAE, *nomen conservandum*

OLIGOMERIS J. Cambessèdes, *nomen conservandum*
O. linifolia (M. H. Vahl) J. Macbride
Sy = Dipetalis subulata K. E. O. Kuntze
Sy = Reseda linifolia M. H. Vahl

RHAMNACEAE, *nomen conservandum*

ADOLPHIA C. Meisner
A. infesta (K. Kunth) C. Meisner
Sy = Ceanothus infestus K. Kunth
Sy = Collectia multiflora A. P. de Candolle

BERCHEMIA A. P. de Candolle, *nomen conservandum*
B. scandens (J. Hill) K. Koch

CEANOTHUS C. Linnaeus
C. americanus C. Linnaeus var. ***pitcheri*** J. Torrey & A. Gray
C. fendleri A. Gray
Sy = C. fendleri var. venosus W. Trelease
Sy = C. fendleri var. viridis A. Gray
Sy = C. subsericeus P. Rydberg
C. greggii A. Gray var. ***greggii***
Sy = C. greggii subsp. greggii
Sy = C. greggii var. orbiculatus E. Kelso
C. herbaceus C. Rafinesque-Schmaltz
Sy = C. herbaceus var. pubescens (J. Torrey & A. Gray *ex* S. Watson) L. Shinners

COLUBRINA L. C. Richard *ex* A. T. de Brongniart, *nomen conservandum*
C. greggii S. Watson
* ***C. stricta*** G. Engelmann *ex* M. C. Johnston [**TOES: V**]
C. texensis (J. Torrey & A. Gray) A. Gray var. ***texensis***
Sy = Rhamnus texensis J. Torrey & A. Gray

CONDALIA A. Cavanilles, *nomen conservandum*
C. ericoides (A. Gray) M. C. Johnston
Sy = Microrhamnus ericoides A. Gray
* ***C. hookeri*** M. C. Johnston var. ***edwardsiana*** (V. Cory) M. C. Johnston [**TOES: V**]
C. hookeri M. C. Johnston var. ***hookeri***
Sy = C. obovata W. Hooker
C. spathulata A. Gray
C. viridis I. M. Johnston
Sy = C. viridis var. reedii V. Cory
C. warnockii M. C. Johnston var. ***warnockii***

FRANGULA P. Miller
F. betulifolia E. Greene subsp. ***betulifolia***
Sy = Rhamnus betulifolia E. Greene
Sy = R. californica J. Eschscholtz var. betulifolia (E. Greene) W. Trelease
Sy = R. purshiana A. P. de Candolle var. betulifolia (E. Greene) V. Cory
F. caroliniana (T. Walter) A. Gray
Sy = Rhamnus caroliniana T. Walter
Sy = R. caroliniana var. mollis M. Fernald

HOVENIA C. Thunberg
H. dulcis C. Thunberg [**cultivated**]

KARWINSKIA J. Zuccarini
K. humboldtiana (J. A. Schultes) J. Zuccarini
Sy = Rhamnus humboldtiana J. A. Schultes

PALIURUS P. Miller
P. spina-christi P. Miller [**cultivated**]
Sy = Rhamnus paliurus C. Linnaeus

RHAMNUS C. Linnaeus
R. lanceolata F. Pursh subsp. ***lanceolata***
R. serrata F. von Humboldt & A. Bonpland *ex* J. A. Schultes var. ***serrata***
Sy = R. fasciculata E. Greene
Sy = R. smithii subsp. fasciculata (E. Greene) C. Wolf

SAGERETIA A. T. de Brongniart
S. thea (P. Osbeck) M. C. Johnston [**cultivated**]
S. wrightii S. Watson

ZIZIPHUS P. Miller
Z. obtusifolia (W. Hooker *ex* J. Torrey & A. Gray) A. Gray var. ***obtusifolia***

Sy = Condalia obtusifolia (W. Hooker *ex* J. Torrey & A. Gray) A. Weberbauer
Sy = Rhamnus obtusifolia W. Hooker *ex* J. Torrey & A. Gray
Sy = Ziziphus lycioides A. Gray
Z. zizyphus (C. Linnaeus) G. Karsten
Sy = Z. jujuba P. Miller

ROSACEAE, *nomen conservandum*

AGRIMONIA C. Linnaeus
* ***A. incisa*** J. Torrey & A. Gray [**TOES: V**]
A. microcarpa C. Wallroth
Sy = A. pubescens C. Wallroth var. microcarpa (C. Wallroth) H. Ahles
Sy = A. pumila G. H. Muhlenberg *ex* N. Britton & A. Brown
A. parviflora W. Aiton
A. rostellata C. Wallroth

AMELANCHIER F. Medikus
A. arborea (F. Michaux) M. Fernald var. ***arborea***
Sy = A. oblongifolia (J. Torrey & A. Gray) M. Römer
A. denticulata (K. Kunth) W. Koch
Sy = Cotoneaster denticulata K. Kunth
Sy = Malacomeles denticulata (K. Kunth) G. Engelmann
A. utahensis B. Köhne subsp. ***utahensis***
Sy = A. alnifolia (T. Nuttall) T. Nuttall *ex* M. Römer var. utahensis (B. Köhne) M. E. Jones
Sy = A. crenata E. Greene
Sy = A. rubescens E. Greene

APHANES C. Linnaeus
A. microcarpa (P. Boissier & G. Reuter) W. Rothmaler
Sy = Alchemilla microcarpa P. Boissier & G. Reuter

ARONIA F. Medikus, *nomen conservandum*
A. arbutifolia (C. Linnaeus) C. Persoon
Sy = A. arbutifolia var. glabra S. Elliott
Sy = Pyrus arbutifolia (C. Linnaeus) C. von Linné
Sy = Sorbus arbutifolia (C. Linnaeus) G. Heynhold

CERCOCARPUS K. Kunth
C. montanus C. Rafinesque-Schmaltz var. ***argenteus*** (P. Rydberg) F. Martin
Sy = C. argenteus P. Rydberg
C. montanus C. Rafinesque-Schmaltz var. ***glaber*** (S. Watson) F. Martin
Sy = C. betuloides T. Nuttall
Sy = C. betuloides var. multiflorus (P. Rydberg) W. Jepson
C. montanus C. Rafinesque-Schmaltz var. ***paucidentatus*** (S. Watson) F. Martin
Sy = C. breviflorus A. Gray
Sy = C. breviflorus var. eximius C. Schneider
Sy = C. paucidentatus (S. Watson) N. Britton

CHAENOMELES J. Lindley, *nomen conservandum*
C. speciosa (R. Sweet) T. Nakai [**cultivated**]
Sy = C. lagenaria (J. Loiseleur-Deslongchamps) G. Koidzumi

COTONEASTER F. Medikus
C. dammeri C. Schneider [**cultivated**]
C. glaucophylla A. Franchet [**cultivated**]
C. horizontalis J. Decaisne [**cultivated**]
C. lacteus W. W. Smith [**cultivated**]
C. microphylla N. Wallich *ex* J. Lindley [**cultivated**]
Sy = C. congestus J. G. Baker

CRATAEGUS C. Linnaeus
C. anamesa C. Sargent
Sy = C. antiplasta C. Sargent
C. berberifolia J. Torrey & A. Gray
Sy = C. berberifolia var. edita (C. Sargent) E. Kruschke
Sy = C. crocina C. Beadle
Sy = C. edita C. Sargent
Sy = C. edura C. Beadle
Sy = C. fera C. Beadle
Sy = C. tersa C. Beadle
Sy = C. torva C. Beadle
C. brachyacantha C. Sargent & G. Engelmann
C. brazoria C. Sargent
C. calpodendron (F. Ehrhart) F. Medikus
Sy = C. calpodendron var. gigantea E. Kruschke
Sy = C. calpodendron var. globosa (C. Sargent) E. Palmer
Sy = C. calpodendron var. hispida (C. Sargent) E. Palmer
Sy = C. calpodendron var. hispidula (C. Sargent) E. Palmer
Sy = C. calpodendron var. microcarpa (A. Chapman) E. Palmer
Sy = C. calpodendron var. mollicula (C. Sargent) E. Palmer
Sy = C. calpodendron var. obesa (W. Ashe) E. Palmer
Sy = C. chapmanii (C. Beadle) W. Ashe
Sy = C. fontanesiana (E. Spach) E. von Steudel

Sy = C. globosa C. Sargent
Sy = C. whittakeri C. Sargent
C. coccinioides W. Ashe [**cultivated**]
C. columbiana T. Howell
C. crus-galli C. Linnaeus
Sy = C. acutifolia C. Sargent
Sy = C. barrettiana C. Sargent
Sy = C. bushii C. Sargent
Sy = C. canby C. Sargent
Sy = C. cherokeënsis C. Sargent
Sy = C. cocksii C. Sargent
Sy = C. crus-galli var. barrettiana (C. Sargent) E. Palmer
Sy = C. crus-galli var. bellica (C. Sargent) E. Palmer
Sy = C. crus-galli var. capillata C. Sargent
Sy = C. crus-galli var. exigua (C. Sargent) W. Eggleston
Sy = C. crus-galli var. leptophylla (C. Sargent) E. Palmer
Sy = C. crus-galli var. macra (C. Beadle) E. Palmer
Sy = C. crus-galli var. oblongata C. Sargent
Sy = C. crus-galli var. pachyphylla (C. Sargent) E. Palmer
Sy = C. crus-galli var. pyracanthifolia W. Aiton
Sy = C. pyracanthoides C. Beadle
Sy = C. sabineana W. Ashe
C. engelmannii C. Sargent
Sy = C. engelmannii var. sinistra (C. Beadle) E. Palmer
Sy = C. sublobulata C. Sargent
C. greggiana W. Eggleston
C. marshallii W. Eggleston
Sy = C. apiifolia A. Michaux
C. mollis G. Scheele
Sy = C. albicans W. Ashe
Sy = C. arkansana C. Sargent
Sy = C. berlandieri C. Sargent
Sy = C. brachyphylla C. Sargent
Sy = C. cibaria C. Beadle
Sy = C. gravida C. Beadle
Sy = C. induta C. Sargent
Sy = C. invisa C. Sargent
Sy = C. lacera C. Sargent
Sy = C. limaria C. Sargent
Sy = C. mollis var. dumetosa (C. Sargent) E. Kruschke
Sy = C. mollis var. gigantea E. Kruschke
Sy = C. mollis var. incisifolia E. Kruschke
Sy = C. mollis var. sera (C. Sargent) W. Eggleston
Sy = C. noelensis C. Sargent
Sy = C. quercina W. Ashe
C. opaca W. Hooker & G. Arnott
C. pearsonii W. Ashe
C. phaenopyrum (C. von Linné) F. Medikus [**cultivated**]
Sy = C. cordata W. Aiton
Sy = C. populifolia T. Walter
Sy = C. youngii C. Sargent
C. poliophylla C. Sargent
C. reverchonii C. Sargent
Sy = C. discolor C. Sargent
Sy = C. reverchonii var. discolor (C. Sargent) E. Palmer
Sy = C. reverchonii var. stevensiana (C. Sargent) E. Palmer
C. rivularis T. Nuttall
Sy = C. douglasii J. Lindley var. rivularis (T. Nuttall) C. Sargent
C. spathulata A. Michaux
C. stenosepala C. Sargent
C. sutherlandensis C. Sargent
C. texana S. Buckley
C. tracyi W. Ashe *ex* W. Eggleston
Sy = C. montivaga C. Sargent
C. turnerorum M. Enquist
Sy = C. secreta J. Phipps
C. uniflora O. von Münchhausen
Sy = C. bisulcata W. Ashe
Sy = C. choriophylla C. Sargent
Sy = C. gregalis C. Beadle
Sy = C. raleighensis W. Ashe
C. viburnifolia C. Sargent
C. viridis C. Linnaeus var. ***desertorum*** (C. Sargent) T. Keeney & M. Enquist
C. viridis C. Linnaeus var. ***viridis***
Sy = C. abbreviata C. Sargent
Sy = C. amicalis C. Sargent
Sy = C. atrorubens W. Ashe
Sy = C. blanda C. Sargent
Sy = C. glabrius C. Sargent
Sy = C. glabriuscula C. Sargent
Sy = C. ingens C. Beadle
Sy = C. interior C. Beadle
Sy = C. micrantha C. Sargent
Sy = C. velutina C. Sargent
Sy = C. viridis var. interior (C. Beadle) E. Palmer
Sy = C. viridis var. lanceolata (C. Sargent) E. Palmer
Sy = C. viridis var. lutensis (C. Sargent) E. Palmer
Sy = C. viridis var. lutescens E. Palmer *ex* H. Gleason

Sy = C. viridis var. velutina (C. Sargent) E. Palmer
C. warneri C. Sargent

DUCHESNEA J. E. Smith
D. indica (A. Andrzejowski) W. Focke
Sy = Fragaria indica A. Andrzejowski

ERIOBOTRYA J. Lindley
E. deflexa T. Nakai **[cultivated]**
E. japonica (C. Thunberg) J. Lindley **[cultivated]**
Sy = Mespilus japonica C. Thunberg

EXOCHORDA J. Lindley
E. racemosa (J. Lindley) A. Rehder **[cultivated]**
Sy = E. grandiflora (W. Hooker) J. Lindley

FALLUGIA S. Endlicher
F. paradoxa (D. Don) S. Endlicher *ex* J. Torrey
Sy = F. mexicana W. Walpers
Sy = Geum cercocarpoides A. P. de Candolle
Sy = Sieversia paradoxa D. Don

FRAGARIA C. Linnaeus
F. chiloënsis (C. Linnaeus) P. Miller **[cultivated]**
F. vesca C. Linnaeus subsp. ***bracteata*** (A. A. Heller) G. Staudt
Sy = F. bracteata A. A. Heller
Sy = F. vesca var. bracteata (A. A. Heller) R. Davis
F. virginiana A. Duchesne subsp. ***grayana*** (P. Rydberg) G. Staudt
Sy = F. grayana P. Vilmorin *ex* J. Gay
Sy = F. virginiana var. illinoënsis A. Gray
F. virginiana A. Duchesne subsp. ***virginiana***
Sy = F. virginiana var. australis P. Rydberg
Sy = F. virginiana var. canadensis (A. Michaux) O. Farwell
Sy = F. virginiana var. ovalis (J. Lehmann) R. Davis
F. × ananassa A. Duchesne [***chiloënsis* × *virginiana***] **[cultivated]**

GEUM C. Linnaeus
G. canadense N. von Jacquin var. ***camporum*** (P. Rydberg) M. Fernald & C. Weatherby
Sy = G. camporum P. Rydberg
G. canadense N. von Jacquin var. ***texanum*** M. Fernald & C. Weatherby
G. vernum (C. Rafinesque-Schmaltz) J. Torrey & A. Gray
Sy = Stylypus vernus C. Rafinesque-Schmaltz

HOLODISCUS (K. Koch) C. Maximowicz, *nomen conservandum*
H. dumosus (T. Nuttall *ex* W. Hooker) A. A. Heller
Sy = H. discolor (F. Pursh) C. Maximowicz var. dumosus (T. Nuttall *ex* W. Hooker) C. Maximowicz *ex* J. Coulter
Sy = Sericotheca dumosa (T. Nuttall) P. Rydberg
Sy = Spiraea dumosa T. Nuttall *ex* W. Hooker

KERRIA A. P. de Candolle
K. japonica (C. Linnaeus) A. P. de Candolle **[cultivated]**

MALUS P. Miller
M. angustifolia (W. Aiton) A. Michaux var. ***angustifolia***
Sy = Pyrus angustifolia W. Aiton
M. floribunda P. von Siebold *ex* L. B. Van Houtte **[cultivated]**
Sy = M. pulcherrima (P. von Siebold) T. Makino
Sy = Pyrus floribunda E. von Kirchner
Sy = P. pulcherrima P. Ascherson & K. Graebner
M. ioënsis (A. Wood) N. Britton var. ***ioënsis***
Sy = M. ioënsis var. bushii A. Rehder
Sy = M. ioënsis var. palmeri A. Rehder
Sy = Pyrus ioënsis (A. Wood) L. Bailey
M. ioënsis (A. Wood) N. Britton var. ***texensis***
Sy = Pyrus ioënsis (A. Wood) L. Bailey var. texana (A. Rehder) L. Bailey

PETROPHYTON P. Rydberg
P. caespitosum (T. Nuttall) P. Rydberg var. ***caespitosum***
Sy = Eriogynia caespitosa (J. Torrey & A. Gray) S. Watson
Sy = Spiraea caespitosa T. Nuttall *ex* J. Torrey & A. Gray
Sy = S. caespitosa var. elatius (T. Nuttall) I. Tidestrom

PHOTINIA J. Lindley
P. serratifolia (R. Desfontaines) C. Kalkman **[cultivated]**
Sy = P. serrulata J. Lindley

P. glabra (C. Thunberg) C. Maximowicz ×
P. serratifolia (R. Desfontaines) C. Kalkman **[cultivated]**
Sy = P. cv. "Fraseri"

PHYSOCARPUS (J. Cambessèdes) C. Maximowicz, *nomen conservandum*
P. monogynus (J. Torrey) J. Coulter
Sy = Opalaster monogynus (J. Torrey) K. E. O. Kuntze
Sy = Spiraea monogyna J. Torrey

PORTERANTHUS N. Britton *ex* J. K. Small
P. stipulatus (G. H. Muhlenberg *ex* C. von Willdenow) N. Britton
Sy = Gillenia stipulata (G. H. Muhlenberg *ex* C. von Willdenow) T. Nuttall

POTENTILLA C. Linnaeus
P. norvegica C. Linnaeus subsp. ***norvegica***
Sy = P. norvegica var. hirsuta (A. Michaux) J. Lehmann
Sy = P. norvegica var. labradorica (J. Lehmann) M. Fernald
P. paradoxa T. Nuttall
Sy = P. nicolletii (S. Watson) E. Sheldon
Sy = P. supina C. Linnaeus subsp. paradoxa (T. Nuttall) J. Soják
P. recta C. Linnaeus
Sy = P. recta var. obscura (C. Nestler) W. Koch
Sy = P. recta var. pilosa (C. von Willdenow) C. von Ledebour
Sy = P. recta var. sulphurea (J. de Lamarck & A. P. de Candolle) J. Peyritsch
P. rivalis T. Nuttall var. ***millegrana*** (G. Engelmann *ex* J. Lehmann) S. Watson
Sy = P. leucocarpa P. Rydberg
Sy = P. millegrana G. Engelmann *ex* J. Lehmann
P. simplex A. Michaux
Sy = P. simplex var. argyrisma M. Fernald
Sy = P. simplex var. calvescens M. Fernald
P. tabernaemontani P. Ascherson **[cultivated]**
Sy = P. crantzii (H. von Crantz) G. Beck *ex* K. Fritsch
Sy = P. crantzii var. hirta J. M. Lange
Sy = P. flabellifolia W. Hooker *ex* J. Torrey & A. Gray var. hirta (J. M. Lange) J. Boivin
Sy = P. maculata P. Pourret de Figeac

PRUNUS C. Linnaeus
P. angustifolia H. Marshall var. ***angustifolia***
Sy = P. angustifolia subsp. varians W. Wight
Sy = P. angustifolia var. varians W. Wight & U. Hedrick
P. armeniaca C. Linnaeus **[cultivated]**
P. caroliniana (P. Miller) W. Aiton
Sy = Laurocerasus caroliniana (P. Miller) M. Römer
P. cerasifera F. Ehrhart **[cultivated]**
P. cerasus C. Linnaeus **[cultivated]**
P. domestica C. Linnaeus **[cultivated]**
P. glandulosa C. Thunberg **[cultivated]**
P. gracilis G. Engelmann & A. Gray
Sy = P. normalis (J. Torrey & A. Gray) J. K. Small
P. havardii (W. Wight) S. Mason
Sy = Amygdalus havardii W. Wight
P. mexicana S. Watson
Sy = P. mexicana var. flutonensis (C. Sargent) C. Sargent
Sy = P. mexicana var. polyandra (C. Sargent) C. Sargent
P. minutiflora G. Engelmann
Sy = Amygdalus minutiflora (G. Engelmann) W. Wight
P. munsoniana W. Wight & U. Hedrick
* ***P. murrayana*** E. Palmer **[TOES: V]**
P. persica (C. Linnaeus) A. Batsch var. ***persica*** **[cultivated]**
Sy = Amygdalus persica C. Linnaeus
Sy = Persica vulgaris P. Miller
P. rivularis G. Scheele
Sy = P. reverchonii C. Sargent
P. serotina F. Ehrhart var. ***eximia*** (J. K. Small) E. Little
Sy = P. eximia J. K. Small
Sy = P. serotina subsp. eximia (J. K. Small) R. McVaugh
P. serotina F. Ehrhart var. ***rufula*** (E. Wooton & P. Standley) R. McVaugh
Sy = Padus rufula E. Wooton & P. Standley
Sy = Prunus virens (E. Wooton & P. Standley) R. McVaugh var. rufula (E. Wooton & P. Standley) C. Sargent
P. serotina F. Ehrhart var. ***serotina***
Sy = P. serotina subsp. serotina
P. serotina F. Ehrhart var. ***virens*** (E. Wooton & P. Standley) F. Shreve
Sy = P. parksii V. Cory
Sy = P. serotina subsp. virens (E. Wooton & P. Standley) R. McVaugh
Sy = P. virens (E. Wooton & P. Standley) F. Shreve
P. texana F. Dietrich
P. umbellata S. Elliott var. ***umbellata***
Sy = P. mitis C. Beadle

Sy = P. tarda C. Sargent
Sy = P. umbellata var. tarda (C. Sargent) W. Wight
P. virginiana C. Linnaeus var. ***demissa*** (T. Nuttall) J. Torrey
Sy = P. demissa (T. Nuttall) W. Walpers
Sy = P. virginiana subsp. demissa (T. Nuttall) Taylor & B. MacBryde
P. virginiana C. Linnaeus var. ***virginiana***
Sy = Padus virginiana (C. Linnaeus) P. Miller
P. yedoënsis J. Matsumura **[cultivated]**

PURSHIA A. P. de Candolle *ex* J. Poiret
P. ericifolia (J. Torrey *ex* A. Gray) J. Henrickson
Sy = Cowania ericifolia J. Torrey *ex* A. Gray

PYRACANTHA M. Römer
P. coccinea M. Römer **[cultivated]**
Sy = Cotoneaster pyracantha (C. Linnaeus) E. Spach
P. fortuneana (C. Maximowicz) H. L. Li **[cultivated]**
Sy = P. crenatiserrata (H. Hance) A. Rehder
P. koidzumii (B. Hayata) A. Rehder **[cultivated]**

PYRUS C. Linnaeus
P. calleryana J. Decaisne **[cultivated]**
P. communis C. Linnaeus **[cultivated]**

RAPHIOLEPIS J. Lindley
R. indica (C. Linnaeus) J. Lindley **[cultivated]**

ROSA C. Linnaeus
R. arkansana T. Porter var. ***arkansana***
Sy = R. lunellii E. Greene
Sy = R. rydbergii E. Greene
R. arkansana T. Porter var. ***suffulta*** (E. Greene) T. Cockerell
Sy = R. suffulta E. Greene
Sy = R. suffulta var. relicta (E. Erlanson) C. Deam
R. banksiae R. Brown **[cultivated]**
R. bracteata J. Wendland **[cultivated]**
R. carolina C. Linnaeus var. ***carolina***
Sy = R. carolina var. glandulosa (F. Crépin) O. Farwell
Sy = R. carolina var. grandiflora (J. G. Baker) A. Rehder
Sy = R. carolina var. obovata (C. Rafinesque-Schmaltz) C. Deam
R. chinense N. von Jacquin **[cultivated]**
R. damascena P. Miller **[cultivated]**
R. eglanteria C. Linnaeus **[cultivated]**
Sy = R. rubiginosa C. Linnaeus
R. foliolosa T. Nuttall *ex* J. Torrey & A. Gray **[cultivated]**
Sy = R. ignota L. Shinners
R. gallica C. Linnaeus var. ***gallica*** **[cultivated]**
R. gallica C. Linnaeus var. ***officinalis*** C. Thory **[cultivated]**
R. laevigata A. Michaux
R. micrantha W. Borrer *ex* J. E. Smith
R. moschata J. Herrmann **[cultivated]**
R. multiflora C. Thunberg *ex* J. Murray
Sy = R. cathayensis (A. Rehder & Percy Wilson) L. Bailey
R. odorata (G. Andréanszky) R. Sweet **[cultivated]**
R. palustris H. Marshall **[cultivated]**
Sy = R. floridana P. Rydberg
Sy = R. lancifolia J. K. Small
Sy = R. palustris var. dasistema (P. Rafinesque-Schmaltz) E. Palmer & J. Steyermark
R. rugosa C. Thunberg **[cultivated]**
R. setigera A. Michaux var. ***setigera***
Sy = R. setigera var. serena E. Palmer & J. Steyermark
R. setigera A. Michaux var. ***tomentosa*** J. Torrey & A. Gray
* ***R. stellata*** E. Wooton subsp. ***mirifica*** (E. Greene) W. Lewis var. ***erlansoniae*** W. Lewis **[TOES: V]**
R. stellata E. Wooton subsp. ***stellata*** var. ***stellata***
Sy = R. tomentosa J. E. Smith
Sy = R. tomentosa var. globulosa G. Rouy
R. wichuaraiana F. Crépin **[cultivated]**
R. woodsii J. Lindley var. ***woodsii***
Sy = R. fendleri F. Crépin
Sy = R. woodsii var. adenosepala (E. Wooton & P. Standley) W. Martin & C. Hutchins
Sy = R. woodsii var. fendleri (F. Crépin) P. Rydberg
Sy = R. woodsii var. hypoleuca (E. Wooton & P. Standley) W. Martin & C. Hutchins
R. × ***anemonoides*** A. Rehder [***laevigata*** × ***odorata***] **[cultivated]**
R. × ***dilecta*** A. Rehder **[cultivated]**
R. × ***rehderiana*** B. Blackburn [***chinensis*** × ***multiflora***] **[cultivated]**

RUBUS C. Linnaeus
R. aboriginum P. Rydberg
Sy = R. almus (L. Bailey) L. Bailey
Sy = R. austrinus L. Bailey
Sy = R. bollianus L. Bailey
Sy = R. clair-brownii L. Bailey
Sy = R. decor L. Bailey

Sy = R. foliaceus L. Bailey
Sy = R. ignarus L. Bailey
Sy = R. ricei L. Bailey
R. apogaeus L. Bailey
Sy = R. exlex L. Bailey
Sy = R. lassus L. Bailey
Sy = R. lundelliorum L. Bailey
Sy = R. uncus L. Bailey
R. argutus J. Link
Sy = R. abundiflorus L. Bailey
Sy = R. betulifolius J. K. Small
Sy = R. floridensis L. Bailey
Sy = R. floridus L. Trattinnick
Sy = R. incisifrons L. Bailey
Sy = R. louisianus A. Berger
Sy = R. penetrans L. Bailey
Sy = R. rhodophyllus P. Rydberg
R. arvensis L. Bailey
Sy = R. saepescandens L. Bailey
R. bifrons L. Vest *ex* L. Trattinnick [**cultivated**]
R. flagellaris C. von Willdenow
Sy = R. alacer L. Bailey
Sy = R. arundelanus W. H. Blanchard
Sy = R. ascendens W. H. Blanchard
Sy = R. ashei L. Bailey
Sy = R. bonus L. Bailey
Sy = R. camurus L. Bailey
Sy = R. clausenii L. Bailey
Sy = R. connixus L. Bailey
Sy = R. cordialis L. Bailey
Sy = R. dissitiflorus M. Fernald
Sy = R. enslenii L. Trattinnick
Sy = R. exemptus L. Bailey
Sy = R. frustratus L. Bailey
Sy = R. geophilus W. H. Blanchard
Sy = R. longipes M. Fernald
Sy = R. maltei L. Bailey
Sy = R. neonefrens L. Bailey
Sy = R. occultus L. Bailey
Sy = R. sailori L. Bailey
Sy = R. serenus L. Bailey
Sy = R. subuniflorus P. Rydberg
Sy = R. tracyi L. Bailey
R. lucidus P. Rydberg
Sy = R. nessianus L. Bailey
R. oklahomus L. Bailey
Sy = R. summotus L. Bailey
R. persistens P. Rydberg
Sy = R. angustus L. Bailey
Sy = R. arrectus L. Bailey
Sy = R. harperi L. Bailey
Sy = R. zoae L. Bailey
R. riograndis L. Bailey
Sy = R. duplaris L. Shinners
Sy = R. trivialis A. Michaux var. duplaris (L. Shinners) W. Mahler

SANGUISORBA C. Linnaeus
S. annua (T. Nuttall *ex* W. Hooker) J. Torrey & A. Gray
Sy = Poteridium annum (T. Nuttall *ex* W. Hooker) E. Spach
S. minor J. Scopoli subsp. ***muricta*** (E. Spach) G. Nordborg [**cultivated**]
Sy = Poteridium polygamum F. von Waldstein & P. Kitaibel

SPIRAEA C. Linnaeus
S. cantoniensis J. de Loureiro [**cultivated**]
S. japonica C. von Linné [**cultivated**]
Sy = S. callosa C. Thunberg
S. prunifolia P. von Siebold & J. Zuccairini [**cultivated**]
Sy = S. prunifolia var. plena C. Schneider
S.* × *bumalda F. Burvenich [***albiflora* × *japonica***] [**cultivated**]
S.* × *vanhouttei (P. L. Briot) É. Carrière [***cantoniensis* × *trilobata***] [**cultivated**]

VAUQUELINIA J. F. Corrêa *ex* F. von Humboldt & A. Bonpland
V. angustifolia P. Rydberg
Sy = V. corymbosa F. von Humboldt & A. Bonpland subsp. angustifolia (P. Rydberg) W. Hess & J. Henrickson

RUBIACEAE, *nomen conservandum*

ASPERULA C. Linnaeus, *nomen conservandum*
A. arvensis C. Linnaeus

BOUVARDIA R. A. Salisbury
B. ternifolia (A. Cavanilles) D. von Schlechtendal
Sy = B. glaberrima G. Engelmann
Sy = Ixora ternifolia A. Cavanilles

CEPHALANTHUS C. Linnaeus
C. occidentalis C. Linnaeus var. ***californicus*** G. Bentham
Sy = C. occidentalis C. Linnaeus var. pubescens C. Rafinesque-Schmaltz
C. salicifolius F. von Humboldt & A. Bonpland
Sy = C. occidentalis C. Linnaeus var. salicifolius A. Gray
Sy = C. peroblongus H. Wernham

CHIOCOCCA P. Browne

*C. **alba*** (C. Linnaeus) A. Hitchcock
Sy = C. macrocarpa M. Martens & H. Galeotti
Sy = C. racemosa C. Linnaeus
Sy = Lonicera alba C. Linnaeus

CRUCIATA P. Miller

*C. **pedemontana*** (C. Bellardi) F. Ehrendorfer
Sy = Galium pedemontanum (C. Bellardi) C. Allioni
Sy = Vallantia pedemontana C. Bellardi

DIODIA C. Linnaeus

D. teres T. Walter var. ***teres***
Sy = Diodella teres (T. Walter) J. K. Small
Sy = Diodia teres var. setifera M. Fernald & L. Griscom

D. virginiana C. Linnaeus var. ***latifolia*** J. Torrey & A. Gray

D. virginiana C. Linnaeus var. ***virginiana***
Sy = D. harperi J. K. Small
Sy = D. hirsuta F. Pursh
Sy = D. tetragona T. Walter

GALIUM C. Linnaeus

G. aparine C. Linnaeus
Sy = G. agreste C. Wallroth var. echinospermon C. Wallroth
Sy = G. aparine subsp. spurium P. K. Dusén
Sy = G. aparine var. echinospermum C. Wallroth
Sy = G. aparine var. intermedium (E. Merrill) J. Briquet
Sy = G. aparine var. minor W. Hooker
Sy = G. aparine var. spurium C. Wimmer & H. Grabowski
Sy = G. aparine var. vaillantii (A. P. de Candolle) W. Koch
Sy = G. spurium C. Linnaeus
Sy = G. vallantii A. P. de Candolle

G. boreale C. Linnaeus
Sy = G. boreale subsp. septentrionale (J. J. Römer & J. A. Schultes) H. Hara
Sy = G. boreale subsp. septentrionale (J. J. Römer & J. A. Schultes) H. Iltis
Sy = G. boreale var. hyssopifolium (G. F. Hoffmann) A. P. de Candolle
Sy = G. boreale var. intermedium A. P. de Candolle
Sy = G. boreale var. linearifolium P. Rydberg
Sy = G. septentrionale J. J. Römer & J. A. Schultes
Sy = G. utahense A. Eastwood

G. circaezans A. Michaux var. ***circaezans***
Sy = G. circaezans var. glabellum N. Britton
Sy = G. circaezans var. glabrum N. Britton
Sy = G. rotundifolium C. Linnaeus var. circaezans (A. Michaux) K. E. O. Kuntze

G. circaezans A. Michaux var. ***hypomalacum*** M. Fernald

* ***G. correllii*** L. Dempster **[TOES: V]**

G. hispidulum A. Michaux
Sy = Bataprine hispidula (A. Michaux) J. Nieuwland
Sy = Galium carolinianum F. Dietrich
Sy = G. hispidulum F. Pursh
Sy = G. peregrina (T. Walter) N. Britton, E. Sterns, & J. Poggenburg

G. mexicanum K. Kunth subsp. ***flexicum*** L. Dempster

G. mexicanum K. Kunth subsp. ***mexicanum***

G. microphyllum A. Gray
Sy = G. nitens M. E. Jones
Sy = Relbunium microphyllum (A. Gray) W. Hemsley

G. obtusum J. Bigelow subsp. ***obtusum***
Sy = G. obtusum var. ramosum H. Gleason

G. orizabense W. Hemsley subsp. ***laevicaule*** (C. Weatherby & S. F. Blake) L. Dempster
Sy = G. pilosum W. Aiton var. laevicaule C. Weatherby & S. F. Blake

G. pilosum W. Aiton var. ***pilosum*** W. Aiton
Sy = G. puncticulosum A. Michaux var. pilosum (W. Aiton) A. P. de Candolle

G. pilosum W. Aiton var. ***puncticulosum*** (A. Michaux) J. Torrey & A. Gray
Sy = G. punctatum C. Persoon
Sy = G. puncticulosum A. Michaux
Sy = G. purpureum T. Walter
Sy = G. walteri J. F. Gmelin

G. proliferum A. Gray
Sy = G. proliferum var. subnudum J. Greenman
Sy = G. virgatum T. Nuttall var. diffusum A. Gray

G. texense A. Gray
Sy = G. californicum W. Hooker & G. Arnott var. texanum J. Torrey & A. Gray
Sy = G. texanum (G. Torrey & A. Gray) K. Wiegand, non G. Scheele

G. tinctorium (C. Linnaeus) J. Scopoli
Sy = G. tinctorium subsp. floridanum (K. Wiegand) C. Puff
Sy = G. tinctorium var. diversifolium W. Wight
Sy = G. tinctorium var. floridanum K. Wiegand

G. trifidum C. Linnaeus subsp. ***subbiflorum*** (K. Wiegand) P. Rydberg

Sy = G. claytonii A. Michaux var. subbiflorum K. Wiegand
Sy = G. subbiflorum (K. Wiegand) P. Rydberg
Sy = G. trifidum var. pusillum A. Gray
Sy = G. trifidum var. subbiflorum K. Wiegand

G. triflorum A. Michaux
Sy = G. brachiatum F. Pursh
Sy = G. flaviflorum A. A. Heller
Sy = G. pennsylvanicum W. Barton
Sy = G. triflorum var. asprelliforme M. Fernald
Sy = G. triflorum var. viridiflorum A. P. de Candolle

G. uncinulatum A. P. de Candolle
Sy = G. uncinulatum var. obstipum (D. von Schlechtendal) S. Watson

G. uniflorum A. Michaux
Sy = Bataprine uniflora (A. Michaux) J. Nieuwland

G. virgatum T. Nuttall
Sy = G. texanum G. Scheele
Sy = G. virgatum var. leiocarpum J. Torrey & A. Gray

G. wrightii A. Gray
Sy = G. frankiniense D. Correll
Sy = G. rothrockii A. Gray
Sy = G. wrightii var. rothrockii (A. Gray) F. Ehrendorfer *ex* R. J. Ferris

GARDENIA J. Ellis, *nomen conservandum*

G. angusta (C. Linnaeus) E. Merrill **[cultivated]**
Sy = G. florida C. Linnaeus
Sy = G. grandiflora J. de Loureiro
Sy = G. jasminoides J. Ellis
Sy = G. radicans C. Thunberg

HAMELIA N. von Jacquin

H. patens N. von Jacquin **[cultivated]**
Sy = H. erecta N. von Jacquin

HEDYOTIS C. Linnaeus

H. angulata F. Fosberg *ex* L. Shinners
Sy = H. nigricans (J. de Lamarck) F. Fosberg var. angulata (F. Fosberg *ex* L. Shinners) W. Lewis
Sy = H. nigricans var. parviflora (A. Gray) W. Lewis

* ***H. butterwickiae*** (E. Terrell) G. Nesom **[TOES: V]**
Sy = Houstonia butterwickiae E. Terrell

H. intricata F. Fosberg
Sy = Houstonia fasciculata A. Gray, non A. Bertolini

* ***H. mullerae*** F. Fosberg **[TOES: V]**
Sy = Houstonia mullerae (F. Fosberg) E. Terrell

H. nigricans (J. de Lamarck) F. Fosberg var. ***austrotexana*** B. L. Turner

H. nigricans (J. de Lamarck) F. Fosberg var. ***nigricans***
Sy = Chamisme angustifolia (A. Michaux) J. Nieuwland
Sy = Gentiana nigricans J. de Lamarck
Sy = Hedyotis nigricans var. filifolia (A. Chapman) L. Shinners
Sy = H. nigricans var. rigidiuscula (A. Gray) L. Shinners
Sy = H. nigricans var. scabra (S. Watson) F. Fosberg
Sy = H. salina (A. A. Heller) L. Shinners
Sy = Houstonia angustifolia A. Michaux
Sy = H. angustifolia var. filifolia (A. Chapman) A. Gray
Sy = H. filifolia (A. Chapman) J. K. Small
Sy = H. nigricans (J. de Lamarck) M. Fernald
Sy = H. rigidiuscula (A. Gray) E. Wooton & P. Standley
Sy = H. salina A. A. Heller
Sy = H. tenuis J. K. Small
Sy = Oldenlandia angustifolia (A. Michaux) A. Gray
Sy = O. angustifolia var. filifolia A. Chapman

H. nigricans (J. de Lamarck) F. Fosberg var. ***papillacea*** B. L. Turner

H. pooleana B. L. Turner

HOUSTONIA C. Linnaeus

H. acerosa (A. Gray) A. Gray *ex* G. Bentham & J. Hooker subsp. ***acerosa***
Sy = Hedyotis acerosa A. Gray var. acerosa
Sy = Mallostoma acerosum W. Hemsley
Sy = Oldenlandia acerosa (A. Gray) A. Gray

H. acerosa (A. Gray) A. Gray *ex* G. Bentham & J. Hooker subsp. ***polypremoides*** (A. Gray) E. Terrell
Sy = Hedyotis acerosa A. Gray var. biglovii (J. Greenman) W. Lewis
Sy = H. polypremoides (A. Gray) L. Shinners
Sy = Houstonia polypremoides A. Gray
Sy = H. polypremoides var. bigelovii J. Greenman

H. correllii (W. Lewis) E. Terrell
* Sy = Hedyotis correllii W. Lewis **[TOES: IV]**

H. croftiae N. Britton & H. Rusby
Sy = Hedyotis croftiae (N. Britton & H. Rusby) L. Shinners

H. humifusa (A. Gray) A. Gray
Sy = Hedyotis humifusa A. Gray
Sy = Oldenlandia humifusa (A. Gray) A. Gray
H. longifolia J. Gaertner var. ***tenuifolia*** (T. Nuttall) A. Wood
Sy = Hedyotis longifolia (J. Gaertner) W. Hooker var. tenuifolia (T. Nuttall) J. Torrey & A. Gray
Sy = H. nuttalliana F. Fosberg
Sy = Houstonia purpurea C. Linnaeus var. tenuifolia (T. Nuttall) F. Fosberg
Sy = H. tenuifolia T. Nuttall
H. micrantha (L. Shinners) E. Terrell
Sy = Hedyotis australis W. Lewis & D. Moore
Sy = H. crassifolia C. Rafinesque-Schmaltz var. micrantha L. Shinners
H. parviflora J. Holzinger *ex* J. Greenman
Sy = Hedyotis greenmanii F. Fosberg *ex* L. Shinners
H. purpurea C. Linnaeus var. ***purpurea***
Sy = Hedyotis purpurea (C. Linnaeus) J. Torrey & A. Gray
Sy = H. purpurea var. pubescens N. Britton
H. pusilla J. Schöpf
Sy = Hedyotis crassifolia C. Rafinesque-Schmaltz
H. rosea (C. Rafinesque-Schmaltz) E. Terrell
Sy = Hedyotis rosea C. Rafinesque-Schmaltz
Sy = H. tayloriae F. Fosberg
H. rubra A. Cavanilles
Sy = Hedyotis rubra (A. Cavanilles) A. Gray
Sy = Houstonia saxicola A. Eastwood
Sy = Oldenlandia rubra (A. Cavanilles) A. Gray
H. subviscosa (C. Wright *ex* A. Gray) A. Gray
Sy = Hedyotis subviscosa (C. Wright *ex* A. Gray) L. Shinners
Sy = Oldenlandia subviscosa C. Wright *ex* A. Gray
H. wrightii A. Gray
Sy = Hedyotis wrightii (A. Gray) F. Fosberg
Sy = Houstonia cervantesii K. Kunth
Sy = H. pygmaea C. H. Muller & M. T. Muller
Sy = H. pygmaea J. J. Römer & J. A. Schultes
Ma = Hedyotis rosea *auct.*, non C. Rafinesque-Schmaltz: Hatch et al. 1990

IXORA C. Linnaeus
I. acuminata W. Roxburgh [**cultivated**]
I. chinensis J. de Lamarck [**cultivated**]
I. coccinea C. Linnaeus [**cultivated**]
I. fulgens W. Roxburgh [**cultivated**]

MANETTIA J. Mutis *ex* C. Linnaeus, *nomen conservandum*
M. cordifolia K. von Martius [**cultivated**]

MITCHELLA C. Linnaeus
M. repens C. Linnaeus

MITRACARPUS J. Zuccarini
M. breviflorus A. Gray
M. hirtus (C. Linnaeus) A. P. de Candolle
Sy = M. villosus (O. Swartz) A. von Chamisso & D. von Schlechtendal *ex* A. P. de Candolle
Sy = Spermacoce hirta C. Linnaeus

OLDENLANDIA C. Linnaeus
O. boscii (A. P. de Candolle) A. Chapman
Sy = Hedyotis boscii A. P. de Candolle
O. corymbosa C. Linnaeus
Sy = Hedyotis corymbosa (C. Linnaeus) J. de Lamarck
O. uniflora C. Linnaeus
Sy = Hedyotis fasciculata A. Bertoloni
Sy = H. uniflora (C. Linnaeus) J. de Lamarck
Sy = H. uniflora var. fasciculata (A. Bertoloni) W. Lewis
Sy = Oldenlandia fasciculata (A. Bertoloni) J. K. Small

PAEDERIA C. Linnaeus, *nomen conservandum*
P. foetida C. Linnaeus [**cultivated**]
Sy = P. scandans (J. de Loureiro) E. Merrill

PENTAS G. Bentham
P. lanceolata (P. Forsskål) M. Deflers [**cultivated**]

PENTODON C. F. Hochstetter
P. pentandrus (K. Schumann) G. Vatke
Sy = Hedyotis pentandra K. Schumann
Sy = Pentodon halei (J. Torrey & A. Gray) A. Gray

RANDIA C. Linnaeus
R. rhagocarpa P. Standley

RICHARDIA C. Linnaeus
R. brasiliensis G. Gómes
R. scabra C. Linnaeus
R. tricocca (J. Torrey & A. Gray) P. Standley
Sy = Crusea tricocca (J. Torrey & A. Gray) A. A. Heller
Sy = Diodia tricocca J. Torrey & A. Gray

SHERARDIA C. Linnaeus
S. arvensis C. Linnaeus

SPERMACOCE C. Linnaeus
S. floridana I. Urban
Sy = S. keyensis J. K. Small
Sy = S. tenuior C. Linnaeus var. floridana (I. Urban) R. Long
S. glabra A. Michaux
S. tenuior C. Linnaeus
Sy = S. riparia A. von Chamisso & D. von Schlechtendal
S. verticillata C. Linnaeus
Sy = Borreria verticillata (C. Linnaeus) G. Meyer

RUTACEAE, *nomen conservandum*

AMYRIS P. Browne
A. madrensis S. Watson
A. texana (S. Buckley) Percy Wilson
Sy = A. parvifolia A. Gray
Sy = Zanthoxylum texanum S. Buckley

CHOISYA K. Kunth
C. dumosa (J. Torrey) A. Gray var. **dumosa**
Sy = Astrophyllum dumosum J. Torrey

CITRUS C. Linnaeus
C. aurantifolia (G. Christmann) W. Swingle **[cultivated]**
Sy = Limonia aurantifolia G. Christmann
C. aurantium C. Linnaeus **[cultivated]**
C. limon (C. Linnaeus) N. Burman **[cultivated]**
C. maxima (N. Burman) E. Merrill var. **maxima** **[cultivated]**
Sy = C. grandis (C. Linnaeus) P. Osbeck
C. medica C. Linnaeus **[cultivated]**
C. reticulata F. M. Blanco **[cultivated]**
C. sinensis (C. Linnaeus) P. Osbeck **[cultivated]**
Sy = C. aurantium C. Linnaeus var. sinensis C. Linnaeus
C. × limonia P. Osbeck [**limon** × **reticulata**] **[cultivated]**
Sy = C. limonum J. Risso
C. × paradisi (C. Linnaeus) J. Macfadyen [**maxima** × **sinensis**] **[cultivated]**
Sy = C. cv. "Ruby" (redblush) **[state fruit]**
Sy = C. maxima (N. Burman) E. Merrill var. uvacarpa E. Merrill & H. A. Lee

COLEONEMA F. Bartling & H. Wendland
C. pulchrum W. Hooker **[cultivated]**

ESENBECKIA K. Kunth
E. berlandieri H. Baillon *ex* W. Hemsley
Sy = E. acapulcensis J. Rose
Sy = E. ovata T. Brandegee
* Sy = E. runyonii C. Morton **[TOES: V]**

FORTUNELLA W. Swingle
F. crassifolia W. Swingle **[cultivated]**
F. japonica (C. Thunberg) W. Swingle **[cultivated]**
F. margarita (J. de Loureiro) W. Swingle **[cultivated]**

HELIETTA L. Tulasne
H. parvifolia (A. Gray *ex* W. Hemsley) G. Bentham
Sy = Ptelea parvifolia A. Gray *ex* W. Hemsley

PONCIRUS C. Rafinesque-Schmaltz
P. trifoliata (C. Linnaeus) C. Rafinesque-Schmaltz
Sy = Citrus trifoliata C. Linnaeus

PTELEA C. Linnaeus
P. trifoliata C. Linnaeus subsp. **angustifolia** (G. Bentham) V. Bailey var. **angustifolia**
Sy = P. angustifolia G. Bentham
Sy = P. jucunda E. Greene
Sy = P. neomexicana E. Greene
Sy = P. trifoliata var. angustifolia (G. Bentham) M. E. Jones
P. trifoliata C. Linnaeus subsp. **angustifolia** (G. Bentham) V. Bailey var. **persicifolia** (E. Greene) V. Bailey
P. trifoliata C. Linnaeus subsp. **polyadenia** (E. Greene) V. Bailey
Sy = P. monticola E. Greene
P. trifoliata C. Linnaeus subsp. **trifoliata** var. **mollis** J. Torrey & A. Gray
Sy = P. tomentosa C. Rafinesque-Schmaltz
P. trifoliata C. Linnaeus subsp. **trifoliata** var. **trifoliata**
Sy = P. baldwinii J. Torrey & A. Gray
Sy = P. microcarpa J. K. Small
Sy = P. serrata J. K. Small
Sy = P. trifoliata var. deamiana J. Nieuwland

RUTA C. Linnaeus
R. chalepensis C. Linnaeus
R. graveolens C. Linnaeus

THAMNOSMA J. Torrey & J. Frémont
T. texana (A. Gray) J. Torrey
Sy = Rutosma purpurea E. Wooton & P. Standley

Sy = R. texana A. Gray
Sy = Thamnosma aldrichii B. Tharp

TRIPHASIA J. de Loureiro
T. trifolia (N. Burman) Percy Wilson

ZANTHOXYLUM C. Linnaeus
Z. clava-herculis C. Linnaeus
Sy = Z. macrophyllum T. Nuttall
Z. fagara (C. Linnaeus) C. Sargent
Sy = Schinus fagara C. Linnaeus
Z. hirsutum S. Buckley
Sy = Z. carolinianum var. fruticosum A. Gray
Sy = Z. clava-herculis var. fruticosum (A. Gray) S. Watson
* ***Z. parvum*** L. Shinners [**TOES: V**]

SALICACEAE, *nomen conservandum*

POPULUS C. Linnaeus
P. alba C. Linnaeus [**cultivated**]
Sy = P. alba var. bolleana F. Lauche
Sy = P. alba var. pyramidalis A. von Bunge
* ***P. angustifolia*** E. James [**TOES: V**]
Sy = P. balsamifera C. Linnaeus var. angustifolia (E. James) S. Watson
Sy = P. canadensis C. Moench var. angustifolia (E. James) A. Wesmael
P. balsamifera C. Linnaeus subsp. ***balsamifera*** [**cultivated**]
Sy = P. candicans W. Aiton
Sy = P. tacamahacca P. Miller
P. deltoides W. Barton *ex* H. Marshall subsp. ***deltoides***
Sy = P. deltoides var. angulata (W. Aiton) C. Sargent
Sy = P. deltoides var. missouriensis (A. C. Henry) A. C. Henry
Sy = P. deltoides var. pilosa (C. Sargent) G. Sudworth
Sy = P. deltoides var. virginiana (Fougeroux de Bondaroy) G. Sudworth
P. deltoides W. Barton *ex* H. Marshall subsp. ***monilifera*** (W. Aiton) J.Eckenwalder
Sy = Monilistus monilifera (W. Aiton) C. Rafinesque-Schmaltz *ex* B. Jackson
Sy = Populus deltoides var. occidentalis P. Rydberg
Sy = P. occidentalis (P. Rydberg) N. Britton *ex* P. Rydberg
Sy = P. sargentii L.-A. Dode
Sy = P. sargentii var. texana (C. Sargent) D. Correll
Sy = P. texana C. Sargent
P. deltoides W. Barton *ex* H. Marshall subsp. ***wislizenii*** (S. Watson) J. Eckenwalder
Sy = P. deltoides var. wislizenii (S. Watson) R. Dorn
Sy = P. fremontii S. Watson var. wislizenii S. Watson
Sy = P. wislizeni (S. Watson) C. Sargent
P. fremontii S. Watson subsp. ***mesetae*** J. Eckenwalder
Sy = P. fremontii var. mesetae (J. Eckenwalder) E. Little
Ma = P. arizonica *auct.*, non C. Sargent: American authors
P. nigra C. Linnaeus
Sy = P. dilatata W. Aiton
Sy = P. italica (J. Du Roi) C. Moench
Sy = P. nigra var. italica J. Du Roi
P. tremuloides A. Michaux
Sy = P. tremula C. Linnaeus subsp. tremuloides (A. Michaux) A. Löve & D. Löve
Sy = P. tremuloides var. aurea (I. Tidestrom) F. Daniels
Sy = P. tremuloides var. cercidiphylla (N. Britton) G. Sudworth
Sy = P. tremuloides var. intermedia F. Victorin
Sy = P. tremuloides var. rhomboidea F. Victorin
Sy = P. tremuloides var. vancouveriana (W. Trelease) C. Sargent
P. × ***acuminata*** P. Rydberg [***angustifolia*** × ***deltoides***]
P. × ***hinkleyana*** D. Correll [***angustifolia*** × ***fremontii***]

SALIX C. Linnaeus
S. amygdaloides N. Andersson
Sy = S. amygdaloides var. wrightii (N. Andersson) C. Schneider
Sy = S. nigra H. Marshall var. amygdaloides (N. Andersson) N. Andersson
Sy = S. wrightii N. Andersson
S. babylonica C. Linnaeus [**cultivated**]
S. caprea C. Linnaeus [**cultivated**]
S. caroliniana A. Michaux
Sy = S. amphibia J. K. Small
Sy = S. harbisonii C. Schneider
Sy = S. longipes R. Shuttleworth *ex* N. Andersson
Sy = S. longipes var. pubescens N. Andersson
Sy = S. longipes var. venulosa (N. Andersson) C. Schneider

Sy = S. longipes var. wardii (M. Bebb) C. Schneider
Sy = S. nigra H. Marshall var. longipes (R. Shuttleworth *ex* N. Andersson) M. Bebb
Sy = S. nigra var. wardii M. Bebb
Sy = S. occidentalis L. Bosc var. longipes (R. Shuttleworth *ex* N. Andersson) M. Bebb
Sy = S. pitcheriana J. Barratt
Sy = S. wardii (M. Bebb) M. Bebb

S. exigua T. Nuttall
Sy = S. exigua subsp. interior (W. Rowlee) A. Cronquist
Sy = S. exigua var. angustissima (N. Andersson) J. Reveal & C. Broome
Sy = S. exigua var. exterior (M. Fernald) C. F. Reed
Sy = S. exigua var. luteosericea (P. Rydberg) C. Schneider
Sy = S. exigua var. nevadensis (S. Watson) C. Schneider
Sy = S. exigua var. pedicellata (N. Andersson) A. Cronquist
Sy = S. exigua var. stenophylla (P. Rydberg) C. Schneider
Sy = S. exigua var. virens W. Rowlee
Sy = S. linearifolia P. Rydberg

S. fragilis C. Linnaeus

S. gooddingii C. Ball
Sy = S. gooddingii var. variabilis C. Ball

S. humilis H. Marshall
Sy = S. humilis var. angustifolia (J. Barratt) N. Andersson
Sy = S. humilis var. grandifolia (J. Barratt) N. Andersson
Sy = S. humilis var. hyporhysa M. Fernald
Sy = S. humilis var. keweenawensis O. Farwell
Sy = S. humilis var. rigidiuscula (N. Andersson) B. Robinson & M. Fernald

S. interior W. Rowlee
Sy = S. interior var. angustissima (N. Andersson) W. A. Dayton
Sy = S. interior var. pedicellata (N. Andersson) C. Ball

S. lasiolepis G. Bentham var. ***lasiolepis***
Sy = S. lasiolepis var. bakeri (K. von Seemen) C. Ball
Sy = S. lasiolepis var. bracelinae C. Ball
Sy = S. lasiolepis var. fallax M. Bebb
Sy = S. lasiolepis var. nivaria W. Jepson
Sy = S. lasiolepis var. sandbergii (P. Rydberg) C. Ball

S. matsudana G. Koidzumi [**cultivated**]

S. nigra H. Marshall
Sy = S. nigra var. altissima C. Sargent
Sy = S. nigra var. brevifolia N. Andersson
Sy = S. nigra var. falcata (F. Pursh) J. Torrey
Sy = S. nigra var. lindheimeri C. Schneider
Sy = S. nigra var. longifolia N. Andersson
Sy = S. nigra var. marginata (C. F. Wimmer *ex* N. Andersson) N. Andersson

S. taxifolia K. Kunth
Sy = S. taxifolia var. lejocarpa N. Andersson
Sy = S. taxifolia var. limitanea I. M. Johnston
Sy = S. taxifolia var. microphylla (D. von Schlechtendal & A. von Chamisso) C. Schneider
Sy = S. taxifolia var. seriocarpa N. Andersson

S.* × *blanda N. Andersson [***babylonica* × *fragilis***] [**cultivated**]
Sy = S. elegantissima K. Koch

SANTALACEAE, *nomen conservandum*

COMANDRA T. Nuttall

C. umbellata (C. Linnaeus) T. Nuttall subsp. ***pallida*** (A. L. de Candolle) M. Piehl
Sy = C. pallida A. L. de Candolle
Sy = C. umbellata var. pallida (A. P. de Candolle) M. E. Jones

SAPINDACEAE, *nomen conservandum*

CARDIOSPERMUM C. Linnaeus

C. corindum C. Linnaeus
Sy = C. corindum var. villosum (P. Miller) L. Radlkofer
Sy = C. keyense J. K. Small

* ***C. dissectum*** (S. Watson) L. Radlkofer [**TOES: V**]
Sy = Urvillea dissecta S. Watson

C. halicacabum C. Linnaeus

CUPANIOPSIS L. Radlkofer

C. anacardioides L. Radlkofer [**cultivated**]

KOELREUTERIA E. Laxmann

K. apiculata E. Laxmann [**cultivated**]

K. bipinnata A. Franchet [**cultivated**]

K. elegans (B. Seemann) A. C. Smith subsp. ***formosana*** (B. Hayata) F. G. Meyer [**cultivated**]
Sy = K. formosana B. Hayata
Sy = K. vitiensis A. C. Smith

K. paniculata E. Laxmann [**cultivated**]
Sy = K. paullinoides L'Héritier de Brutelle
Sy = Sapindus chinensis J. Murray

SAPINDUS C. Linnaeus
S. *saponaria* C. Linnaeus var. ***drummondii*** (W. Hooker & G. Arnott) L. Benson
Sy = S. drummondii W. Hooker & G. Arnott
Sy = S. marginatus J. Coulter

SERJANIA P. Miller
S. *brachycarpa* A. Gray
S. *incisa* J. Torrey

UNGNADIA S. Endlicher
U. *speciosa* S. Endlicher

URVILLEA K. Kunth
U. *ulmacea* K. Kunth
Sy = U. mexicana A. Gray

SAPOTACEAE, *nomen conservandum*

CHRYSOPHYLLUM C. Linnaeus
C. *oliviformi* C. Linnaeus [**cultivated**]
Sy = Cynodendron oliviforme (C. Linnaeus) C. Baehni

SIDEROXYLON C. Linnaeus
S. *celastrinum* (K. Kunth) T. Pennington
Sy = Bumelia angustifolia T. Nuttall
Sy = B. celastrina K. Kunth
Sy = B. celastrina var. angustifolia (T. Nuttall) R. Long
Sy = B. schottii N. Britton
S. *lanuginosum* A. Michaux subsp. ***oblongifolium*** (T. Nuttall) T. Pennington
Sy = Bumelia lanuginosa (A. Michaux) C. Persoon subsp. oblongifolium (T. Nuttall) A. Cronquist var. albicans C. Sargent
Sy = B. lanuginosa var. albicans C. Sargent
Sy = B. lanuginosa var. oblongifolia (T. Nuttall) R. B. Clark
Sy = B. oblongifolia T. Nuttall
Sy = Sideroxylon lanuginosum subsp. albicans (C. Sargent) J. Kartesz & K. Gandhi
S. *lanuginosum* A. Michaux subsp. ***rigidum*** (A. Gray) T. Pennington
Sy = Bumelia lanuginosa (A. Michaux) C. Persoon subsp. rigida (A. Gray) A. Cronquist
Sy = B. lanuginosa subsp. rigida var. texana (S. Buckley) A. Cronquist
Sy = B. lanuginosa var. rigida A. Gray
Sy = B. spinosa A. P. de Candolle, *sensu* S. Watson
Sy = B. texana S. Buckley
S. *lycioides* C. Linnaeus
Sy = Bumelia lycioides (C. Linnaeus) C. Persoon
Sy = B. smallii R. B. Clark

SARRACENIACEAE, *nomen conservandum*

SARRACENIA C. Linnaeus
S. *alata* A. Wood
Sy = S. sledgei J. MacFarlane

SAURURACEAE, *nomen conservandum*

ANEMOPSIS W. Hooker & G. Arnott
A. *californica* (T. Nuttall) W. Hooker & G. Arnott
Sy = A. californica var. subglabra L. Kelso

HOUTTUYNIA C. Thunberg, *nomen conservandum*
H. *cordata* C. Thunberg [**cultivated**]
Sy = Polypara cochinchinensis J. de Loureiro
Sy = P. cordata (C. Thunberg) H. Buek

SAURURUS C. Linnaeus
S. *cernuus* C. Linnaeus

SAXIFRAGACEAE, *nomen conservandum*

HEUCHERA C. Linnaeus
H. *americana* C. Linnaeus var. ***americana***
Sy = H. americana var. brevipetala C. Rosendal, F. Butters, & O. Lakela
Sy = H. americana var. calycosa (J. K. Small) C. Rosendal, F. Butters, & O. Lakela
Sy = H. americana var. heteradenia M. Fernald
Sy = H. americana var. subtruncata M. Fernald
H. *rubescens* J. Torrey var. ***rubescens***
H. *rubescens* J. Torrey var. ***versicolor*** (E. Greene) M. Stewart
Sy = H. leptomeria E. Greene
Sy = H. villosa A. Michaux var. intermedia C. Rosendal, F. Butters, & O. Lakela
Sy = H. villosa var. macrorhiza (J. K. Small) C. Rosendal, F. Butters, & O. Lakela

LEPUROPETALON S. Elliott
L. *spathulatum* S. Elliott
Sy = L. amplexifolium C. Sternberg
Sy = Pyxidanthera spathulata G. H. Muhlenberg

PARNASSIA C. Linnaeus
* ***P. asarifolia*** E. Ventenat [**TOES: IV**]
P. grandifolia A. P. de Candolle

SAXIFRAGA C. Linnaeus
S. stolonifera N. Meerburg [**cultivated**]
Sy = S. sarmentosa C. Linnaeus
S. texana S. Buckley
Sy = Micranthes texana (S. Buckley) J. K. Small
Sy = Saxifraga reevesii V. Cory

SCROPHULARIACEAE, *nomen conservandum*

AGALINIS C. Rafinesque-Schmaltz, *nomen conservandum*
A. aspera (D. Douglas *ex* G. Bentham) N. Britton
Sy = Gerardia aspera D. Douglas *ex* G. Bentham
* ***A. auriculata*** (A. Michaux) S. F. Blake [**TOES: V**]
Sy = Aureolaria auriculata (A. Michaux) O. Farwell
Sy = Gerardia auriculata A. Michaux
Sy = Otophylla auriculata (A. Michaux) J. K. Small
Sy = Tomanthera auriculata (A. Michaux) C. Rafinesque-Schmaltz
A. calycina F. Pennell
Sy = Gerardia calycina (F. Pennell) F. Pennell
A. densiflora (G. Bentham) S. F. Blake
Sy = Gerardia densiflora G. Bentham
Sy = Otophylla densiflora (G. Bentham) J. K. Small
Sy = Tomanthera densiflora (G. Bentham) F. Pennell
A. edwardsiana F. Pennell var. ***glabra*** F. Pennell
Sy = Gerardia edwardsiana (F. Pennell) F. Pennell
Sy = G. edwardsiana subsp. glabra (F. Pennell) F. Pennell
Sy = G. edwardsiana var. glabra (F. Pennell) F. Pennell
A. fasciculata (S. Elliott) C. Rafinesque-Schmaltz
Sy = A. fasciculata var. peninsularis F. Pennell
Sy = A. georgiana (C. Boyton) F. Pennell
Sy = A. purpurea (C. Linnaeus) F. Pennell var. racemulosa (F. Pennell) J. Boivin
Sy = Gerardia fasciculata S. Elliott
Sy = G. fasciculata subsp. peninsularis (F. Pennell) F. Pennell
A. gattingeri (J. K. Small) J. K. Small *ex* N. Britton
Sy = Gerardia gattingeri J. K. Small
A. heterophylla (T. Nuttall) J. K. Small *ex* N. Britton
Sy = Gerardia heterophylla T. Nuttall
A. homalantha F. Pennell
Sy = Gerardia homalantha (F. Pennell) F. Pennell
A. maritima (C. Rafinesque-Schmaltz) C. Rafinesque-Schmaltz var. ***grandiflora*** (G. Bentham) L. Shinners
Sy = A. spiciflora (G. Engelmann) F. Pennell
Sy = Gerardia maritima C. Rafinesque-Schmaltz subsp. grandiflora (G. Bentham) F. Pennell
Sy = G. maritima var. grandiflora G. Bentham
* ***A. navasotensis*** M. Dubrule & J. Canne-Hilliker [**TOES: V**]
A. oligophylla F. Pennell
Sy = A. keyensis F. Pennell
Sy = A. oligophylla var. pseudophylla F. Pennell
Ma = Gerardia aphylla *auct.*, non T. Nuttall: American authors
A. pulchella F. Pennell
Sy = Gerardia pulcherrima F. Pennell
A. purpurea (C. Linnaeus) F. Pennell
Sy = A. purpurea var. carteri F. Pennell
Sy = Gerardia purpurea C. Linnaeus
Sy = G. purpurea subsp. parvula F. Pennell
Sy = G. purpurea var. carteri (F. Pennell) F. Pennell
A. strictifolia (G. Bentham) F. Pennell
Sy = Gerardia strictifolia G. Bentham
A. tenuifolia (M. H. Vahl) C. Rafinesque-Schmaltz var. ***leucanthera*** (C. Rafinesque-Schmaltz) F. Pennell
Sy = A. tenuifolia subsp. leucanthera (C. Rafinesque-Schmaltz) F. Pennell
Sy = Gerardia leucanthera C. Rafinesque-Schmaltz
Sy = G. tenuifolia M. H. Vahl subsp. leucanthera (C. Rafinesque-Schmaltz) F. Pennell
Sy = G. tenuifolia var. leucanthera (C. Rafinesque-Schmaltz) L. Shinners
A. viridis (J. K. Small) F. Pennell
Sy = Gerardia viridis J. K. Small

ANTIRRHINUM C. Linnaeus
A. majus C. Linnaeus [**cultivated**]

AUREOLARIA C. Rafinesque-Schmaltz
A. flava (C. Linnaeus) O. Farwell var. ***flava***
Sy = Agalinis flava (C. Linnaeus) J. Boivin
Sy = Aureolaria calycosa (K. Mackenzie & B. Bush) F. Pennell
Sy = A. dispersa (J. K. Small) F. Pennell
Sy = A. flava subsp. reticulata (C. Rafinesque-Schmaltz) F. Pennell

Sy = A. flava var. reticulata (C. Rafinesque-Schmaltz) F. Pennell
Sy = Gerardia dispersa (J. K. Small) K. Schumann
A. flava (C. Linnaeus) O. Farwell var. ***macrantha*** F. Pennell
Sy = Agalinis flava (C. Linnaeus) J. Boivin var. macrantha F. Pennell
Sy = Gerardia flava C. Linnaeus var. macrantha (F. Pennell) M. Fernald
A. grandiflora (G. Bentham) F. Pennell var. ***cinerea*** F. Pennell
Sy = A. grandiflora subsp. cinerea (F. Pennell) F. Pennell
Sy = A. grandiflora var. cinerea (F. Pennell) V. Cory
A. grandiflora (G. Bentham) F. Pennell var. ***grandiflora***
Sy = Agalinis grandiflora (G. Bentham) S. F. Blake
Sy = Dasistoma grandiflora (G. Bentham) A. Wood
Sy = Gerardia grandiflora G. Bentham
A. grandiflora (G. Bentham) F. Pennell var. ***serrata*** (J. Torrey *ex* G. Bentham) F. Pennell
Sy = Agalinis grandiflora (G. Bentham) S. F. Blake var. serrata (J. Torrey *ex* G. Bentham) S. F. Blake
Sy = Aureolaria grandiflora subsp. serrata (J. Torrey *ex* G. Bentham) F. Pennell
Sy = Dasistoma serrata (J. Torrey *ex* G. Bentham) J. K. Small
Sy = Gerardia grandiflora G. Bentham var. serrata (J. Torrey) B. Robinson
A. pectinata (T. Nuttall) F. Pennell
Sy = A. pectinata subsp. eurycarpa (F. Pennell) F. Pennell
Sy = A. pectinata subsp. floridana F. Pennell
Sy = A. pectinata var. floridana (F. Pennell) F. Pennell
Sy = A. pectinata var. ozarkensis F. Pennell
Sy = A. pectinata var. transcendens F. Pennell
Sy = Gerardia pectinata (T. Nuttall) G. Bentham

BACOPA J. Aublet, *nomen conservandum*
B. caroliniana (T. Walter) B. Robinson
Sy = Hydrotrida caroliniana (T. Walter) J. K. Small
B. monnieri (C. Linnaeus) F. Pennell
Sy = Bramia monnieri (C. Linnaeus) E. Drake del Castillo
Sy = Lysimachia monnieri C. Linnaeus
B. repens (O. Swartz) R. Wettstein
Sy = Gratiola repens O. Swartz
Sy = Macuillamia repens (O. Swartz) F. Pennell
B. rotundifolia (A. Michaux) R. Wettstein
Sy = B. nobsiana F. Mason
Sy = B. simulans M. Fernald
Sy = Bramia rotundifolia (A. Michaux) N. Britton
Sy = Herpestis rotundifolia F. Pursh
Sy = Hydranthelium rotundifolium (A. Michaux) F. Pennell
Sy = Macuillamia rotundifolia (A. Michaux) C. Rafinesque-Schmaltz
Sy = Monniera rotundifolia A. Michaux
Sy = Ranapalus rotundifolius F. Pennell

BELLARDIA C. Allioni
B. trixago (C. Linnaeus) C. Allioni
Sy = Bartsia trixago C. Linnaeus

BUCHNERA C. Linnaeus
B. americana C. Linnaeus
Sy = B. breviflora F. Pennell
Sy = B. floridana M. Gandoger

CASTILLEJA J. Mutis *ex* C. von Linné
C. genevievana G. Nesom
C. indivisa G. Engelmann
C. integra A. Gray var. ***integra***
* Sy = C. elongata F. Pennell [**TOES**: V]
C. lanata A. Gray
C. mexicana (W. Hemsley) A. Gray
Sy = C. tortifolia F. Pennell
C. purpurea (T. Nuttall) G. Don var. ***citrina*** (F. Pennell) L. Shinners
Sy = C. citrina F. Pennell
Sy = C. labiata F. Pennell
C. purpurea (T. Nuttall) G. Don var. ***lindheimeri*** (A. Gray) L. Shinners
Sy = C. lindheimeri A. Gray
Sy = C. mearnsii F. Pennell
Sy = C. williamsii F. Pennell
C. purpurea (T. Nuttall) G. Don var. ***purpurea***
C. rigida A. Eastwood
Sy = C. latebracteata F. Pennell
C. sessiliflora F. Pursh
C. wootonii P. Standley
* Sy = C. ciliata F. Pennell [**TOES**: V]
C. indivisa G. Engelmann × ***C. purpurea*** (T. Nuttall) G. Don var. ***purpurea***
C. lanata A. Gray × ***C. sessiliflora*** F. Pursh

COLLINSIA T. Nuttall
C. violacea T. Nuttall

CORDYLANTHUS T. Nuttall *ex* G. Bentham
C. wrightii A. Gray subsp. ***wrightii***
Sy = C. wrightii var. pauciflorus T. Kearney & R. Peebles

DASISTOMA C. Rafinesque-Schmaltz
D. macrophylla (T. Nuttall) C. Rafinesque-Schmaltz
Sy = Afzelia macrophylla (T. Nuttall) K. E. O. Kuntze
Sy = Seymeria macrophylla T. Nuttall

DIGITALIS C. Linnaeus
D. purpurea C. Linnaeus [**cultivated**]

EPIXIPHIUM (A. Gray) P. Munz
E. wislizeni (G. Engelmann *ex* A. Gray) P. Munz
Sy = Antirrhinum wislizeni (G. Engelmann *ex* A. Gray) I. Tidestrom
Sy = Asarina wislizeni (G. Engelmann *ex* A. Gray) F. Pennell
Sy = Maurandya wislizenii G. Engelmann *ex* A. Gray

GRATIOLA C. Linnaeus
G. brevifolia C. Rafinesque-Schmaltz
G. flava M. Leavenworth
G. neglecta J. Torrey
Sy = G. neglecta var. glaberrima M. Fernald
G. pilosa A. Michaux
Sy = G. pilosa var. epilis F. Pennell
Sy = Sophronanthe pilosa (A. Michaux) J. K. Small
Sy = Tragiola pilosa (A. Michaux) J. K. Small & F. Pennell
G. virginiana C. Linnaeus var. ***virginiana***
Sy = G. sphaerocarpa S. Elliott

KICKXIA C. Dumortier
K. elatine (C. Linnaeus) C. Dumortier
Sy = Antirrhinum elatine C. Linnaeus
Sy = Linaria elatine (C. Linnaeus) P. Miller

LEUCOPHYLLUM F. von Humboldt & A. Bonpland
L. candidum I. M. Johnston
Sy = L. violaceum F. Pennell
L. frutescens (J. Berlandier) I. M. Johnston var. ***frutescens***
Sy = L. texanum G. Bentham
Sy = Terania frutescens J. Berlandier
L. laevigatum P. Standley [**cultivated**]
L. minus A. Gray

LEUCOSPORA T. Nuttall
L. multifida (A. Michaux) T. Nuttall
Sy = Conobea multifida (A. Michaux) G. Bentham

LIMNOPHILA R. Brown, *nomen conservandum*
L. sessiliflora (M. H. Vahl) K. von Blume [**federal noxious weed**]

LINARIA P. Miller
L. vulgaris P. Miller
Sy = Antirrhinum linaria C. Linnaeus
Sy = Linaria linaria (C. Linnaeus) G. Karsten

LINDERNIA C. Allioni
L. dubia (C. Linnaeus) F. Pennell var. ***anagallidea*** (A. Michaux) T. Cooperrider
Sy = Gratiola anagallidea A. Michaux
Sy = Ilysanthes anagallidea C. Rafinesque-Schmaltz
Sy = I. inequalis (T. Walter) F. Pennell
Sy = Lindernia anagallidea (A. Michaux) F. Pennell
L. dubia (C. Linnaeus) F. Pennell var. ***dubia***
Sy = Gratiola dubia C. Linnaeus
Sy = Ilysanthes dubia (C. Linnaeus) J. Barnhart
Sy = I. gratioloides G. Bentham
Sy = Lindernia dubia var. major (F. Pursh) F. Pennell
Sy = L. dubia var. riparia (C. Rafinesque-Schmaltz) M. Fernald

MAURANDYA C. Ortega
M. antirrhiniflora F. von Humboldt & A. Bonpland *ex* C. von Willdenow subsp. ***antirrhiniflora***
Sy = Antirrhinum antirrhiniflora (F. von Humboldt & A. Bonpland *ex* C. Willdenow) C. Hitchcock
Sy = A. maurandioides A. Gray
Sy = Asarina antirrhiniflora (F. von Humboldt & A. Bonpland *ex* C. Willdenow) F. Pennell
Sy = Maurandella antirrhiniflora (F. von Humboldt & A. Bonpland *ex* C. Willdenow) W. Rothmaler

MAZUS J. de Loureiro
M. pumilus (N. Burman) C. van Steenis
Sy = M. japonicus (C. Thunberg) K. E. O. Kuntze

MECARDONIA H. Ruiz López & J. Pavón
M. acuminata (T. Walter) J. K. Small var. ***acuminata***

Sy = Bacopa acuminata (T. Walter) B. Robinson
Sy = Pagesia acuminata (T. Walter) F. Pennell
M. procumbens (P. Miller) J. K. Small
Sy = M. dianthera F. Pennell
Sy = M. peduncularis (G. Bentham) J. K. Small
Sy = M. tenuis J. K. Small
Ma = M. vandellioides *auct.*, non (K. Kunth) F. Pennell: Texas authors

MICRANTHEMUM A. Michaux, *nomen conservandum*
M. umbrosum (J. F. Gmelin) S. F. Blake
Sy = Globifera umbrosa J. F. Gmelin

MIMULUS C. Linnaeus
M. alatus W. Aiton
* ***M. dentilobus*** B. Robinson & M. Fernald **[TOES: V]**
Sy = M. parvulus E. Wooton & P. Standley
M. glabratus K. Kunth var. ***jamesii*** (J. Torrey & A. Gray *ex* G. Bentham) A. Gray
Sy = M. glabratus var. fremontii (G. Bentham) A. Grant
Sy = M. jamesii J. Torrey & A. Gray *ex* G. Bentham
Sy = M. jamesii var. fremontii G. Bentham
M. ringens C. Linnaeus var. ***ringens***
Sy = M. minthodes E. Greene
Sy = M. pallidus R. A. Salisbury
Sy = M. ringens var. congesta O. Farwell
Sy = M. ringens var. minthodes (E. Greene) A. Grant
M. rubellus A. Gray
Sy = M. gratioloides P. Rydberg

NUTTALLANTHUS D. Sutton
N. canadensis (C. Linnaeus) D. Sutton
Sy = Antirrhinum canadense C. Linnaeus
Sy = Linaria canadensis (C. Linnaeus) C. Dumortier
N. texanus (G. Scheele) D. Sutton
Sy = Linaria canadensis (C. Linnaeus) C. Dumortier var. texana (G. Scheele) F. Pennell
Sy = L. texana G. Scheele

PARENTUCELLIA D. Viviani
P. viscosa (C. Linnaeus) C. Caruel

PAULOWNIA P. von Siebold & J. Zuccarini **[cultivated]**
P. tomentosa (C. Thunberg) P. von Siebold & J. Zuccarini *ex* E. von Steudel

PEDICULARIS C. Linnaeus
P. canadensis C. Linnaeus subsp. ***canadensis*** var. ***dobbsii*** M. Fernald

PENSTEMON C. Schmidel
* ***P. alamosensis*** F. Pennell & G. Nisbet **[TOES: V]**
P. albidus T. Nuttall
P. ambiguus J. Torrey var. ***ambiguus***
Sy = Leiostemon ambiguus E. Greene
Sy = L. purpureum C. Rafinesque-Schmaltz
P. ambiguus J. Torrey var. ***laevissimus*** (D. Keck) H. Holmgren
Sy = P. ambiguus subsp. laevissimus D. Keck
P. arkansanus F. Pennell
Sy = P. multicaulis F. Pennell
Sy = P. pallidus J. K. Small subsp. arkansanus (F. Pennell) A. Bennett
Sy = P. wherryi F. Pennell
P. australis J. K. Small subsp. ***laxiflorus*** (F. Pennell) A. Bennett
Sy = P. australis var. ameles F. Crosswhite
Sy = P. laxiflorus F. Pennell
P. baccharifolius W. Hooker
Sy = P. baccharifolius var. schaffneri W. Hemsley
P. barbatus (A. Cavanilles) A. Roth subsp. ***barbatus***
Sy = Chelone barbata A. Cavanilles
Sy = Penstemon barbatus var. barbatus
Sy = P. barbatus var. puberulus A. Gray
P. barbatus (A. Cavanilles) A. Roth subsp. ***torreyi*** (G. Bentham) D. Keck
Sy = P. barbatus var. torreyi (G. Bentham) A. Gray
Sy = P. torreyi G. Bentham
P. buckleyi F. Pennell
* ***P. cardinalis*** E. Wooton & P. Standley subsp. ***regalis*** (A. Nelson) G. Nisbet & R. Jackson **[TOES: V]**
Sy = P. regalis A. Nelson
P. cobaea T. Nuttall
Sy = P. cobaea var. purpureus F. Pennell
P. dasyphyllus A. Gray
P. digitalis T. Nuttall *ex* J. Sims
P. fendleri J. Torrey & A. Gray
P. grandiflorus T. Nuttall
Sy = P. bradburii F. Pursh
P. guadalupensis A. A. Heller
Sy = P. guadalupensis var. ernstii (F. Pennell) V. Cory
P. havardii A. Gray
P. jamesii G. Bentham
Sy = P. brevibarbatus F. Crosswhite
Sy = P. similis A. Nelson

P. laevigatus W. Aiton subsp. ***digitalis*** (T. Nuttall *ex* J. Sims) A. Bennett
P. murrayanus W. Hooker
P. ramosus F. Crosswhite
Ma = P. lanceolatus *auct.*, non G. Bentham: American authors
P. tenuis J. K. Small
P. thurberi J. Torrey
Sy = Leiostemon thurberi E. Greene
Sy = Penstemon ambiguus J. Torrey var. thurberi A. Gray
Sy = P. scoparius A. Nelson
Sy = P. thurberi var. anestius J. Reveal & J. Beatley
P. triflorus A. A. Heller subsp. ***integrifolius*** F. Pennell
Sy = P. helleri J. K. Small
Sy = P. triflorus var. integrifolius (F. Pennell) V. Cory
P. triflorus A. A. Heller subsp. ***triflorus***
P. tubiflorus T. Nuttall var. ***tubiflorus***
P. wrightii W. Hooker

RUSSELIA N. von Jacquin
R. equisetiformis D. von Schlechtendal & A. von Chamisso [**cultivated**]
Sy = R. juncea J. Zuccarini

SCHWALBEA C. Linnaeus
* ***S. americana*** C. Linnaeus [**TOES: V**]
Sy = S. americana var. australis (F. Pennell) J. Reveal & C. Broome
Sy = S. australis F. Pennell

SCOPARIA C. Linnaeus
S. dulcis C. Linnaeus

SCROPHULARIA C. Linnaeus
S. marilandica C. Linnaeus

SEYMERIA F. Pursh, *nomen conservandum*
S. bipinnatisecta B. Seemann
Sy = Afzelia havardii F. Pennell
Sy = Seymeria havardii (F. Pennell) F. Pennell
S. cassioides (J. F. Gmelin) S. F. Blake
Sy = Afzelia cassioides J. F. Gmelin
S. scabra A. Gray
S. texana (A. Gray) F. Pennell

STEMODIA C. Linnaeus, *nomen conservandum*
S. lanata M. Sessé y Lacasta & J. Mociño *ex* G. Bentham
Sy = Erinus tomentosus P. Miller
Sy = Herpestis tomentosa D. von Schlechtendal & A. von Chamisso
Sy = Stemodia tomentosa (P. Miller) J. Greenman & C. Thompson, non (W. Roxburgh) G. Don
Sy = Stemodiacra tomentosa (P. Miller) K. E. O. Kuntze
S. schottii J. Holzinger
Sy = S. purpusii T. Brandegee

VERBASCUM C. Linnaeus
V. blattaria C. Linnaeus
Sy = Blattaria alba P. Miller
Sy = B. vulgaris J. Fourreau
Sy = Verbascum blattaria var. albiflorum D. Don
Sy = V. glabrum P. Miller
V. thapsus C. Linnaeus
V. virgatum J. Stokes
Sy = Blattaria virgata J. Fourreau

VERONICA C. Linnaeus
V. agrestis C. Linnaeus
Sy = V. polita E. Fries
V. americana L. Schweinitz *ex* G. Bentham
Sy = V. americana var. crassula P. Rydberg
Sy = V. beccabunga C. Linnaeus var. americana C. Rafinesque-Schmaltz
Sy = V. crenatifolia E. Greene
Sy = V. oxylobula E. Greene
V. anagallis-aquatica C. Linnaeus
Sy = V. anagallis C. Linnaeus
Sy = V. catenata F. Pennell
Sy = V. glandifera F. Pennell
Sy = V. micromera E. Wooton & P. Standley
V. arvensis C. Linnaeus
V. peregrina C. Linnaeus subsp. ***peregrina***
Sy = V. peregrina var. peregrina
V. peregrina C. Linnaeus subsp. ***xalapensis*** (K. Kunth) F. Pennell
Sy = V. peregrina var. xalapensis (K. Kunth) F. Pennell
Sy = V. sherwoodii M. Peck
Sy = V. xalapensis K. Kunth
V. persica J. Poiret
Sy = V. buxbaumii M. Tenore
Sy = V. diffusa C. Rafinesque-Schmaltz
Sy = V. precox C. Rafinesque-Schmaltz
Sy = V. tournefortii K. Gmelin

VERONICASTRUM L. Heister *ex* P. Fabricius
V. virginicum (C. Linnaeus) O. Farwell

Sy = Leptandra virginica (C. Linnaeus) T. Nuttall
Sy = Veronica virginica C. Linnaeus

SIMAROUBACEAE, *nomen conservandum*

AILANTHUS R. Desfontaines, *nomen conservandum*
A. altissima (P. Miller) W. Swingle
Sy = A. glandulosa R. Desfontaines

CASTELA P. Turpin, *nomen conservandum*
C. erecta P. Turpin subsp. ***texana*** (J. Torrey & A. Gray) J. Rose
Sy = C. erecta subsp. texana (J. Torrey & A. Gray) A. Cronquist
Sy = C. texana (J. Torrey & A. Gray) J. Rose
Sy = C. tortuosa F. Liebmann
Sy = Castelaria texana (J. Torrey & A. Gray) J. K. Small

HOLACANTHA A. Gray
H. stewartii C. H. Muller
Sy = Castela stewartii (C. H. Muller) R. Moran & R. Felger

SOLANACEAE, *nomen conservandum*

BOUCHETIA M. Dunal
B. erecta A. P. de Candolle
Sy = Salpiglossis erecta (A. P. de Candolle) W. D'Arcy

BRUGMANSIA C. Persoon
B. suaveolens (F. von Humboldt & A. Bonpland *ex* C. von Willdenow) B. von Berchtold & K. Presl **[cultivated]**
Sy = Datura suaveolens F. von Humboldt & A. Bonpland *ex* C. von Willdenow

BRUNFELSIA C. Linnaeus (orthographic *conservandum), nomen conservandum*
B. americana C. Linnaeus **[cultivated]**
B. australis G. Bentham **[cultivated]**
B. latifolia G. Bentham **[cultivated]**

CALIBRACHOA P. de la Llave & J. de Lexarza
C. parviflora (A. L. de Jussieu) W. D'Arcy
Sy = Petunia integrifolia (W. Hooker) H. Schinz & A. Thellung
Sy = P. parviflora A. L. de Jussieu

CAPSICUM C. Linnaeus
C. annuum C. Linnaeus var. ***annuum*** **[cultivated]**
Hn = C. annuum var. annuum cv. "jalapeño"
[Note: The 74th state legislature designated the jalapeño as the **state pepper**]
C. annuum C. Linnaeus var. ***aviculare*** (J. Dierbach) W. D'Arcy & W. Eshbaugh
Sy = C. annuum var. glabriusculum (M. Dunal) C. Heiser & B. Pickersgill
Sy = C. annuum var. minus (K. Fingerhuth) L. Shinners
Sy = C. minimum P. Miller
C. annuum C. Linnaeus var. ***grossum*** O. Sendtner **[cultivated]**
C. annuum C. Linnaeus var. ***longum*** O. Sendtner **[cultivated]**
C. baccatum C. Linnaeus var. ***pendulum*** C. von Willdenow **[cultivated]**
C. frutescens C. Linnaeus **[cultivated]**

CESTRUM C. Linnaeus
C. aurantiacum J. Lindley **[cultivated]**
C. diurnum C. Linnaeus var. ***diurnum*** **[cultivated]**
C. nocturnum C. Linnaeus **[cultivated]**
Sy = C. nocturnum var. mexicanum O. E. Shulz
C. parqui L'Héritier de Brutelle **[cultivated]**

CHAMAESARACHA (A. Gray) G. Bentham
C. conoides (M. Moricand *ex* M. Dunal) N. Britton
Sy = Solanum conoides M. Moricand
C. coronopus (M. Dunal) A. Gray
Sy = Saracha coronopus A. Gray
Sy = Solanum coronopus M. Dunal
C. crenata P. Rydberg
C. edwardsiana J. Averett
C. pallida J. Averett
C. sordida (M. Dunal) A. Gray
C. villosa P. Rydberg

DATURA C. Linnaeus
D. candida (C. Persoon) W. Safford **[cultivated]**
D. ferox C. Linnaeus
D. inoxia P. Miller
D. metel C. Linnaeus
Sy = D. fastuosa C. Linnaeus
D. quercifolia K. Kunth
D. stramonium C. Linnaeus
Sy = D. stramonium var. tatula (C. Linnaeus) J. Torrey
Sy = D. tatula C. Linnaeus
D. wrightii E. von Regel
Sy = D. metel var. quinquecuspida J. Torrey
Ma = D. meteloides *auct.*, non M. Dunal: American authors

HUNZIKERIA W. D'Arcy
H. texana (J. Torrey) W. D'Arcy
Sy = Browallia texana J. Torrey
Sy = Leptoglossis texana (J. Torrey) A. Gray
Sy = Nierembergia viscosa J. Torrey

LYCIUM C. Linnaeus
L. berlandieri M. Dunal var. ***berlandieri***
L. berlandieri M. Dunal var. ***parviflorum*** (A. Gray) A. Terracciano
Sy = L. berlandieri var. bervilobum C. Hitchcock
Sy = L. parviflorum A. Gray
L. carolinianum T. Walter var. ***quadrifidum*** (M. Dunal) C. Hitchcock
L. pallidum J. Miers var. ***pallidum***
L. puberulum A. Gray var. ***berberioides*** (D. Correll) C. Chiang
Sy = L. berberioides D. Correll
L. puberulum A. Gray var. ***puberulum***
L. texanum D. Correll
L. torreyi A. Gray
Sy = L. torreyi var. filiforme M. E. Jones

LYCOPERSICON P. Miller
L. esculentum P. Miller var. ***cerastiforme*** (M. Dunal) F. Alefeld **[cultivated]**
Sy = L. cerastiforme M. Dunal
Sy = L. esculentum subsp. galenii (P. Miller) L. Luckwill
Sy = L. esculentum var. leptophyllum (M. Dunal) W. D'Arcy
Sy = L. lycopersicum (C. Linnaeus) G. Karsten *ex* O. Farwell var. cerasiforme (M. Dunal) F. Alefeld
L. esculentum P. Miller var. ***esculentum*** **[cultivated]**, *nomen conservandum*
Sy = L. lycopersicum (C. Linnaeus) G. Karsten *ex* O. Farwell
Sy = Solanum lycopersicum C. Linnaeus

MARGARANTHUS D. von Schlechtendal
M. solanaceus D. von Schlechtendal
Sy = M. lemmonii A. Gray
Sy = M. purpurascens P. Rydberg
Sy = M. tenuis J. Miers

NECTOUXIA K. Kunth
N. formosa K. Kunth

NICOTIANA C. Linnaeus
N. glauca R. Graham
N. longiflora A. Cavanilles
N. plumbaginifolia D. Viviani
N. repanda C. von Willdenow *ex* J. Lehmann
N. tabacum C. Linnaeus
N. trigonophylla M. Dunal var. ***trigonophylla***

NIEREMBERGIA H. Ruiz López & J. Pavón
N. hippomanica J. Miers var. ***coerulea*** (J. Miers) A. Millan **[cultivated]**

PETUNIA A. L. de Jussieu
P. axillaris (J. de Lamarck) N. Britton, E. Sterns, & J. Poggenburg **[cultivated]**
Sy = P. hybrida P. L. de Vilmorin
Sy = P. × atkinsiana D. Don *ex* J. Loudon

PHYSALIS C. Linnaeus
P. acutifolia (J. Miers) N. Sandwith
Sy = Chamaesaracha physaloides E. Greene
Sy = Physalis wrightii A. Gray
P. angulata C. Linnaeus
Sy = P. angulata var. lanceifolia (C. Nees von Esenbeck) U. Waterfall
Sy = P. angulata var. pendula (P. Rydberg) U. Waterfall
Sy = P. lanceifolia C. Nees von Esenbeck
Sy = P. pendula P. Rydberg
P. cinerascens (M. Dunal) A. Hitchcock var. ***cinerascens***
Sy = P. mollis T. Nuttall var. cinerascens (M. Dunal) A. Gray
Sy = P. pensylvanica C. Linnaeus var. cinerascens M. Dunal
Sy = P. viscosa C. Linnaeus var. cinerascens (M. Dunal) U. Waterfall
P. cinerascens (M. Dunal) A. Hitchcock var. ***spathulifolia*** (J. Torrey) J. Sullivan
Sy = P. lanceolata A. Michaux var. spathulifolia J. Torrey
Sy = P. viscosa var. spathulifolia (J. Torrey) A. Gray
P. cordata P. Miller
Sy = P. barbadensis N. von Jacquin var. glabra (A. Michaux) M. Fernald
Sy = P. barbadensis var. obscura (A. Michaux) P. Rydberg
Sy = P. pubescens C. Linnaeus var. glabra (A. Michaux) U. Waterfall
P. hederifolia A. Gray var. ***comata*** (P. Rydberg) U. Waterfall
Sy = P. comata P. Rydberg
Sy = P. rotundata P. Rydberg

P. hederifolia A. Gray var. ***fendleri*** (A. Gray) A. Cronquist
Sy = P. fendleri (A. Gray) A. Cronquist
Sy = P. fendleri var. cordifolia (A. Gray) U. Waterfall
P. hederifolia A. Gray var. ***hederifolia***
Sy = P. hederifolia var. puberula A. Gray
P. heterophylla C. Nees von Esenbeck var. ***heterophylla***
Sy = P. ambigua (A. Gray) N. Britton
Sy = P. heterophylla var. ambigua (A. Gray) P. Rydberg
Sy = P. heterophylla var. clavipes M. Fernald
Sy = P. heterophylla var. nyctaginea (M. Dunal) P. Rydberg
Sy = P. heterophylla var. villosa U. Waterfall
P. heterophylla C. Nees von Esenbeck var. ***rowellii*** E. Stanford
P. hispida (U. Waterfall) A. Cronquist
Sy = P. longifolia T. Nuttall var. hispida (U. Waterfall) J. Steyermark
Sy = P. pumila T. Nuttall subsp. hispida (U. Waterfall) Hinton
Sy = P. virginiana var. hispida U. Waterfall
P. ixocarpa F. Brotero *ex* J. Hornemann var. ***immaculata*** (U. Waterfall) J. Kartesz & K. Gandhi
Sy = P. philadelphica J. de Lamarck var. immaculata U. Waterfall
P. longifolia T. Nuttall var. ***longifolia***
Sy = P. rigida C. Pollard & C. Ball
Sy = P. virginiana P. Miller var. sonorae (J. Torrey) U. Waterfall
P. longifolia T. Nuttall var. ***subglabrata*** (K. Mackenzie & B. Bush) A. Cronquist
Sy = P. macrophysa P. Rydberg
Sy = P. subglabrata K. Mackenzie & B. Bush
Sy = P. virginiana var. subglabrata (K. Mackenzie & B. Bush) U. Waterfall
P. missouriensis K. Mackenzie & B. Bush
Sy = P. pubescens C. Linnaeus var. missouriensis (K. Mackenzie & B. Bush) U. Waterfall
P. mollis T. Nuttall var. ***mollis***
Sy = P. viscosa var. mollis (T. Nuttall) U. Waterfall
P. mollis T. Nuttall var. ***variovestita*** (U. Waterfall) J. Sullivan
Sy = P. variovestita U. Waterfall
P. pruinosa C. Linnaeus
Sy = P. pubescens C. Linnaeus var. grisea U. Waterfall
P. pubescens C. Linnaeus var. ***integrifola*** (M. Dunal) U. Waterfall
P. pubescens C. Linnaeus var. ***pubescens***
Sy = P. barbadensis N. von Jacquin
Sy = P. floridana P. Rydberg
P. pumila T. Nuttall
P. subulata P. Rydberg var. ***neomexicana*** (P. Rydberg) U. Waterfall *ex* J. Kartesz & K. Gandhi
Sy = P. neomexicana P. Rydberg
P. turbinata F. Medikus
P. virginiana P. Miller var. ***texana*** (P. Rydberg) U. Waterfall
Sy = P. texana P. Rydberg
P. virginiana P. Miller var. ***virginiana***
Sy = P. intermedia P. Rydberg
Sy = P. monticola C. Mohr

QUINCULA C. Rafinesque-Schmaltz
Q. lobata (J. Torrey) C. Rafinesque-Schmaltz
Sy = Physalis lobata J. Torrey
Sy = P. lobata var. albiflora U. Waterfall
Sy = P. sabeana S. Buckley

SALPICHROA J. Miers
S. origanifolia (J. de Lamarck) H. Baillon
Sy = Perizoma rhomboidea (J. Gillies & W. Hooker) J. K. Small
Sy = Salpichroa rhomboidea (J. Gillies & W. Hooker) J. Miers

SCHIZANTHUS H. Ruiz López & J. Pavón
S. pinnatus H. Ruiz López & J. Pavón

SOLANUM C. Linnaeus
S. americanum P. Miller
Sy = S. americanum var. nodiflorum (N. von Jacquin) J. Edmonds
Sy = S. americanum var. patulum (C. Linnaeus) J. Edmonds
Sy = S. nigrum C. Linnaeus var. americanum (P. Miller) E. Schulz
Sy = S. nigrum var. virginicum C. Linnaeus
Sy = S. nodiflorum N. von Jacquin
S. campechiense C. Linnaeus
Sy = S. guanicense I. Urban
S. capsicastrum J. Link *ex* J. Schauer
S. capsicoides C. Allioni
Sy = S. ciliatum J. de Lamarck
Ma = S. aculeatissimum *auct.*, non N. von Jacquin: *sensu* O. E. Schulz
S. carolinense C. Linnaeus var. ***carolinense***

S. citrullifolium A. Braun var. ***citrullifolium***
S. citrullifolium A. Braun var. ***setigerum*** H. Bartlett
S. davisense M. Whalen
S. dimidiatum C. Rafinesque-Schmaltz
Sy = S. perplexum J. K. Small
Sy = S. torreyi A. Gray
S. diphyllum C. Linnaeus
S. douglasii M. Dunal
Sy = S. arizonicum F. Pursh
Sy = S. nigrum C. Linnaeus var. douglasii (M. Dunal) A. Gray
S. elaeagnifolium A. Cavanilles
Sy = S. flavidum J. Torrey
Sy = S. leprosum C. Ortega
Sy = S. roemerianaum G. Scheele
Sy = S. texense G. Engelmann & A. Gray
S. erianthum D. Don
Sy = S. erianthum var. adulterinum (W. Hamilton *ex* G. Don) J. G. Baker & J. H. Simmonds
Ma = S. verbascifolium *auct.*, non C. Linnaeus
S. fendleri A. Gray *ex* J. Torrey var. ***fendleri***
S. fendleri A. Gray *ex* J. Torrey var. ***texense*** D. Correll
S. heterodoxum M. Dunal var. ***setigeroides*** M. Whalen
S. interius P. Rydberg
S. jamesii J. Torrey
S. jasminoides J. Paxton **[cultivated]**
* ***S. leptosepalum*** D. Correll **[TOES: V]**
S. melongena C. Linnaeus **[cultivated]**
S. pseudocapsicum C. Linnaeus
S. ptycanthum M. Dunal
Ma = S. americanum *auct.*, non P. Miller: Texas authors
Ma = S. nigrum *auct.*, non C. Linnaeus: Texas authors
S. rantonnetii É. Carrière **[cultivated]**
S. rostratum M. Dunal
Sy = Androcera lobata T. Nuttall
Sy = A. rostrata (M. Dunal) P. Rydberg
Sy = Solanum heterandrum F. Pursh
S. sarrachoides O. Sendtner
Sy = S. villosum P. Miller
S. seaforthianum G. Andrásovszky **[cultivated]**
S. sisymbriifolium J. de Lamarck
S. tenuipes H. Bartlett var. ***latisectum*** M. Whalen
S. tenuipes H. Bartlett var. ***tenuipes***
S. triflorum T. Nuttall
S. triquetrum A. Cavanilles
Sy = S. lindheimerianum G. Scheele
S. tuberosum C. Linnaeus **[cultivated]**

SPHENOCLEACEAE, *nomen conservandum*

SPHENOCLEA J. Gaertner, *nomen conservandum*
S. zeylanica J. Gaertner

STERCULIACEAE, *nomen conservandum*

AYENIA C. Linnaeus
A. filiformis S. Watson
* ***A. limitaris*** C. Cristobal **[TOES: III]**
Sy = Nephropetalum pringlei B. Robinson & J. Greenman
A. microphylla A. Gray
A. pilosa C. Cristobal
Sy = A. pusilla C. Linnaeus

FIRMIANA G. Marsili
F. simplex (C. Linnaeus) W. Wight **[cultivated]**
Sy = F. platanifolia (C. von Linné) H. Schott & S. Endlicher
Sy = Sterculia platanifolia C. von Linné

HERMANNIA C. Linnaeus
H. texana A. Gray

MELOCHIA C. Linnaeus
M. corchorifolia C. Linnaeus
M. pyramidata C. Linnaeus var. ***pyramidata***
M. tomentosa C. Linnaeus
Sy = M. arida J. Rose
Sy = M. plicata J. Presl
Sy = M. speciosa S. Watson
Sy = M. tomentosa var. frutescens (N. von Jacquin) A. P. de Candolle

WALTHERIA C. Linnaeus
W. indica C. Linnaeus
Sy = W. americana C. Linnaeus
Sy = W. detonsa A. Gray
Sy = W. indica var. americana (C. Linnaeus) R. Brown *ex* Hosaka
Sy = W. pyrolifolia A. Gray

STYRACACEAE, *nomen conservandum*

HALESIA J. Ellis *ex* C. Linnaeus, *nomen conservandum*
H. diptera J. Ellis
Sy = H. diptera var. magniflora R. Godfrey

STYRAX C. Linnaeus
S. americanus J. de Lamarck
Sy = S. americanus var. pulverulentus (A. Michaux) J. Perkins *ex* A. Rehder
Sy = S. pulverulentus A. Michaux

S. grandifolius W. Aiton
S. platanifolius G. Engelmann *ex* J. Torrey var. ***platanifolius***
S. platanifolius G. Engelmann *ex* J. Torrey var. ***stellatus*** V. Cory
* ***S. texanus*** V. Cory **[TOES: I]**
* ***S. youngiae*** V. Cory **[TOES: III]**

SYMPLOCACEAE, *nomen conservandum*

SYMPLOCOS N. von Jacquin
S. tinctoria (C. Linnaeus) L'Héritier de Brutelle
Sy = S. tinctoria var. ashei T. Harbison
Sy = S. tinctoria var. pygmaea M. Fernald

TAMARICACEAE, *nomen conservandum*

TAMARIX C. Linnaeus
T. aphylla (C. Linnaeus) G. Karsten
Sy = T. articulata M. H. Vahl
Sy = T. furas F. Buchanan-Hamilton *ex* J. Royle
Sy = T. orientalis P. Forsskål
Sy = T. usneoides E. Meyer
T. chinensis J. de Loureiro
Sy = T. pentandra P. von Pallas
T. gallica C. Linnaeus
Sy = T. africana J. Poiret
T. parviflora A. P. de Candolle
T. ramosissima C. von Ledebour

THEACEAE, *nomen conservandum*

CAMELLIA C. Linnaeus
C. japonica C. Linnaeus **[cultivated]**
Sy = Thea japonica (C. Linnaeus) H. Baillon
C. sasanqua C. Thunberg **[cultivated]**
Sy = Thea miyagii G. Koidz
Sy = T. sasanqua (C. Thunberg) L. Noisette
C. vernalis (T. Makino) T. Makino **[cultivated]**

CLEYERA C. Thunberg
C. japonica C. Thunberg **[cultivated]**

FRANKLINIA W. Bartram *ex* H. Marshall
F. alatamaha W. Bartram *ex* H. Marshall **[cultivated]**
Sy = Gordonia alatamaha (W. Bartram *ex* H. Marshall) C. Sargent
Sy = G. pubescens L'Héritier de Brutelle

GORDONIA J. Ellis, *nomen conservandum*
G. lasianthus (C. Linnaeus) J. Ellis **[cultivated]**

STEWARTIA C. Linnaeus
S. malacodendron C. Linnaeus

TERNSTROEMIA J. Mutis *ex* C. von Linné, *nomen conservandum*
T. gymnanthera (R. Wight & G. Arnott) T. A. Sprague **[cultivated]**
Sy = Cleyera gymnanthera R. Wight & G. Arnott

THYMELAEACEAE, *nomen conservandum*

DAPHNE J. de Tournefort *ex* C. Linnaeus
D. cannabina N. Wallich **[cultivated]**
Sy = D. odora D. Don
D. cneorum C. Linnaeus **[cultivated]**

TILIACEAE, *nomen conservandum*

CORCHORUS C. Linnaeus
C. hirtus C. Linnaeus var. ***glabellus*** A. Gray
C. orinocensis K. Kunth
Sy = C. hirtus C. Linnaeus var. orionocensis (K. Kunth) K. Schumann

GREWIA C. Linnaeus
G. caffra C. Meisner **[cultivated]**

TILIA C. Linnaeus
T. americana C. Linnaeus var. ***caroliniana*** (P. Miller) L. G. Castiglioni
Sy = T. australis J. K. Small
Sy = T. caroliniana P. Miller
Sy = T. floridana J. K. Small
Sy = T. georgiana C. Sargent
Sy = T. leptophylla (E. Ventenat) J. K. Small
Sy = T. leucocarpa W. Ashe
Sy = T. littoralis C. Sargent
Sy = T. porracea W. Ashe
Sy = T. pubescens W. Aiton

TROPAEOLACEAE, *nomen conservandum*

TROPAEOLUM C. Linnaeus
T. majus C. Linnaeus **[cultivated]**

TURNERACEAE, *nomen conservandum*

TURNERA C. Linnaeus
T. diffusa C. von Willdenow *ex* J. A. Schultes var. ***aphrodisiaca*** (L. F. Ward) I. Urban
Sy = T. aphrodisiaca L. F. Ward
T. ulmifolia C. Linnaeus **[cultivated]**

ULMACEAE, *nomen conservandum*

CELTIS C. Linnaeus
- ***C. laevigata*** C. von Willdenow var. ***laevigata***
  - Sy = C. laevigata var. anomala C. Sargent
  - Sy = C. laevigata var. brachyphylla C. Sargent
- ***C. laevigata*** C. von Willdenow var. ***reticulata*** (J. Torrey) L. Benson
  - Sy = C. occidentalis C. Linnaeus var. reticulata (J. Torrey) C. Sargent
  - Sy = C. reticulata J. Torrey
  - Sy = C. reticulata var. vesita C. Sargent
- ***C. lindheimeri*** G. Engelmann *ex* K. Koch
- ***C. occidentalis*** C. Linnaeus var. ***occidentalis***
  - Sy = C. occidentalis var. crassifolia (J. de Lamarck) A. Gray
- ***C. pallida*** J. Torrey
  - Sy = C. spinosa K. Sprengel var. pallida (J. Torrey) M. C. Johnston
  - Sy = C. tala J. Gillies var. pallida (J. Torrey) J. Planchon
  - Sy = Momisia pallida (J. Torrey) J. Planchon
- ***C. tenuifolia*** T. Nuttall
  - Sy = C. georgiana J. K. Small
  - Sy = C. laevigata C. von Willdenow var. smallii C. Sargent
  - Sy = C. smallii C. Beadle
  - Sy = C. tenuifolia T. Nuttall var. georgiana (J. K. Small) M. Fernald & B. Schubert
  - Sy = C. tenuifolia var. soperi J. Boivin

PLANERA J. F. Gmelin
- ***P. aquatica*** (T. Walter) J. F. Gmelin

ULMUS C. Linnaeus
- ***U. alata*** A. Michaux
- ***U. americana*** C. Linnaeus
  - Sy = U. americana var. floridana (A. Chapman) E. Little
  - Sy = U. floridana A. Chapman
- ***U. crassifolia*** T. Nuttall
- ***U. parvifolia*** N. von Jacquin [**cultivated**]
  - Sy = U. chinensis C. Persoon
- ***U. pumila*** C. Linnaeus [**cultivated**]
- ***U. rubra*** G. H. Muhlenberg
  - Sy = U. fulva A. Michaux

ZELKOVA E. Spach, *nomen conservandum*
- ***Z. serrata*** (C. Thunberg) T. Makino [**cultivated**]

URTICACEAE, *nomen conservandum*

BOEHMERIA N. von Jacquin
- ***B. cylindrica*** (C. Linnaeus) O. Swartz
  - Sy = B. austrina J. K. Small
  - Sy = B. cylindrica var. drummondiana (H. Weddell) H. Weddell
  - Sy = B. cylindrica var. scabra T. Porter
  - Sy = Urtica cylindrica C. Linnaeus
- ***B. nivea*** (C. Linnaeus) J. Gaudin
  - Sy = Ramium niveum (C. Linnaeus) J. K. Small
  - Sy = Urtica nivea C. Linnaeus

PARIETARIA C. Linnaeus
- ***P. floridana*** T. Nuttall
  - Sy = P. nummularia J. K. Small
- ***P. judaica*** C. Linnaeus
  - Sy = P. diffusa F. Mertens & W. Koch
- ***P. pensylvanica*** G. H. Muhlenberg *ex* C. von Willdenow var. ***obtusa*** (P. Rydberg *ex* J. K. Small) L. Shinners
  - Sy = P. obtusa P. Rydberg *ex* J. K. Small
  - Sy = P. occidentalis P. Rydberg
- ***P. pensylvanica*** G. H. Muhlenberg *ex* C. von Willdenow var. ***pensylvanica***

PILEA J. Lindley, *nomen conservandum*
- ***P. microphylla*** (C. Linnaeus) F. Liebmann
  - Sy = Parietaria microphylla C. Linnaeus
- ***P. pumila*** (C. Linnaeus) A. Gray var. ***deamii*** (J. Lunell) M. Fernald

URTICA C. Linnaeus
- ***U. chamaedryoides*** F. Pursh var. ***chamaedryoides***
- ***U. chamaedryoides*** var. ***runyonii*** D. Correll
- ***U. dioica*** C. Linnaeus subsp. ***gracilis*** (W. Aiton) N. Selander
  - Sy = U. californica E. Greene
  - Sy = U. cardiophylla P. Rydberg
  - Sy = U. dioica var. angustifolia D. von Schlechtendal
  - Sy = U. dioica var. californica (E. Greene) C. Hitchcock
  - Sy = U. dioica var. gracilis (W. Aiton) C. Hitchcock
  - Sy = U. dioica var. lyallii (S. Watson) C. Hitchcock
  - Sy = U. dioica var. procera (G. H. Muhlenberg *ex* C. von Willdenow) H. Weddell
  - Sy = U. gracilis W. Aiton
  - Sy = U. strigosissima P. Rydberg
  - Sy = U. viridis P. Rydberg
- ***U. gracilenta*** E. Greene
- ***U. urens*** C. Linnaeus

VALERIANACEAE, *nomen conservandum*

VALERIANA C. Linnaeus
- ***V. arizonica*** A. Gray

Sy = V. acutiloba P. Rydberg var. ovata (P. Rydberg) A. Nelson
Sy = V. ovata P. Rydberg
* ***V. texana*** J. Steyermark **[TOES: V]**

VALERIANELLA P. Miller
V. amarella (F. Lindheimer *ex* G. Engelmann) T. Krok
Sy = Fedia amarella F. Lindheimer *ex* G. Engelmann
V. florifera L. Shinners
V. radiata (C. Linnaeus) P. Dufresne
Sy = V. radiata var. fernaldii S. Dyal
Sy = V. radiata var. missouriensis S. Dyal
Sy = V. stenocarpa (G. Engelmann) T. Krok var. parviflora S. Dyal
V. stenocarpa (G. Engelmann *ex* A. Gray) S. Dyal
* ***V. texana*** S. Dyal **[TOES: V]**
V. woodsiana (J. Torrey & A. Gray) W. Walpers

VERBENACEAE, *nomen conservandum*

ALOYSIA A. L. de Jussieu
A. gratissima (J. Gillies & W. Hooker) N. Troncoso var. ***gratissima***
Sy = A. lycioides A. von Chamisso
Sy = Lippia lycioides (A. von Chamisso) E. von Steudel
A. gratissima (J. Gillies & W. Hooker) N. Troncoso var. ***schulziae*** (P. Standley) L. Benson
Sy = A. lycioides var. schulziae (P. Standley) H. Moldenke
A. macrostachya (J. Torrey) H. Moldenke
Sy = Lippia macrostachya (J. Torrey) S. Watson
A. triphylla (L'Héritier de Brutelle) N. Britton **[cultivated]**
Sy = A. citriodora C. Ortega *ex* C. Persoon
Sy = Lippia citriodora K. Kunth
A. wrightii A. A. Heller *ex* L. Abrams
Sy = Lippia wrightii A. Gray *ex* J. Torrey

BOUCHEA A. von Chamisso, *nomen conservandum*
B. linifolia A. Gray
B. prismatica (C. Linnaeus) K. E. O. Kuntze
Sy = B. prismatica var. brevirostra M. Grenzebach
Sy = B. prismatica var. longirostra M. Grenzebach
Sy = Verbena prismatica C. Linnaeus
B. spathulata J. Torrey

CALLICARPA C. Linnaeus
C. americana C. Linnaeus
Sy = C. americana var. lactea F. von Mueller

CARYOPTERIS A. von Bunge
C. × clandonensis A. Rehder [***incana × mongholica***] **[cultivated]**

CITHAREXYLUM C. Linnaeus
C. berlandieri B. Robinson
C. brachyanthum (A. Gray) A. Gray
Sy = C. brachyanthum var. glabrum C. Hitchcock & H. Moldenke
* Sy = C. spathulatum H. Moldenke & C. Lundell **[TOES: V]**
Sy = Lycium brachyanthum A. Gray

CLERODENDRUM C. Linnaeus
C. bungei E. von Steudel **[cultivated]**
Sy = C. foetidum A. von Bunge
C. indicum (C. Linnaeus) K. E. O. Kuntze **[cultivated]**
Sy = Siphonanthus indicus C. Linnaeus
C. thompsoniane I. Balfour **[cultivated]**

DURANTA C. Linnaeus
D. erecta C. Linnaeus **[cultivated]**
Sy = D. erecta var. alba (M. T. Masters) J. A. Caro
Sy = D. repens C. Linnaeus
Sy = D. repens var. alba (M. T. Masters) L. Bailey
Sy = D. repens var. macrophylla (R. Desfontaines) H. Moldenke

GLANDULARIA J. F. Gmelin
G. bipinnatifida (T. Nuttall) T. Nuttall var. ***bipinnatifida***
Sy = Verbena ambrosiifolia P. Rydberg *ex* J. K. Small
Sy = V. bipinnatifida T. Nuttall
Sy = V. bipinnatifida var. latilobata L. Perry
Sy = V. ciliata G. Bentham var. ciliata
Sy = V. ciliata var. longidentata L. Perry
Sy = V. ciliata var. pubera (E. Greene) L. Perry
Sy = V. pubera E. Greene
G. bipinnatifida (T. Nuttall) T. Nuttall var. ***brevispicata*** R. Umber
Sy = Verbena bipinnatifida T. Nuttall var. brevispicata (R. Umber) H. Moldenke
G. canadensis (C. Linnaeus) T. Nuttall
Sy = Verbena canadensis (C. Linnaeus) T. Nuttall var. atroviolacea H. Dermen
Sy = V. canadensis var. compacta H. Dermen
Sy = V. canadensis var. drummondii (J. Lindley) E. Baxter
Sy = V. canadensis var. grandiflora (F. Haage & F. Schmidt) H. Moldenke

Sy = V. canadensis var. lambertii (J. Sims) A. Thellung
Sy = V. × oklahomensis H. Moldenke
G. delticola (J. K. Small *ex* L. Perry) R. Umber
Sy = Verbena cameronensis L. Davis
Sy = V. delticola J. K. Small *ex* L. Perry
Sy = V. lundelliorum H. Moldenke
G. elegans (K. Kunth) R. Umber var. ***asperata*** (L. Perry) R. Umber
Sy = Verbena elegans K. Kunth var. asperata L. Perry
G. polyantha R. Umber
Sy = Verbena polyantha (R. Umber) H. Moldenke
G. pulchella (R. Sweet) N. Troncoso
Sy = G. tenuisecta (J. Briquet) J. K. Small
Sy = Verbena pulchella R. Sweet
Sy = V. pulchella var. gracilior (N. Troncoso) L. Shinners
Sy = V. tenuisecta J. Briquet var. alba H. Moldenke
G. pumila (P. Rydberg) R. Umber
Sy = Verbena inconspicua E. Greene
Sy = V. pumila P. Rydberg
G. quandrangulata (A. A. Heller) R. Umber
Sy = Verbena quadrangulata A. A. Heller
G. racemosa (H. Eggert) R. Umber
Sy = Verbena racemosa H. Eggert
G. tumidula (L. Perry) R. Umber
Sy = Verbena tumidula L. Perry
G. vercunda R. Umber
Sy = Verbena vercunda (R. Umber) H. Moldenke
G. wrightii (A. Gray) R. Umber
Sy = Verbena wrightii A. Gray

LANTANA C. Linnaeus
L. achyranthifolia R. Desfontaines
Sy = L. macropoda J. Torrey
Sy = L. macropodioides J. Greenman
L. camara C. Linnaeus
Sy = L. aculeata C. Linnaeus
Sy = L. camara var. aculeata (C. Linnaeus) H. Moldenke
Sy = L. camara var. flava (F. Medikus) H. Moldenke
Sy = L. camara var. hybrida (W. Neubert) H. Moldenke
Sy = L. camara var. mista (C. Linnaeus) L. Bailey
Sy = L. camara var. mutabilis (W. Hooker) L. Bailey
Sy = L. camara var. nivea (E. Ventenat) L. Bailey
Sy = L. camara var. sanguinea (F. Medikus) L. Bailey
L. canescens K. Kunth
Sy = Goniostachyum citrosum J. K. Small
Sy = Lantana microcephala A. Richard
L. montevidensis (K. Sprengel) J. Briquet
Sy = L. sellowiana J. Link & C. Otto
Sy = Lippia montevidensis K. Sprengel
L. urticoides A. von Hayek
Sy = L. notha H. Moldenke
Sy = L. scorta H. Moldenke
Sy = L. urticoides var. hispidula H. Moldenke
Ma = L. horrida *auct.*, non K. Kunth: *sensu* H. Moldenke and Texas authors
L. velutina M. Martens & H. Galeotti
Sy = Camara velutina K. E. O. Kuntze
Sy = Lantana frutilla H. Moldenke

LIPPIA C. Linnaeus
L. alba (P. Miller) N. Brown
Sy = Lantana alba P. Miller
Sy = Lippia geminata K. Kunth
L. graveolens K. Kunth
Sy = Goniostachyum graveolens (K. Kunth) J. K. Small
Sy = Lippia berlandier J. Schauer

PHYLA J. de Loureiro
P. cuneifolia (J. Torrey) E. Greene
Sy = Lippia cuneifolia (J. Torrey) E. von Steudel
Sy = Zapania cuneifolia J. Torrey
P. lanceolata (A. Michaux) E. Greene
Sy = Lippia lanceolata A. Michaux
Sy = Zapania lanceolata A. L. de Jussieu
P. nodiflora (C. Linnaeus) E. Greene
Sy = Lippia incisa (J. K. Small) I. Tidestrom
Sy = L. nodiflora (J. de Lamarck) A. Michaux
Sy = Phyla incisa J. K. Small
Sy = P. nodiflora var. incisa (J. K. Small) H. Moldenke
Sy = P. nodiflora var. texensis H. Moldenke
Sy = Verbena nodiflora C. Linnaeus
Sy = Zapania nodiflora J. de Lamarck var. rosea D. Don
P. strigulosa (M. Martens & H. Galeotti) H. Moldenke var. ***sericea*** (K. E. O. Kuntze) H. Moldenke
Sy = P. strigulosa var. parviflora (H. Moldenke) H. Moldenke
P. strigulosa (M. Martens & H. Galeotti) H. Moldenke var. ***strigulosa***
Sy = Lippia strigulosa M. Martens & H. Galeotti

PRIVA M. Adanson
P. lappulacea (C. Linnaeus) C. Persoon

STYLODON C. Rafinesque-Schmaltz
S. carneus (F. Medikus) H. Moldenke
Sy = S. carolinensis (T. Walter) J. K. Small
Sy = Verbena carnea F. Medikus

TETRACLEA A. Gray
T. coulteri A. Gray
Sy = T. angustifolia E. Wooton & P. Standley
Sy = T. coulteri var. angustifolia (E. Wooton & P. Standley) A. Nelson & J. Macbride

VERBENA C. Linnaeus
V. bonariensis C. Linnaeus
V. bracteata M. Lagasca y Segura & J. Rodríguez
Sy = V. bracteosa A. Michaux
Sy = V. bracteosa var. brevibracteata A. Gray
Sy = V. imbricata E. Wooton & P. Standley
Sy = V. prostrata G. Savi
V. brasiliensis J. Velloso de Miranda
V. canescens K. Kunth
Sy = V. canescens var. roemeriana (G. Scheele) L. Perry
V. cloverae H. Moldenke var. ***cloverae***
Sy = V. cloverae var. lilacina H. Moldenke
V. halei J. K. Small
Sy = V. leucanthemifolia E. Greene
Sy = V. officinalis C. Linnaeus subsp. halei (J. K. Small) S. Barber
V. hastata C. Linnaeus
Sy = V. hastata var. scabra H. Moldenke
V. hybrida A. Voss *ex* T. Rümpler
V. litoralis K. Kunth
V. macdougalii A. A. Heller
V. menthifolia G. Bentham
V. neomexicana (A. Gray) J. K. Small var. ***hirtella*** L. Perry
V. neomexicana (A. Gray) J. K. Small var. ***neomexicana***
Sy = V. canescens K. Kunth var. neomexicana A. Gray
V. neomexicana (A. Gray) J. K. Small var. ***xylopoda*** L. Perry
V. perennis E. Wooton
V. plicata E. Greene var. ***degeneri*** H. Moldenke
V. plicata E. Greene var. ***plicata***
V. rigida K. Sprengel
Sy = V. rigida var. lilacina (W. Harrow) H. Moldenke
V. runyonii H. Moldenke
V. scabra M. H. Vahl
V. simplex J. Lehmann
Sy = V. angustifolia A. Michaux
V. stricta E. Ventenat
V. urticifolia C. Linnaeus var. ***leiocarpa*** L. Perry & M. Fernald
V. urticifolia C. Linnaeus var. ***urticifolia***
V. xutha J. Lehmann

VITEX C. Linnaeus
V. agnus-castus C. Linnaeus var. ***agnus-castus*** **[cultivated]**
V. agnus-castus C. Linnaeus var. ***caerulea*** A. Rehder **[cultivated]**
V. negundo C. Linnaeus var. ***heterophylla*** (A. Franchet) A. Rehder **[cultivated]**
Sy = V. negundo var. incisa (J. de Lamarck) C. B. Clarke
V. negundo C. Linnaeus var. ***intermedia*** (S. Ji Pei) H. Moldenke **[cultivated]**

VIOLACEAE, *nomen conservandum*

HYBANTHUS N. von Jacquin, *nomen conservandum*
H. verticillatus (C. Ortega) H. Baillon var. ***platyphyllus*** (A. Gray) V. Cory & H. Parks
H. verticillatus (C. Ortega) H. Baillon var. ***verticillatus***
Sy = Calceolaria verticillata (C. Ortega) K. E. O. Kuntze
Sy = Hybanthus linearis (J. Torrey) L. Shinners
Sy = Viola verticillata C. Ortega

VIOLA C. Linnaeus
V. bicolor F. Pursh
Sy = V. kitaibeliana J. A. Schultes var. rafinesquei M. Fernald
Sy = V. rafinesquei E. Greene
V. cornuta C. Linnaeus **[cultivated]**
* ***V. guadalupensis*** A. M. Powell & B. Wauer **[TOES: V]**
V. lanceolata C. Linnaeus subsp. ***lanceolata***
V. lanceolata C. Linnaeus subsp. ***vittata*** (E. Greene) N. Russell
Sy = V. lanceolata var. vittata (E. Greene) C. Weatherby & L. Griscom
V. odorata C. Linnaeus **[cultivated]**
V. palmata C. Linnaeus var. ***trilobata*** (L. Schweinitz) F. Gingins de la Sarraz *ex* A. P. de Candolle
Sy = V. angellae C. Pollard
Sy = V. esculenta S. Elliott
Sy = V. lovelliana E. Brainerd
Sy = V. triloba L. Schweinitz

V. pedata C. Linnaeus
Sy = V. pedata var. concolor H. Holm
Sy = V. pedata var. lineariloba A. P. de Candolle
Sy = V. pedata var. ranunculifolia A. P. de Candolle
Sy = V. populifolia E. Greene
Sy = V. variabilis E. Greene
V. primulifolia C. Linnaeus subsp. ***primulifolia***
Sy = V. primulifolia subsp. villosa (A. Eaton) N. Russell
Sy = V. primulifolia var. acuta (J. Bigelow) J. Torrey & A. Gray
Sy = V. primulifolia var. villosa A. Eaton
V. pubescens W. Aiton var. ***pubescens***
Sy = V. eriocarpon (T. Nuttall) L. Schweinitz
Sy = V. pensylvanica A. Michaux
Sy = V. pubescens var. eriocarpa (L. Schweinitz) N. Russell
V. sagittata W. Aiton var. ***sagittata***
Sy = V. emarginata (T. Nuttall) J. Le Conte
Sy = V. sagittata var. emarginata T. Nuttall
Sy = V. sagittata var. subsagittata (E. Greene) C. Pollard
Sy = V. subsagittata E. Greene
V. sororia C. von Willdenow var. ***missouriensis*** (E. Greene) L. McKinney
Sy = V. missouriensis E. Greene
V. sororia C. von Willdenow var. ***sororia***
Sy = V. affinis J. Le Conte var. langloisii (E. Greene) L. Griscom
Sy = V. asarifolia F. Pursh
Sy = V. langloisii E. Greene
Sy = V. nephrophylla E. Greene
Sy = V. papilionacea F. Pursh
Sy = V. pratincola E. Greene
V. tricolor C. Linnaeus [**cultivated**]
V. villosa T. Walter
Sy = V. alabamensis C. Pollard
Sy = V. palmata C. Linnaeus var. villosa (T. Walter) B. Robinson
Sy = V. sororia T. Nuttall, non C. von Willdenow
V. walteri H. House
V.* × *wittrockiana H. Gams [***altaica* × *lutea***] [**cultivated**]

VISCACEAE

ARCEUTHOBIUM F. Marschall von Bieberstein, *nomen conservandum*
A. divaricatum G. Engelmann *ex* L. Wheeler
Sy = Razoumofskya divaricata (G. Engelmann) F. Coville
A. douglasii G. Engelmann *ex* L. Wheeler
Sy = Razoumofskya douglasii (G. Engelmann) K. E. O. Kuntze
A. vaginatum (C. von Willdenow) J. Presl subsp. ***cryptopodum*** (G. Engelmann) F. G. Hawksworth, O. Swartz, & D. Wiens
Sy = A. cryptopodum G. Engelmann
Sy = Razoumofskya cryptopodum (G. Engelmann) F. G. Hawksworth & D. Weins

PHORADENDRON T. Nuttall
P. bolleanum (B. Seemann) A. Eichler subsp. ***bolleanum***
P. densum J. Torrey *ex* W. Trelease
Sy = P. bolleanum (B. Seemann) A. Eichler var. densum (J. Torrey *ex* W. Trelease) F. Fosberg
P. hawksworthii (D. Wiens) D. Wiens
Sy = P. bolleanum (B. Seemann) A. Eichler subsp. hawksworthii D. Wiens
P. juniperinum G. Engelmann subsp. ***juniperinum***
Sy = P. juniperinum var. ligatum (W. Trelease) F. Fosberg
Sy = P. ligatum W. Trelease
P. leucarpum (C. Rafinesque-Schmaltz) J. Reveal & M. C. Johnston
Sy = P. eatonii W. Trelease
Sy = P. flavescens T. Nuttall
Sy = P. flavescens var. orbiculatum (G. Engelmann) G. Engelmann
Sy = P. macrotomum W. Trelease
Sy = P. serotinum (C. Rafinesque-Schmaltz) M. C. Johnston
Sy = Viscum leucarpum C. Rafinesque-Schmaltz, non Phoradendron leucocarpum J. Paczoski
P. macrophyllum (G. Engelmann) T. Cockerell subsp. ***cockerellii*** (W. Trelease) D. Wiens
Sy = P. cockerellii W. Trelease
P. macrophyllum (G. Engelmann) T. Cockerell subsp. ***macrophyllum***
Sy = P. coloradense W. Trelease
Sy = P. flavescens T. Nuttall *ex* G. Engelmann var. macrophyllum G. Engelmann
Sy = P. longispicum W. Trelease
Sy = P. tomentosum (A. P. de Candolle) G. Engelmann *ex* A. Gray subsp. macrophyllum (G. Engelmann) D. Wiens
Sy = P. tomentosum var. macrophyllum (G. Engelmann) L. Benson
P. tomentosum (A. P. de Candolle) G. Engelmann *ex* A. Gray

Sy = P. flavescens T. Nuttall *ex* G. Engelmann var. pubescens G. Engelmann *ex* A. Gray
Sy = P. serotinum (C. Rafinesque-Schmaltz) M. C. Johnston var. pubescens (G. Engelmann *ex* A. Gray) M. C. Johnston
P. villosum (T. Nuttall) T. Nuttall subsp. ***coryae*** (W. Trelease) D. Wiens
Sy = P. coryae W. Trelease
Sy = P. havardianum W. Trelease

VITACEAE, *nomen conservandum*

AMPELOPSIS A. Michaux
A. arborea (C. Linnaeus) B. Köhne
Sy = A. bipinnata A. Michaux
Sy = Cissus arborea (C. Linnaeus) C. Des Moulins
A. cordata A. Michaux
Sy = Cissus ampelopsis C. Persoon

CAYRATIA A. L. de Jussieu, *nomen conservandum*
C. japonica (C. Thunberg) F. Gagnepain [**cultivated**]

CISSUS C. Linnaeus
C. incisa C. Des Moulins
Sy = Vitis incisa T. Nuttall *ex* J. Torrey & A. Gray

PARTHENOCISSUS J. Planchon, *nomen conservandum*
P. heptaphylla (S. Buckley) N. Britton *ex* J. K. Small
Sy = Ampelopsis heptaphylla S. Buckley
Sy = Parthenocissus texana A. Rehder
Sy = Psedera heptaphylla (S. Buckley) A. Rehder
Sy = Vitis heptaphylla (S. Buckley) N. Britton
P. quinquefolia (C. Linnaeus) J. Planchon var. ***quinquefolia***
Sy = Ampelopsis quinquefolia (C. Linnaeus) A. Michaux
Sy = Hedera quinquefolia C. Linnaeus
Sy = Parthenocissus hirsuta (F. Pursh) K. Graebner
Sy = P. inserta (A. Kerner von Marilaun) K. Fritsch
Sy = P. quinquefolia var. hirsuta (F. Pursh) J. Planchon
Sy = P. quinquefolia var. saintpaulii (B. Köhne *ex* K. Graebner) A. Rehder
Sy = Vitis quinquefolia (C. Linnaeus) J. de Lamarck
P. tricuspidata (P. von Siebold & J. Zuccarini) J. Planchon [**cultivated**]
Sy = Ampelopsis tricuspidata P. von Siebold & J. Zuccarini
P. vitacea (E. Knerr) A. Hitchcock

VITIS C. Linnaeus
V. acerifolia C. Rafinesque-Schmaltz
Sy = V. longii W. Prince var. longii
Sy = V. longii var. microsperma (T. Munson) L. Bailey
Sy = V. nuevo-mexican J. Lemmon *ex* T. Munson
Sy = V. solonis J. Planchon
Sy = V. solonis var. microsperma T. Munson
V. aestivalis A. Michaux var. ***aestivalis***
Sy = V. lincecumii S. Buckley var. glauca T. Munson
Sy = V. lincecumii var. lactea J. K. Small
Sy = V. rufotomentosa J. K. Small
Sy = V. smalliana L. Bailey
V. aestivalis A. Michaux var. ***lincecumii*** (S. Buckley) T. Munson
Sy = V. lincecumii S. Buckley
V. arizonica G. Engelmann var. ***arizonica***
V. arizonica G. Engelmann var. ***glabra*** T. Munson
V. cinerea (G. Engelmann) G. Engelmann *ex* P. M. Millardet var. ***cinerea***
Sy = V. aestivalis A. Michaux var. canescens G. Engelmann
Sy = V. aestivalis var. cinerea G. Engelmann
Sy = V. cinerea var. canescens (G. Engelmann) L. Bailey
V. cinerea (G. Engelmann) G. Engelmann *ex* P. M. Millardet var. ***helleri*** (L. Bailey) M. Moore
Sy = V. berlandieri J. Planchon
Sy = V. cordifolia A. Michaux var. helleri L. Bailey
Sy = V. helleri (L. Bailey) J. K. Small
V. monticola S. Buckley
Sy = V. aestivalis A. Michaux var. monticola (S. Buckley) G. Engelmann
Sy = V. foexana J. Planchon
Sy = V. montana S. Buckley *ex* G. L. Foëx
Sy = V. texana T. Munson
Ma = V. champinii *auct.*, non J. Planchon, in part
V. mustangensis S. Buckley
Sy = V. candicans G. Engelmann *ex* A. Gray
Sy = V. mustangensis var. diversa (L. Bailey) L. Shinners
V. palmata M. H. Vahl
Sy = V. rubra A. Michaux *ex* J. Planchon

V. riparia A. Michaux
Sy = V. riparia var. praecox G. Engelmann *ex* L. Bailey
Sy = V. riparia var. syrticola (M. Fernald & K. Wiegand) M. Fernald
Sy = V. vulpina C. Linnaeus subsp. riparia (A. Michaux) R. Clausen
Sy = V. vulpina var. praecox (G. Engelmann *ex* L. Bailey) L. Bailey
Sy = V. vulpina var. syrticola M. Fernald & K. Wiegand
V. rotundifolia A. Michaux var. ***rotundifolia***
Sy = Muscadinia rotundifolia (A. Michaux) J. K. Small
V. rupestris G. Scheele
Sy = V. rupestris var. dissecta H. Eggert *ex* L. Bailey
V. vinifera C. Linnaeus [**cultivated**]
V. vulpina C. Linnaeus
Sy = V. cordifolia J. de Lamarck
Sy = V. cordifolia var. foetida G. Engelmann
Sy = V. cordifolia var. sempervirens T. Munson
Sy = V. illex L. Bailey
V. × ***champinii*** J. Planchon [***mustangensis*** × ***rupestris***]
V. × ***doaniana*** T. Munson *ex* P. Viala [***mustangensis*** × ***acerifolia***]

ZYGOPHYLLACEAE, *nomen conservandum*

GUAJACUM C. Linnaeus
G. angustifolium G. Engelmann
Sy = Porlieria angustifolia (G. Engelmann) A. Gray

KALLSTROEMIA J. Scopoli
K. californica (S. Watson) A. Vail
Sy = K. brachystylis A. Vail
Sy = K. californica var. brachystylis (A. Vail) T. Kearney & R. Peebles
Sy = Tribulus californicus S. Watson
K. grandiflora J. Torrey *ex* A. Gray
Sy = Tribulus grandiflorus G. Hooker & W. Hooker *ex* J. Brewer & S. Watson
K. hirsutissima A. Vail *ex* J. K. Small
K. maxima (C. Linnaeus) W. Hooker & G. Arnott
K. parviflora J. Norton
Sy = K. intermedia P. Rydberg
Sy = K. latevirens J. Thornber
* ***K. perennans*** B. L. Turner [**TOES: V**]

LARREA A. Cavanilles, *nomen conservandum*
L. tridentata (M. Sessé y Lacasta & J. Mociño *ex* A. P. de Candolle) F. Coville
Sy = Covillea glutinosa (G. Engelmann) P. Rydberg
Sy = C. tridentata A. Vail
Sy = Larrea glutinosa G. Engelmann
Sy = L. mexicana M. Moricand
Sy = L. tridentata var. arenaria L. Benson
Sy = Zygophyllum tridentatum M. Sessé y Lacasta & J. Mociño *ex* A. P. de Candolle
Ma = L. divaricata *auct.*, non A. Cavanilles

PEGANUM C. Linnaeus
P. harmala C. Linnaeus
P. mexicanum A. Gray

TRIBULUS C. Linnaeus
T. cistoides C. Linnaeus
Sy = Kallstroemia cistoides S. Endlicher
Sy = Tribulus terrestris C. Linnaeus var. cistoides D. Oliver
T. terrestris C. Linnaeus

ZYGOPHYLLUM C. Linnaeus
Z. fabago C. Linnaeus
Ma = Z. fabago var. brachycarpum *auct.*, non P. Boissier

CLASS: LILIOPSIDA

ACORACEAE

ACORUS C. Linneaus
A. americanus (C. Rafinesque-Schmaltz) C. Rafinesque-Schmaltz [**cultivated**]
Sy = A. calamus C. Linnaeus var. americanus (C. Rafinesque-Schmaltz) H. D. Wulff

AGAVACEAE, *nomen conservandum*

AGAVE C. Linnaeus
A. americana C. Linnaeus subsp. ***americana*** var. ***americana*** [**cultivated**]
Sy = A. americana var. marginata W. Trelease
Sy = A. americana var. medico-picta W. Trelease
Sy = A. complicata W. Trelease *ex* E. Ochoterana
Sy = A. felina W. Trelease
Sy = A. gracilispina G. Engelmann *ex* W. Trelease
A. americana C. Linnaeus subsp. ***protamericana*** H. Gentry

A. glomeruliflora (G. Engelmann) A. Berger
Sy = A. chisoënsis C. H. Muller
A. gracilipes W. Trelease
A. havardiana W. Trelease
A. lechuguilla J. Torrey
Sy = A. lophantha C. Schiede var. poselgeri (J. Salm-Reiffersheid-Dyck) A. Berger
Sy = A. multilineata J. Baker
Sy = A. poselgeri J. Salm-Reiffersheid-Dyck
A. lophantha C. Schiede
Sy = A. heteracantha J. Zuccarini
Sy = A. univittata A. Haworth
Sy = A. vittata E. von Regel
A. neomexicana E. Wooton & P. Standley
A. scabra J. Salm-Reifferscheid-Dyck subsp. ***scabra***
Sy = A. asperrima G. Jacobi
Sy = A. caeciliana A. Berger

CORDYLINE P. Commerson *ex* A. L. de Jussieu, *nomen conservandum*
C. fruticosa (C. Linnaeus) A. Chevalier [**cultivated**]
Sy = Convallaria fruticosa C. Linnaeus
Sy = Cordyline terminalis (C. Linnaeus) K. Kunth
Sy = Dracaena terminalis (C. Linnaeus) C. Linnaeus

DASYLIRION J. Zuccarini
D. heteracanthum I. M. Johnston
D. leiophyllum G. Engelmann *ex* W. Trelease
D. texanum G. Scheele
D. wheeleri S. Watson

HESPERALOË G. Engelmann
H. funifera (K. Koch) W. Trelease
H. parviflora (J. Torrey) J. Coulter var. ***engelmannii*** (E. Krauskopf) W. Trelease
H. parviflora (J. Torrey) J. Coulter var. ***parviflora***

MANFREDA R. A. Salisbury
* ***M. longiflora*** (J. Rose) S. Verhoek-Williams [**TOES: IV**]
Sy = Polianthes runyonii L. Shinners
Sy = Runyonia longiflora J. Rose
M. maculosa (W. Hooker) J. Rose
Sy = Agave maculosa W. Hooker
Sy = Polianthes maculosa (W. Hooker) L. Shinners
M. sileri S. Verhoek-Williams
M. variegata (G. Jacobi) J. Rose
Sy = Agave variegata G. Jacobi
Sy = Polianthes variegata (G. Jacobi) L. Shinners
M. virginica (C. Linnaeus) R. A. Salisbury *ex* J. Rose
Sy = Agave lata L. Shinners
Sy = A. virginica C. Linnaeus
Sy = Polianthes lata (L. Shinners) L. Shinners
Sy = P. virginica (C. Linnaeus) L. Shinners

NOLINA A. Michaux
* ***N. arenicola*** D. Correll [**TOES: V**]
N. erumpens (J. Torrey) S. Watson
Ma = N. microcarpa S. Watson: Texas authors
N. lindheimeriana (G. Scheele) S. Watson
N. micrantha I. M. Johnston
N. texana S. Watson var. ***compacta*** (W. Trelease) I. M. Johnston
N. texana S. Watson var. ***texana***

PHORMIUM J. R. Forster & J. G. Forster
P. tenax J. R. Forster & J. G. Forster [**cultivated**]

SANSEVIERIA C. Thunberg (orthographic *conservandum*), *nomen conservandum*
S. hyacinthoides (C. Linnaeus) G. Druce [**cultivated**]
Sy = Aloë hyacinthoides C. Linnaeus
Sy = Sansevieria guineënsis (C. Linnaeus) C. von Willdenow
Sy = S. thrysiflora V. Petagna
S. trifasciata D. Prain [**cultivated**]

YUCCA C. Linnaeus
Y. aloifolia C. Linnaeus [**cultivated**]
Y. arkansana W. Trelease
Sy = Y. angustissima G. Engelmann *ex* W. Trelease var. mollis G. Engelmann
Sy = Y. glauca T. Nuttall var. mollis G. Engelmann
Y. baccata J. Torrey var. ***baccata***
Sy = Y. baccata var. vespertina S. McKelvey
Y. campestris S. McKelvey
Y. constricta S. Buckley
Y. elata (G. Engelmann) G. Engelmann var. elata
Sy = Y. angustissima G. Engelmann *ex* W. Trelease var. elata G. Engelmann
Y. faxoniana (W. Trelease) C. Sargent
Ma = Y. carnerosana *auct.*, non (W. Trelease) S. McKelvey
Y. filamentosa C. Linnaeus
Sy = Y. flaccida A. Haworth
Sy = Y. smalliana M. Fernald
Y. glauca T. Nuttall var. ***glauca***
Sy = Y. angustifolia F. Pursh
Y. gloriosa C. Linnaeus [**cultivated**]

Y. louisianensis W. Trelease
Sy = Y. arkansana W. Trelease var. paniculata S. McKelvey
Sy = Y. freemanii L. Shinners
* ***Y. necopina*** L. Shinners [**TOES: V**]
Y. pallida S. McKelvey
Sy = Y. rupicola W. Trelease
Y. recurvifolia R. A. Salisbury [**cultivated**]
Y. reverchonii W. Trelease
Y. rostrata G. Engelmann *ex* W. Trelease
Sy = Y. rostrata var. linearis W. Trelease
Sy = Y. thompsoniana W. Trelease
Y. rupicola G. Scheele
Sy = Y. tortifolia J. Lindley & G. Engelmann
Y. tenuistyla W. Trelease
Y. torreyi J. Shafer
Sy = Y. baccata J. Torrey var. macrocarpa J. Torrey
Sy = Y. macrocarpa (J. Torrey) F. Coville, non V. Cory
Y. treculeana É. Carrière
Sy = Y. canaliculata W. Hooker
Sy = Y. macrocarpa V. Cory, non (J. Torrey) F. Coville
Sy = Y. treculeana var. canaliculata (W. Hooker) W. Trelease
Sy = Y. treculeana var. succulenta S. McKelvey

ALISMATACEAE, *nomen conservandum*

ALISMA C. Linnaeus
A. subcordatum C. Rafinesque-Schmaltz
Sy = A. parviflorum F. Pursh
Sy = A. plantago-aquatica C. Linnaeus var. parviflorum (F. Pursh) J. Torrey

ECHINODORUS L. C. Richard *ex* G. Engelmann
E. beteroi (K. Sprengel) N. Fassett
Sy = Alisma berteroi K. Sprengel
Sy = A. rostratum T. Nuttall
Sy = Echinodorus beteroi var. lanceolatus (G. Engelmann) N. Fassett
Sy = E. rostratus (T. Nuttall) G. Engelmann *ex* A. Gray
Sy = E. rostratus var. lanceolatus G. Engelmann *ex* S. Watson & J. Coulter
E. cordifolius (C. Linnaeus) A. Grisebach subsp. ***fluitans*** (N. Fassett) R. Haynes & L. B. Holm-Nielsen
E. tenellus (K. von Martius) F. Buchenau
Sy = Alisma tenellum K. von Martius
Sy = Echinodorus parvulus G. Engelmann *ex* A. Gray
Sy = E. tenellus var. parvulus (G. Engelmann) N. Fassett
Sy = Helianthium parvulum (G. Engelmann) J. K. Small

SAGITTARIA C. Linnaeus
S. brevirostra K. Mackenzie & B. Bush
Sy = S. engelmanniana J. G. Smith subsp. brevirostra (K. Mackenzie & B. Bush) C. Bogin
S. cuneata E. Sheldon
Sy = S. arifolia T. Nuttall *ex* J. G. Smith
Sy = S. paniculata J. Blankinship
S. graminea A. Michaux subsp. ***graminea***
Sy = S. graminea var. graminea
Sy = S. stolonifera G. Engelmann & A. Gray
S. lancifolia C. Linnaeus subsp. ***lancifolia***
S. lancifolia C. Linnaeus subsp. ***media*** (P. Micheli) C. Bogin
Sy = S. falcata F. Pursh
Sy = S. lancifolia var. falcata (F. Pursh) J. G. Smith
Sy = S. lancifolia var. media P. Micheli
S. latifolia C. von Willdenow
Sy = S. latifolia var. obtusa (G. Engelmann) K. Wiegand
Sy = S. latifolia var. pubescens (G. H. Muhlenberg *ex* T. Nuttall) J. G. Smith
Sy = S. pubescens G. H. Muhlenberg *ex* T. Nuttall
Sy = S. sagittifolia C. Linnaeus var. pubescens (G. H. Muhlenberg) J. Torrey
S. longiloba G. Engelmann *ex* J. Torrey
Sy = S. greggii J. G. Smith
Sy = S. saggitifolia C. Linnaeus var. mexicana M. Martens & H. Galeotti
S. montevidensis A. von Chamisso & D. von Schlechtendal subsp. ***calycina*** (G. Engelmann) C. Bogin
Sy = Lophotocarpus calycinus (G. Engelmann) J. G. Smith
Sy = Sagittaria calycina G. Engelmann var. calycina
Sy = S. calycina var. fluitans G. Engelmann
Sy = S. calycina var. maxima G. Engelmann
S. papillosa F. Buchenau
S. platyphylla (G. Engelmann) J. G. Smith
Sy = S. graminea A. Michaux var. platyphylla G. Engelmann
Sy = S. mohrii J. G. Smith

ARACEAE, *nomen conservandum*

ARISAEMA K. von Martius
A. *dracontium* (C. Linnaeus) H. Schott
Sy = Muricauda dracontium (C. Linnaeus) J. K. Small
A. *triphyllum* (C. Linnaeus) H. Schott subsp. ***quinatum*** (S. Buckley) D. Huttleston
Sy = A. quinatum (S. Buckley) H. Schott
A. *triphyllum* (C. Linnaeus) H. Schott subsp. ***triphyllum***
Sy = A. acuminatum J. K. Small
Sy = A. atrorubens (W. Aiton) K. von Blume

ARUM C. Linnaeus
A. *italicum* P. Miller [**cultivated**]

CALADIUM E. Ventenat
C. *bicolor* (W. Aiton) E. Ventenat [**cultivated**]
Sy = Arum bicolor W. Aiton
Sy = Cyrtospadix bicolor (W. Aiton) N. Britton & Percy Wilson

COLOCASIA H. Schott, *nomen conservandum*
C. *esculenta* (C. Linnaeus) H. Schott [**cultivated**]
Sy = Arum esculentum C. Linnaeus
Sy = Colocasia esculenta var. antiquorum (H. Schott) E. Hubbard & A. Rehder
Sy = C. esculenta var. aquatilis J. Hasskarl

MONSTERA M. Adanson, *nomen conservandum*
M. *deliciosa* F. Liebmann [**cultivated**]

ORONTIUM C. Linnaeus
O. *aquaticum* C. Linnaeus

PELTANDRA C. Rafinesque-Schmaltz, *nomen conservandum*
P. *virginica* (C. Linnaeus) H. Schott & S. Endlicher
Sy = P. tharpii F. Barkley

PHILODENDRON H. Schott, *nomen conservandum*
P. *bipinnifolium* H. Schott [**cultivated**]
Ma = P. panduriforme *auct.*, non (K. Kunth) K. Kunth
P. *giganteum* H. Schott [**cultivated**]
P. *selloum* K. Koch [**cultivated**]

PISTIA C. Linnaeus
P. *stratiotes* C. Linnaeus

XANTHOSOMA H. Schott
X. *sagittifolium* (C. Linnaeus) H. Schott

ZANTEDESCHIA K. Sprengel, *nomen conservandum*
Z. *aethiopica* (C. Linnaeus) K. Sprengel [**cultivated**]
Sy = Calla aethiopica C. Linnaeus

ARECACEAE, *nomen conservandum*

ACOELORRAPHE H. Wendland
A. *wrightii* (A. Grisebach & H. Wendland) H. Wendland *ex* O. Beccari [**cultivated**]
Sy = A. arborescens (C. Sargent) O. Beccari
Sy = Paurotis wrightii (A. Grisebach & H. Wendland) N. Britton

ARECASTRUM O. Beccari
A. *romanzoffianum* O. Beccari [**cultivated**]
Sy = Cocos plumosa W. Hooker

ARENGA J. de Houtton Labillardière, *nomen conservandum*
A. *engleri* O. Beccari [**cultivated**]

BUTIA O. Beccari
B. *capitata* O. Beccari [**cultivated**]

CHAMAEROPS C. Linnaeus
C. *humilis* C. Linnaeus [**cultivated**]

CHRYSALIDOCARPUS H. Wendland
C. *lutescens* (Bory de Saint-Vincent) H. Wendland [**cultivated**]
Sy = Areca lutescens

COCOS C. Linnaeus
C. *nucifera* C. Linnaeus [**cultivated**]

ERYTHEA S. Watson
E. *armata* S. Watson [**cultivated**]
Sy = Brahea armata S. Watson
Sy = Glaucothea armata O. F. Cook

HOWEA O. Beccari
H. *belmoreana* O. Beccari [**cultivated**]

JUBAEA K. Kunth
J. *spectabilis* K. Kunth [**cultivated**]

LIVISTONA R. Brown
L. *australis* K. von Martius [**cultivated**]
Sy = Corypha australis

L. chinensis (N. von Jacquin) R. Brown *ex* K. von Martius [**cultivated**]
Sy = L. oliviformis (J. Hasskarl) K. von Martius
Sy = Saribus oliviformis J. Hasskarl

PHOENIX C. Linnaeus
P. canariensis S. Chabaud [**cultivated**]
P. dactylifera C. Linnaeus [**cultivated**]
P. loureiri K. Kunth [**cultivated**]
P. reclinata N. von Jacquin [**cultivated**]
P. roebelenii J. O'Brien [**cultivated**]
P. sylvestris W. Roxburgh [**cultivated**]

RHAPIDOPHYLLUM H. Wendland & C. Drude *ex* C. Drude
R. hystrix (F. Pursh) H. Wendland & C. Drude *ex* C. Drude [**cultivated**]

RHAPIS C. von Linné
R. excelsa A. C. Henry [**cultivated**]
R. humilis K. von Blume [**cultivated**]

ROYSTONEA O. F. Cook
R. regia (K. Kunth) O. F. Cook [**cultivated**]

SABAL M. Adanson
S. etonia W. Swingle *ex* G. Nash [**cultivated**]
Sy = S. miamiensis S. Zona
* ***S. mexicana*** K. von Martius [**TOES: IV**]
Sy = S. texana (O. F. Cook) O. Beccari
S. minor (N. von Jacquin) C. Persoon
Sy = S. louisiana (J. Darby) M. Bomhard
S. palmetto (T. Walter) C. Loddiges *ex* J. A. Schultes & J. H. Schultes [**cultivated**]
Sy = Corypha palmetto T. Walter
Sy = Inodes schwarzii O. F. Cook
Sy = Sabal jamesiana J. K. Small
S. umbraculifera K. von Martius [**cultivated**]
* ***S. mexicana*** K. von Martius × ***S. minor*** (N. von Jacquin) C. Persoon [**TOES: V**]

SERENOA J. Hooker
S. repens (W. Bartram) J. K. Small [**cultivated**]
Sy = Brahea serrulata (A. Michaux) H. Wendland
Sy = Corypha repens W. Bartram
Sy = Serenoa serrulata (A. Michaux) D. Nicholson

TRACHYCARPUS H. Wendland
T. fortunei H. Wendland [**cultivated**]
Sy = Chamaerops excelsa
T. martianus H. Wendland [**cultivated**]

WASHINGTONIA H. Wendland, *nomen conservandum*
W. filifera (J. J. Linden *ex* É. André) H. Wendland [**cultivated**]
W. robusta H. Wendland [**cultivated**]
Sy = W. gracilis S. Parish
Sy = W. sonorae S. Watson

BROMELIACEAE, *nomen conservandum*

HECHTIA J. Klotzsch
H. glomerata J. Zuccarini
Sy = H. gamopetala C. Mez
Sy = H. ghiesbreghtii C. Lemaire
Sy = H. morreniana C. Mez
H. texensis S. Watson
Sy = H. scariosa L. B. Smith

TILLANDSIA C. Linnaeus
* ***T. baileyi*** J. Rose *ex* J. K. Small [**TOES: V**]
T. recurvata (C. Linnaeus) C. Linnaeus
Sy = Diaphoranthema recurvata (C. Linnaeus) J. Beer
Sy = D. uniflora J. Beer
Sy = Renealmia recurvata C. Linnaeus
Sy = Tillandsia uniflora K. Kunth
T. usneoides (C. Linnaeus) C. Linnaeus
Sy = Dendropogon usneoides (C. Linnaeus) C. Rafinesque
Sy = Renealmia usneoides C. Linnaeus

BURMANNIACEAE, *nomen conservandum*

APTERIA T. Nuttall
A. aphylla (T. Nuttall) J. Barnhart *ex* J. K. Small
Sy = A. aphylla var. hymenanthera (F. Miquel) F. Jonker

BURMANNIA C. Linnaeus
B. biflora C. Linnaeus
B. capitata (J. F. Gmelin) K. von Martius
Sy = Vogelia capitata T. Walter *ex* J. F. Gmelin

CANNACEAE, *nomen conservandum*

CANNA C. Linnaeus
C. flaccida R. A. Salisbury [**cultivated**]
C. glauca C. Linnaeus [**cultivated**]
C. indica C. Linnaeus [**cultivated**]
Sy = C. coccinea P. Miller
Sy = C. edulis J. Ker-Gawler
C. × ***generalis*** L. Bailey [***glauca*** × ***indica***] [**cultivated**]

C. × ***orchiodes*** L. Bailey [***flaccida*** × unknown] [**cultivated**]

COMMELINACEAE, *nomen conservandum*

CALLISIA P. Löfling
C. micrantha (J. Torrey) D. Hunt
Sy = Phodina micrantha (J. Torrey) D. Hunt
Sy = Tradescantia micrantha J. Torrey
C. repens (N. von Jacquin) C. Linnaeus
Sy = C. fragrans (J. Lindley) A. Woods
Sy = Hapalanthus repens N. von Jacquin
Sy = Spironema robbinsii Sanvalle

COMMELINA C. Linnaeus
C. caroliniana T. Walter
C. communis C. Linnaeus var. ***communis***
Sy = C. willdenowii K. Kunth
C. dianthifolia A. Delile var. ***dianthifolia***
C. diffusa N. L. Burnman
Sy = C. gigas J. K. Small
Sy = C. longicaulis N. von Jacquin
C. erecta C. Linnaeus var. ***angustifolia*** (A. Michaux) M. Fernald
Sy = C. angustifolia A. Michaux
Sy = C. nashii J. K. Small
Sy = C. swingleana G. Nash
C. erecta C. Linnaeus var. ***deamiana*** M. Fernald
Sy = C. erecta var. greenei N. Fassett
C. erecta C. Linnaeus var. ***erecta***
Sy = C. elegans K. Kunth
Sy = C. saxicola J. K. Small
C. virginica C. Linnaeus
Sy = C. deficiens W. Hooker
Sy = C. longifolia A. Michaux

MURDANNIA J. Royle, *nomen conservandum*
M. nudiflora (C. Linnaeus) J. Brenan
Sy = Aneilema nudicaule (N. L. Burman) G. Don
Sy = A. nudiflorum (C. Linnaeus) K. Kunth
Sy = Commelina nudiflora C. Linnaeus

TINANTIA M. Scheidweiler, *nomen conservandum*
T. anomala (J. Torrey) C. B. Clarke
Sy = Commeliantia anomala (J. Torrey) B. Tharp
Sy = Commelina anomala (J. Torrey) A. Woods
Sy = Tradescantia anomala J. Torrey

TRADESCANTIA C. Linnaeus
T. brevifolia (J. Torrey) J. Rose
Sy = Setcreasea brevifolia (J. Torrey) R. Pilger
Sy = Tradescantia leiandra J. Torrey var. brevifolia J. Torrey
T. buckleyi (I. M. Johnston) D. Hunt
Sy = Setcreasea buckleyi I. M. Johnston
T. edwardsiana B. Tharp & F. Barkley
T. fluminensis J. Velloso de Miranda [**cultivated**]
Sy = T. albiflora K. Kunth
T. gigantea J. Rose
T. hirsutiflora B. Bush
Sy = T. australis B. Bush
Sy = T. eglandulosa B. Bush
T. humilis J. Rose
T. leiandra J. Torrey var. ***glandulosa*** (D. Correll) K. Gandhi
Sy = Setcreasea leiandra (J. Torrey) R. Pilger var. glandulosa D. Correll
T. leiandra J. Torrey var. ***leiandra***
Sy = Setcreasea leiandra (J. Torrey) R. Pilger
T. occidentalis (N. Britton) B. Smyth var. ***occidentalis***
Sy = T. virginiana C. Linnaeus var. occidentalis N. Britton
T. ohiensis C. Rafinesque-Schmaltz
Sy = T. foliosa J. K. Small
Sy = T. incarnata J. K. Small
Sy = T. ohiensis var. foliosa (J. K. Small) D. T. MacRoberts
Sy = T. reflexa C. Rafinesque-Schmaltz
T. pallida (J. Rose) D. Hunt [**cultivated**]
Sy = Setcreasea purpurea B. K. Boom
T. reverchonii B. Bush
T. subacaulis B. Bush
Sy = T. texana B. Bush
T. tharpii E. Anderson & R. Woodson
T. wrightii J. Rose & B. Bush var. ***glandulopubescens*** B. L. Turner
T. wrightii J. Rose & B. Bush var. ***wrightii***
T. × ***diffusa*** B. Bush [***humilis*** × ***occidentalis***]
Sy = T. pedicellata R. Celerier

COSTACEAE

COSTUS C. Linnaeus
C. cuspidata A. Garden [**cultivated**]
C. speciosus (K. König) J. E. Smith [**cultivated**]

CYMODOCEACEAE, *nomen conservandum*

CYMODOCEA K. König, *nomen conservandum*
C. filiformis (F. Kutzing) D. Correll
Sy = C. manatorum P. Ascherson
Sy = Syringodium filiforme K. E. O. Kuntze

HALODULE S. Endlicher
H. *beaudettei* (C. den Hartog) C. den Hartog
Sy = Diplanthera beaudettei C. den Hartog
Sy = Halodule wrightii P. Ascherson

CYPERACEAE, *nomen conservandum*

BOLBOSCHOENUS (P. Ascherson) E. Palla
B. fluviatilis (J. Torrey) J. Soják. NOTE: S. G. Smith suggests this taxon is in the Texas panhandle but he has seen no specimen, nor have we.
Sy = Schoenoplectus fluviatilis (J. Torrey) M. T. Strong
Sy = Scirpus fluviatilis (J. Torrey) A. Gray
B. maritimus (C. Linnaeus) E. Palla
Sy = B. paludosus (A. Nelson) K. Soó von Bere
Sy = Schoenoplectus maritimus (C. Linnaeus) K. Lye
Sy = Scirpus maritimus C. Linnaeus var. maritimus
Sy = S. maritimus var. paludosus (A. Nelson) G. Kükenthal
Sy = S. paludosus A. Nelson
B. robustus (F. Pursh) J. Soják
Sy = Schoenoplectus robustus (F. Pursh) M. T. Strong
Sy = Scirpus maritimus C. Linnaeus var. macrostachyus A. Michaux
Sy = S. robustus F. Pursh

BULBOSTYLIS K. Kunth, *nomen conservandum*
B. barbata (C. Rottbøll) C. B. Clarke
Sy = Scirpus barbatus C. Rottbøll
Sy = Stenophyllus barbatus (C. Rottbøll) N. Britton
B. capillaris (C. Linnaeus) K. Kunth *ex* C. B. Clarke
Sy = B. capillaris var. crebra M. Fernald
Sy = B. capillaris var. isopoda M. Fernald
Sy = Fimbristylis capillaris (C. Linnaeus) A. Gray
Sy = Scirpus capillaris C. Linnaeus
Sy = Stenophyllus capillaris (C. Linnaeus) N. Britton
B. ciliatifolia (S. Elliott) M. Fernald var. ***ciliatifolia***
Sy = Scirpus ciliatifolius S. Elliott
Sy = Stenophyllus ciliatifolius (S. Elliott) C. Mohr
B. ciliatifolia (S. Elliott) M. Fernald var. ***coarctata*** (S. Elliott) R. Kral
Sy = B. coarctata (S. Elliott) C. Mohr
Sy = Scirpus coarctatus S. Elliott
Sy = Stenophyllus coarctatus (S. Elliott) N. Britton
B. juncoides (M. H. Vahl) G. Kükenthal
Sy = B. fendleri C. B. Clarke
Sy = B. juncoides var. ampliceps G. Kükenthal
Sy = Fimbristylis juncoides (M. H. Vahl) Brother Alain
Sy = Schoenus juncoides M. H. Vahl

CAREX C. Linnaeus
C. absondita K. Mackenzie
Sy = C. absondita var. glauca (A. Chapman) M. Fernald
Sy = C. absondita var. rostellata M. Fernald
Sy = C. magnifolia K. Mackenzie
C. agrostoides K. Mackenzie
Ma = C. praegracilis *auct.*, non W. Boott: Texas authors, in part
C. alata J. Torrey
Sy = C. albolutescens L. Schweinitz var. alata (J. Torrey) G. Kükenthal
Sy = C. straminea C. von Willdenow var. alata (J. Torrey) L. Bailey
Sy = Kolerma alata (J. Torrey) C. Rafinesque-Schmaltz
C. albicans C. von Willdenow *ex* K. Sprengel var. ***australis*** (L. Bailey) J. Rettig
Sy = C. emmonsii C. Dewey var. australis (L. Bailey) J. Rettig
Sy = C. physorhyncha F. Liebmann *ex* E. von Steudel
Sy = C. varia G. H. Muhlenberg *ex* C. von Willdenow var. australis L. Bailey
C. albolutescens L. Schweinitz
C. albula H. Allan [**cultivated**]
C. amphibola E. von Steudel
Sy = C. grisea G. Wahlenberg var. amphibola (E. von Steudel) G. Kükenthal
C. annectens (E. Bicknell) E. Bicknell var. ***annectens***
Sy = C. vulpinoidea A. Michaux var. annectens (E. Bicknell) O. Farwell
C. arkansana L. Bailey
Sy = C. rosea C. Schkuhr *ex* C. von Willdenow var. arkansana L. Bailey
C. athrostachya S. Olney
Sy = C. athrostachya var. minor S. Olney
Sy = C. tenuirostris S. Olney
Ma = C. praegracilis *auct.*, non W. Boott: Texas authors, in part
C. atlantica L. Bailey subsp. ***atlantica***
Sy = C. atlantica var. incomperta (E. Bicknell) F. J. Hermann
Sy = C. echinata J. Murray var. conferta (A. Chapman) L. Bailey
Sy = C. incomperta E. Bicknell

C. atlantica L. Bailey subsp. ***capillacea*** (L. Bailey) A. Reznicek
Sy = C. atlantica var. capillacea (L. Bailey) A. Cronquist
Sy = C. delicatula E. Bicknell
Sy = C. howei K. Mackenzie
Sy = C. howei var. capillacea (L. Bailey) M. Fernald
Sy = C. interior L. Bailey var. capillacea L. Bailey
Sy = C. mohriana K. Mackenzie
C. aurea T. Nuttall
Sy = C. aurea var. androgyna S. Olney
Sy = C. aurea var. celsa L. Bailey
Sy = C. garberi M. Fernald var. bifaria M. Fernald
Sy = C. hassei L. Bailey
C. austrina (J. K. Small) K. Mackenzie
Sy = C. muehlenbergii C. Schkuhr *ex* C. von Willdenow var. australis S. Olney
Sy = C. muehlenbergii var. austrina J. K. Small
C. basiantha E. von Steudel
Ma = C. willdenowii S. Schkuhr *ex* C. von Willdenow: American authors
C. bicknellii N. Britton var. ***opaca*** F. J. Hermann
C. blanda C. Dewey
Sy = C. anceps G. H. Muhlenberg *ex* C. von Willdenow var. blanda (C. Dewey) F. Boott
Sy = C. bulbosa J. Böckeler
Sy = C. laxiflora J. de Lamarck var. blanda (C. Dewey) F. Boott
Sy = C. truncata J. Böckeler
Sy = Deweya blanda (C. Dewey) C. Rafinesque-Schmaltz
Ma = Carex gracilescens *auct.*, non E. von Steudel: Texas authors
C. brevior (C. Dewey) K. Mackenzie *ex* J. Lunell
Sy = C. festucacea C. Schkuhr *ex* C. von Willdenow var. brevior (C. Dewey) M. Fernald
Sy = C. straminea C. von Willdenow *ex* C. Schkuhr var. brevior C. Dewey
C. bromoides C. Schkuhr *ex* C. von Willdenow var. ***bromoides***
Sy = Loncoperis bromoides C. Rafinesque-Schmaltz
C. bulbostylis K. Mackenzie
Sy = C. amphibola E. von Steudel var. globosa (L. Bailey) L. Bailey
Sy = C. grisea G. Wahlenberg var. globosa L. Bailey
C. bushii K. Mackenzie
Sy = C. caroliniana L. Schweinitz var. cuspidata (C. Dewey) L. Shinners
Sy = C. triceps A. Michaux var. longicuspis G. Kükenthal
C. caroliniana L. Schweinitz
Sy = C. gynandra L. Schweinitz var. caroliniana (L. Schweinitz) S. Olney
Sy = C. smithii T. Porter, non I. Tausch
C. cephalophora G. H. Muhlenberg *ex* C. von Willdenow
Sy = C. cephalophora var. bracteosa O. Farwell
Sy = Diemisa cephalophora (G. H. Muhlenberg *ex* C. von Willdenow) C. Rafinesque-Schmaltz
C. cherokeënsis L. Schweinitz
Sy = C. baazasana E. von Steudel [typographic error]
Sy = C. brazasana E. von Steudel
Sy = C. christyana W. Boott
Sy = C. recurva G. H. Muhlenberg, non W. Hudson, non C. Schkuhr
Sy = Edritria recurva (G. H. Muhlenberg) C. Rafinesque-Schmaltz
C. comosa F. Boott
Sy = C. furcata S. Elliott, non P. Lapeyrouse
Sy = C. pseudo-cyperus C. Linnaeus var. americana C. F. Hochstetter
Sy = C. pseudo-cyperus var. comosa (F. Boott) W. Boott
Sy = C. pseudo-cyperus var. furcata G. Kükenthal
C. complanata J. Torrey & W. Hooker
Sy = C. bolliana J. Böckeler
Sy = C. triceps A. Michaux, non F. von Schrank
Sy = Facolos complanata (J. Torrey & W. Hooker) C. Rafinesque-Schmaltz
C. corrugata M. Fernald
Sy = C. rugata M. Fernald, non J. Ohwi
C. crebriflora K. Wiegand
C. crinita J. de Lamarck var. ***brevicrinis*** M. Fernald
C. crus-corvi R. Shuttleworth *ex* G. Kunze
Sy = C. bayardii M. Fernald
Sy = C. halei C. Dewey
Sy = C. sicaeformis F. Boott
C. davisii L. Schweinitz & J. Torrey
Sy = C. albirostris C. B. Clarke
Sy = Loxotrema davisii (L. Schweinitz & J. Torrey) C. Rafinesque-Schmaltz
C. debilis A. Michaux var. ***debilis***
C. decomposita G. H. Muhlenberg
Sy = C. nuttallii L. Schweinitz
Sy = C. paniculata C. Linnaeus var. decomposita (G. H. Muhlenberg) C. Dewey
C. digitalis C. Schkuhr *ex* C. von Willdenow var. ***asymmetrica*** M. Fernald

*C. **digitalis*** C. Schkuhr *ex* C. von Willdenow var. ***macropoda*** M. Fernald
*C. **eburnea*** F. Boott
Sy = C. alba J. Scopoli var. setifolia C. Dewey
*C. **edwardsiana*** E. Bridges & S. Orzell
Ma = C. oligocarpa *auct.*, non C. Schkuhr *ex* C. von Willdenow: Texas authors, in part
*C. **emoryi*** C. Dewey
Sy = C. stricta J. de Lamarck var. elongata (J. Böckeler) H. Gleason
Sy = C. stricta var. emoryi (C. Dewey) L. Bailey
*C. **festucacea*** C. Schkuhr *ex* C. von Willdenow
Sy = C. straminea C. von Willdenow var. festucacea (C. Schkuhr *ex* C. von Willdenow) E. Tuckerman
*C. **fissa*** K. Mackenzie var. ***fissa***
*C. **flacca*** J. von Schreber [**cultivated**]
Sy = C. glauca J. Scopoli
*C. **flaccosperma*** C. Dewey
Sy = C. grisea G. Wahlenberg var. mutica (J. Torrey & A. Gray) J. Carey
Sy = C. laxiflora J. de Lamarck var. mutica J. Torrey & A. Gray
Sy = C. microsperma E. von Steudel, non G. Wahlenberg
Sy = C. xanthosperma C. Dewey
*C. **frankii*** K. Kunth
Sy = C. atherodes J. Frank
Sy = C. aurolensis E. von Steudel
Sy = C. involucrata J. Böckeler
*C. **geophila*** K. Mackenzie
Sy = C. pityophila K. Mackenzie
*C. **gigantea*** E. Rudge
Sy = C. gigantea var. grandis (L. Bailey) O. Farwell
Sy = C. grandis L. Bailey
Sy = C. lacustris C. von Willdenow var. gigantea (E. Rudge) F. Pursh
*C. **glaucescens*** S. Elliott
Sy = C. rufidula E. von Steudel
Sy = C. verrucosa G. H. Muhlenberg var. glaucescens (S. Elliott) A. Wood
Sy = Edritria glaucescens (S. Elliott) C. Rafinesque-Schmaltz
*C. **glaucodea*** E. Tuckerman *ex* S. Olney
Sy = C. flaccosperma C. Dewey var. glaucodea (E. Tuckerman *ex* S. Olney) G. Kükenthal
*C. **granularis*** G. H. Muhlenberg *ex* C. von Willdenow
Sy = C. granularis var. haleana (S. Olney) T. Porter
Sy = C. granularis var. recta C. Dewey
Sy = C. rectior K. Mackenzie
Sy = Deweya granularis (G. H. Muhlenberg *ex* C. von Willdenow) C. Rafinesque-Schmaltz
*C. **gravida*** L. Bailey
Sy = C. gravida var. lunelliana (K. Mackenzie) F. J. Hermann
Sy = C. lunelliana K. Mackenzie
*C. **grisea*** G. Wahlenberg
Sy = C. amphibola E. von Steudel var. turgida M. Fernald
Sy = C. turgida (M. Fernald) J. Moore
Sy = Manochlaenia grisea (G. Wahlenberg) K. Börner *ex* F. Fedde & C. Schuster
* *C. **hyalina*** F. Boott [**TOES: V**]
Sy = C. straminea C. von Willdenow var. hyalina (F. Boott) F. Boott
*C. **hyalinolepis*** E. von Steudel
Sy = C. impressa (S. Wright) K. Mackenzie
Sy = C. lacustris C. von Willdenow var. laxiflora C. Dewey
Sy = C. riparia M. Curtis var. impressa S. Wright
*C. **hystericina*** G. H. Muhlenberg *ex* C. von Willdenow
*C. **intumescens*** E. Rudge
Sy = C. folliculata C. Linnaeus var. major K. Kunth
Sy = C. intumescens var. fernaldii L. Bailey
*C. **joorii*** L. Bailey
*C. **lanuginosa*** A. Michaux
Sy = C. filiformis C. Linnaeus var. latifolia J. Böckeler
Sy = C. filiformis var. lanuginosa (A. Michaux) N. Britton, E. Sterns, & J. Poggenburg
Sy = C. lasiocarpa F. Ehrhart var. lanuginosa (A. Michaux) G. Kükenthal
Sy = C. lasiocarpa var. latifolia (J. Böckeler) C. Gilly
*C. **lativena*** S. D. Jones & G. D. Jones
*C. **leavenworthii*** C. Dewey
Sy = C. cephalophora G. H. Muhlenberg *ex* C. von Willdenow var. angustifolia F. Boott
Sy = C. cephalophora var. leavenworthii (C. Dewey) G. Kükenthal
*C. **leptalea*** G. Wahlenberg subsp. ***harperi*** (M. Fernald) W. Stone
Sy = C. harperi M. Fernald
Sy = C. leptalea var. harperi (M. Fernald) C. Weatherby & L. Griscom
Ma = C. leptalea var. leptalea *auct.*, non G. Wahlenberg: Texas authors
*C. **lonchocarpa*** C. von Willdenow
Sy = C. folliculata C. Linnaeus var. australis L. Bailey
Sy = C. smalliana K. Mackenzie

C. longii K. Mackenzie
C. louisianica L. Bailey
C. lupuliformis H. Sartwell *ex* C. Dewey
C. lupulina G. H. Muhlenberg *ex* C. von Willdenow
Sy = C. lupulina var. pedunculata A. Gray
C. lurida G. Wahlenberg
C. meadii C. Dewey
Sy = C. tetanica C. Schkuhr var. meadii (C. Dewey) L. Bailey
C. mesochorea K. Mackenzie
Sy = C. cephalophora G. H. Muhlenberg *ex* C. von Willdenow var. mesochorea (K. Mackenzie) H. Gleason
Sy = C. mediterranea K. Mackenzie, non C. B. Clarke
C. microdonta J. Torrey & W. Hooker
Sy = C. alveata W. Boott
Sy = C. microdonta var. latifolia L. Bailey
Sy = C. roemeriana G. Scheele
Sy = C. wrightii C. Dewey
C. microrhyncha K. Mackenzie
C. morrowii F. Boott var. ***morrowii*** **[cultivated]**
C. muehlenbergii C. Schkuhr *ex* C. von Willdenow var. ***enervis*** F. Boott
Sy = C. muhlenbergii var. enervis [orthographic error]
Sy = C. onusta K. Mackenzie
Sy = C. plana K. Mackenzie
C. muehlenbergii C. Schkuhr *ex* C. von Willdenow var. ***muehlenbergii***
Sy = C. cephalophora G. H. Muhlenberg *ex* C. von Willdenow var. anomala O. Farwell
Sy = C. muhlenbergii var. muhlenbergii [orthographic error]
Sy = C. mühlenbergii var. mühlenbergii [orthographic variant]
Sy = C. pinetorum C. von Willdenow
Sy = C. piniaria L. Bosc
Sy = Vignea muehlenbergii (C. Schkuhr *ex* C. von Willdenow) C. Rafinesque-Schmaltz
C. muriculata F. J. Hermann
C. nigromarginata L. Schweinitz var. ***floridana*** (L. Schweinitz) G. Kükenthal
Sy = C. floridana L. Schweinitz
C. nigromarginata L. Schweinitz var. ***nigromarginata***
C. occidentalis L. Bailey
Sy = C. muricata C. Linnaeus var. americana L. Bailey
Sy = C. neomexicana K. Mackenzie
Ma = C. praegracilis *auct.*, non W. Boott: Texas authors, in part
C. oklahomensis K. Mackenzie
Sy = C. stipata G. H. Muhlenberg *ex* C. von Willdenow var. oklahomensis (K. Mackenzie) H. Gleason
C. oxylepis J. Torrey & W. Hooker
Sy = C. familiaris E. von Steudel
C. ozarkana P. Rothrock & A. Reznicek
C. perdentata S. D. Jones
C. planostachys G. Kunze
Sy = C. alpestris var. tripla C. Dewey
Sy = C. halleriana I. Asso subsp. planostachys (G. Kunze) G. Kükenthal
C. reniformis (L. Bailey) J. K. Small
Sy = C. straminea C. von Willdenow var. reniformis L. Bailey
C. retroflexa G. H. Muhlenberg *ex* C. von Willdenow
Sy = C. rosea C. Schkuhr *ex* C. von Willdenow var. retroflexa (G. H. Muhlenberg *ex* C. von Willdenow) J. Torrey
Sy = Diemisa retroflexa (G. H. Muhlenberg *ex* C. von Willdenow) C. Rafinesque-Schmaltz
C. rosea C. Schkuhr *ex* C. von Willdenow
Sy = C. convoluta K. Mackenzie
Sy = C. flaccidula E. von Steudel
Sy = C. rosea var. pusilla C. H. Peck & W. E. Howe
C. socialis R. Mohlenbrock & J. Schwegman
C. striatula A. Michaux
Sy = C. anceps G. H. Muhlenberg *ex* C. von Willdenow var. striatula (A. Michaux) J. Carey
Sy = C. ignota C. Dewey
Sy = C. laxiflora J. de Lamarck var. angustifolia C. Dewey
Sy = C. laxiflora var. michauxii L. Bailey
Sy = C. laxiflora var. striatula (A. Michaux) J. Carey
C. stricta J. de Lamarck
Sy = C. stricta var. strictior (C. Dewey) J. Carey
Sy = C. strictior C. Dewey
C. styloflexa S. Buckley
Sy = C. laxiflora J. de Lamarck var. styloflexa (S. Buckley) W. Boott
Sy = C. protracia E. von Steudel
Sy = C. styloflexa var. remotiflora K. Wiegand
C. tenax A. Chapman
Sy = C. chapmanii H. Sartwell
Sy = C. dasycarpa G. H. Muhlenberg var. tenax (A. Chapman) G. Kükenthal
C. tetrastachya G. Scheele
Sy = C. brittoniana L. Bailey

Sy = C. straminea C. von Willdenow var. maxima L. Bailey
Sy = C. straminea var. porepens G. Kükenthal
Sy = C. wrightii S. Olney, non C. Dewey
C. texensis (J. Torrey) L. Bailey
Sy = C. retroflexa G. H. Muhlenberg *ex* C. von Willdenow var. texensis (J. Torrey) M. Fernald
C. triangularis J. Böckeler
Sy = C. vulpinoidea A. Michaux var. drummondiana J. Böckeler
Sy = C. vulpinoidea var. triangularis (J. Böckeler) G. Kükenthal
C. tribuloides G. Wahlenberg
Sy = C. lagopodioides C. Schkuhr
Sy = C. scoparia C. Schkuhr var. lagopodioides (C. Schkuhr) J. Torrey
C. typhina A. Michaux
Sy = C. squarrosa C. Linnaeus var. typhina (A. Michaux) T. Nuttall
C. verrucosa G. H. Muhlenberg
Sy = C. macrokolea E. von Steudel
C. vulpinoidea A. Michaux var. ***vulpinoidea***
Sy = C. setacea C. Dewey
Sy = C. vulpiniformis E. Tuckerman
Sy = Vignea setacea (C. Dewey) C. Rafinesque-Schmaltz

CLADIUM P. Browne
C. mariscoides (G. H. Muhlenberg) J. Torrey
Sy = Mariscus mariscoides (G. H. Muhlenberg) K. E. O. Kuntze
C. mariscus (C. Linnaeus) J. Pohl subsp. ***jamaicense*** (H. von Crantz) G. Kükenthal
Sy = C. jamaicense H. von Crantz
Sy = C. leptostachum C. Nees von Esenbeck & F. Meyen
Sy = Mariscus jamaicensis (H. von Crantz) N. Britton
Sy = Schoenus mariscus C. Linnaeus

CYPERUS C. Linnaeus
C. acuminatus J. Torrey & W. Hooker *ex* J. Torrey
Sy = C. acuminatus var. cyrtolepis (J. Torrey & W. Hooker *ex* J. Torrey) G. Kükenthal
Sy = C. cyrtolepis J. Torrey & W. Hooker
C. albostriatus H. A. Schrader [**cultivated**]
C. articulatus C. Linnaeus
Sy = C. articulatus var. nodosus (F. von Humboldt & A. Bonpland *ex* C. von Willdenow) G. Kükenthal
Sy = C. corymbosus C. Rottbøll var. corymbosus
Sy = C. corymbosus var. subnodosus (C. Nees von Esenbeck & F. Meyen *ex* C. Nees von Esenbeck) G. Kükenthal
Sy = C. subnodosus C. Nees von Esenbeck & F. Meyen *ex* C. Nees von Esenbeck
C. bipartitus J. Torrey
Sy = C. lagunetto E. von Steudel
Sy = C. niger H. Ruiz López & J. Pavón var. rivularis (K. Kunth) V. Grant
Sy = C. rivularis K. Kunth
Sy = C. rivularis subsp. lagunetto (E. von Steudel) G. Kükenthal
Sy = Pycreus rivularis (K. Kunth) E. Palla
* ***C. cephalanthus*** J. Torrey & W. Hooker [**TOES: V**]
Sy = C. laetus K. Kunth var. cephalanthus (J. Torrey & W. Hooker) G. Kükenthal
Sy = C. rigens J. Presl subsp. cephalanthus (J. Torrey & W. Hooker) T. Pedersen
Sy = Mariscus cephalanthus (J. Torrey & W. Hooker) C. B. Clarke
C. compressus C. Linnaeus
Sy = C. compressus var. pectiniformis (J. J. Römer & J. A. Schultes) C. B. Clarke
Sy = C. pectinatus W. Roxburgh, non M. H. Vahl
Sy = C. pectiniformis J. J. Römer & J. A. Schultes
C. croceus M. H. Vahl
Sy = C. baldwinii J. Torrey
Sy = C. multiflorus (N. Britton) J. K. Small
Sy = Mariscus bracheilema E. von Steudel
Ma = Cyperus globulosus *auct.*, non J. Aublet: American authors
C. difformis C. Linnaeus
Sy = C. lateriflorus J. Torrey
C. digitatus W. Roxburgh subsp. ***digitatus***
Sy = C. bourgaei C. B. Clarke *ex* C. Lundell, non C. B. Clarke
Sy = C. digitatus var. mexicanus (F. Liebmann) G. Kükenthal
Sy = C. mexicanus F. Liebmann
Ma = C. giganteus, *auct.*, non M. H. Vahl: Texas authors
C. drummondii J. Torrey & W. Hooker
Sy = C. virens A. Michaux var. drummondii (J. Torrey & W. Hooker) G. Kükenthal
C. echinatus (C. Linnaeus) A. Wood
Sy = C. ovularis (A. Michaux) J. Torrey
Sy = C. ovularis var. sphaericus J. Böckeler
C. elegans C. Linnaeus

Sy = C. trachynotus J. Torrey
Sy = C. viscosus O. Swartz
C. entrerianus J. Böckeler
Sy = C. luzulae (C. Linnaeus) C. Rottbøll *ex* A. Retzius var. entrerianus (J. Böckeler) M. Barros
Sy = C. tucumanensis J. Böckeler
C. eragrostis J. de Lamarck
Sy = C. eragrostis var. compactus (N. Desvaux) G. Kükenthal
Sy = C. monandrus A. Roth
Sy = C. serrulatus S. Watson
Sy = C. vegetus C. von Willdenow
C. erythrorhizos G. H. Muhlenberg
Sy = Chlorocyperus erythrorhizos E. Palla
Sy = Cyperus chrysokerkos E. von Steudel
Sy = C. cupreus J. Presl & K. Presl
Sy = C. erythrorhizos var. cupreus (J. Presl & K. Presl) G. Kükenthal
Sy = C. halei J. Torrey *ex* N. Britton
Sy = C. occidentalis J. Torrey
Sy = C. washingtonensis M. Gandoger
C. esculentus C. Linnaeus var. ***esculentus***
Sy = C. esculentus C. Linnaeus var. leptostachyus J. Böckeler
C. esculentus C. Linnaeus var. ***macrostachyus*** J. Böckeler
Sy = C. esculentus var. sprucei C. B. Clarke
Sy = C. lutescens J. Torrey & W. Hooker
C. fendlerianus J. Böckeler var. ***debilis*** (N. Britton) G. Kükenthal
Sy = Chlorocyperus mexicanus E. Palla
Sy = C. fendelerianus var. leucolepis (J. Böckeler) G. Kükenthal
Sy = C. rubsyi N. Britton
Sy = C. schweinitzii J. Torrey var. debilis N. Britton
Sy = C. sphaerolepis J. Böckeler
C. fendlerianus J. Böckeler var. ***fendlerianus***
Sy = Chlorocyperus fendlerianus (J. Böckeler) E. Palla
Sy = Cyperus fendlerianus var. leucolepis (J. Böckeler) G. Kükenthal
Sy = C. fendlerianus var. major G. Kükenthal
Sy = C. leucolepis J. Böckeler
Sy = C. sphaerolepis J. Böckeler
Sy = Mariscus fendlerianus (J. Böckeler) T. Koyama
C. filiculmis M. H. Vahl
Sy = C. martindalei N. Britton
C. flavescens C. Linnaeus var. ***flavescens***
Sy = C. flavescens var. poiformis (F. Pursh) M. Fernald
Sy = C. nieuwlandii M. Geise
Sy = Pycreus flavescens (C. Linnaeus) H. G. L. Reichenbach
C. flavicomus A. Michaux
Sy = C. albomarginatus K. Martens & H. A. Schrader
Sy = C. sabulosus (K. von Martius & H. A. Schrader *ex* C. Nees von Esenbeck) E. von Steudel
Sy = Pycreus albomarginatus K. von Martius & H. A. Schrader *ex* C. Nees von Esenbeck
Sy = P. flavicomus (A. Michaux) C. Adams
Sy = P. sabulosus K. von Martius & H. A. Schrader *ex* C. Nees von Esenbeck
C. floribundus (G. Kükenthal) J. R. Carter & S. D. Jones (***ined.***)
Sy = C. uniflorus J. Torrey & W. Hooker var. floribundus G. Kükenthal
C. fugax F. Liebmann
Sy = C. inconspicuus F. Liebmann
Sy = C. polystachyos C. Rottbøll f. fugax (F. Liebmann) G. Kükenthal
Sy = C. polystachyos f. inconspicuus (F. Liebmann) G. Kükenthal
Sy = C. tenellus J. Presl & K. Presl, non C. tenellus C. von Linné
Sy = Pycreus fugax (F. Liebmann) C. D. Adams
C. grayoides R. Mohlenbrock
Sy = C. grayioides R. Mohlenbrock (orthographic error)
C. haspan C. Linnaeus
Sy = C. aphyllus M. H. Vahl
Sy = C. autumnalis M. H. Vahl
Sy = C. efoliatus J. Böckeler
Sy = C. haspan var. americanus J. Böckeler
Sy = C. juncoides J. de Lamarck
C. hystricinus M. Fernald
Sy = C. retrofractus (C. Linnaeus) J. Torrey var. hystricinus (M. Fernald) G. Kükenthal
C. involucratus C. Rottbøll [**cultivated**]
Sy = C. alternifolius C. Linnaeus subsp. flabelliformis (C. Rottbøll) G. Kükenthal
Sy = C. flabelliformis C. Rottbøll
Sy = C. onustus E. von Steudel
Ma = C. alternifolius *auct.*, non C. Linnaeus: American authors
C. iria C. Linnaeus
Sy = C. panicoides J. de Lamarck
C. laevigatus C. Linnaeus
Sy = C. careyi N. Britton
Sy = C. distachyos C. Allioni
Sy = C. mucronatus C. Rottbøll

Sy = C. neokunthianus G. Kükenthal

C. lanceolatus J. Poiret

Sy = C. ambiguus F. Liebmann

Sy = C. densus J. Link

Sy = C. lanceolatus var. compositus J. Presl & K. Presl

Sy = C. olfersianus K. Kunth

C. lentiginosus C. Millspaugh & M. A. Chase

Sy = C. tenuis O. Swartz var. lentiginosus (C. Millspaugh & M. A. Chase) G. Kükenthal

C. lupulinus (K. Sprengel) B. G. Marcks subsp. ***lupulinus***

Sy = Mariscus cyperiformis (G. H. Muhlenberg) J. Torrey

Sy = Scirpus cyperiformis G. H. Muhlenberg

Sy = S. lupulinus K. Sprengel

C. macrocephalus F. Liebmann var. ***eggersii*** (J. Böckeler) S. D. Jones, J. Wipff, & J. R. Carter [***ined.***]

Sy = C. eggersii J. Böckeler

Sy = Torulinium eggersii (J. Böckeler) C. B. Clarke

Sy = T. macrocephalum (F. Liebmann) T. Koyama var. eggersii (J. Böckeler) C. Adams

C. macrocephalus F. Liebmann var. ***macrocephalus***

Sy = C. oxycarioides N. Britton

Sy = Torulinium macrocephalum (F. Liebmann) T. Koyama

C. niger H. Ruiz López & J. Pavón

Sy = C. diandrus J. Torrey var. capitatus N. Britton

Sy = C. melanostachyus K. Kunth

Sy = C. melanostachyus var. robustus F. Liebmann

Sy = C. niger var. capitatus (N. Britton) H. O'Neill

Sy = C. niger var. castaneus (F. Pursh) G. Kükenthal

Sy = Pycreus niger (H. Ruiz López & J. Pavón) G. Cufodontis

C. ochraceus M. H. Vahl

C. odoratus C. Linnaeus var. ***engelmannii*** (E. von Steudel) J. R. Carter, S. D. Jones, & J. Wipff [***ined.***]

Sy = C. engelmannii E. von Steudel

C. odoratus C. Linnaeus var. ***ferruginescens*** (J. Böckeler) S. D. Jones, J. R. Carter, & J. Wipff [***ined.***]

Sy = C. ferruginescens J. Böckeler

Sy = C. speciosus M. H. Vahl var. ferruginescens (J. Böckeler) N. Britton

C. odoratus C. Linnaeus var. ***odoratus***

Sy = C. acicularis H. A. Schrader *ex* C. Nees von Esenbeck

Sy = C. ferax L. C. Richard subsp. speciosus (N. Britton) G. Kükenthal

Sy = C. huarmensis (K. Kunth) M. C. Johnston

Sy = C. longispicatus J. Norton

Sy = C. michauxicanus J. A. Schultes

Sy = C. odoratus var. acicularis (H. A. Schrader *ex* C. Nees von Esenbeck) H. O'Neill

Sy = C. odoratus var. squarrosus (N. Britton) G. Kükenthal

Sy = C. speciosus M. H. Vahl

Sy = C. speciosus var. squarrosus N. Britton

Sy = Mariscus huarmensis K. Kunth

Sy = Torulinium confertum N. Desvaux *ex* W. Hamilton

Sy = T. michauxicanus (J. A. Schultes) C. B. Clarke

Sy = T. odoratum (C. Linnaeus) S. Hooper

* ***C. onerosus*** M. C. Johnston [**TOES: V**]

C. oxylepis C. Nees von Esenbeck *ex* E. von Steudel

C. pallidicolor (G. Kükenthal) G. Tucker subsp. ***ammophilous*** S. D. Jones, J. Wipff, & J. R. Carter [**ined.**]

Ma = C. aggregatus *auct.*, non (C. von Willdenow) S. Endlicher: American authors

Ma = C. huarmensis (K. Kunth) M. C. Johnston: *sensu* Correll & Johnston 1970, Gould 1975

C. papyrus C. Linnaeus [**cultivated**]

Sy = Papyrus antiquorum C. von Willdenow

C. phaeolepis H. Chermezon [**cultivated**]

C. plukenetii M. Fernald

Ma = C. retrofractus *auct.*, non (C. Linnaeus) J. Torrey: Texas authors, in part

C. polystachyos C. Rottbøll

Sy = C. brizaeus J. Presl & K. Presl

Sy = C. holosericeus J. Link

Sy = C. microdontus J. Torrey

Sy = C. olidus M. H. Vahl

Sy = C. polystachyos var. holosericeus (J. Link) T. Koyama

Sy = C. polystachyos var. laxiflorus G. Bentham

Sy = C. polystachyos var. leptostachyus J. Böckeler

Sy = C. polystachyos var. texensis (J. Torrey) M. Fernald

Sy = Pycreus polystachyos (C. Rottbøll) A. Palisot de Beauvois

C. prolifer J. de Lamarck [**cultivated**]

Sy = C. aequalis M. H. Vahl

Sy = C. papyroides J. Poiret

C. prolixus K. Kunth
Sy = C. amplissimus E. von Steudel
C. pseudothyrsiflorus (G. Kükenthal) J. R. Carter & S. D. Jones [***ined.***]
Sy = C. uniflorus J. Torrey & W. Hooker var. pseudothyrsiflorus G. Kükenthal
C. pseudovegetus E. von Steudel var. ***pseudovegetus***
C. reflexus M. H. Vahl var. ***fraternus*** (K. Kunth) K. E. O. Kuntze
C. reflexus M. H. Vahl var. ***reflexus***
C. retroflexus S. Buckley var. ***pumilus*** (N. Britton) J. R. Carter & S. D. Jones [***ined.***]
Sy = C. subuniflorus N. Britton
Sy = C. uniflorus J. Torrey & W. Hooker var. pumilus N. Britton
Sy = Mariscus subuniflorus (N. Britton) T. Koyama
C. retroflexus S. Buckley var. ***retroflexus***
Sy = C. uniflorus J. Torrey & W. Hooker, non C. Thunberg
Sy = C. uniflorus var. retroflexus (S. Buckley) G. Kükenthal
Sy = Mariscus uniflorus (J. Torrey & W. Hooker) E. von Steudel
C. retrofractus (C. Linnaeus) J. Torrey
Sy = C. dipsaciformis M. Fernald
Sy = C. retrofractus var. dipsaciformis (M. Fernald) G. Kükenthal
Sy = Mariscus retrofractus (C. Linnaeus) M. H. Vahl
Sy = Scirpus retrofractus C. Linnaeus
C. retrorsus A. Chapman
Sy = C. cylindricus (S. Elliott) N. Britton, non A. Chapman
Sy = C. globulosus J. Aublet var. robustus (J. Böckeler) L. Shinners
Sy = C. ovularis (A. Michaux) J. Torrey var. cylindricus (S. Elliott) J. Torrey
Sy = C. pollardii N. Britton *ex* J. K. Small
Sy = C. retrorsus var. cylindricus (S. Elliott) M. Fernald & L. Griscom
Sy = C. retrorsus var. robustus (J. Böckeler) G. Kükenthal
Sy = Mariscus cylindricus S. Elliott
C. rotundus C. Linnaeus
C. schweinitzii J. Torrey
Sy = Chlorocyperus schweinitzii E. Palla
Sy = Mariscus schweinitzii (J. Torrey) T. Koyama
C. seslerioides K. Kunth
Sy = C. andinus E. Palla *ex* G. Kükenthal
Sy = C. perpusillus J. Böckeler
Sy = C. seslerioides var. minor G. Kükenthal
Sy = Scirpus hahnii C. B. Clarke
C. setigerus J. Torrey & W. Hooker
Sy = C. hallii N. Britton
C. spectabilis J. Link
Sy = C. buckleyi N. Britton *ex* J. Coulter
Sy = C. mohrii N. Britton *ex* C. B. Clarke
Sy = C. parryi N. Britton *ex* C. B. Clarke
Sy = C. scaberrimus C. Nees von Esenbeck
C. squarrosus C. Linnaeus var. ***runyonii*** (H. O'Neill) S. D. Jones & J. K. Wipff
Sy = C. aristatus C. Rottbøll var. runyonii H. O'Neill
C. squarrosus C. Linnaeus var. ***squarrosus***
Sy = C. aristatus var. inflexus (G. H. Muhlenberg) J. Böckeler *ex* G. Kükenthal
Sy = C. falciculosus F. Liebmann
Sy = C. inflexus G. H. Muhlenberg
Sy = Dichostylis aristatus E. Palla
Ma = Cyperus aristatus C. Rottbøll: American authors
C. strigosus C. Linnaeus var. ***strigosus***
C. surinamensis C. Rottbøll
Sy = C. barrancae M. E. Jones
Sy = C. bipontini J. Böckeler
Sy = C. denticulatus H. A. Schrader *ex* J. A. Schultes
Sy = C. surinamensis var. lutescens J. Böckeler
C. thyrsiflorus F. Junghuhn
Sy = C. dissitiflorus J. Torrey
Ma = C. hermaphroditus (N. von Jacquin) P. Standley
Ma = C. tenuis *auct.*, non O. Swartz: Texas authors
C. virens A. Michaux var. ***virens***
Sy = Scirpus reticulatus J. de Lamarck
C. grayoides R. Mohlenbrock × ***C. filiculmis*** M. H. Vahl
C. lupulinus (K. Sprengel) B. G. Marcks subsp. ***lupulinus*** × ***C. schweinitzii*** J. Torrey
Sy = C. houghtonii J. Torrey var. uberior G. Kükenthal
Sy = C. × mesochorus M. Geise

DULICHIUM C. Persoon
D. arundinaceum (C. Linnaeus) N. Britton
Sy = Cyperus arundinaceum C. Linnaeus
Sy = C. spathaceus C. Linnaeus
Sy = Dulichium spathaceum (L. C. Richard) C. Persoon
Sy = Schoenus spathaceus L. C. Richard

ELEOCHARIS R. Brown
E. ***acicularis*** (C. Linnaeus) J. J. Römer & J. A. Schultes var. ***acicularis***
Sy = Clavula acicularis (C. Linnaeus) C. Dumortier
Sy = Eleocharis acicularis var. occidentalis H. Svenson
Sy = E. nervata H. Svenson
Sy = E. reverchonii H. Svenson
Sy = Isolepis acicularis (C. Linnaeus) J. J. Römer & J. A. Schultes
Sy = Scirpus acicularis C. Linnaeus
E. ***acutangula*** (W. Roxburgh) J. A. Schultes
Sy = E. fistulosa (J. Poiret) J. A. Schultes
Sy = Scirpus acutangulus W. Roxburgh
Sy = S. fistulosus J. Poiret
E. ***acutisquamata*** S. Buckley
E. ***albida*** J. Torrey
E. ***atropurpurea*** (A. Retzius) J. Presl & K. Presl
Sy = E. multiflora A. Chapman
Sy = Scirpus atropurpureus A. Retzius
E. ***austrotexana*** M. C. Johnston
E. ***baldwinii*** (J. Torrey) A. Chapman
Sy = E. capillacea K. Kunth
Sy = E. prolifera J. Torrey
* *E.* ***brachycarpa*** H. Svenson [**TOES**: **V**]
E. ***brittonii*** H. Svenson *ex* J. K. Small
Sy = E. microcarpa J. Torrey var. brittonii (H. Svenson *ex* J. K. Small) H. Svenson
E. ***cellulosa*** J. Torrey
E. ***compressa*** W. Sullivant
Sy = E. acuminata (G. H. Muhlenberg) C. Nees von Esenbeck
* *E.* ***cylindrica*** S. Buckley [**TOES**: **III**]
E. ***elongata*** A. Chapman
E. ***engelmannii*** E. von Steudel
Sy = E. engelmannii var. monticola (M. Fernald) H. Svenson
Sy = E. engelmannii var. robusta M. Fernald
Sy = E. monticola M. Fernald
Sy = E. monticola var. pallida H. St. John
Sy = E. obtusa (C. von Willdenow) J. A. Schultes var. detonsa (A. Gray) D. Drapalik & R. Mohlenbrock
Sy = E. ovata (A. Roth) J. J. Römer & J. A. Schultes var. engelmannii (E. Steudel) N. Britton
E. ***equisetoides*** (S. Elliott) J. Torrey
Sy = E. elliottii A. Dietrich
Sy = Scirpus equisetoides S. Elliott
E. ***fallax*** C. Weatherby
Sy = E. ambigens M. Fernald
E. ***flavescens*** (J. Poiret) I. Urban var. ***flavescens***
Sy = E. flavescens var. fuscescens (G. Kükenthal) H. Svenson
Sy = E. ochreata *auct.*, non (C. Nees von Esenbeck) E. von Steudel: Mexican authors
Sy = E. praticola N. Britton
Sy = Heleocharis flaccida (H. G. L. Reichenbach) I. Urban
Sy = H. flaccida var. fuscescens G. Kükenthal
Sy = Scirpus flaccidus (H. G. L. Reichenbach *ex* K. Sprengel) I. Urban
Sy = S. flavescens J. Poiret
E. ***geniculata*** (C. Linnaeus) J. J. Römer & J. A. Schultes
Sy = E. capitata (C. Linnaeus) R. Brown var. dispar (E. Hill) M. Fernald
Sy = Scirpus geniculatus C. Linnaeus
Ma = Eleocharis caribaea *auct.*, non (C. Rottbøll) S. F. Blake
Ma = E. caribaea var. dispar *auct.*, non (E. Hill) S. F. Blake
Ma = Scirpus caribaeus *auct.*, non C. Rottbøll
E. ***interstincta*** (M. H. Vahl) J. J. Römer & J. A. Schultes
Sy = Scirpus interstinctus M. H. Vahl
E. ***lanceolata*** M. Fernald
Sy = E. obtusa (C. von Willdenow) J. A. Schultes var. lanceolata (M. Fernald) C. Gilly
E. ***melanocarpa*** J. Torrey
E. ***microcarpa*** J. Torrey var. ***microcarpa***
Sy = E. torreyana J. Böckeler
E. ***minima*** K. Kunth var. ***minima***
Sy = E. durandii J. Böckeler
E. ***montana*** (K. Kunth) J. J. Römer & J. A. Schultes
Sy = E. montana var. nodulosa (A. Roth) H. Svenson
Sy = E. nodulosa (A. Roth) J. A. Schultes
Sy = Scirpus montanus K. Kunth
Sy = S. nodulosus A. Roth
E. ***montevidensis*** K. Kunth
Sy = E. arenicola J. Torrey *ex* G. Engelmann
Sy = Limnochloa montevidensis C. Nees von Esenbeck
Sy = Trichophyllum arenicolum H. House
E. ***obtusa*** (C. von Willdenow) J. A. Schultes
Sy = E. diandra C. Wright
Sy = E. macounii M. Fernald
Sy = E. obtusa var. ellipsoidalis M. Fernald
Sy = E. obtusa var. gigantea (C. B. Clarke) M. Fernald

Sy = E. obtusa var. jejuna M. Fernald
Sy = E. obtusa var. peasei H. Svenson
Sy = E. ovata (A. Roth) J. J. Römer & J. A. Schultes var. obtusa (C. von Willdenow) G. Kükenthal *ex* C. Skottsberg

E. olivacea J. Torrey
Sy = E. flaccida (H. G. L. Reichenbach) I. Urban var. olivacea (J. Torrey) M. Fernald & L. Griscom
Sy = E. flavescens (J. Poiret) I. Urban var. olivacea (J. Torrey) H. Gleason

E. palmeri H. Svenson

E. palustris (C. Linnaeus) J. J. Römer & J. A. Schultes
Sy = E. calva J. Torrey var. australis (C. Nees von Esenbeck) H. St. John
Sy = E. glaucescens (C. von Willdenow) J. A. Schultes
Sy = E. macrostachya N. Britton
Sy = E. palustris var. australis C. Nees von Esenbeck
Sy = E. palustris var. major O. W. Sonder
Sy = E. perlong M. Fernald & A. Brackett
Sy = E. smallii N. Britton var. major (O. W. Sonder) Seymour
Sy = E. xyridiformis M. Fernald & A. Brackett
Sy = Scirpus palustris C. Linnaeus

E. parvula (J. J. Römer & J. A. Schultes) J. Link *ex* M. Bluff, C. Nees von Esenbeck, & J. Schauer var. ***anachaeta*** (J. Torrey) H. Svenson

E. quadrangulata (A. Michaux) J. J. Römer & J. A. Schultes var. ***crassior*** M. Fernald

E. radicans (A. Dietrich) K. Kunth
Sy = E. acicularis (C. Linnaeus) J. J. Römer & J. A. Schultes var. radicans (A. Dietrich) N. Britton
Sy = E. lindheimeri H. Svenson
Sy = Eliogiton radicans A. Dietrich
Sy = Scirpus radicans J. Poiret, non C. Schkuhr

E. rostellata (J. Torrey) J. Torrey
Sy = E. rostellata var. congdonii W. Jepson
Sy = E. rostellata var. occidentalis S. Watson
Sy = Scirpus rostellatus J. Torrey
Sy = Trichophyllum rostellatum O. Farwell

E. tenuis (C. von Willdenow) J. A. Schultes var. ***verrucosa*** (H. Svenson) H. Svenson
Sy = E. capitata (C. Linnaeus) R. Brown var. verrucosa H. Svenson
Sy = E. verrucosa (H. Svenson) L. Harms

E. tortilis (J. Link) J. A. Schultes

E. tuberculosa (A. Michaux) J. J. Römer & J. A. Schultes
Sy = E. simplex (S. Elliott) A. Dietrich
Sy = E. tuberculosa var. pubnicoënsis M. Fernald

E. vivipara J. Link
Sy = E. curtisii J. K. Small

E. wolfii (A. Gray) A. Gray *ex* N. Britton
Sy = Scirpus wolfii A. Gray

FIMBRISTYLIS M. H. Vahl, *nomen conservandum*

F. annua (C. Allioni) J. J. Römer & J. A. Schultes
Sy = F. alamosana M. Fernald
Sy = F. baldwiniana (J. A. Schultes) J. Torrey
Sy = F. laxa M. H. Vahl
Sy = Scirpus annuus C. Allioni

F. autumnalis (C. Linnaeus) J. J. Römer & J. A. Schultes
Sy = F. autumnalis var. mucronulata (A. Michaux) M. Fernald
Sy = F. geminata (C. Nees von Esenbeck) K. Kunth
Sy = Scirpus autumnalis C. Linnaeus

F. caroliniana (J. de Lamarck) M. Fernald
Sy = F. harperi N. Britton
Sy = Scirpus carolinianus J. de Lamarck

F. castanea (A. Michaux) M. H. Vahl
Sy = Scirpus castaneus A. Michaux

F. decipiens R. Kral

F. dichotoma (C. Linnaeus) M. H. Vahl
Sy = F. diphylla (A. Retzius) M. H. Vahl
Sy = Scirpus dichotomus C. Linnaeus

F. miliacea (C. Linnaeus) M. H. Vahl
Sy = Scirpus miliaceus C. Linnaeus

F. puberula (A. Michaux) M. H. Vahl *ex* J. K. Small & N. Britton var. ***interior*** (N. Britton) R. Kral
Sy = F. interior N. Britton

F. puberula (A. Michaux) M. H. Vahl *ex* J. K. Small & N. Britton var. ***puberula***
Sy = F. anomala J. Böckeler
Sy = F. castanea (A. Michaux) M. H. Vahl var. puberula (A. Michaux) N. Britton
Sy = F. drummondii (J. Torrey & W. Hooker *ex* J. Torrey) J. Böckeler
Sy = F. puberula var. drummondii (J. Torrey & W. Hooker *ex* J. Torrey) L. F. Ward

F. tomentosa M. H. Vahl
Sy = F. pilosa M. H. Vahl

F. vahlii (J. de Lamarck) J. Link
Sy = F. congesta J. Torrey
Sy = Scirpus vahlii J. de Lamarck

FUIRENA C. Rottbøll

F. breviseta (F. Coville) F. Coville
Sy = F. squarrosa A. Michaux var. breviseta F. Coville

F. bushii R. Kral
Sy = F. ciliata B. Bush
F. longa A. Chapman
F. pumila (J. Torrey) K. Sprengel
Sy = F. squarrosa A. Michaux var. pumila J. Torrey
Sy = F. torreyana L. Beck
F. scirpoidea A. Michaux
Sy = Scirpus scirpoideus (A. Michaux) T. Koyama
F. simplex M. H. Vahl var. ***aristulata*** (J. Torrey) R. Kral
Sy = F. squarrosa A. Michaux var. aristulata J. Torrey
F. simplex M. H. Vahl var. ***simplex***
F. squarrosa A. Michaux
Sy = F. hispida S. Elliott

ISOLEPIS R. Brown
I. carinatus J. Hooker & G. Arnott *ex* J. Torrey
Sy = I. koilolepis E. von Steudel
Sy = Scirpus carinatus (J. Hooker & G. Arnott *ex* J. Torrey) A. Gray
Sy = S. koilolepis (E. von Steudel) H. Gleason
I. cernua J. J. Römer & J. A. Schultes
Sy = Scirpus cernuus M. H. Vahl var. californicus (J. Torrey) A. Beetle
I. molesta (M. C. Johnston) S. G. Smith
Sy = Scirpus molestus M. C. Johnston

KYLLINGA C. Rottbøll, *nomen conservandum*
K. brevifolia C. Rottbøll
Sy = Cyperus brevifolius (C. Rottbøll) S. Endlicher *ex* J. Hasskarl
K. odorata M. H. Vahl var. ***odorata***
Sy = Cyperus sesquiflorus (J. Torrey) J. Mattfeld & G. Kükenthal *ex* G. Kükenthal
Sy = Kyllinga sesquiflora J. Torrey
K. pumila A. Michaux
Sy = Cyperus densicaespitosus J. Mattfeld & G. Kükenthal *ex* G. Kükenthal
Sy = C. densicaespitosus var. major (C. Nees von Esenbeck) G. Kükenthal
Sy = C. tenuifolia E. von Steudel
Sy = C. tenuifolius (E. von Steudel) J. Dandy

LIPOCARPHA R. Brown, *nomen conservandum*
L. aristulata (F. Coville) G. Tucker
Sy = Hemicarpha aristulata (F. Coville) B. Smyth
Sy = H. micrantha (M. H. Vahl) F. Pax var. aristulata F. Coville
L. micrantha (M. H. Vahl) G. Tucker
Sy = Hemicarpha micrantha (M. H. Vahl) F. Pax
Sy = H. micrantha var. drummondii (C. Nees von Esenbeck) S. Friedland
Sy = Isolepis micranthus J. J. Römer & J. A. Schultes
Sy = Scirpus micranthus M. H. Vahl

OXYCARYUM C. Nees von Esenbeck
O. cubense (E. Poeppig & K. Kunth) K. Lye
Sy = Anosporum cubense (E. Poeppig & K. Kunth) J. Böckeler var. gracile J. Böckeler
Sy = Oxycaryum schomburgkianum C. Nees von Esenbeck
Sy = Scirpus cubensis E. Poeppig & K. Kunth
Sy = S. cubensis var. gracilis (J. Böckeler) A. Beetle

RHYNCHOSPORA M. H. Vahl (orthographic *conservandum*), *nomen conservandum*
R. caduca S. Elliott
Sy = R. patula A. Gray
R. capillacea J. Torrey
Sy = R. capillacea var. leviseta E. Hill *ex* A. Gray
Sy = R. smallii N. Britton
R. capitellata (A. Michaux) M. H. Vahl
Sy = R. glomerata (C. Linnaeus) M. H. Vahl var. capitellata (A. Michaux) G. Kükenthal
Sy = R. glomerata var. leptocarpa A. Chapman *ex* N. Britton
Sy = R. glomerata var. minor N. Britton
Sy = R. leptocarpa (A. Chapman *ex* N. Britton) J. K. Small
Sy = Schoenus capitellatus A. Michaux
R. cephalantha A. Gray
Sy = R. axillaris (J. de Lamarck) N. Britton
Sy = R. cephalantha var. pleiocephala M. Fernald & S. Gale
R. chalarocephala M. Fernald & S. Gale
Sy = R. chalarocephala var. angusta S. Gale
R. colorata (C. Linnaeus) H. Pfeiffer
Sy = Dichromena colorata (C. Linnaeus) A. Hitchcock
Sy = Rhynchospora drummondiana E. von Steudel
Sy = R. stellata (J. de Lamarck) A. Grisebach
Sy = Schoenus coloratus C. Linnaeus
Sy = S. stellatus J. de Lamarck
R. corniculata (J. de Lamarck) A. Gray
Sy = R. corniculata var. interior M. Fernald
R. debilis S. Gale
Sy = R. fascicularis (A. Michaux) M. H. Vahl var. trichodes A. Chapman
Ma = R. trichodes *auct.*, non C. B. Clarke
R. divergens A. Chapman *ex* M. Curtis

R. elliottii D. Dietrich
Sy = R. schoenoides (S. Elliott) A. Wood
R. fascicularis (A. Michaux) M. H. Vahl
Sy = R. fascicularis subsp. fascicularis
Sy = R. fascicularis var. fascicularis
Sy = Schoenus fascicularis A. Michaux
R. filifolia A. Gray
Sy = R. fuscoides C. B. Clarke
R. globularis (A. Chapman) J. K. Small var. ***globularis***
Sy = R. cymosa S. Elliott var. globularis A. Chapman
Sy = R. obliterata S. Gale
R. globularis (A. Chapman) J. K. Small var. ***pinetorum*** (N. Britton & J. K. Small *ex* J. K. Small) S. Gale
Sy = R. pinetorum N. Britton & J. K. Small *ex* J. K. Small
R. glomerata (C. Linnaeus) M. H. Vahl
Sy = R. cymosa S. Elliott
Sy = R. glomerata var. angusta S. Gale
Sy = R. glomerata var. paniculata A. Chapman
Sy = Schoenus glomeratus C. Linnaeus
R. gracilenta A. Gray
Sy = R. gracilenta var. diversifolia M. Fernald
Sy = R. drummondiana J. Böckeler
R. grayi K. Kunth
Sy = R. distans S. Elliott, non M. H. Vahl
Sy = Schoenus distans G. H. Muhlenberg
R. harveyi W. Boott
Sy = R. earlei N. Britton *ex* J. K. Small
Sy = R. planckii N. Britton *ex* J. K. Small
R. indianolensis J. K. Small
R. inexpansa (A. Michaux) M. H. Vahl
Sy = Schoenus inexpansus A. Michaux
R. inundata (W. Oakes) M. Fernald
R. latifolia (W. Baldwin *ex* S. Elliott) W. Thomas
Sy = Dichromena latifolia W. Baldwin *ex* S. Elliott
R. macra (C. B. Clarke) J. K. Small
Sy = R. alba (C. Linnaeus) M. H. Vahl var. macra C. B. Clarke *ex* N. Britton
R. macrostachya J. Torrey *ex* A. Gray
Sy = R. macrostachya var. colpophila M. Fernald & S. Gale
R. microcarpa W. Baldwin *ex* A. Gray
Sy = R. edisoniana N. Britton *ex* J. K. Small
* ***R. miliacea*** (J. de Lamarck) A. Gray **[TOES: IV]**
Sy = Schoenus miliaceus J. de Lamarck
R. mixta N. Britton
Sy = R. prolifera J. K. Small
R. nitens (M. H. Vahl) A. Gray
Sy = Psilocarya nitens (M. H. Vahl) A. Wood
Sy = P. portoricensis N. Britton
R. nivea J. Böckeler
Sy = Dichromena nivea (J. Böckeler) J. Böckeler *ex* N. Britton
R. oligantha A. Gray
R. perplexa N. Britton var. ***perplexa***
R. plumosa S. Elliott
Sy = R. penniseta A. Grisebach
Sy = R. semiplumosa A. Gray
R. pusilla A. Chapman *ex* M. Curtis
Sy = R. bruneri N. Britton
R. rariflora (A. Michaux) S. Elliott
Sy = Schoenus rariflorus A. Michaux
R. recognita (S. Gale) R. Kral
Sy = R. globularis (A. Chapman) J. K. Small var. recognita S. Gale
R. scirpoides (J. Torrey) A. Gray
Sy = Psilocarya corymbifera (C. Wright) N. Britton
Sy = P. corymbiformis G. Bentham
Sy = P. scirpoides J. Torrey
Sy = P. scirpoides var. grimesii M. Fernald & L. Griscom
R. stenophylla A. Chapman *ex* M. Curtis
R. tracyi N. Britton
Sy = Ceratoschoenus capitatus A. Chapman

SCHOENOPLECTUS (H. G. L. Reichenbach) E. Palla, *nomen conservandum*
S. acutus (G. H. Muhlenberg *ex* J. Bigelow) A. Löve & D. Löve var. ***acutus***
Sy = Scirpus acutus G. H. Muhlenberg *ex* J. Bigelow
Sy = S. lacustris C. Linnaeus var. condensatus C. H. Peck
Sy = S. validus M. H. Vahl var. condensatus (C. H. Peck) A. Beetle
S. acutus (G. H. Muhlenberg *ex* J. Bigelow) A. Löve & D. Löve var. ***occidentalis*** (S. Watson) S. G. Smith
Sy = Scirpus acutus G. H. Muhlenberg *ex* J. Bigelow var. occidentalis (S. Watson) A. Beetle
Sy = S. lacustris C. Linnaeus var. occidentalis S. Watson
Sy = S. malheurensis L. F. Henderson
Sy = S. occidentalis (S. Watson) M. A. Chase
Sy = S. rubiginosus A. Beetle
S. americanus (C. Persoon) A. von Volkart *ex* H. Schinz & R. Keller
Sy = Scirpus americanus C. Persoon

Sy = S. americanus subsp. monophyllus (J. Presl & K. Presl) T. Koyama
Sy = S. olneyi A. Gray
Ma = S. chilensis *auct.*, non C. Nees von Esenbeck & G. Meyer *ex* K. Kunth

S. californicus (C. von Meyer) J. Soják
Sy = Elytrospermum californicum C. von Meyer
Sy = Scirpus californicus (C. von Meyer) E. von Steudel
Sy = S. riparius J. Presl & K. Presl

S. deltarum (A. E. Schuyler) J. Soják
Sy = Scirpus deltarum A. E. Schuyler

S. erectus (J. Poiret) E. Palla *ex* J. Raynal
Sy = Scirpus erectus J. Poiret
Sy = S. erismanae A. E. Schuyler
Sy = S. wilkensii A. E. Schuyler

S. etuberculatus (E. von Steudel) J. Soják
Sy = Scirpus etuberculatus (E. von Steudel) K. E. O. Kuntze

S. heterochaetus (M. A. Chase) J. Soják
Sy = Scirpus heterochaetus M. A. Chase

S. pungens (M. H. Vahl) E. Palla var. ***longispicatus*** (N. Britton) S. G. Smith
Sy = S. americanus C. Persoon subsp. longispicatus (N. Britton) J. Soják
Sy = S. pungens subsp. longispicatus (N. Britton) A. Löve & D. Löve
Sy = Scirpus americanus C. Persoon var. longispicatus N. Britton
Sy = S. longispicatus (N. Britton) B. Smyth
Sy = S. pungens var. longispicatus (N. Britton) A. Cronquist
Sy = S. solispicatus J. Lunell

S. saximontanus (M. Fernald) J. Raynal
Sy = Scirpus bergsonii A. E. Schuyler
Sy = S. saximontanus M. Fernald
Sy = S. supinus C. Linnaeus var. saximontanus (M. Fernald) T. Koyama

S. tabernaemontani (K. Gmelin) E. Palla
Sy = S. validus (M. H. Vahl) A. Löve & D. Löve
Sy = Scirpus glaucus J. E. Smith
Sy = S. lacustris C. Linnaeus subsp. tabernaemontani (K. Gmelin) A. Löve & D. Löve
Sy = S. lacustris C. Linnaeus subsp. tabernaemontani (K. Gmelin) J. Syme
Sy = S. tabernaemontani K. Gmelin
Sy = S. validus M. H. Vahl
Sy = S. validus var. creber M. Fernald

SCHOENUS C. Linnaeus

S. nigricans C. Linnaeus

SCIRPUS C. Linnaeus, *nomen conservandum*

S. atrovirens C. von Willdenow

S. cyperinus (C. Linnaeus) K. Kunth
Sy = S. cyperinus var. condensatus M. Fernald
Sy = S. cyperinus var. eriophorum (A. Michaux) K. E. O. Kuntze
Sy = S. cyperinus var. laxus (A. Gray) A. Beetle
Sy = S. cyperinus var. pelius M. Fernald
Sy = S. cyperinus var. rubricosus (M. Fernald) C. Gilly
Sy = S. eriophorum A. Michaux
Sy = S. rubricosus M. Fernald

* ***S. divaricatus*** S. Elliott [**TOES: IV**]

S. georgianus R. Harper
Sy = S. atrovirens C. von Willdenow var. georgianus (R. Harper) M. Fernald

S. pallidus (N. Britton) M. Fernald
Sy = S. atrovirens C. von Willdenow var. pallidus N. Britton

S. pendulus G. H. Muhlenberg
Ma = S. lineatus *auct.*, non A. Michaux: Texas authors

SCLERIA P. J. Bergius

S. baldwinii (J. Torrey) E. von Steudel
Sy = S. costata (N. Britton) J. K. Small

S. ciliata A. Michaux var. ***ciliata***
Sy = S. ciliata var. elliottii (A. Chapman) M. Fernald
Sy = S. elliottii A. Chapman

S. ciliata A. Michaux var. ***curtissii*** (N. Britton) J. Kessler
Sy = S. pauciflora G. H. Muhlenberg *ex* C. von Willdenow var. curtissii (N. Britton) J. Fairey
Sy = S. curtissii N. Britton

S. ciliata A. Michaux var. ***glabra*** (A. Chapman) J. Fairey
Sy = S. brittonii E. Core *ex* J. K. Small
Sy = S. pauciflora G. H. Muhlenberg *ex* C. von Willdenow var. glabra A. Chapman

S. georgiana E. Core
Sy = S. gracilis S. Elliott

S. hirtella O. Swartz
Sy = S. distans J. Poiret
Sy = S. doradoënsis N. Britton
Sy = S. interrupta L. C. Richard
Sy = S. lindleyana C. B. Clarke
Sy = S. nutans K. Kunth

S. oligantha A. Michaux

S. pauciflora G. H. Muhlenberg *ex* C. von Willdenow
Sy = S. pauciflora var. kansana M. Fernald
S. reticularis A. Michaux
Sy = S. muhlenbergii E. von Steudel
Sy = S. reticularis var. pubescens N. Britton
Sy = S. reticularis var. pumila N. Britton
Sy = S. setacea J. Poiret
Sy = S. stevensiana N. Britton
S. triglomerata A. Michaux
Sy = S. flaccida E. von Steudel
Sy = S. minor (N. Britton) W. Stone
Sy = S. nitida G. H. Muhlenberg *ex* C. von Willdenow
S. verticillata G. H. Muhlenberg *ex* C. von Willdenow

DIOSCOREACEAE, *nomen conservandum*

DIOSCOREA C. Linnaeus
D. bulbifera C. Linnaeus [**cultivated**]
D. oppositifolia C. Linnaeus [**cultivated**]
Sy = D. batatas J. Decaisne
D. quaternata (T. Walter) J. F. Gmelin
D. villosa C. Linnaeus
Sy = D. hirticaulis H. Bartlett
Sy = D. paniculata A. Michaux

ERIOCAULACEAE, *nomen conservandum*

ERIOCAULON C. Linnaeus
E. aquaticum (J. Hill) G. Druce
Sy = E. septangulare W. Withering
* ***E. compressum*** J. de Lamarck var. ***compressum*** [**TOES: V**]
E. decangulare C. Linnaeus
Sy = E. decangulare var. latifolium A. Chapman
Sy = E. decangulare var. minor H. Moldenke
* ***E. kornickianum*** H. Van Heurck & J. Müller of Aargau [**TOES: V**]
E. texense F. Körnicke

LACHNOCAULON K. Kunth
L. anceps (T. Walter) T. Morong
Sy = L. floridanum J. K. Small
Sy = L. glabrum F. Körnicke
L. digynum F. Körnicke

HYDROCHARITACEAE, *nomen conservandum*

EGERIA J. Planchon
E. densa J. Planchon
Sy = Anacharis densa (J. Planchon) J. Marie-Victorin
Sy = Elodea densa (J. Planchon) J. Caspary
Sy = Philotria densa (J. Planchon) J. K. Small

HALOPHILA L. Aubert du Petit-Thouars
H. engelmannii P. Ascherson

HYDRILLA L. C. Richard
H. verticillata (C. von Linné) J. Royle [**federal noxious weed**]

LIMNOBIUM L. C. Richard
L. spongia (L. Bosc) L. C. Richard *ex* E. von Steudel

OTTELIA C. Persoon
O. alismoides (C. Linnaeus) C. Persoon [**federal noxious weed**]

THALASSIA J. Banks & D. Solander *ex* K. König
T. testudinum J. Banks & D. Solander *ex* K. König
Ma = Posidonia oceanica *auct.*, non K. König: Texas authors

VALLISNERIA C. Linnaeus
V. americana A. Michaux
Ma = V. spiralis *auct.*, non C. Linnaeus: American authors

HYPOXIDACEAE, *nomen conservandum*

HYPOXIS C. Linnaeus
H. curtissii J. Rose
Sy = H. erecta C. Linnaeus var. leptocarpa G. Engelmann & A. Gray
Sy = H. hirsuta var. leptocarpa (G. Engelmann & A. Gray) A. Brackett
Sy = H. leptocarpa (G. Engelmann & A. Gray) J. K. Small
H. hirsuta (C. Linnaeus) F. Coville
Sy = H. rigida A. Chapman
H. micrantha C. Pollard
H. sessilis C. Linnaeus
Sy = H. longii M. Fernald
H. wrightii (J. G. Baker) A. Brackett

IRIDACEAE, *nomen conservandum*

ACIDANTHERA C. F. Hochstetter
A. bicolor C. F. Hochstetter [**cultivated**]

ALOPHIA W. Herbert
A. drummondii (R. Graham) R. Foster
Sy = Herbertia drummondii (R. Graham) J. K. Small
Ma = Eustylis purpurea *auct.*, non (W. Herbert) G. Engelmann & A. Gray: Texas authors

BELAMCANDA M. Adanson, *nomen conservandum*
B. chinensis (C. Linnaeus) A. P. de Candolle **[cultivated]**

CROCOSMIA J. Planchon
C. × *crocosmiiflora* (V. Lemoine *ex* E. Morris) N. Brown [***aurea*** × ***pottsii***] **[cultivated]**
Sy = Tritonia × crocosmiiflora (V. Lemoine *ex* E. Morris) G. E. Nichols

CROCUS C. Linnaeus
C. angustifolius R. Weston **[cultivated]**
Sy = C. susianus J. Ker-Gawler
C. vernus (C. Linnaeus) J. Hill **[cultivated]**
Sy = C. vernus var. neapolitanus J. Ker-Gawler

FREESIA C. Ecklon *ex* F. Klatt, *nomen conservandum*
F. corymbosa (N. Burman) N. E. Brown **[cultivated]**
Sy = F. refracta (N. von Jacquin) C. Ecklon *ex* F. Klatt

GLADIOLUS C. Linnaeus
G. communis C. Linnaeus subsp. ***byzantinus*** (P. Miller) A. Hamilton **[cultivated]**
Sy = G. byzantinus P. Miller
G. tristis W. Herbert **[cultivated]**
G. × *gandavensis* L. B. Van Houtte [***cardinalis*** × ***natlcnsis***] **[cultivated]**
Sy = G. × hortulanus L. Bailey

HERBERTIA R. Sweet
H. lahue (J. Molina) P. Goldblatt subsp. ***caerulea*** (W. Herbert) P. Goldblatt
Sy = Trifurcia lahue (J. Molina) P. Goldblatt subsp. caerulea (W. Herbert) P. Goldblatt
Ma = Alophia drummondii *auct.*, non (R. Graham) R. Foster: Texas authors

IRIS C. Linnaeus
I. brevicaulis C. Rafinesque-Schmaltz
I. cristata D. Solander **[cultivated]**
I. fulva J. Ker-Gawler
I. germanica C. Linnaeus **[cultivated]**
I. hexagona T. Walter var. ***flexicaulis*** (J. K. Small) R. Foster
I. laevigata F. von Fischer *ex* F. von Fischer & C. von Meyer **[cultivated]**
Sy = I. kaempferi P. von Siebold *ex* C. Lemaire
I. pallida J. de Lamarck **[cultivated]**
I. pseudacorus C. Linnaeus **[cultivated]**
I. reticulata F. Marschall von Bieberstein **[cultivated]**
I. setosa P. von Pallas *ex* J. Link **[cultivated]**
I. sibirica C. Linnaeus **[cultivated]**
I. spuria C. Linnaeus subsp. ***ochroleuca*** (C. Linnaeus) W. Dykes **[cultivated]**
Sy = I. ochroleuca C. Linnaeus
I. tingitana P. Boissier & G. Reuter **[cultivated]**
Sy = Xiphion tingitana (P. Boissier & G. Reuter) G. Rodionenko
I. virginica C. Linnaeus var. ***shrevei*** (J. K. Small) E. Anderson
Sy = I. shrevei J. K. Small
I. xiphium C. Linnaeus **[cultivated]**
Sy = Xiphion vernum F. von Schrank
Sy = X. vulgare P. Miller

IXIA C. Linnaeus, *nomen conservandum*
I. maculata C. Linnaeus **[cultivated]**

MORAEA P. Miller *ex* C. Linnaeus
M. bicolor E. von Steudel **[cultivated]**
M. iridioides J. Gaertner **[cultivated]**

NEMASTYLIS T. Nuttall
N. geminiflora T. Nuttall
Sy = N. acuta W. Herbert
N. nuttallii C. Pickering *ex* R. Foster
N. tenuis (W. Herbert) G. Bentham subsp. ***pringlei*** (S. Watson) P. Goldblatt
Sy = N. pringlei S. Watson
Sy = N. tenuis var. pringlei (S. Watson) R. Foster

NEOMARICA T. A. Sprague
N. gracilis (W. Herbert) T. A. Sprague **[cultivated]**
N. longifolia **[cultivated]**

SISYRINCHIUM C. Linnaeus
S. albidum C. Rafinesque-Schmaltz
S. angustifolium P. Miller
Sy = S. bermudiana C. Linnaeus
S. atlanticum E. Bicknell
S. biforme E. Bicknell
Sy = S. dimorphum R. Oliver
S. campestre E. Bicknell
Sy = S. campestre var. kansanum E. Bicknell
S. cernuum (E. Bicknell) T. Kearney
Sy = S. powellii B. Warnock

S. chilense W. Hooker
Sy = S. ensigerum E. Bicknell
Sy = S. scabrum A. von Chamisso & D. von Schlechtendal
S. demissum E. Greene
Sy = S. demissum var. amethystinum (E. Bicknell) T. Kearney & R. Peebles
S. iridifolium K. Kunth **[cultivated]**
Sy = S. laxum C. Otto *ex* J. Sims
S. langloisii E. Greene
Sy = S. pruinosum E. Bicknell
S. minus G. Engelmann & A. Gray
S. montanum E. Greene var. ***montanum***
Sy = S. heterocarpum E. Bicknell
Sy = S. strictum E. Bicknell
S. rosulatum E. Bicknell
Sy = S. exile E. Bicknell
S. sagittiferum E. Bicknell

SPARAXIS J. Ker-Gawler
S. grandiflora (D. Delaroche) J. Ker-Gawler **[cultivated]**

WATSONIA P. Miller, *nomen conservandum*
W. borbonica (P. Pourret) P. Goldblatt **[cultivated]**
W. marginata (C. von Linné) J. Ker-Gawler **[cultivated]**
W. meriana (C. Linnaeus) P. Miller **[cultivated]**

JUNCACEAE, *nomen conservandum*

JUNCUS C. Linnaeus
J. acuminatus A. Michaux
J. balticus C. von Willdenow var. ***littoralis*** G. Engelmann
Sy = J. arcticus C. von Willdenow subsp. littoralis (G. Engelmann) O. Hultén
J. brachycarpus G. Engelmann
J. brachyphyllus K. Wiegand
J. bufonius C. Linnaeus var. ***bufonius***
Sy = J. bufonius var. congestus G. Wahlenberg
Sy = J. bufonius var. occidentalis F. J. Hermann
J. capitatus C. von Weigel
J. coriaceus K. Mackenzie
Sy = J. setaceus F. Rostkovius
J. debilis A. Gray
J. dichotomus S. Elliott
Sy = J. dichotomus var. platyphyllus K. Wiegand
Sy = J. platyphyllus (K. Wiegand) M. Fernald
Sy = J. tenuis C. von Willdenow var. dichotomus (S. Elliott) A. Wood
Sy = J. tenuis var. platyphyllus (K. Wiegand) F. J. Hermann
J. diffusissimus S. Buckley
J. dudleyi K. Wiegand
Sy = J. tenuis C. von Willdenow var. dudleyi (K. Wiegand) F. J. Hermann
J. effusus C. Linnaeus var. ***solutus*** M. Fernald & K. Wiegand
Sy = J. effusus subsp. solutus (M. Fernald & K. Wiegand) R.-L. Hämet-Ahti
J. elliottii A. Chapman var. ***elliottii***
J. filipendulus S. Buckley
J. interior K. Wiegand var. ***arizonicus*** (K. Wiegand) F. J. Hermann
Sy = J. arizonicus G. Vasey
J. interior K. Wiegand var. ***interior***
J. interior K. Wiegand var. ***neomexicanus*** (K. Wiegand) F. J. Hermann
Sy = J. neomexicanus K. Wiegand
J. marginatus F. Rostkovius var. ***marginatus***
Sy = J. aristulatus A. Michaux
J. marginatus F. Rostkovius var. ***setosus*** F. Coville
Sy = J. biflorus S. Elliott
Sy = J. marginatus var. biflorus (S. Elliott) G. Engelmann
Sy = J. setosus (F. Coville) J. K. Small
J. megacephalus M. Curtis
J. mexicanus C. von Willdenow *ex* J. A. Schultes & J. H. Schultes
Sy = J. balticus C. von Willdenow var. mexicanus (C. von Willdenow *ex* J. A. Schultes & J. H. Schultes) K. E. O. Kuntze
Sy = J. compressus K. Kunth
J. nodatus F. Coville
Sy = J. robustus (G. Engelmann) F. Coville
J. nodosus C. Linnaeus var. ***meridianus*** F. J. Hermann
J. nodosus C. Linnaeus var. ***nodosus***
J. patens E. Meyer **[cultivated]**
J. polycephalus A. Michaux
J. repens A. Michaux
J. roemerianus G. Scheele
J. saximontanus A. Nelson
Sy = J. brunnescens P. Rydberg
Sy = J. ensifolius J. Wikstrom var. brunnescens (P. Rydberg) A. Cronquist
Sy = J. parous P. Rydberg
Sy = J. saximontanus var. robustior M. Peck
J. scirpoides J. de Lamarck
J. tenuis C. von Willdenow
Sy = J. tenuis var. anthelatus K. Wiegand
Sy = J. tenuis var. multicornis E. Meyer

Sy = J. tenuis var. williamsii M. Fernald
J. texanus (G. Engelmann) F. Coville
J. torreyi F. Coville
Sy = J. megacephalus A. Wood, non W. Curtis
Sy = J. nodosus C. Linnaeus var. megacephalus J. Torrey
J. trigonocarpus E. von Steudel
J. validus F. Coville var. ***fascinatus*** M. C. Johnston
J. validus F. Coville var. ***validus***
Sy = J. crassifolius F. Buchenau
Sy = J. platycephalus A. Michaux

LUZULA A. P. de Candolle, *nomen conservandum*
L. bulbosa (A. Wood) B. Smyth
Sy = L. campestris (C. Linnaeus) A. P. de Candolle var. bulbosa A. Wood
L. echinata (J. K. Small) F. J. Hermann var. ***mesochorea*** F. J. Hermann

LEMNACEAE, *nomen conservandum*

LEMNA C. Linnaeus
L. aequinoctialis F. Welwitch
Sy = L. angolensis F. Welwitch *ex* C. Hegelmaier
Sy = L. paucicostata C. Hegelmaier
Sy = L. trinervis (C. Austin) J. K. Small
L. gibba C. Linnaeus
Sy = Lenticula gibba (C. Linnaeus) C. Moench
Sy = Telmatophace gibba M. Schleiden
L. minuta K. Kunth
Sy = L. minuscula W. Herter
Sy = L. valdiviana R. Philippi var. abbreviata C. Hegelmaier
Sy = L. valdiviana var. minima C. Hegelmaier
L. obscura (C. Austin) E. Daubs
Sy = L. minor C. Linnaeus var. colorata C. Hegelmaier
Sy = L. minor var. obscura C. Austin
Ma = L. minor *auct.* non C. Linnaeus: Texas authors, in part
L. perpusilla J. Torrey
Sy = L. perpusilla var. trinervis C. Austin
L. turionifera E. Landolt
Ma = L. minor *auct.* non C. Linnaeus: Texas authors, in part
L. valdiviana R. Philippi
Sy = L. cyclostasa (S. Elliott) C. Thompson
Sy = L. torreyi C. Austin
Sy = L. valdiviana var. pellucida C. Hegelmaier
Sy = L. valdiviana var. platyclades C. Hegelmaier
Ma = L. trisulca *auct.*, non C. Linnaeus: Texas authors

SPIRODELA M. Schleiden
S. polyrhiza (C. Linnaeus) M. Schleiden
Sy = Lemna polyrhiza C. Linnaeus
Sy = Lenticula polyrhiza J. de Lamarck
Sy = Spirodela polyrhiza var. masonii E. Daubs
Sy = Telmatophace polyrhiza D. Godron
S. punctata (G. Meyer) C. Thompson
Sy = Lemna punctata G. Meyer
Sy = Spirodela oligorrhiza (F. Kurtz) C. Hegelmaier

WOLFFIA J. Horkel *ex* M. Schleiden, *nomen conservandum*
W. brasiliensis H. Weddell
Sy = Bruniera punctata (A. Grisebach) J. Nieuwland
Sy = Wolffia papulifera C. Thompson
Sy = W. punctata A. Grisebach
W. columbiana G. Karsten
Sy = Bruniera columbiana (G. Karsten) J. Nieuwland

WOLFFIELLA C. Hegelmaier
W. gladiata (J. D. Smith *ex* C. Hegelmaier) C. Hegelmaier
Sy = Wolffia floridana (J. D. Smith *ex* C. Hegelmaier) C. Thompson
Sy = W. gladiata J. D. Smith *ex* C. Hegelmaier
W. lingulata (C. Hegelmaier) C. Hegelmaier
Sy = Wolffia lingulata C. Hegelmaier
W. oblonga (R. Philippi) C. Hegelmaier
Sy = Lemna oblonga R. Philippi

LILIACEAE, *nomen conservandum*

AGAPANTHUS L'Héritier de Brutelle
A. praecox C. von Willdenow subsp. ***orientalis*** (F. M. Leighton) F. M. Leighton [**cultivated**]
Sy = A. orientalis F. M. Leighton

ALETRIS C. Linnaeus
A. aurea T. Walter
A. farinosa C. Linnaeus

ALLIUM C. Linnaeus
A. ampeloprasum C. Linnaeus
A. canadense C. Linnaeus var. ***canadense***
Sy = A. acetabulum (C. Rafinesque-Schmaltz) L. Shinners
Sy = A. canadense var. ovoideum O. Farwell

Sy = A. canadense var. robustum O. Farwell
Sy = A. continuum J. K. Small
A. canadense C. Linnaeus var. ***ecristatum*** (M. E. Jones) M. Ownbey
Sy = A. canadense subsp. ecristatum (M. E. Jones) H. Traub & M. Ownbey
Sy = A. reticulatum G. Don var. ecristatum M. E. Jones
A. canadense C. Linnaeus var. ***fraseri*** M. Ownbey
Sy = A. canadense subsp. fraseri (M. Ownbey) H. Traub & M. Ownbey
Sy = A. fraseri (M. Ownbey) L. Shinners
Sy = A. lavandulare J. Bates var. fraseri (M. Ownbey) L. Shinners
A. canadense C. Linnaeus var. ***hyacinthoides*** (B. Bush) M. Ownbey & H. Aase
Sy = A. canadense subsp. hyacinthoides (B. Bush) H. Traub & M. Ownbey
Sy = A. hyacinthoides B. Bush
A. canadense C. Linnaeus var. ***mobilense*** (E. von Regel) M. Ownbey
Sy = A. arenicola J. K. Small
Sy = A. canadense subsp. mobilense (E. von Regel) M. Ownbey
Sy = A. microscordion J. K. Small
Sy = A. mobilense E. von Regel
Sy = A. mutabile A. Michaux
Sy = A. zenobine V. Cory
A. cepa C. Linnaeus var. ***cepa*** **[cultivated]**
A. cernuum A. Roth var. ***neomexicanum*** (P. Rydberg) J. Macbride
Sy = A. neomexicanum P. Rydberg
A. coryi M. E. Jones
A. drummondii E. von Regel
Sy = A. helleri J. K. Small
Sy = A. nuttallii S. Watson
* ***A. elmendorfii*** M. E. Jones *ex* M. Ownbey **[TOES: V]**
A. geyeri S. Watson
Sy = A. dictyotum E. Greene
Sy = A. funiculosum A. Nelson
Sy = A. geyeri subsp. tenerum (M. E. Jones) H. Traub & M. Ownbey
Sy = A. geyeri var. graniferum L. Henderson
Sy = A. geyeri var. tenerum M. E. Jones
Sy = A. rubrum G. Osterhout
Sy = A. rydbergii F. Macbride
Sy = A. sabulicola G. Osterhout
A. glandulosum J. Link & C. Otto
Sy = A. rhizomatum E. Wooton & P. Standley
A. gomphrenoides P. Boissier & T. von Heldreich **[cultivated]**
Sy = A. ascalonicum Bory de Saint-Vincent & L. Chaubard
A. kunthii G. Don
Sy = A. scaposum G. Bentham
Sy = Schoenoprasum lineare K. Kunth
A. macropetalum P. Rydberg
Sy = A. deserticola (M. E. Jones) E. Wooton & P. Standley
A. neapolitanum D. Cirillo **[cultivated]**
Sy = A. inodorum W. Aiton
Sy = Nothoscordum inodorum (W. Aiton) G. E. Nichols
A. perdulce S. Fraser var. ***perdulce***
A. perdulce S. Fraser var. ***sperryi*** M. Ownbey
A. porrum C. Linnaeus **[cultivated]**
A. runyonii M. Ownbey
A. sativum C. Linnaeus **[cultivated]**
A. schoenoprasum C. Linnaeus **[cultivated]**
A. stellatum T. Nuttall *ex* J. Ker-Gawler

ALOË C. Linnaeus
A. barbadensis P. Miller **[cultivated]**

ALSTROEMERIA C. Linnaeus
A. psittacina J. Lehmann **[cultivated]**
A. pulchella C. von Linné **[cultivated]**

× AMACRINUM A. Garden [***Amaryllis*** × ***Crinum***]
× ***A. nemoria-corsii*** (A. Ragionieri) H. Moore [***Amaryllis bella-donna*** × ***Crinum moorei***] **[cultivated]**
Sy = × A. cv. "howardii"

AMARYLLIS C. Linnaeus, *nomen conservandum*
A. bella-donna C. Linnaeus **[cultivated]**

ANDROSTEPHIUM J. Torrey
A. coeruleum (G. Scheele) E. Greene
Sy = Milla coerulea G. Scheele

ASPARAGUS C. Linnaeus
A. cochinchinensis (J. de Loureiro) E. Merrill **[cultivated]**
Sy = A. falcatus C. Thunberg
Sy = A. lucidus J. Lindley
Sy = Melanthium cochinchinense J. de Loureiro
A. densiflorus (K. Kunth) J. Jessop **[cultivated]**
Sy = A. sprengeri E. von Regel
A. officinalis C. Linnaeus **[cultivated]**
A. retrofractus C. Linnaeus **[cultivated]**
A. setaceus (K. Kunth) J. Jessop **[cultivated]**
Sy = A. plumosus J. G. Baker

ASPHODELUS C. Linnaeus
A. *fistulosus* C. Linnaeus [**federal noxious weed**]

ASPIDISTRA J. Ker-Gawler
A. *elatior* K. von Blume [**cultivated**]
Sy = Plectogyne variegata J. Link

CALOCHORTUS F. Pursh
C. *kennedyi* T. Porter

CAMASSIA J. Lindley, *nomen conservandum*
C. *angusta* (G. Engelmann & A. Gray) J. Blankinship
C. *scilloides* (C. Rafinesque-Schmaltz) V. Cory
Sy = C. esculenta (C. Rafinesque-Schmaltz) V. Cory
Sy = Quamasia hyacinthina (C. Rafinesque-Schmaltz) N. Britton

CHIONODOXA P. Boissier
C. *luciliae* P. Boissier [**cultivated**]

CHLIDANTHUS W. Herbert
C. *fragrans* W. Herbert [**cultivated**]

CLIVIA J. Lindley
C. *miniata* E. von Regel [**cultivated**]

COOPERIA W. Herbert
C. *drummondii* W. Herbert
Sy = Zephyranthes brazosensis (W. Herbert) H. Traub
C. *jonesii* V. Cory
Sy = Zephyranthes jonesii (V. Cory) H. Traub
C. *pedunculata* W. Herbert
Sy = Zephyranthes drummondii D. Don
C. *smallii* E. Alexander
Sy = Zephyranthes smallii (E. Alexander) H. Traub
C. *traubii* W. Hayward

CRINUM C. Linnaeus
C. *americanum* C. Linnaeus subsp. ***americanum***
C. *americanum* C. Linnaeus subsp. ***traubii*** (H. Moldenke) L. Hannibal
Sy = C. strictum W. Herbert var. traubii H. Moldenke
C. *asiaticum* C. Linnaeus [**cultivated**]
C. *bulbispermum* (N. Burman) E. Milne-Redhead & H. G. Schweickerdt [**cultivated**]
Sy = C. longifolium (C. Linnaeus) C. Thunberg
C. *zeylanicum* (C. Linnaeus) C. Linnaeus [**cultivated**]
Sy = Amaryllis zeylanicum C. Linnaeus
Sy = Crinum latifolium C. Linnaeus var. zeylanicum (C. Linnaeus) J. Hooker

ECHEANDIA C. Ortega
* **E. *chandleri*** (J. Greenman & C. Thompson) M. C. Johnston [**TOES: V**]
Sy = Anthericum chandleri J. Greenman & C. Thompson
E. *flavescens* (J. A. Schultes & J. H. Schultes) R. Cruden
Sy = Anthericum flavescens J. A. Schultes & J. H. Schultes
Ma = A. torreyi *auct.*, non J. G. Baker: Texas authors, in part
E. *reflexa* (A. Cavanilles) J. Rose
Sy = Anthericum reflexum A. Cavanilles
Ma = A. torreyi *auct.*, non J. G. Baker: Texas authors, in part

ERYTHRONIUM C. Linnaeus
E. *albidum* T. Nuttall
E. *americanum* J. Ker-Gawler subsp. ***americanum***
E. *mesochoreum* E. Knerr
Sy = E. albidum T. Nuttall var. coloratum E. Sterns
Sy = E. albidum var. mesochoreum (E. Knerr) H. Rickett
E. *rostratum* W. Wolf
E. *umblicatum* J. Parks & J. Hardin

EUCHARIS J. Planchon, *nomen conservandum*
E. *grandiflora* J. Planchon *ex* J. J. Linden [**cultivated**]

GALANTHUS C. Linnaeus
G. *nivalis* C. Linnaeus [**cultivated**]

GALTONIA J. Decaisne
G. *candicans* J. Decaisne [**cultivated**]

GLORIOSA C. Linnaeus
G. *rothschildiana* J. O'Brien [**cultivated**]
G. *superba* C. Linnaeus [**cultivated**]

HABRANTHUS W. Herbert
H. *tubispathus* (L'Héritier de Brutelle) H. Traub
Sy = H. texanus (W. Herbert) E. von Steudel
Sy = Zephyranthes texana W. Herbert

HAEMANTHUS J. de Tournefort *ex* C. Linnaeus
H. coccineus C. Linnaeus [**cultivated**]
H. katharinae J. G. Baker [**cultivated**]

HEMEROCALLIS C. Linnaeus
H. fulva (C. Linnaeus) C. Linnaeus [**cultivated**]
Sy = H. fulva var. kwanso E. von Regel
Ma = H. lilioalphodelus *auct.*, non C. Linnaeus: Hatch et al. 1990

HIPPEASTRUM W. Herbert, *nomen conservandum*
H. bifidum J. G. Baker [**cultivated**]
H. puniceum (J. de Lamarck) K. E. O. Kuntze [**cultivated**]
Sy = Amaryllis punicea J. de Lamarck

HOSTA L. Trattinnick, *nomen conservandum*
H. undulata L. Bailey [**cultivated**]

HYACINTHUS C. Linnaeus
H. orientalis C. Linnaeus [**cultivated**]

HYMENOCALLIS R. A. Salisbury
H. calathina W. Nicholson [**cultivated**]
H. caroliniana (C. Linnaeus) W. Herbert
Sy = H. choctawensis H. Traub
Sy = H. coronaria (J. Le Conte) K. Kunth
Sy = H. georgiana H. Traub
Sy = H. occidentalis K. Kunth
Sy = H. palusvirensis H. Traub
Sy = Pancratium caroliniatum C. Linnaeus
H. eulae L. Shinners
H. latifolia (P. Miller) M. Römer [**cultivated**]
Sy = H. caymanensis W. Herbert
Sy = H. collieri J. K. Small
Sy = H. keyensis J. K. Small
Sy = H. kimballiae J. K. Small
Sy = Pancratium latifolium P. Miller
H. liriosme (C. Rafinesque-Schmaltz) L. Shinners
H. rotata (J. Ker-Gawler) W. Herbert [**cultivated**]
Sy = H. bidentata J. K. Small
Sy = H. traubii H. Moldenke
Sy = Pancratium rotatum J. Ker-Gawler

IPHEION C. Rafinesque-Schmaltz
I. uniflorum (R. Graham) C. Rafinesque-Schmaltz [**cultivated**]
Sy = Brodiaea uniflora (R. Graham) A. Engler
Sy = Triteleia uniflora (R. Graham) J. Lindley

KNIPHOFIA C. Moench, *nomen conservandum*
K. uvaria (C. Linnaeus) L. Oken [**cultivated**]

LEUCOJUM C. Linnaeus
L. aestivum C. Linnaeus [**cultivated**]
L. vernum C. Linnaeus [**cultivated**]

LILIUM C. Linnaeus
L. candidum C. Linnaeus [**cultivated**]
L. formosanum A. Wallace [**cultivated**]
L. henryi J. G. Baker [**cultivated**]
L. lancifolium C. Thunberg [**cultivated**]
Sy = L. tigrinum J. Ker-Gawler
L. longiflorum C. Thunberg [**cultivated**]
L. michauxii J. Poiret
Sy = L. carolinianum A. Michaux, non C. Bosc *ex* J. de Lamarck
Sy = L. fortunofulgidum M. Roane & A. C. Henry
Sy = L. superbum C. Linnaeus var. carolinianum (A. Michaux) A. Chapman
L. philadelphicum C. Linnaeus var. ***andinum*** (T. Nuttall) J. Ker-Gawler
Sy = L. andinum T. Nuttall
Sy = L. montanum A. Nelson
Sy = L. philadelphicum var. montanum (A. Nelson) E. Wherry
Sy = L. umbellatum F. Pursh
L. philippinense J. G. Baker [**cultivated**]
L. regale E. H. Wilson [**cultivated**]
L. speciosum C. Thunberg [**cultivated**]
L. superbum C. Linnaeus [**cultivated**]
L. tenuifolium F. von Fischer [**cultivated**]
L. × hollandicum J. Bergmans [**cultivated**]

LIRIOPE J. de Loureiro
L. muscari (J. Decaisne) L. Bailey [**cultivated**]
L. spicatum J. de Loureiro [**cultivated**]

LYCORIS W. Herbert
L. aurea (L'Héritier de Brutelle) W. Herbert [**cultivated**]
Sy = Amaryllis aurea L'Héritier de Brutelle
Sy = Lycoris africana M. Römer
Sy = Nerine aurea (L'Héritier de Brutelle) P. Bury
L. radiata (L'Héritier de Brutelle) W. Herbert var. ***radiata*** [**cultivated**]
Sy = Amaryllis radiata L'Héritier de Brutelle
Sy = Nerine japonica F. Miquel
L. squamigera C. Maximowicz [**cultivated**]
L. traubii W. Hayward [**cultivated**]

MAIANTHEMUM G. Weber *ex* F. Wiggers, *nomen conservandum*

M. racemosum (C. Linnaeus) J. Link subsp. ***amplexicaule*** (T. Nuttall) J. La Frankie
Sy = M. amplexicaule (T. Nuttall) W. A. Weber
Sy = M. racemosum var. amplexicaule (T. Nuttall) R. Dorn
Sy = Smilacina amplexicaulis T. Nuttall
Sy = S. amplexicaulis var. glabra J. Macbride
Sy = S. amplexicaulis var. jenkinsii J. Boivin
Sy = S. amplexicaulis var. ovata J. Boivin
Sy = S. racemosa (C. Linnaeus) R. Desfontaines var. amplexicaulis (T. Nuttall) S. Watson
Sy = S. racemosa var. brachystyla G. Henderson
Sy = S. racemosa var. glabra (J. Macbride) H. St. John
Sy = S. racemosa var. jenkinsii (J. Boivin) J. Boivin
Sy = Vagnera amplexicaulis (T. Nuttall) E. Greene
Sy = V. amplexicaulis var. glabra (J. Macbride) L. Abrams

MELANTHIUM C. Linnaeus
M. virginicum C. Linnaeus
Sy = M. dispersum J. K. Small
Sy = Veratrum virginicum (C. Linnaeus) W. T. Aiton

MILLA A. Cavanilles
M. biflora A. Cavanilles

MUSCARI P. Miller
M. armeniacum M. Leichtlin *ex* J. G. Baker **[cultivated]**
M. neglectum G. Gussone *ex* M. Tenore **[cultivated]**
Sy = M. atlanticum P. Boissier & G. Reuter
Sy = M. racemosum (C. Linnaeus) J. de Lamarck & A. P. de Candolle

NARCISSUS C. Linnaeus
N. jonquilla C. Linnaeus **[cultivated]**
N. poeticus C. Linnaeus **[cultivated]**
N. pseudonarcissus C. Linnaeus **[cultivated]**
Sy = N. cyclamineus A. P. de Candolle
N. tazetta C. Linnaeus **[cultivated]**
N. × incomparabilis P. Miller [***poeticus* × *pseudonarcissus***] **[cultivated]**
N. × medioluteus P. Miller [***poeticus* × *tazetta***] **[cultivated]**
Sy = N. biflorus W. Curtis
N. × odorus C. Linnaeus [***jonquilla* × *pseudonarcissus***] **[cultivated]**

NERINE W. Herbert, *nomen conservandum*
N. samiensis W. Herbert **[cultivated]**

NOTHOSCORDUM K. Kunth, *nomen conservandum*
N. bivalve (C. Linnaeus) N. Britton
Sy = Allium bivalve C. Linnaeus
Sy = A. striatum N. von Jacquin
Sy = Ornithogalum bivalve C. Linnaeus
N. texanum M. E. Jones

OPHIOPOGON J. Ker-Gawler
O. jaburan (P. von Siebold) C. Loddiges **[cultivated]**
Sy = Flueggea jaburan K. Kunth
Sy = Slateria jaburan P. von Siebold
O. japonicus C. von Linné **[cultivated]**
Sy = Convallaria japonica C. von Linné
Sy = Ophiopogon japonicus var. minor C. Thunberg
O. planiscapus T. Nakai **[cultivated]**
Sy = O. japonicus C. von Linné var. wallichianus C. Maximowicz

ORNITHOGALUM C. Linnaeus
O. umbellatum C. Linnaeus **[cultivated]**

POLYGONATUM P. Miller
* ***P. biflorum*** (T. Walter) S. Elliott **[TOES: V]**
Sy = P. biflorum var. commutatum (J. A. Schultes & J. H. Schultes) T. Morong
Sy = P. commutatum (J. A. Schultes & J. H. Schultes) A. Dietrich
Ma = P. canaliculatum *auct.*, non (G. H. Muhlenberg *ex* C. von Willdenow) F. Pursh
P. cobrense (E. Wooton & P. Standley) R. Gates
Sy = Salamonia cobrensis E. Wooton & P. Standley

SCHOENOCAULON A. Gray
S. drummondii A. Gray
S. texanum G. Scheele

SCHOENOLIRION J. Torrey *ex* E. Durand, *nomen conservandum*
S. croceum (A. Michaux) A. Wood
S. wrightii H. Sherman
Sy = Oxytria texana (G. Scheele) C. Pollard
Sy = Schoenolirion texanum (G. Scheele) A. Gray

SCILLA C. Linnaeus
S. hispanica P. Miller **[cultivated]**
Sy = S. campanulata W. Aiton
S. siberica A. Haworth *ex* H. Andrews **[cultivated]**

SPREKELIA L. Heister
S. formosissima (C. Linnaeus) W. Herbert **[cultivated]**

STENANTHIUM (A. Gray) K. Kunth, *nomen conservandum*
  S. gramineum (J. Ker-Gawler) T. Morong var. ***robustum*** (S. Watson) M. Fernald
    Sy = S. robustum S. Watson

STERNBERGIA F. von Waldstein & P. Kitaibel
  S. lutea (C. Linnaeus) J. Ker-Gawler *ex* K. Sprengel **[cultivated]**

TIGRIDIA A. L. Jussieu
  T. pavonia (C. von Linné) A. P. de Candolle **[cultivated]**

TOFIELDIA W. Hudson
  T. racemosa (T. Walter) N. Britton, E. Sterns, & J. Poggenburg
    Sy = Melanthium racemosum T. Walter
    Sy = Triantha racemosa (T. Walter) J. K. Small

TRILLIUM C. Linnaeus
* ***T. gracile*** J. Freeman **[TOES: V]**
    Ma = T. ludoviciana *auct.*, non T. Harbison
* ***T. recurvatum*** L. Beck **[TOES: III]**
  T. texanum S. Buckley
*   Sy = T. pusillum A. Michaux var. texanum (S. Buckley) J. Reveal & C. Bromme **[TOES: V]**
    Sy = T. pusillum var. texanum (S. Buckley) C. F. Reed
  T. viridescens T. Nuttall

TULBAGHIA C. Linnaeus, *nomen conservandum*
  T. fragrans F. Verdoorn **[cultivated]**
  T. violacea W. Harvey **[cultivated]**

TULIPA C. Linnaeus
  T. acuminata M. H. Vahl *ex* J. Hornemann **[cultivated]**
  T. clusiana A. P. de Candolle **[cultivated]**
  T. greigi E. von Regel **[cultivated]**
  T. kaufmanniana E. von Regel **[cultivated]**
  T. praestans A. Garden **[cultivated]**
  T. saxatilis F. Sieber *ex* K. Sprengel **[cultivated]**
  T. sylvestris C. Linnaeus **[cultivated]**
  T. × gesneriana C. Linnaeus **[cultivated]**

UVULARIA C. Linnaeus
* ***U. perfoliata*** C. Linnaeus, *nomen conservandum* **[TOES: V]**

ZEPHYRANTHES W. Herbert, *nomen conservandum*
  Z. candida (J. Lindley) W. Herbert **[cultivated]**
    Sy = Amaryllis candida J. Lindley
  Z. citrina J. G. Baker **[cultivated]**
    Sy = Z. aurea (H. Ruiz López & J. Pavón) J. G. Baker
    Sy = Z. eggersiana I. Urban
  Z. grandiflora J. Lindley **[cultivated]**
    Sy = Amaryllis carinata K. Sprengel
    Sy = Atamosco carinata (W. Herbert) Percy Wilson
    Sy = Zephyranthes carinata W. Herbert
  Z. longifolia W. Hemsley
    Sy = Atamosco longifolia (W. Hemsley) T. Cockerell
  Z. pulchella J. G. Smith
  Z. refugiensis F. B. Jones

ZIGADENUS A. Michaux
  Z. densus (L. Desrousseaux) M. Fernald
    Sy = Tracyanthus angustifolius (A. Michaux) J. K. Small
  Z. elegans F. Pursh subsp. ***elegans***
    Sy = Anticlea elegans (F. Pursh) P. Rydberg
    Sy = Zigadenus alpinus J. Blankinship
  Z. glaberrimus A. Michaux
  Z. leimanthoides A. Gray
    Sy = Oceanorus leimanthoides (A. Gray) J. K. Small
  Z. nuttallii (A. Gray) S. Watson
    Sy = Amianthium nuttallii A. Gray
    Sy = Toxicoscordion nuttallii (A. Gray) P. Rydberg

LIMNOCHARITACEAE

HYDROCLEYS L. C. Richard
  H. nymphoides (F. von Humboldt & A. Bonpland *ex* C. von Willdenow) F. Buchenau

MARANTACEAE, *nomen conservandum*

THALIA C. Linnaeus
  T. dealbata J. Fraser *ex* W. Roscoe
    Sy = T. barbata J. K. Small

MAYACACEAE, *nomen conservandum*

MAYACA J. Aublet
  M. fluviatilis J. Aublet
    Sy = M. aubletii A. Michaux

MUSCACEAE, *nomen conservandum*

MUSA C. Linnaeus
M. ensete J. F. Gmelin [**cultivated**]
M. sumatrana O. Beccari [**cultivated**]
M. × paradisiaca C. Linnaeus [**acuminata × balbisiana**] [**cultivated**]
Sy = M. ornata W. Roxburgh var. normalis K. E. O. Kuntze
Sy = M. sapientum C. Linnaeus

RAVENALA M. Adanson
R. madagascariensis P. Sonnerat [**cultivated**]

NAJADACEAE, *nomen conservandum*

NAJAS C. Linnaeus
N. guadalupensis (K. Sprengel) P. Magnus
Sy = Caulinia guadalupensis K. Sprengel
Sy = Najas flexilis (C. von Willdenow) F. Rostkovius & F. Schmidt var. guadalupensis A. Braun
N. marina C. Linnaeus
Sy = N. major C. Allioni
Sy = N. major var. gracilis W. R. Dudley
Sy = N. marina var. californica A. Rendle
Sy = N. marina var. mexicana A. Rendle
Sy = N. marina var. recurvata W. R. Dudley

ORCHIDACEAE, *nomen conservandum*

BLETILLA H. G. Reichenbach, *nomen conservandum*
B. striata (C. Thunberg) H. G. Reichenbach [**cultivated**]

CALOPOGON R. Brown *ex* W. T. Aiton, *nomen conservandum*
C. oklahomensis D. H. Goldman
Ma = C. barbatus *auct.*, non (O. Swartz) O. Ames: Texas authors [**Toes: IV**]
C. tuberosus (C. Linnaeus) N. Britton, E. Sterns, & J. Poggenburg var. **tuberosus**
Sy = C. pulchellus (R. A. Salisbury) R. Brown *ex* W. T. Aiton
Sy = C. pulchellus var. latifolius (H. St. John) J. Boivin
Sy = C. tuberosus var. latifolius (H. St. John) J. Boivin
Sy = Limodorum tuberosum C. Linnaeus

CLEISTES L. C. Richard *ex* J. Lindley
C. bifaria (M. Fernald) P. Catling & K. Gregg
Sy = Cleistes divaricata (C. Linnaeus) O. Ames var. bifaria M. Fernald
Ma = Cleistes divaricata *auct.*, non (C. Linnaeus) O. Ames: Texas authors

CORALLORHIZA A. Gagnebin de la Ferrière
C. maculata (C. Rafinesque-Schmaltz) C. Rafinesque-Schmaltz
Sy = Cladorhiza maculata C. Rafinesque-Schmaltz
Sy = Corallorhiza hortensis W. Suksdorf
Sy = C. leimbachiana W. Suksdorf
Sy = C. maculata var. flavida (M. Peck) T. Cockerell
Sy = C. maculata var. immaculata M. Peck
Sy = C. maculata var. intermedia O. Farwell
Sy = C. maculata var. occidentalis (J. Lindley) T. Cockerell
Sy = C. maculata var. punicea H. Bartlett
Sy = C. multiflora T. Nuttall
C. odontorhiza (C. von Willdenow) T. Nuttall
Sy = C. micrantha A. Chapman
C. striata J. Lindley var. **striata**
Sy = C. ochroleuca P. Rydberg
Sy = C. striata var. flavida T. K. Todsen & T. A. Todsen
Sy = C. striata var. ochroleuca (P. Rydberg) L. Magrath
C. wisteriana S. Conrad
Sy = C. odontorhiza A. Chapman
Sy = C. unguiculata C. Rafinesque-Schmaltz

CYPRIPEDIUM C. Linnaeus
* **C. kentuckiense** C. F. Reed [**TOES: IV**]
C. pubescens C. von Willdenow
Sy = C. calceolus C. Linnaeus var. pubescens (C. von Willdenow) D. Correll
Sy = C. parviflorum R. A. Salisbury var. pubescens (C. von Willdenow) J. Knight

DEIREGYNE F. Schlechter
D. confusa L. Garay
Sy = Spiranthes confusa (L. Garay) J. Kartesz & K. Gandhi
Ma = S. durangensis O. Ames & C. Schweinitz

DICHROMANTHUS L. Garay
D. cinnabarinus (P. de la Llave & J. de Lexarza) L. Garay

Sy = Neottia cinnabarina P. de la Llave & J. de Lexarza
Sy = Spiranthes cinnabarina (P. de la Llave & J. de Lexarza) W. Hemsley
Sy = Stenorrhynchus cinnabarinus (P. de la Llave & J. de Lexarza) J. Lindley

EPIPACTIS J. Zinn, *nomen conservandum*
E. gigantea D. Douglas *ex* W. Hooker
Sy = Amesia gigantea A. Nelson & J. Macbride
Sy = Helleborine gigantea G. Druce
Sy = Limodorum giganteum K. E. O. Kuntze
Sy = Peramium giganteum J. Coulter
Sy = Serapias gigantea A. Eaton

HABENARIA C. von Willdenow
H. quinqueseta (A. Michaux) A. Eaton var. ***quinqueseta***
H. repens T. Nuttall
Sy = Platanthera repens (T. Nuttall) A. Wood

HEXALECTRIS C. Rafinesque-Schmaltz
H. grandiflora (A. Richard & H. Galeotti) L. O. Williams
Sy = Corallorhiza grandiflora A. Richard & H. Galeotti
Sy = Hexalectris mexicana J. Greenman
Sy = Neottia grandiflora (A. Richard & H. Galeotti) K. E. O. Kuntze
H. nitida L. O. Williams
* ***H. revoluta*** D. Correll **[TOES: V]**
H. spicata (T. Walter) J. Barnhart var. ***arizonica*** (S. Watson) P. Catling & V. Engel
Sy = Corallorhiza arizonica S. Watson
H. spicata (T. Walter) J. Barnhart var. ***spicata***
Sy = Arethusa spicata T. Walter
Sy = Bletia aphylla T. Nuttall
Sy = Corallorhiza arizonica S. Watson
Sy = C. spicata (T. Walter) I. Tidestrom
Sy = Hexalectris aphylla (T. Nuttall) C. Rafinesque-Schmaltz
H. warnockii O. Ames & D. Correll

ISOTRIA C. Rafinesque-Schmaltz
* ***I. verticillata*** (G. H. Muhlenberg *ex* C. von Willdenow) C. Rafinesque-Schmaltz **[TOES: IV]**
Sy = Arethusa verticillata G. H. Muhlenberg *ex* C. von Willdenow
Sy = Odonectis verticillata (G. H. Muhlenberg *ex* C. von Willdenow) C. Rafinesque-Schmaltz
Sy = Pogonia verticillata (G. H. Muhlenberg *ex* C. von Willdenow) T. Nuttall

LISTERA R. Brown *ex* W. T. Aiton, *nomen conservandum*
L. australis J. Lindley
Sy = Bifolium australe (J. Lindley) J. Nieuwland
Sy = Diphryllum australe (J. Lindley) K. E. O. Kuntze
Sy = Ophrys australis (J. Lindley) H. House

MALAXIS D. Solander *ex* O. Swartz
M. macrostachya (J. de Lexarza) K. E. O. Kuntze
Sy = M. ophioglossoides G. H. Muhlenberg *ex* C. von Willdenow
Sy = M. soulei L. O. Williams
Sy = Microstylis macrostachya J. de Lexarza
Sy = Ophrys macrostachya J. de Lexarza
M. unifolia A. Michaux
Sy = Achroanthes unifolia (A. Michaux) C. Rafinesque-Schmaltz
Sy = Microstylis unifolia (A. Michaux) N. Britton, E. Sterns, & J. Poggenburg
M. wendtii G. Salazar
Ma = M. ehrenbergii *auct.*, non (H. G. Reichenbach) K. E. O. Kuntze: Texas authors

PHAIUS J. de Loureiro
P. tankervilliae (J. Banks) K. von Blume **[cultivated]**
Sy = Bletia tankervilliae (J. Banks) R. Brown
Sy = Limodorum tankervilliae (W. T. Aiton) J. Banks

PLATANTHERA L. C. Richard, *nomen conservandum*
P. blephariglottis (C. von Willdenow) J. Lindley var. ***conspicua*** (G. Nash) C. Luer
Sy = Blephariglottis conspicua (G. Nash) J. K. Small
Sy = Habenaria blephariglottis (C. von Willdenow) W. Hooker var. conspicua (G. Nash) O. Ames
Sy = H. conspicua G. Nash
P. chapmanii (J. K. Small) C. Luer
Sy = Blephariglottis chapmanii J. K. Small
Sy = Habenaria chapmanii (J. K. Small) O. Ames
P. ciliaris (C. Linnaeus) J. Lindley
Sy = Blephariglottis ciliaris (C. Linnaeus) P. Rydberg
Sy = B. flaviflora C. Rafinesque-Schmaltz

Sy = Habenaria ciliaris (C. Linnaeus) R. Brown
Sy = Orchis ciliaris C. Linnaeus
P. clavellata (A. Michaux) C. Luer
Sy = Denslovia clavellata (A. Michaux) P. Rydberg
Sy = Gymnadeniopsis clavellata (A. Michaux) P. Rydberg
Sy = Habenaria clavellata (A. Michaux) K. Sprengel
Sy = H. clavellata var. ophioglossoides M. Fernald
Sy = Orchis clavellata A. Michaux
Sy = Peristylus clavellatus (A. Michaux) F. Kränzlin
P. cristata (A. Michaux) J. Lindley
Sy = Blephariglottis cristata (A. Michaux) P. Rydberg
Sy = Habenaria cristata (A. Michaux) R. Brown *ex* W. T. Aiton
Sy = Orchis cristata A. Michaux
P. flava (C. Linnaeus) J. Lindley var. ***flava***
Sy = Habenaria flava (C. Linnaeus) R. Brown
Sy = H. scutellata (T. Nuttall) F. Morris
Sy = Orchis flava C. Linnaeus
Sy = Perularia bidentata (S. Elliott) J. K. Small
Sy = P. flava (C. Linnaeus) O. Farwell
Sy = P. scutellata (T. Nuttall) J. K. Small
P. integra (T. Nuttall) A. Gray *ex* L. Beck
Sy = Gymnadeniopsis integra (T. Nuttall) P. Rydberg
Sy = Habenaria integra (T. Nuttall) K. Sprengel
Sy = Orchis integra T. Nuttall
* ***P. lacera*** (A. Michaux) G. Don var. ***lacera*** **[TOES: IV]**
Sy = Blephariglottis lacera (A. Michaux) O. Farwell
Sy = Fimbriella lacera (A. Michaux) O. Farwell
Sy = Habenaria lacera (A. Michaux) C. Loddiges
Sy = Orchis lacera A. Michaux
* ***P. nivea*** (T. Nuttall) C. Luer **[TOES: IV]**
Sy = Gymnadenia nivea (T. Nuttall) J. Lindley
Sy = Gymnadeniopsis nivea (T. Nuttall) P. Rydberg
Sy = Habenaria nivea (T. Nuttall) K. Sprengel
Sy = Orchis nivea T. Nuttall
Sy = Peristylus niveus (T. Nuttall) F. Kränzlin

POGONIA A. L. de Jussieu
P. ophioglossoides (C. Linnaeus) A. L. de Jussieu
Sy = Arethusa ophioglossoides C. Linnaeus
Sy = Pogonia ophioglossoides var. brachypogon M. Fernald

PONTHIEVA R. Brown *ex* W. T. Aiton
P. racemosa (T. Walter) C. Mohr
Sy = Arethusa racemosa T. Walter

SCHIEDEELA F. Schlechter
S. parasitica (A. Richard & H. Galeotti) F. Schlechter
Sy = Spiranthes parasitica A. Richard & H. Galeotti

SPIRANTHES L. C. Richard, *nomen conservandum*
S. brevilabris J. Lindley var. ***brevilabris***
Sy = S. gracilis var. brevilabris (J. Lindley) D. Correll
S. brevilabris J. Lindley var. ***floridana*** (E. Wherry) C. Luer
Sy = Ibidium floridanum E. Wherry
Sy = Spiranthes gracilis var. floridana (J. Bigelow) (E. Wherry) D. Correll
S. cernua (C. Linnaeus) L. C. Richard
Sy = Ibidium cernuum (C. Linnaeus) H. House
Sy = Ophrys cernua C. Linnaeus
Sy = Spiranthes cernua var. incurva O. Jennings
Sy = Triorchis cernua (C. Linnaeus) J. Nieuwland
S. lacera (C. Rafinesque-Schmaltz) C. Rafinesque-Schmaltz var. ***gracilis*** (J. Bigelow) C. Luer
Sy = Gyrostachys gracilis (J. Bigelow) K. E. O. Kuntze
Sy = Ibidium gracile (J. Bigelow) H. House
Sy = Neottia gracilis J. Bigelow
Sy = Spiranthes gracilis (J. Bigelow) L. Beck
Sy = Triochris gracilis (J. Bigelow) J. Nieuwland
S. laciniata (J. K. Small) O. Ames
Sy = Gyrostachys laciniata J. K. Small
Sy = Ibidium laciniatum (J. K. Small) H. House
Sy = Triochris laciniata (J. K. Small) H. House
S. longilabris J. Lindley
Sy = Ibidium longilabre (J. Lindley) H. House
S. magnicamporum C. Sheviak
S. odorata (T. Nuttall) J. Lindley
Sy = Gyrostachys odorata (T. Nuttall) K. E. O. Kuntze
Sy = Ibidium odoratum (T. Nuttall) H. House
Sy = Neottia odorata T. Nuttall
Sy = Spiranthes cernua (C. Linnaeus) L. C. Richard var. odorata (T. Nuttall) D. Correll
Sy = Triorchis odorata (T. Nuttall) J. Nieuwland

S. ovalis J. Lindley var. ***ovalis***
Sy = Gyrostachys ovalis (J. Lindley) K. E. O. Kuntze
Sy = Ibidium ovale (J. Lindley) H. House
Sy = Spiranthes parviflora (A. Chapman) O. Ames
Sy = S. smallii F. Schlechter
Sy = Triorchis ovalis (J. Lindley) J. Nieuwland
* ***S. parksii*** D. Correll [**TOES: I**]
S. praecox (T. Walter) S. Watson
Sy = Gyrostachys praecox (T. Walter) K. E. O. Kuntze
Sy = Ibidium praecox (T. Walter) H. House
Sy = Limodorum praecox T. Walter
Sy = Triorchis praecox (T. Walter) J. Nieuwland
S. tuberosa C. Rafinesque-Schmaltz
Sy = S. grayi O. Ames
Sy = S. simplex A. Gray
Sy = S. tuberosa C. Rafinesque-Schmaltz var. grayi (O. Ames) M. Fernald
S. vernalis G. Engelmann & A. Gray
Sy = Ibidium vernalis (G. Engelmann & A. Gray) H. House
Sy = Triorchis vernalis (G. Engelmann & A. Gray) H. House

STENORRHYNCHOS L. C. Richard *ex* K. Sprengel
S. michuacanus (P. de la Llave & J. de Lexarza) F. Schlechter
Sy = Neottia michuacana P. de la Llave & J. de Lexarza
Sy = Spiranthes michuacana (P. de la Llave & J. de Lexarza) W. Hemsley

TIPULARIA T. Nuttall
T. discolor (F. Pursh) T. Nuttall
Sy = Limodorum unifolium G. H. Muhlenberg
Sy = Orchis discolor F. Pursh
Sy = Tipularia unifolia (G. H. Muhlenberg) N. Britton, E. Sterns, & J. Poggenburg

TRIPHORA T. Nuttall
T. trianthophora (O. Swartz) P. Rydberg
Sy = Arethusa trianthophoros O. Swartz
Sy = Pogonia trianthophorus (O. Swartz) N. Britton, E. Sterns, & J. Poggenburg
Sy = Triphora trianthophora var. schaffneri W. Camp

ZEUXINE J. Lindley, *nomen conservandum*
Z. strateumatica (C. Linnaeus) F. Schlechter

POACEAE, *nomen conservandum*

AEGILOPS C. Linnaeus
A. cylindrica N. Host
Sy = Cylindropyrum cylindricum (N. Host) Á. Löve
Sy = Triticum cylindricum (N. Host) V. de Cesati

× **AGROPOGON** P. Fournier [***Agrostis*** × ***Polypogon***]
× ***A. littoralis*** (J. E. Smith) C. Hubbard [***Agrostis stolonifera*** × ***Polypogon monspeliensis***]
Sy = Polypogon interruptus K. Kunth
Sy = P. littoralis J. E. Smith
Ma = P. lutosus *auct.*, non (J. Poiret) A. Hitchcock: Hitchcock (1935)

AGROPYRON J. Gaertner
A. cristatum (C. Linnaeus) J. Gaertner
Sy = A. desertorum (F. von Fischer) J. A. Schultes
Sy = A. pectiniforme J. J. Römer & J. A. Schultes
Sy = Bromus cristatus C. Linnaeus

AGROSTIS C. Linnaeus
A. avenacea J. F. Gmelin
Sy = A. retrofracta C. von Willdenow
A. elliottiana J. A. Schultes
Sy = A. exigua G. Thurber
A. exarata K. von Trinius
A. hyemalis (T. Walter) N. Britton, E. Sterns, & J. Poggenburg
Sy = A. hyemalis var. scabra (C. von Willdenow) S. Blomquist
Sy = A. scabra C. von Willdenow
Sy = Cornucopiae hyemalis T. Walter
A. perennans (T. Walter) E. Tuckerman
Sy = A. perennans var. aestivalis G. Vasey
Sy = A. perennans var. elata (F. Pursh) A. Hitchcock
Sy = Cornucopiae perennans T. Walter
A. stolonifera C. Linnaeus
Sy = A. gigantea A. Roth
Sy = A. palustris W. Hudson
Sy = A. stolonifera var. palustris (W. Hudson) O. Farwell
Ma = A. alba *auct.*, non C. Linnaeus: American authors

AIRA C. Linnaeus
A. caryophyllea C. Linnaeus var. ***capillaris*** A. Mutel

Sy = A. capillaris N. Host., non G. Savi
Sy = A. elegans C. von Willdenow *ex* J. J. Römer & J. A. Schultes
Sy = A. elegantissima C. Schur
A. caryophyllea C. Linnaeus var. ***caryophyllea***

ALLOLEPIS T. Soderstrom & H. Decker
A. texana (G. Vasey) T. Soderstrom & H. Decker
Sy = Distichlis texana (G. Vasey) F. Lamson-Scribner
Sy = Poa texana G. Vasey

ALOPECURUS C. Linnaeus
A. carolinianus T. Walter
Sy = A. macounii G. Vasey
Sy = A. ramosus J. Poiret
A. myosuroides W. Hudson
Sy = A. agretis C. Linnaeus

ANDROPOGON C. Linnaeus
A. gerardii F. Vitman var. ***chrysocomus*** (G. Nash) M. Fernald
A. gerardii F. Vitman var. ***gerardii***
A. gerardii F. Vitman var. ***paucipilus*** (G. Nash) M. Fernald
Sy = A. hallii E. Hackel
A. glomeratus (T. Walter) N. Britton, E. Sterns, & J. Poggenburg
Sy = A. glomeratus var. glaucopsis (S. Elliott) C. Mohr
Sy = A. glomeratus var. hirsutior (E. Hackel) C. Mohr
Sy = A. glomeratus var. pumilus G. Vasey
Sy = A. glomeratus var. scabriglumis C. Campbell
Sy = Cinna glomerata T. Walter
A. gyrans W. Ashe var. ***gyrans***
Sy = A. campyloracheus G. Nash
Sy = A. subtenuis G. Nash
Ma = A. elliottii *auct.*, non A. Chapman: Texas authors
A. gyrans W. Ashe var. ***stenophyllus*** (E. Hackel) C. Campbell
Sy = A. perangustatus G. Nash
Sy = A. virginicus C. Linnaeus var. stenophyllus (E. Hackel) M. Fernald & L. Griscom
A. spadiceus J. Swallen
A. ternarius A. Michaux var. ***ternarius***
Sy = A. elliottii A. Chapman
A. virginicus C. Linnaeus
Sy = A. capillipes G. Nash
Sy = A. virginicus var. decipiens C. Campbell
Sy = A. virginicus var. glaucus E. Hackel
Sy = A. virginicus var. tetrastachyus (S. Elliott) E. Hackel

ANTHENANTIA A. Palisot de Beauvois
Sy = Anthaenantia (orthographic variant)
Sy = Anthenanthia (orthographic variant)
NOTE: According to Clayton and Renvoize (1986), *Anthenantia* is the etymologically correct version of three alternative spellings given by A. Palisot de Beauvois.
A. rufa (S. Elliott) J. A. Schultes
Sy = Aulaxanthus rufus S. Elliott
A. villosa (A. Michaux) A. Palisot de Beauvois
Sy = Aulaxanthus ciliatus S. Elliott
Sy = Phalaris villosa A. Michaux

ANTHOXANTHUM C. Linnaeus
A. aristatum P. Boissier
A. odoratum C. Linnaeus

ARISTIDA C. Linnaeus
A. adscensionis C. Linnaeus
Sy = A. adscensionis var. abortiva A. Beetle
Sy = A. adscensionis var. modesta E. Hackel
Sy = A. fasciculata J. Torrey
A. arizonica G. Vasey
A. basiramea G. Engelmann *ex* G. Vasey
A. brownii B. Warnock
A. desmantha K. von Trinius & F. Ruprecht
A. dichotoma A. Michaux var. ***dichotoma***
A. divaricata C. von Willdenow
A. gypsophila A. Beetle
A. havardii G. Vasey
Sy = A. barbata E. Fournier
A. lanosa G. H. Muhlenberg *ex* S. Elliott
Sy = A. lanosa var. macera M. Fernald & L. Griscom
A. longespica J. Poiret var. ***geniculata*** (C. Rafinesque-Schmaltz) M. Fernald
Sy = A. geniculata C. Rafinesque-Schmaltz
Sy = A. intermedia F. Scribner & C. Ball
Sy = A. necopina L. Shinners
A. longespica J. Poiret var. ***longespica***
Sy = A. gracilis S. Elliott
A. oligantha A. Michaux
A. palustris (A. Chapman) G. Vasey
Sy = A. virgata K. von Trinius var. palustris A. Chapman
Ma = A. affinis *auct.*, non (J. A. Schultes) K. Kunth

A. pansa E. Wooton & P. Standley var. ***dissita*** (I. M. Johnston) A. Beetle
Sy = A. dissita I. M. Johnston
A. pansa E. Wooton & P. Standley var. ***pansa***
A. purpurascens J. Poiret var. ***purpurascens***
Sy = A. affinis (J. A. Schultes) K. Kunth
Sy = A. purpurascens var. minor G. Vasey
A. purpurascens J. Poiret var. ***virgata*** (K. von Trinius) K. Allred
Sy = A. virgata K. von Trinius
A. purpurea T. Nuttall var. ***fendleriana*** (E. von Steudel) G. Vasey
Sy = A. fendleriana E. von Steudel
A. purpurea T. Nuttall var. ***longiseta*** (E. von Steudel) G. Vasey
Sy = A. longiseta E. von Steudel
Sy = A. longiseta var. rariflora A. Hitchcock
Sy = A. longiseta var. robusta E. Merrill
Sy = A. purpurea var. robusta (E. Merrill) C. Piper
A. purpurea T. Nuttall var. ***nealleyi*** (G. Vasey) K. Allred
Sy = A. glauca (C. Nees von Esenbeck) W. Walpers
Sy = A. nealleyi (G. Vasey) G. Vasey
Sy = A. purpurea var. glauca (C. Nees von Esenbeck) A. Holmgren & N. Holmgren
Sy = A. reverchonii G. Vasey
Sy = A. stricta A. Michaux var. nealleyi G. Vasey
A. purpurea T. Nuttall var. ***purpurea***
Sy = A. purpurea T. Nuttall var. laxiflora E. Merrill
Sy = A. roemeriana G. Scheele
A. purpurea T. Nuttall var. ***wrightii*** (G. Nash) K. Allred
Sy = A. wrightii G. Nash
A. ramosissima G. Engelmann *ex* A. Gray
Sy = A. ramosissima var. chaseana J. Henrard
A. schiedeana K. von Trinius & F. Ruprecht
Sy = A. orcuttiana G. Vasey
A. ternipes A. Cavanilles var. ***gentilis*** (J. Henrard) K. Allred
Sy = A. gentilis J. Henrard
Sy = A. gentilis var. breviaristata J. Henrard
Sy = A. hamulosa J. Henrard
Sy = A. imbricata J. Henrard
Sy = A. ternipes var. hamulosa (J. Henrard) J. Trent
A. ternipes A. Cavanilles var. ***ternipes***

ARTHRAXON A. Palisot de Beauvois
A. hispidus (C. Thunberg) T. Makino var. ***hispidus***
Sy = A. ciliaris A. Palisot de Beauvois
Sy = A. hispidus var. cryptatherus (E. Hackel) M. Honda
Sy = Phalaris hispida C. Thunberg

ARUNDINARIA A. Michaux
A. amabilis F. McClure **[cultivated]**
A. argenteostriata (E. von Regel) P. L. de Vilmorin **[cultivated]**
Sy = Bambusa argenteostriata E. von Regel
Sy = Pleioblastus argenteostriatus (E. von Regel) T. Nakai
A. auricoma A. Freeman-Mitford **[cultivated]**
Sy = A. viridi-striata T. Makino
Sy = Bambusa viridi-striata P. von Siebold *ex* É. André, non E. von Regel
Sy = Pleioblastus viridi-striatus T. Makino
A. chino (A. Franchet & P. Savatier) T. Makino **[cultivated]**
Sy = Bambusa chino A. Franchet & P. Savatier
Sy = Pleioblastus chino (A. Franchet & P. Savatier) T. Makino
A. funghomii F. McClure **[cultivated]**
A. gigantea (T. Walter) G. H. Muhlenberg subsp. ***gigantea***
Sy = Arundo gigantea T. Walter
A. pygmaea (F. Miquel) P. Ascherson & K. Graebner var. ***disticha*** (A. Freeman-Mitford) C. S. Chao & S. A. Renvoize **[cultivated]**
Sy = Arundinaria disticha (A. Freeman-Mitford) E. Pfitzer
Sy = Bambusa disticha A. Freeman-Mitford
Sy = Pleioblastus pygmaeus (F. Miquel) T. Nakai var. distichus (A. Freeman-Mitford) T. Nakai
Sy = Sasa disticha (A. Freeman-Mitford) E. Camus
A. pygmaea (F. Miquel) P. Ascherson & K. Graebner var. ***pygmaea*** **[cultivated]**
Sy = Bambusa pygmaea F. Miquel
Sy = Pleioblastus pygmaeus (F. Miquel) T. Nakai
Sy = Sasa pygmaea (F. Miquel) E. Camus
A. simonii (É. Carrière) A. Rivière & C. Rivière f. ***simonii*** **[cultivated]**
Sy = Arundinaria simonii var. simonii
Sy = Bambusa simonii É. Carrière
Sy = Pleioblastus simonii (É. Carrière) T. Nakai
A. simonii (É. Carrière) A. Rivière & C. Rivière f. ***variegata*** J. Hooker **[cultivated]**
Sy = Arundinaria simonii var. variegata J. Hooker
Sy = Pleioblastus simonii (É. Carrière) T. Nakai f. variegatus (J. Hooker) H. Muroi
A. variegata (P. von Siebold *ex* F. Miquel) T. Makino **[cultivated]**

Sy = Bambusa fortunei L. Van Houtte
Sy = B. variegata P. von Siebold *ex* F. Miquel
Sy = Pleioblastus fortunei (L. Van Houtte) T. Nakai

ARUNDO C. Linnaeus
A. donax C. Linnaeus
Sy = A. donax var. versicolor (P. Miller) J. Stokes [**cultivated**]

AVENA C. Linnaeus
A. fatua C. Linnaeus var. ***fatua***
Sy = A. fatua var. glabrata W. Petermann
Sy = A. fatua var. vilis (C. Wallroth) H. Haussknecht
A. fatua C. Linnaeus var. ***sativa*** (C. Linnaeus) H. Haussknecht [**cultivated**]
Sy = A. byzantina K. Koch
Sy = A. sativa C. Linnaeus

AXONOPUS A. Palisot de Beauvois
A. compressus (O. Swartz) A. Palisot de Beauvois
Sy = Anastrophus compressus (O. Swartz) D. von Schlechtendal *ex* J. Döll
Sy = Milium compressum O. Swartz
A. fissifolius (G. Raddi) J. Kuhlmann
Sy = A. affinis M. A. Chase
Sy = Paspalum fissifolium G. Raddi
A. furcatus (J. Flüggé) A. Hitchcock
Sy = Anastrophus furcatus (J. Flüggé) G. Nash
Sy = Panicum furcatum J. Flüggé

BAMBUSA J. von Schreber, *nomen conservandum*
B. arundinacea (A. Retzius) C. von Willdenow [**cultivated**]
Sy = Bambos arundinacea A. Retzius
Sy = Bambusa surinamensis F. Ruprecht
Sy = B. thouarsii K. Kunth
Sy = B. vulgaris H. A. Schrader *ex* J. Wendland
B. beecheyana W. Munro [**cultivated**]
Sy = Sinocalamus beecheyanus (W. Munro) F. McClure
B. multiplex (J. de Loureiro) E. Räuschel *ex* J. A. Schultes & J. H. Schultes
Sy = Arundinaria glaucescens (C. von Willdenow) A. Palisot de Beauvois
Sy = Arundo multiplex J. de Loureiro
Sy = Bambusa glaucescens (C. von Willdenow) P. von Siebold *ex* W. Munro
Sy = B. nana W. Roxburgh
Sy = Ludolphia glaucescens C. von Willdenow
B. oldhamii W. Munro
Sy = B. oldhami W. Munro (orthographic error)
Sy = Sinocalamus oldhamii (W. Munro) F. McClure
B. textilis F. McClure [**cultivated**]
B. tuldoides W. Munro [**cultivated**]
B. ventricosa F. McClure [**cultivated**]

BLEPHARIDACHNE E. Hackel
B. bigelovii (S. Watson) E. Hackel
Sy = Eremochloë bigelovii S. Watson

BLEPHARONEURON G. Nash
B. tricholepis (J. Torrey) G. Nash
Sy = Vilfa tricholepis J. Torrey

BOTHRIOCHLOA K. E. O. Kuntze
B. alta (A. Hitchcock) J. Henrard
Sy = Andropogon altus A. Hitchcock
B. barbinodis (M. Lagasca y Segura) W. Herter var. ***barbinodis***
Sy = Andropogon barbinodis M. Lagasca y Segura
B. barbinodis (M. Lagasca y Segura) W. Herter var. ***perforata*** (E. Fournier) F. Gould
Sy = Andropogon perforatus K. von Trinius *ex* E. Fournier
B. bladhii (A. Retzius) S. T. Blake subsp. ***bladhii***
Sy = Andropogon bladhii A. Retzius
Sy = A. intermedius R. Brown
Sy = Bothriochloa intermedia (R. Brown) A. Camus
B. caucasica (K. von Trinius) C. Hubbard
Sy = Andropogon caucasicus K. von Trinius
B. edwardsiana (F. Gould) L. Parodi
Sy = Andropogon edwardsianus F. Gould
B. exaristata (G. Nash) J. Henrard
Sy = Amphilophis exaristatus G. Nash
Sy = Andropogon exaristatus (G. Nash) A. Hitchcock
Sy = A. hassleri E. Hackel
Sy = A. saccharoides var. submuticus G. Vasey *ex* E. Hackel
Sy = Bothriochloa hassleri (E. Hackel) A. Cabrera
B. hybrida (F. Gould) F. Gould
Sy = Andropogon hybridus F. Gould
B. ischaemum (C. Linnaeus) Y. Keng var. ***ischaemum***
Sy = Andropogon ischaemum C. Linnaeus

B. ischaemum (C. Linnaeus) Y. Keng var. ***songarica*** (F. von Fischer & C. von Meyer) R. Celarier & J. Harlan
Sy = Andropogon ischaemum (C. Linnaeus) G. Nash var. songaricus (F. Ruprecht *ex* F. von Fischer & C. von Meyer) R. Celarier & J. Harlan
B. laguroides (A. P. de Candolle) W. Herter subsp. ***torreyana*** (E. von Steudel) K. Allred & F. Gould
Sy = Andropogon saccharoides O. Swartz var. torreyanus (E. von Steudel) E. Hackel
Sy = Bothriochloa longipaniculata (F. Gould) K. Allred & F. Gould
Sy = B. saccharoides (O. Swartz) P. Rydberg var. longipaniculata (F. Gould) F. Gould
Sy = B. saccharoides var. torreyana (E. von Steudel) F. Gould
Ma = B. saccharoides *auct.*, non (O. Swartz) P. Rydberg: American authors
B. pertusa (C. Linnaeus) A. Camus
Sy = Andropogon pertusus (C. Linnaeus) C. von Willdenow
Sy = Holcus pertusus C. Linnaeus
B. springfieldii (F. Gould) L. Parodi
Sy = Andropogon springfieldii F. Gould
B. wrightii (E. Hackel) J. Henrard
Sy = Andropogon wrightii E. Hackel

BOUTELOUA M. Lagasca y Segura, *nomen conservandum*
B. aristidoides (K. Kunth) A. Grisebach var. ***aristidoides***
B. barbata M. Lagasca y Segura var. ***barbata***
Sy = B. arenosa G. Vasey
Sy = Chondrosum exile E. Fournier
Sy = C. barbatum (M. Lagasca y Segura) W. Clayton
Sy = C. microstachyum E. Fournier
Sy = C. polystachum G. Bentham
Sy = C. subscorpioides J. K. A. Müller
B. breviseta G. Vasey
Sy = Chondrosum brevisetum (G. Vasey) W. Clayton
B. chondrosioides (K. Kunth) G. Bentham *ex* S. Watson
Sy = Chondrosum humboldtianum K. Kunth
Sy = Dinebra chondrosioides K. Kunth
B. curtipendula (A. Michaux) J. Torrey var. ***caespitosa*** F. Gould & Z. Kapadia **[state grass]**
B. curtipendula (A. Michaux) J. Torrey var. ***curtipendula*** **[state grass]**
Sy = Atheropogon curtipendulus (A. Michaux) E. Fournier
Sy = Chloris curtipendula A. Michaux
B. eriopoda (J. Torrey) J. Torrey
Sy = Chondrosum eriopodum J. Torrey
B. gracilis (C. von Willdenow *ex* K. Kunth) M. Lagasca y Segura *ex* D. Griffiths
Sy = B. gracilis var. stricta (G. Vasey) A. Hitchcock
Sy = B. oligostachya (T. Nuttall) J. Torrey *ex* A. Gray
Sy = Chondrosum gracile C. von Willdenow *ex* K. Kunth
Sy = C. oligostachum (T. Nuttall) J. Torrey
B. hirsuta M. Lagasca y Segura subsp. ***hirsuta***
Sy = Chondrosum hirsutum (M. Lagasca y Segura) K. Kunth
B. hirsuta M. Lagasca y Segura subsp. ***pectinata*** (H. Featherly) J. Wipff & S. D. Jones
Sy = B. hirsuta M. Lagasca y Segura var. pectinata (H. Featherly) V. Cory
Sy = B. pectinata H. Featherly
Sy = Chondrosum pectinatum (H. Featherly) W. Clayton
* ***B. kayi*** B. Warnock **[TOES: V]**
Sy = Chondrosum kayi (B. Warnock) W. Clayton
B. ramosa F. Lamson-Scribner *ex* G. Vasey
B. repens (K. Kunth) F. Lamson-Scribner & E. Merrill
Sy = B. filiformis (E. Fournier) D. Griffiths
Sy = B. heterostega (K. von Trinius) D. Griffiths
Sy = Dinebra repens K. Kunth
B. rigidiseta (E. von Steudel) A. Hitchcock
Sy = Aegopogon rigidisetus E. von Steudel
B. simplex M. Lagasca y Segura
Sy = Chondrosum procumbens N. Desvaux *ex* A. Palisot de Beauvois
B. trifida G. Thurber
Sy = Chondrosum trifidum (G. Thurber) W. Clayton
Sy = C. trinii E. Fournier
B. uniflora G. Vasey var. ***uniflora***
B. warnockii F. Gould & Z. Kapadia

BRACHIARIA (K. von Trinius) A. Grisebach
B. eruciformis (J. E. Smith) A. Grisebach
Sy = Panicum eruciforme J. E. Smith

BRACHYELYTRUM A. Palisot de Beauvois
* ***B. erectum*** (J. von Schreber) A. Palisot de Beauvois var. ***erectum*** **[TOES: IV]**
Sy = B. aristosum (A. Michaux) W. Trelease
Sy = Muhlenbergia erectum J. von Schreber

BRACHYPODIUM A. Palisot de Beauvois
  B. distachyon (C. Linnaeus) A. Palisot de Beauvois
    Sy = Bromus distachyon C. Linnaeus

BRIZA C. Linnaeus
  B. maxima C. Linnaeus [**cultivated**]
  B. media C. Linnaeus [**cultivated**]
  B. minor C. Linnaeus

BROMUS C. Linnaeus
  B. anomalus F. Ruprecht *ex* E. Fournier
    Sy = Bromopsis anomala (F. Ruprecht *ex* E. Fournier) J. Holub
  B. arizonicus (C. Shear) G. Stebbins
    Sy = B. carinatus W. Hooker & G. Arnott var. arizonicus C. Shear
  B. biebersteinii J. J. Römer & J. A. Schultes [**cultivated**]
  B. carinatus W. Hooker & G. Arnott var. ***carinatus***
    Sy = B. marginatus C. Nees von Esenbeck *ex* E. von Steudel
    Sy = B. paniculatus (C. Shear) P. Rydberg
    Sy = B. polyanthus F. Lamson-Scribner
    Sy = Ceratochloa carinata (W. Hooker & G. Arnott) T. Tutin
    Sy = C. polyantha (F. Lamson-Scribner) N. Tzvelev
  B. catharticus M. A. Vahl
    Sy = B. brevis C. Nees von Esenbeck *ex* E. von Steudel
    Sy = B. haenkeanus (J. Presl) K. Kunth
    Sy = B. unioloides (C. von Willdenow) K. Kunth
    Sy = B. willdenowii K. Kunth
    Sy = Ceratochloa unioloides (C. von Willdenow) A. Palisot de Beauvois
    Sy = Festuca unioloides C. von Willdenow
  B. ciliatus C. Linnaeus
    Sy = Bromopsis canadensis (A. Michaux) J. Holub
    Sy = B. ciliata (C. Linnaeus) J. Holub
    Sy = Bromus canadensis A. Michaux
    Sy = B. ciliatus var. genuinus M. Fernald
    Sy = B. ciliatus var. richardsonii (J. Link) J. Boivin
    Sy = B. richardsonii J. Link
  B. diandrus A. Roth
    Sy = Anisantha diandra (A. Roth) T. Tutin *ex* N. Tzvelev
    Sy = Bromus gussonei F. Parlatore
    Sy = B. rigidus A. Roth
    Sy = B. rigidus var. gussonei (F. Parlatore) E. Cosson & M. Durieu de Maisonneuve
  B. hordeaceus C. Linnaeus
    Sy = B. hordeaceus subsp. molliformis (J. Lloyd) R. Maire & M. Weiller
    Sy = B. hordeaceus subsp. pseudothominii P. Smith
    Sy = B. hordeaceus subsp. thominei (C. Hardham *ex* C. Nyman) J. Marie-Victorin & M. Weiller
    Sy = B. molliformis J. Lloyd
    Sy = B. mollis C. Linnaeus
    Sy = B. racemosus C. Linnaeus
  B. inermis F. von Leysser var. ***inermis***
    Sy = Bromopsis inermis (F. von Leysser) J. Holub
  B. japonicus C. Thunberg *ex* J. Murray
    Sy = B. commutatus H. A. Schrader
  B. lanatipes (C. Shear) P. Rydberg
    Sy = B. anomalus F. Ruprecht *ex* E. Fournier var. lanatipes (C. Shear) A. Hitchcock
    Sy = B. lanatipes (C. Shear) J. Holub
    Sy = B. porteri (J. Coulter) G. Nash var. lanatipes C. Shear
  B. lanceolatus A. Roth var. lanuginosa (J. Poiret) J. Dinsmore
    Sy = B. lanuginosus J. Poiret
    Sy = B. macrostachys R. Desfontaines
  B. madritensis C. Linnaeus subsp. ***rubens*** (C. Linnaeus) P. Husnot
    Sy = Anisantha rubens (C. Linnaeus) S. Nevski
    Sy = Bromus rubens C. Linnaeus
  B. porteri (J. Coulter) G. Nash
    Sy = Bromopsis frondosus (C. Shear) J. Holub
    Sy = B. porteri (J. Coulter) J. Holub
    Sy = Bromus frondosus (C. Shear) E. Wooton & P. Standley
    Sy = B. kalmii A. Gray var. porteri J. Coulter
    Sy = B. porteri var. frondosus C. Shear
  B. pubescens G. H. Muhlenberg *ex* C. von Willdenow
    Sy = Bromopsis pubescens (G. H. Muhlenberg *ex* C. von Willdenow) J. Holub
    Ma = Bromus purgans *auct.*, non C. Linnaeus: American authors
  B. secalinus C. Linnaeus
    Sy = B. secalinus var. hirsutus N. Kindberg
    Ma = B. racemosus *auct.*, non C. Linnaeus: Texas authors
  B. sterilis C. Linnaeus, *nomen conservandum*
    Sy = Anisantha sterilis (C. Linnaeus) S. Nevski
  B. tectorum C. Linnaeus var. ***glabratus*** F. Spenner
    Sy = B. tectorum var. nudus F. Klett & K. Richter
  B. tectorum C. Linnaeus var. ***tectorum***
    Sy = Anisantha tectorum (C. Linnaeus) S. Nevski

B. texensis (C. Shear) A. Hitchcock
Sy = Bromopsis texensis (C. Shear) J. Holub
Sy = Bromus purgans C. Linnaeus var. texensis C. Shear

BUCHLOË G. Engelmann, *nomen conservandum*
B. dactyloides (T. Nuttall) G. Engelmann
Sy = Bulbilis dactyloides (T. Nuttall) C. Rafinesque-Schmaltz *ex* K. E. O. Kuntze
Sy = Sesleria dactyloides T. Nuttall

CALAMAGROSTIS M. Adanson
C. arundinacea (C. Linnaeus) A. Roth var. ***brachytricha*** (E. von Steudel) E. Hackel [**cultivated**]
Sy = C. arundinacea subsp. brachytricha (E. von Steudel) N. Tzvelev
Sy = C. brachytricha E. von Steudel

CALAMOVILFA (A. Gray) F. Lamson-Scribner
C. gigantea (T. Nuttall) F. Lamson-Scribner & E. Merrill
Sy = Calamagrostis gigantea T. Nuttall

CATHESTECUM J. Presl
C. erectum G. Vasey & E. Hackel

CENCHRUS C. Linnaeus
C. brownii J. J. Römer & J. A. Schultes
Sy = C. viridis K. Sprengel
C. echinatus C. Linnaeus
Sy = C. echinatus var. hillebrandianus (A. Hitchcock) F. Brown
Sy = C. hillebrandianus A. Hitchcock
C. longispinus (E. Hackel) M. Fernald
C. myosuroides K. Kunth
Sy = Cenchropsis myosuroides (K. Kunth) G. Nash
C. spinifex A. Cavanilles
Sy = C. incertus M. Curtis
Sy = C. parviceps L. Shinners
Sy = C. pauciflorus G. Bentham
Ma = C. carolinianus *auct.*, non T. Walter: J. Kartesz (1994)

CHASMANTHIUM J. Link
C. latifolium (A. Michaux) H. Yates
Sy = Uniola latifolia A. Michaux
C. laxum (C. Linnaeus) H. Yates var. ***laxum***
Sy = Uniola laxa (C. Linnaeus) N. Britton, E. Sterns, & J. Poggenburg
C. laxum (C. Linnaeus) H. Yates var. ***sessiliflorum*** (J. Poiret) J. Wipff & S. D. Jones
Sy = C. laxum (C. Linnaeus) H. Yates subsp. sessiliflorum (J. Poiret) L. C. Clark
Sy = C. sessiliflorum (J. Poiret) H. Yates
Sy = Uniola longifolia F. Lamson-Scribner
Sy = Uniola sessiliflora J. Poiret

CHIMONOBAMBUSA T. Makino
C. marmorea (A. Freeman-Mitford) T. Makino [**cultivated**]
Sy = Bambusa marmorea A. Freeman-Mitford
C. quadrangularis (E. Fenzi) T. Makino [**cultivated**]
Sy = Bambusa quadrangularis E. Fenzi

CHLORIS O. Swartz
C. andropogonoides E. Fournier
C. barbata O. Swartz
Sy = C. inflata J. Link
Sy = C. paraguaiensis E. von Steudel
C. canterai J. Arechavleta var. ***canterai***
C. ciliata O. Swartz
C. cucullata G. Bischoff
C. divaricata R. Brown
Sy = C. cynodontioides B. Balansa
Sy = C. divaricata var. cynodontioides (B. Balansa) M. Lazarides
C. gayana K. Kunth
C. submutica K. Kunth
* ***C. texensis*** G. Nash [**TOES: V**]
C. verticillata T. Nuttall
C. virgata O. Swartz
C. × brevispica G. Nash [***andropogonoides × cucullata***]
C. × subdolichostachya J. K. A. Müller [***cucullata × verticillata***]
Sy = C. latisquamea G. Nash

CHRYSOPOGON K. von Trinius
C. pauciflorus (A. Chapman) G. Bentham *ex* G. Vasey
Sy = Rhaphis pauciflora (A. Chapman) G. Nash

CINNA C. Linnaeus
C. arundinacea C. Linnaeus
Sy = C. arundinacea var. inexpansa M. Fernald & L. Griscom

COIX C. Linnaeus
C. lacryma-jobi C. Linnaeus [**cultivated**]

CORTADERIA O. Stapf, *nomen conservandum*
C. jubata (P. Lemoine) O. Stapf [**cultivated**]
Sy = Gynerium jubatum P. Lemoine

C. selloana (J. A. Schultes & J. H. Schultes) P. Ascherson & K. Graebner [**cultivated**]
Sy = Arundo dioëca K. Sprengel, non J. de Loureiro
Sy = A. selloana J. A. Schultes & J. H. Schultes

COTTEA K. Kunth
C. pappophoroides K. Kunth

CYMBOPOGON K. Sprengel
C. citratus (A. P. de Candolle) O. Stapf [**cultivated**]

CYNODON L. C. Richard, *nomen conservandum*
C. aethiopicus W. Clayton & J. Harlan
C. dactylon (C. Linnaeus) C. Persoon var ***dactylon***
Sy = Capriola dactylon (C. Linnaeus) K. E. O. Kuntze
Sy = Cynodon dactylon var. aridus J. Harlan & J. de Wet
Sy = C. dactylon var. elegans A. Rendle
Sy = C. dactylon var. polevansii (S. Stent) J. Harlan & J. de Wet
Sy = Panicum dactylon C. Linnaeus
C. nlemfuensis A. Vanderyst var. ***nlemfuensis***
C. transvaalensis J. Burtt-Davy
C. × ***magenisii*** R. Hurcombe [***dactylon*** × ***transvaalensis***]

DACTYLIS C. Linnaeus
D. glomerata C. Linnaeus [**cultivated**]

DACTYLOCTENIUM C. von Willdenow
D. aegyptium (C. Linnaeus) A. Palisot de Beauvois
Sy = Cynosurus aegyptius C. Linnaeus

DANTHONIA A. P. de Candolle, *nomen conservandum*
D. sericea T. Nuttall
Sy = D. epilis F. Lamson-Scribner
Sy = D. sericea var. epilis (F. Lamson-Scribner) S. Blomquist
D. spicata (C. Linnaeus) A. Palisot de Beauvois *ex* J. J. Römer & J. A. Schultes
Sy = D. spicata var. longipila F. Lamson-Scribner & E. Merrill
Sy = D. spicata var. pinetorum C. Piper
Sy = D. thermalis F. Lamson-Scribner

DASYOCHLOA E. von Steudel
D. pulchella (K. Kunth) P. Rydberg
Sy = Erioneuron pulchellum (K. Kunth) T. Tateoka
Sy = Tridens pulchellus (K. Kunth) A. Hitchcock

DESMAZERIA C. Dumortier
D. rigida (C. Linnaeus) T. Tutin subsp. ***rigida***
Sy = Catapodium rigidum (C. Linnaeus) C. Hubbard *ex* J. Dony
Sy = Poa rigida C. Linnaeus
Sy = Scleropoa rigida (C. Linnaeus) A. Grisebach

DIARRHENA A. Palisot de Beauvois, *nomen conservandum*
D. obovata (H. Gleason) A. Brandenburg
Sy = D. americana A. Palisot de Beauvois var. obovata H. Gleason

DICHANTHIUM P. Willemet
D. annulatum (P. Forsskål) O. Stapf var. ***annulatum***
Sy = Andropogon annulatus P. Forsskål
D. aristatum (J. Poiret) C. Hubbard
Sy = Andropogon nodosus (P. Willemet) G. Nash
D. sericeum (R. Brown) A. Camus subsp. ***sericeum***
Sy = Andropogon sericeus R. Brown

DIGITARIA A. von Haller, *nomen conservandum*
D. arenicola (J. Swallen) A. Beetle
Sy = D. cognata (J. A. Schultes) R. Pilger var. arenicola (J. Swallen) R. D. Webster
Sy = Leptoloma arenicola J. Swallen
Sy = L. cognatum (J. A. Schultes) M. A. Chase var. arenicola (J. Swallen) F. Gould
D. bicornis (J. de Lamarck) J. J. Römer & J. A. Schultes
Sy = D. diversiflora J. Swallen
Sy = Paspalum bicorne J. de Lamarck
D. californica (G. Bentham) J. Henrard
Sy = Trichachne californica (G. Bentham) M. A. Chase
D. ciliaris (A. Retzius) G. Köler
Sy = D. adscendens (K. Kunth) J. Henrard
Sy = D. sanguinalis (C. Linnaeus) J. Scopoli var. ciliaris (A. Retzius) F. Parlatore
Sy = Panicum ciliare A. Retzius
D. cognata (J. A. Schultes) R. Pilger subsp. ***cognata***
Sy = Leptoloma cognatum (J. A. Schultes) M. A. Chase
D. cognata (J. A. Schultes) R. Pilger subsp. ***pubiflora*** (G. Vasey) J. Wipff
Sy = Panicum autumnale L. Bosc *ex* K. Sprengel var. publiflora G. Vasey
D. filiformis (C. Linnaeus) G. Köler var. ***filiformis***
Sy = Panicum filiforme C. Linnaeus
Sy = Syntherisma filiformis (C. Linnaeus) G. Nash

D. filiformis (C. Linnaeus) G. Köler var. ***villosa*** (T. Walter) M. Fernald
Sy = D. villosa (T. Walter) C. Persoon
Sy = Syntherisma villosa T. Walter
D. hitchcockii (M. A. Chase) T. Stuckert
Sy = Trichachne hitchcockii (M. A. Chase) M. A. Chase
D. insularis (C. Linnaeus) F. Fedde
Sy = Andropogon insulare C. Linnaeus
Sy = Trichachne insularis (C. Linnaeus) C. Nees von Esenbeck
Sy = Valota insularis (C. Linnaeus) M. A. Chase
D. ischaemum (J. von Schreber) J. von Schreber *ex* G. H. Muhlenberg
Sy = Syntherisma ischaemum (J. von Schreber) G. Nash
D. milanjiana (A. Rendle) O. Stapf
D. patens (J. Swallen) J. Henrard
Sy = Trichachne patens J. Swallen
D. sanguinalis (C. Linnaeus) J. Scopoli
Sy = Syntherisma sanguinalis (C. Linnaeus) J. Dulac
D. texana A. Hitchcock
Sy = D. albicoma J. Swallen
Sy = D. runyonii A. Hitchcock
Sy = D. subcalva A. Hitchcock
D. violascens J. Link
Sy = D. chinensis (C. Nees von Esenbeck) A. Camus
Sy = D. ischaemum (J. von Schreber) J. von Schreber *ex* G. H. Muhlenberg var. violascens (J. Link) A. E. Radford
Sy = Syntherisma chinensis (C. Nees von Esenbeck) A. Hitchcock

DISTICHLIS C. Rafinesque-Schmaltz
D. spicata (C. Linnaeus) E. Greene
Sy = D. spicata subsp. stricta (J. Torrey) R. F. Thorne
Sy = D. spicata var. stricta (J. Torrey) F. Lamson-Scribner
Sy = D. stricta (J. Torrey) P. Rydberg
Sy = D. stricta var. dentata (P. Rydberg) C. Hitchcock
Sy = Uniola spicata C. Linnaeus

ECHINOCHLOA A. Palisot de Beauvois, *nomen conservandum*
E. colona (C. Linnaeus) J. Link
Sy = Panicum colonum C. Linnaeus
E. crus-galli (C. Linnaeus) A. Palisot de Beauvois var. ***crus-galli***
Sy = E. pungens (J. Poiret) P. Rydberg var. coarctata M. Fernald & L. Griscom
Sy = Panicum crus-galli C. Linnaeus
E. crus-pavonis (K. Kunth) J. A. Schultes var. ***crus-pavonis***
Sy = Oplismenus crus-pavonis K. Kunth
E. crus-pavonis (K. Kunth) J. A. Schultes var. ***macera*** (K. Wiegand) F. Gould
Sy = E. crus-galli (C. Linnaeus) P. de Beauvois var. zeylayensis (C. Kunth) A. Hitchcock
Sy = E. zelayensis (K. Kunth) J. A. Schultes
E. frumentacea J. Link
Sy = E. crus-galli (C. Linnaeus) A. Palisot de Beauvois var. edulis A. Hitchcock
Sy = E. crus-galli var. frumentacea (J. Link) W. Wight
Sy = Panicum frumentaceum W. Roxburgh, non R. A. Salisbury
E. muricata (A. Palisot de Beauvois) M. Fernald var. ***microstachya*** K. Wiegand
Sy = E. crus-galli (C. Linnaeus) A. Palisot de Beauvois var. mitis (F. Pursh) W. Petermann
Sy = E. microstachya (K. Wiegand) P. Rydberg
Sy = E. muricata var. occidentalis K. Wiegand
Sy = E. occidentalis (K. Wiegand) P. Rydberg
Sy = E. pungens (J. Poiret) P. Rydberg var. microstachya (K. Wiegand) M. Fernald & L. Griscom
Sy = E. pungens var. multiflora (K. Wiegand) M. Fernald & L. Griscom
E. muricata (A. Palisot de Beauvois) M. Fernald var. ***muricata***
Sy = E. muricata var. ludoviciana K. Wiegand
Sy = E. pungens (J. Poiret) P. Rydberg
Sy = E. pungens var. ludoviciana (K. Wiegand) M. Fernald & L. Griscom
E. paludigena K. Wiegand
E. polystachya (K. Kunth) A. Hitchcock var. ***polystachya***
Sy = Oplismenus polystachus K. Kunth
E. walteri (F. Pursh) A. A. Heller
Sy = Panicum walteri F. Pursh

EHRHARTA C. Thunberg, *nomen conservandum*
E. calycina J. E. Smith

ELEUSINE J. Gaertner
E. indica (C. Linnaeus) J. Gaertner subsp. ***indica***
Sy = Cynosurus indicus C. Linnaeus

ELIONURUS F. von Humboldt & A. Bonpland *ex* C. von Willdenow
Sy = Elyonurus (orthographic error)
NOTE: Kunth (1816) gives the etymologically correct form as *Elionurus*.
E. barbiculmis E. Hackel
Sy = E. barbiculmis var. parviflorus F. Lamson-Scribner
E. tripsacoides F. von Humboldt & A. Bonpland *ex* C. von Willdenow var. ***tripsacoides***

ELYMUS C. Linnaeus
E. canadensis C. Linnaeus var. ***canadensis***
Sy = E. canadensis var. brachystachya (F. Lamson-Scribner & C. Ball) O. Farwell
Sy = E. canadensis var. robustus (F. Lamson-Scribner & J. G. Smith) K. Mackenzie & B. Bush
Sy = E. canadensis var. villosus (G. H. Muhlenberg) L. Shinners
Sy = E. canadensis var. wiegandii (M. Fernald) W. M. Bowden
Sy = E. wiegandii M. Fernald
E. canadensis C. Linnaeus var. ***interruptus*** (S. Buckley) G. Church
Sy = E. interruptus S. Buckley
E. elymoides (C. Rafinesque-Schmaltz) G. Swezey subsp. ***brevifolius*** (J. G. Smith) M. Barkworth
Sy = E. elymoides var. brevifolius (J. G. Smith) R. Dorn
Sy = Sitanion elymoides C. Rafinesque-Schmaltz
Sy = S. hystrix (T. Nuttall) J. G. Smith
E. repens (C. Linnaeus) F. Gould
Sy = Agropyron repens (C. Linnaeus) A. Palisot de Beauvois
Sy = Elytrigia repens (C. Linnaeus) S. Nevski
Sy = Triticum repens (C. Linnaeus)
E. trachycaulus (J. Link) F. Gould *ex* L. Shinners
Sy = Agropyron trachycaulum J. Link
Sy = Elymus trachycaulus subsp. subsecundus (J. Link) Á. Löve & D. Löve
E. virginicus C. Linnaeus var. ***virginicus***
Sy = E. glabriflorus (G. Vasey) F. Lamson-Scribner & C. Ball
Sy = E. jejunus (F. Ramaley) P. Rydberg
Sy = E. macgregorii R. E. Brooks
Sy = E. submuticus (W. Hooker) B. Smyth
Sy = E. virginicus var. australis (F. Lamson-Scribner & C. Ball) A. Hitchcock
Sy = E. virginicus var. glabriflorus (G. Vasey) B. Bush
Sy = E. virginicus var. intermedius (G. Vasey) B. Bush
Sy = E. virginicus var. jejunus (F. Ramaley) B. Bush
Sy = E. virginicus var. submuticus W. Hooker

ENNEAPOGON A. Palisot de Beauvois
E. desvauxii A. Palisot de Beauvois
Sy = Pappophorum wrightii S. Watson

ENTEROPOGON C. Nees von Esenbeck
E. chlorideus(J. Presl) W. Clayton
Sy = Chloris chloridea (J. Presl) A. Hitchcock
Sy = Dinebra chloridea J. Presl

ERAGROSTIS N. von Wolf
E. airoides C. Nees von Esenbeck
E. amabilis (C. Linnaeus) R. Wight & G. Arnott *ex* C. Nees von Esenbeck
Sy = E. tenella (C. Linnaeus) A. Palisot de Beauvois *ex* J. J. Römer & J. A. Schultes
Sy = Poa amabilis C. Linnaeus
Sy = P. tenella C. Linnaeus
E. barrelieri J. Daveau
E. capillaris (C. Linnaeus) C. Nees von Esenbeck
E. cilianensis (C. Allioni) F. Vignolo Lutati *ex* E. Janchen
Sy = E. major N. Host
Sy = E. megastachya (G. Köler) J. Link
Sy = Poa cilianensis C. Allioni
E. ciliaris (C. Linnaeus) R. Brown var. ***ciliaris***
Sy = Poa ciliaris C. Linnaeus
E. curtipedicellata S. Buckley
E. curvula (H. A. Schrader) C. Nees von Esenbeck
E. elliottii S. Watson
E. erosa F. Lamson-Scribner
E. glomerata (T. Walter) L. Dewey
Sy = Poa glomerata T. Walter
E. hirsuta (A. Michaux) C. Nees von Esenbeck
Sy = E. hirsuta var. laevivaginata M. Fernald
Sy = Poa hirsuta A. Michaux
E. hypnoides (J. de Lamarck) N. Britton, E. Sterns, & J. Poggenburg
Sy = Poa hypnoides J. de Lamarck
E. intermedia A. Hitchcock var. ***intermedia***
E. lehmanniana C. Nees von Esenbeck
E. lugens C. Nees von Esenbeck
E. mexicana (J. Hornemann) J. Link subsp. ***mexicana***
Sy = E. neomexicana G. Vasey
Sy = Poa mexicana J. Hornemann

E. minor N. Host
Sy = E. eragrostis (C. Linnaeus) A. Palisot de Beauvois
Sy = E. poaeoides A. Palisot de Beauvois
E. palmeri S. Watson
E. pectinacea (A. Michaux) C. Nees von Esenbeck *ex* E. von Steudel var. ***miserrima*** (E. Fournier) J. R. Reeder
Sy = E. arida A. Hitchcock
Sy = E. tephrosanthes J. A. Schultes
E. pectinacea (A. Michaux) C. Nees von Esenbeck *ex* E. von Steudel var. ***pectinacea***
Sy = E. diffusa S. Buckley
Sy = Poa pectinacea A. Michaux
E. pilosa (C. Linnaeus) A. Palisot de Beauvois var. ***perplexa*** (L. Harvey) S. Koch
Sy = E. perplexa L. Harvey
E. pilosa (C. Linnaeus) A. Palisot de Beauvois var. ***pilosa***
Sy = E. multicaulis E. von Steudel
Sy = Poa pilosa C. Linnaeus
E. refracta (G. H. Muhlenberg *ex* S. Elliott) F. Lamson-Scribner
Sy = Poa pectinacea A. Michaux var. refracta (G. H. Muhlenberg) A. Chapman
Sy = P. refracta G. H. Muhlenberg *ex* S. Elliott
E. reptans (A. Michaux) C. Nees von Esenbeck
Sy = Neeragrostis reptans (A. Michaux) E. Nicora
Sy = Poa reptans A. Michaux
E. secundiflora J. Presl subsp. ***oxylepis*** (J. Torrey) S. Koch
Sy = E. beyrichii J. G. Smith
Sy = E. oxylepis (J. Torrey) J. Torrey
E. sessilispica S. Buckley
Sy = Acamptoclados sessilispicus (S. Buckley) G. Nash
E. silveana J. Swallen
E. spectabilis (F. Pursh) E. von Steudel
Sy = E. spectabilis var. sparsihirsuta O. Farwell
Sy = Poa spectabilis F. Pursh
E. spicata G. Vasey
E. superba J. Peyritsch
E. swallenii A. Hitchcock
E. trichocolea E. Hackel & J. Arechavleta var. ***floridana*** (A. Hitchcock) J. Witherspoon
Sy = E. floridana A. Hitchcock
E. trichodes (T. Nuttall) A. Wood var. ***pilifera*** (G. Scheele) M. Fernald
Sy = E. pilifera G. Scheele
E. trichodes (T. Nuttall) A. Wood var. ***trichodes***
Sy = Poa trichodes T. Nuttall
E. trichophora E. Cosson & M. Durieu de Maisonneuve
Sy = E. atherstonei O. Stapf

EREMOCHLOA L. Büse
E. ophiuroides (W. Munro) E. Hackel
Sy = Ischaemum ophiuroides W. Munro

ERIOCHLOA K. Kunth
E. acuminata (J. Presl) K. Kunth var. ***acuminata***
Sy = E. gracilis (E. Fournier) A. Hitchcock
Sy = E. lemmonii G. Vasey & F. Lamson-Scribner var. gracilis (E. Fournier) F. Gould
E. acuminata (J. Presl) K. Kunth var. ***minor*** (G. Vasey) R. Shaw
Sy = E. gracilis (E. Fournier) A. Hitchcock var. minor G. Vasey
Sy = E. punctata (C. Linnaeus) N. Desvaux *ex* W. Hamilton var. minor G. Vasey
Ma = E. lemmonii var. lemmonii *auct.*, non G. Vasey & F. Lamson-Scribner: *sensu* F. Gould
E. contracta A. Hitchcock
Sy = Helopus mollis J. K. A. Müller, non K. Kunth
E. polystachya K. Kunth
Sy = E. subglabra (G. Nash) A. Hitchcock
E. pseudoacrotricha (O. Stapf *ex* A. Thellung) C. Hubbard *ex* S. T. Blake
Sy = E. ramosa (A. Retzius) K. E. O. Kuntze var. pseudoacrotricha O. Stapf *ex* A. Thellung
E. punctata (C. Linnaeus) N. Desvaux *ex* W. Hamilton
Sy = Milium punctatum C. Linnaeus
E. sericea (G. Scheele) W. Munro *ex* G. Vasey
Sy = Paspalum sericeum G. Scheele

ERIONEURON G. Nash
E. avenaceum (K. Kunth) T. Tateoka var. ***avenaceum***
Sy = E. avenaceum var. grandiflorum (G. Vasey) T. Tateoka
Sy = Tridens avenaceus (K. Kunth) A. Hitchcock
Sy = T. grandiflorus (G. Vasey) E. Wooton & P. Standley
Sy = Triodia grandiflora G. Vasey
E. avenaceum (K. Kunth) T. Tateoka var. ***nealleyi*** (G. Vasey) F. Gould
Sy = E. nealleyi (G. Vasey) T. Tateoka
Sy = Tridens nealleyi (G. Vasey) E. Wooton & P. Standley
E. pilosum (S. Buckley) G. Nash var. ***pilosum***
Sy = Tridens pilosus (S. Buckley) A. Hitchcock
Sy = Uralepis pilosa S. Buckley

EUSTACHYS N. Desvaux
E. caribaea (K. Sprengel) W. Herter
Sy = Chloris caribaea K. Sprengel
Sy = Eustachys paspaloides (M. A. Vahl) D. Lanza & G. Mattei subsp. caribaea (K. Sprengel) R. Nowack
E. petraea (O. Swartz) N. Desvaux
Sy = Chloris petraea O. Swartz
E. retusa (M. Lagasca y Segura) K. Kunth
Sy = Chloris argentina (E. Hackel) M. Lillo & L. Parodi

FESTUCA C. Linnaeus
F. arizonica G. Vasey
* *F. ligulata* J. Swallen [**TOES: V**]
F. paradoxa N. Desvaux
Sy = F. nutans J. Biehler
F. rubra C. Linnaeus
F. subverticillata (C. Persoon) E. Alexeev
Sy = F. obtusa J. Biehler
F. versuta W. Beal

GASTRIDIUM A. Palisot de Beauvois
G. phleoides (C. Nees von Esenbeck & F. Meyen) C. Hubbard
Ma = G. ventricosum *auct.*, non (A. Gouan) H. Schinz & A. Thellung: American authors

GLYCERIA R. Brown, *nomen conservandum*
G. septentrionalis A. Hitchcock var. *arkansana* (M. Fernald) J. Steyermark & J. Kucera
Sy = G. arkansana M. Fernald
G. septentrionalis A. Hitchcock var. *septentrionalis*
Sy = Panicularia septentrionalis (A. Hitchcock) E. Bicknell
G. striata (J. de Lamarck) A. Hitchcock
Sy = G. striata var. stricta (F. Lamson-Scribner) O. Hultén

GYMNOPOGON A. Palisot de Beauvois
G. ambiguus (A. Michaux) N. Britton, E. Sterns, & J. Poggenburg
G. brevifolius K. von Trinius

HAINARDIA W. Greuter
H. cylindrica (C. von Willdenow) W. Greuter
Sy = Lepturus cylindricus (C. von Willdenow) K. von Trinius
Sy = Monerma cylindrica (C. von Willdenow) E. Cosson & M. Durieu de Maisonneuve
Sy = Rottboellia cylindrica C. von Willdenow

HAKONECHLOA T. Makino *ex* M. Honda
H. macra (W. Munro) T. Makino [**cultivated**]
Sy = Phragmites macra W. Munro

HEMARTHRIA R. Brown
H. altissima (J. Poiret) O. Stapf & C. Hubbard
Sy = Manisuris altissima (J. Poiret) A. Hitchcock

HESPEROSTIPA
H. comata (K. von Trinius & F. Ruprecht) M. Barkworth subsp. *comata*
Sy = Stipa comata K. von Trinius & F. Ruprecht
Sy = S. comata subsp. intonsa C. Piper
Sy = S. comata var. suksdorfii H. St. John
H. comata (K. von Trinius & F. Ruprecht) M. Barkworth subsp. *intermedia* (F. Lamson-Scribner & F. Tweedy) M. Barkworth
Sy = Stipa comata K. von Trinius & F. Ruprecht var. intermedia F. Lamson-Scribner & F. Tweedy
H. neomexicana (G. Thurber) M. Barkworth
Sy = Stipa neomexicana (G. Thurber) F. Lamson-Scribner
Sy = S. pennata C. Linnaeus var. neomexicana G. Thurber

HETEROPOGON C. Persoon
H. contortus (C. Linnaeus) A. Palisot de Beauvois *ex* J. J. Römer & J. A. Schultes
Sy = Andropogon contortus C. Linnaeus
H. melanocarpus (S. Elliott) G. Bentham
Sy = Andropogon melanocarpus S. Elliott

HILARIA K. Kunth
H. belangeri (E. von Steudel) G. Nash var. *belangeri*
H. swallenii V. Cory

HOLCUS C. Linnaeus, *nomen conservandum*
H. lanatus C. Linnaeus
Sy = Nothoholcus lanatus (C. Linnaeus) G. Nash

HORDEUM C. Linnaeus
H. brachyantherum S. Nevski subsp. *brachyantherum*
Sy = Critesion brachyantherum (S. Nevski) M. Barkworth & D. R. Dewey
Ma = Hordeum nodosum *auct.*, non C. Linnaeus: American authors
H. jubatum C. Linnaeus subsp. *jubatum*
Sy = Critesion jubatum (C. Linnaeus) S. Nevski

H. murinum C. Linnaeus subsp. ***glaucum*** (E. von Steudel) N. Tzvelev
Sy = Critesion murinum (C. Linnaeus) Á. Löve subsp. glaucum (E. von Steudel) W. A. Weber
Sy = Hordeum glaucum E. von Steudel
Sy = H. stebbinsii G. Covas
H. murinum C. Linnaeus subsp. ***leporinum*** (J. Link) G. Arcangeli
Sy = Critesion murinum (C. Linnaeus) Á. Löve subsp. leporinum (J. Link) Á. Löve
Sy = Hordeum leporinum J. Link
H. pusillum T. Nuttall
Sy = Critesion pusillum (T. Nuttall) Á. Löve
Sy = Hordeum pusillum var. pubens A. Hitchcock
H. vulgare C. Linnaeus

IMPERATA D. Cirillo
I. brevifolia G. Vasey
Sy = I. hookeri (F. Ruprecht *ex* N. Andersson) F. Ruprecht *ex* E. Hackel
I. cylindrica (C. Linnaeus) A. Palisot de Beauvois [**cultivated**]
Sy = I. koenigii (A. Retzius) A. Palisot de Beauvois
Sy = Lagurus cylindricus C. Linnaeus
Hn = I. cylindrica cv. "Red Baron"
Hn = I. cylindrica cv. "Rubrum"
NOTE: Only the commercial variety is found in Texas. This red-foliaged ornamental cultivar is diminutive and nonaggressive, and has little chance of escaping. However, it was cultivated from a very aggressive species that is listed in the Federal Noxious Weed list and can occasionally revert to its natural, green-foliaged, aggressive state.

INDOCALAMUS T. Nakai
I. tessellatus (W. Munro) P. Keng [**cultivated**]
Sy = Bambusa tessellata W. Munro
Sy = Sasa tessellatus (W. Munro) T. Makino & K. Shibata

ISCHAEMUM C. Linnaeus
I. rugosum R. A. Salisbury var. ***rugosum*** [**federal noxious weed**]
NOTE: Introduced into southern Texas (weed in a greenhouse), but it is reported to have been eradicated.

KOELERIA C. Persoon
K. macrantha (C. von Ledebour) J. A. Schultes
Sy = K. cristata (C. Linnaeus) C. Persoon
Sy = K. gracilis C. Persoon
Sy = K. nitida T. Nuttall
Ma = K. pyramidata *auct.*, non J. de Lamarck: American authors

LAMARCKIA C. Moench, *nomen conservandum*
L. aurea (C. Linnaeus) C. Moench
Sy = Cynosurus aureus C. Linnaeus

LEERSIA O. Swartz, *nomen conservandum*
L. hexandra O. Swartz
Sy = Homalocenchrus hexandrus (O. Swartz) K. E. O. Kuntze
L. lenticularis A. Michaux
Sy = Homalocenchrus lenticularis (A. Michaux) K. E. O. Kuntze
L. monandra O. Swartz
Sy = Homalocenchrus monandrus (O. Swartz) K. E. O. Kuntze
L. oryzoides (C. Linnaeus) O. Swartz var. ***oryzoides***
Sy = Homalocenchrus oryzoides (C. Linnaeus) J. Pollich
Sy = Leersia oryzoides f. glabra A. Eaton
Sy = Phalaris oryzoides C. Linnaeus
L. virginica C. von Willdenow
Sy = Homalocenchrus virginicus (C. von Willdenow) N. Britton
Sy = Leersia virginica var. ovata (J. Poiret) M. Fernald

LEPTOCHLOA A. Palisot de Beauvois
L. chloridiformis (E. Hackel) L. Parodi
L. dubia (K. Kunth) C. Nees von Esenbeck
Sy = Diplachne dubia (K. Kunth) C. Nees von Esenbeck
L. fascicularis (J. de Lamarck) A. Gray var. ***fascicularis***
Sy = Diplachne acuminata G. Nash
Sy = D. fascicularis (J. de Lamarck) A. Palisot de Beauvois
Sy = Leptochloa acuminata (G. Nash) H. Gleason
L. mucronata (A. Michaux) K. Kunth
Sy = L. attenuata (T. Nuttall) E. von Steudel
Sy = L. filiformis (J. de Lamarck) A. Palisot de Beauvois
Sy = L. filiformis var. attenuata (T. Nuttall) J. Steyermark & J. Kucera
Sy = L. filiformis var. pulchella (F. Lamson-Scribner) A. Beetle
L. nealleyi G. Vasey
L. panicoides (J. Presl) A. Hitchcock
Sy = Diplachne halei G. Nash

Sy = D. panicoides (J. Presl) J. McNeill
Sy = Leptochloa floribunda J. Döll
Sy = Megastachya panicoides J. Presl
L. uninervia (J. Presl) A. Hitchcock & M. A. Chase
Sy = Diplachne uninervia (J. Presl) L. Parodi
L. virgata (C. Linnaeus) A. Palisot de Beauvois
Sy = Cynosurus virgatus C. Linnaeus
Sy = Leptochloa domingensis (N. von Jacquin) K. von Trinius
L. viscida (F. Lamson-Scribner) W. Beal
Sy = Diplachne viscida F. Lamson-Scribner

LEYMUS C. F. Hochstetter
L. arenarius (C. Linnaeus) C. F. Hochstetter **[cultivated]**
Sy = Elymus arenarius C. Linnaeus
Hn = E. arenarius cv. "Findhorn"
Hn = E. arenarius cv. "Glaucus"
L. triticoides (S. Buckley) R. Pilger
Sy = Elymus condensatus J. Presl var. triticoides (S. Buckley) G. Thurber
Sy = E. triticoides S. Buckley
Sy = E. triticoides var. pubescens A. Hitchcock

LIMNODEA L. Dewey
L. arkansana (T. Nuttall) L. Dewey
Sy = Greenia arkansana T. Nuttall

LOLIUM C. Linnaeus
L. arundinaceum (J. von Schreber) S. J. Darbyshire
Sy = Festuca arundinacea J. von Schreber
Sy = F. elatior C. Linnaeus
Sy = F. elatior subsp. arundinacea (J. von Schreber) L. J. Čelakovský
Sy = F. elatior var. arundinacea (J. von Schreber) C. F. Wimmer
L. perenne C. Linnaeus var. ***aristatum*** C. von Willdenow
Sy = L. multiflorum J. de Lamarck
Sy = L. perenne subsp. multiflorum (J. de Lamarck) T. Husnot
Sy = L. perenne var. italicum (R. Brown) D. Parnell
L. perenne C. Linnaeus var. ***perenne***
L. pratense (W. Hudson) S. J. Darbyshire
Sy = Festuca elatior C. Linnaeus subsp. pratensis (W. Hudson) E. Hackel
Sy = F. elatior var. pratensis (W. Hudson) A. Gray
Sy = F. pratensis W. Hudson
L. rigidum J. Gaudin
Sy = L. lepturoides P. Boissier
Sy = L. loliaceum (Bory de Saint-Vincent & L. Chaubard) H. Handel-Mazzetti
Sy = L. perenne C. Linnaeus subsp. rigidum (J. Gaudin) Á. Löve & D. Löve
Sy = L. perenne var. rigidum (J. Gaudin) E. Cosson & M. Durieu de Maisonneauve
Sy = L. rigidum subsp. lepturoides (P. Boissier) F. Sennen & A. Mauricio
Sy = L. rigidum var. rottbollioides T. von Heldreich *ex* P. Boissier
Sy = L. strictum C. Presl
L. temulentum C. Linnaeus subsp. ***temulentum***
Sy = L. arvense W. Withering
Sy = L. multiflorum J. de Lamarck var. ramosum (G. Gussone) F. Parlatore
Sy = L. temulentum var. arvense (W. Withering) S. Liljeblad
Sy = L. temulentum var. leptochaeton A. Braun
Sy = L. temulentum var. macrochaeton A. Braun
Sy = L. temulentum var. ramosum G. Gussone

LUZIOLA A. L. de Jussieu
L. fluitans (A. Michaux) E. Terrell & H. Robinson
Sy = Hydrochloa caroliniensis A. Palisot de Beauvois

LYCURUS K. Kunth
L. phleoides K. Kunth
L. setosus (T. Nuttall) C. Reeder
Sy = Pleopogon setosus T. Nuttall

MELICA C. Linnaeus
M. bulbosa C. Geyer *ex* T. Porter & J. Coulter
Sy = Bromelica bulbosa (C. Geyer *ex* T. Porter & J. Coulter) W. A. Weber
Sy = Melica bulbosa var. intonsa (C. Piper) M. Peck
M. montezumae C. Piper
M. mutica T. Walter
M. nitens (F. Lamson-Scribner) T. Nuttall *ex* C. Piper
M. porteri F. Lamson-Scribner var. ***laxa*** W. Boyle
M. porteri F. Lamson-Scribner var. ***porteri***

MELINIS A. Palisot de Beauvois
M. repens (C. von Willdenow) G. Zizka subsp. ***repens***
Sy = Rhynchelytrum repens (C. von Willdenow) C. Hubbard

MICROCHLOA R. Brown
M. kunthii N. Desvaux
Sy = Paspalum tenuissimum M. E. Jones

MICROSTEGIUM C. Nees von Esenbeck
M. vimineum (K. von Trinius) A. Camus var. **_vimineum_**
Sy = Andropogon vimineum K. von Trinius
Sy = Eulalia viminea (K. von Trinius) K. E. O. Kuntze
Sy = E. viminea var. variabilis K. E. O. Kuntze
Sy = Microstegium vimineum var. imberbe (C. Nees von Esenbeck) M. Honda

MISCANTHUS N. Andersson
M. cv. **"Giganteus"** [**cultivated**]
Ma = M. floridulus (J. Houtton de Labillardière) O. Warburg *ex* K. Schumann & K. Lauterbach
M. sinensis N. Andersson var. **_sinensis_** [**cultivated**]
Sy = M. sinensis var. gracillimus A. Hitchcock
Sy = M. sinensis var. variegatus W. Beal
Sy = M. sinensis var. zebrinus (G. Nicholson) W. Beal
M. transmorrisonensis B. Hayata [**cultivated**]

MNESITHEA K. Kunth
M. cylindrica (A. Michaux) R. de Koning & M. Sosef
Sy = Coelorachis cylindrica (A. Michaux) G. Nash
Sy = Manisuris cylindrica A. Michaux
M. granularis (C. Linnaeus) R. de Koning & M. Sosef
Sy = Hackelochloa granularis (C. Linnaeus) K. E. O. Kuntze
M. rugosa (A. Michaux) R. de Koning & M. Sosef
Sy = Coelorachis rugosa (T. Nuttall) G. Nash
Sy = Manisuris rugosa (T. Nuttall) K. E. O. Kuntze

MONANTHOCHLOË G. Engelmann
M. littoralis G. Engelmann

MONROA J. Torrey
Sy = Munroa [orthographic variant]
M. squarrosa (T. Nuttall) J. Torrey
Sy = Crypsis squarrosa T. Nuttall
Sy = Munroa squarrosa var. flaccuosa G. Vasey *ex* W. Beal

MUHLENBERGIA J. von Schreber
M. andina (T. Nuttall) A. Hitchcock
M. arenacea (S. Buckley) A. Hitchcock
M. arenicola S. Buckley
M. asperifolia (C. Nees von Esenbeck & F. Meyen) L. Parodi
Sy = Sporobolus asperifolius (C. Nees von Esenbeck & F. Meyen *ex* C. von Trinius) C. Nees von Esenbeck
M. brevis C. Goodding
M. bushii R. Pohl
Sy = M. brachyphylla B. Bush
M. capillaris (J. de Lamarck) K. von Trinius var. **_capillaris_**
M. capillaris (J. de Lamarck) K. von Trinius var. **_filipes_** (M. Curtis) A. Chapman *ex* W. Beal
Sy = M. filipes M. Curtis
M. crispiseta A. Hitchcock
M. depauperata F. Lamson-Scribner
M. dubia E. Fournier *ex* W. Hemsley
M. dumosa F. Lamson-Scribner [**cultivated**]
M. eludens C. Reeder
M. emersleyi G. Vasey
M. expansa (J. Poiret) K. von Trinius
Sy = Agrostis trichopodes S. Elliott
Sy = Muhlenbergia capillaris (J. de Lamarck) K. von Trinius var. trichopodes (S. Elliott) G. Vasey
M. fragilis J. Swallen
M. frondosa (J. Poiret) M. Fernald
Sy = Agrostis frondosa J. Poiret
Ma = M. mexicana *auct.*, non (C. Linnaeus) K. von Trinius: Hitchcock (1935)
M. glabrifloris F. Lamson-Scribner
M. glauca (C. Nees von Esenbeck) C. Mez
M. lindheimeri A. Hitchcock
M. metcalfei M. E. Jones
M. mexicana (C. Linnaeus) K. von Trinius
Sy = Agrostis mexicana C. Linnaeus
Sy = Muhlenbergia mexicana var. filiformis (C. von Willdenow) F. Lamson-Scribner
M. minutissima (E. von Steudel) J. Swallen
M. montana (T. Nuttall) A. Hitchcock
M. parviglumis G. Vasey
M. pauciflora S. Buckley
M. polycaulis F. Lamson-Scribner
M. porteri F. Lamson-Scribner *ex* W. Beal
M. pungens G. Thurber
M. racemosa (A. Michaux) N. Britton, E. Sterns, & J. Poggenburg
Sy = Agrostis racemosa A. Michaux
M. repens (J. Presl) A. Hitchcock
M. reverchonii G. Vasey & F. Lamson-Scribner
Sy = M. reverchoni G. Vasey & F. Lamson-Scribner (orthographic error)
M. rigens (G. Bentham) A. Hitchcock
Sy = Epicampes rigens G. Bentham
M. rigida (K. Kunth) K. Kunth

M. schreberi J. F. Gmelin
Sy = M. palustris F. Lamson-Scribner
Sy = M. schreberi var. palustris (F. Lamson-Scribner) F. Lamson-Scribner
M. setifolia G. Vasey
M. sobolifera (G. H. Muhlenberg *ex* C. von Willdenow) K. von Trinius var. ***setigera*** F. Lamson-Scribner
Sy = M. setigera (F. Lamson-Scribner) B. Bush
Sy = M. sobolifera f. setigera (F. Lamson-Scribner) C. Deam
M. sobolifera (G. H. Muhlenberg *ex* C. von Willdenow) K. von Trinius var. ***sobolifera***
Sy = Agrostis sobolifera G. H. Muhlenberg *ex* C. von Willdenow
M. sylvatica J. Torrey *ex* A. Gray
Sy = Agrostis diffusa G. H. Muhlenberg, non N. Host
Sy = A. sylvatica J. Torrey, non W. Hudson
Sy = Muhlenbergia diffusa O. Farwell, non C. von Willdenow
Sy = M. sylvatica f. attenuata (F. Lamson-Scribner) E. Palmer & J. Steyermark
Sy = M. sylvatica var. robusta M. Fernald
M. tenuifolia (K. Kunth) K. Kunth
Sy = M. monticola S. Buckley
M. texana S. Buckley
M. torreyi (K. Kunth) A. Hitchcock *ex* B. Bush
M. utilis (J. Torrey) A. Hitchcock
M. villiflora A. Hitchcock var. ***villosa*** (J. Swallen) C. Morden
Sy = M. villosa J. Swallen
M. × curtisetosa (F. Lamson-Scribner) B. Bush [***frondosa × schreberi***]
Sy = M. schreberi J. F. Gmelin subsp. curtisetosa F. Lamson-Scribner
M. × involuta J. Swallen [***reverchonii*** × unknown]

NASSELLA N. Desvaux
N. leucotricha (K. von Trinius & F. Ruprecht) R. Pohl
Sy = Stipa leucotricha K. von Trinius & F. Ruprecht
N. tenuissima (K. von Trinius) M. Barkworth
Sy = Stipa tenuissima K. von Trinius

OPLISMENUS A. Palisot de Beauvois, *nomen conservandum*
O. hirtellus (C. Linnaeus) A. Palisot de Beauvois
Sy = O. hirtellus subsp. setarius (J. de Lamarck) C. Mez
Sy = O. setarius (J. de Lamarck) J. J. Römer & J. A. Schultes

ORYZA C. Linnaeus
O. sativa C. Linnaeus var. ***fatua*** D. Prain
O. sativa C. Linnaeus var. ***sativa*** **[cultivated]**

OTATEA (F. McClure & E. W. Smith) C. Calderón & T. Soderstrom
O. acuminata (W. Munro) C. Calderón & T. Soderstrom subsp. ***aztecorum*** (F. McClure & E. W. Smith) R. Guzmán **[cultivated]**
Sy = O. aztecorum (F. McClure & E. W. Smith) C. Calderón & T. Soderstrom
Sy = Yushania aztecorum F. McClure & E. W. Smith

PANICUM C. Linnaeus
P. aciculare N. Desvaux *ex* J. Poiret var. ***aciculare***
Sy = Dichanthelium aciculare (N. Desvaux *ex* J. Poiret) F. Gould & C. A. Clark
Sy = Panicum aciculare subsp. aciculare
Sy = P. aciculare subsp. neuranthum (A. Grisebach) R. Freckmann & M. Lelong
Sy = P. neuranthum A. Grisebach
P. aciculare N. Desvaux *ex* J. Poiret var. ***angustifolium*** (S. Elliott) J. Wipff & S. D. Jones
Sy = Dichanthelium aciculare (N. Desvaux *ex* J. Poiret) F. Gould & C. A. Clark var. ramosum (A. Grisebach) G. Davidse
Sy = D. angustifolium (S. Elliott) F. Gould
Sy = Panicum aciculare N. Desvaux *ex* J. Poiret subsp. angustifolium (S. Elliott) R. Freckmann and M. Lelong
Sy = P. aciculare subsp. fusiforme (A. Hitchcock) R. Freckmann and M. Lelong
Sy = P. aciculare var. arenicoloides (W. Ashe) A. Beetle
Sy = P. angustifolium S. Elliott
Sy = P. arenicoloides W. Ashe
Sy = P. fusiforme A. Hitchcock
Sy = P. neuranthum A. Grisebach var. ramosum A. Grisebach
Sy = P. nitidum J. de Lamarck var. angustifolium (S. Elliott) A. Gray
P. acuminatum O. Swartz var. ***acuminatum***
Sy = Dichanthelium acuminatum (O. Swartz) F. Gould & C. A. Clark
Sy = D. acuminatum var. thurowii (F. Lamson-Scribner & J. G. Smith) F. Gould & C. A. Clark
Sy = D. lanuginosum (S. Elliott) F. Gould
Sy = D. sabulorum (J. de Lamarck) F. Gould & C. A. Clark var. thinium (A. Hitchcock & M. A. Chase) F. Gould & C. A. Clark

Sy = D. villosissimum (G. Nash) R. Freckmann var. praecocius (A. Hitchcock & M. A. Chase) R. Freckmann
Sy = Panicum acuminatum subsp. acuminatum
Sy = P. acuminatum subsp. fasciculatum (J. Torrey) R. Freckmann & M. Lelong
Sy = P. acuminatum var. implicatum (F. Lamson-Scribner) A. Beetle
Sy = P. lanuginosum S. Elliott
Sy = P. ovale S. Elliott subsp. praecocius (A. Hitchcock & M. A. Chase) R. Freckmann & M. Lelong
Sy = P. praecocius A. Hitchcock & M. A. Chase

P. acuminatum O. Swartz var. ***consanguineum*** (K. Kunth) J. Wipff & S. D. Jones
Sy = Dichanthelium consanguineum (K. Kunth) F. Gould & C. A. Clark
Sy = D. ovale (S. Elliott) F. Gould & C. A. Clark
Sy = Panicum commutatum J. A. Schultes var. consanguineum (K. Kunth) W. Beal
Sy = P. consanguineum K. Kunth
Sy = P. ovale S. Elliott
Sy = P. villosum S. Elliott, non J. de Lamarck

P. acuminatum O. Swartz var. ***densiflorum*** (E. L. Rand & J. Redfield) M. Lelong
Sy = Dichanthelium acuminatum var. densiflorum (E. L. Rand & J. Redfield) F. Gould & C. A. Clark
Sy = Panicum nitidum J. de Lamarck var. densiflorum E. L. Rand & J. Redfield
Sy = P. spretum J. A. Schultes subsp. spretum

P. acuminatum var. ***lindheimeri*** (G. Nash) A. Beetle
Sy = Dichanthelium acuminatum (O. Swartz) F. Gould & C. A. Clark var. lindheimeri (G. Nash) F. Gould & C. A. Clark
Sy = D. lanuginosum (S. Elliott) F. Gould var. lindheimeri (G. Nash) R. Freckmann
Sy = D. lindheimeri (G. Nash) F. Gould
Sy = Panicum acuminatum subsp. lindheimeri (G. Nash) R. Freckmann & M. Lelong
Sy = P. lanuginosum S. Elliott var. lindheimeri (G. Nash) M. Fernald
Sy = P. lindheimeri G. Nash

P. acuminatum O. Swartz var. ***longiligulatum*** (G. Nash) M. Lelong
Sy = Dichanthelium acuminatum (O. Swartz) F. Gould & C. A. Clark var. longiligulatum (G. Nash) F. Gould & C. A. Clark
Sy = D. acuminatum var. wrightianum (F. Lamson-Scribner) F. Gould & C. A. Clark
Sy = D. leucothrix (G. Nash) R. Freckmann
Sy = D. longiligulatum (G. Nash) R. Freckmann
Sy = D. wrightianum (F. Lamson-Scribner) R. Freckmann
Sy = Panicum acuminatum var. leucothrix (G. Nash) M. Lelong
Sy = P. leucothrix G. Nash
Sy = P. longiligulatum G. Nash
Sy = P. pilatum J. Swallen
Sy = P. spretum J. A. Schultes subsp. leucothrix (G. Nash) R. Freckmann & M. Lelong
Sy = P. spretum J. A. Schultes subsp. longiligulatum (G. Nash) R. Freckmann & M. Lelong
Sy = P. wrightianum F. Lamson-Scribner

P. acuminatum O. Swartz var. ***villosum*** (A. Gray) A. Beetle
Sy = Dichanthelium acuminatum (O. Swartz) F. Gould & C. A. Clark var. villosum (A. Gray) F. Gould & C. A. Clark
Sy = D. ovale (S. Elliott) F. Gould & C. A. Clark var. addisonii (G. Nash) F. Gould & C. A. Clark
Sy = Panicum nitidum J. de Lamarck var. villosum A. Gray
Sy = P. ovale S. Elliott subsp. pseudopubescens (G. Nash) R. Freckmann & M. Lelong
Sy = P. ovale subsp. villosissimum (G. Nash) R. Freckmann & M. Lelong
Sy = P. ovale var. pseudopubescens (G. Nash) M. Lelong
Sy = P. ovale var. villosum (A. Gray) M. Lelong
Sy = P. pseudopubescens G. Nash
Sy = P. villosissimum G. Nash
Sy = P. villosissimum var. pseudopubescens (G. Nash) M. Fernald

P. amarum S. Elliott var. ***amarulum*** (A. Hitchcock & M. A. Chase) P. Palmer
Sy = P. amarulum A. Hitchcock & M. A. Chase
Sy = P. amarum subsp. amarulum (A. Hitchcock & M. A. Chase) R. Freckmann & M. Lelong

P. amarum S. Elliott var. ***amarum***
Sy = P. amarum subsp. amarum

P. anceps A. Michaux var. ***anceps***
Sy = P. anceps subsp. anceps

P. anceps A. Michaux var. ***rhizomatum*** (A. Hitchcock & M. A. Chase) M. Fernald
Sy = P. anceps subsp. rhizotum (A. Hitchcock & M. A. Chase) R. Freckmann & M. Lelong
Sy = P. rhizomatum A. Hitchcock & M. A. Chase

P. antidotale A. Retzius

P. bergii J. Arechavleta
Sy = P. pilcomayense E. Hackel

P. boscii J. Poiret
Sy = Dichanthelium boscii (J. Poiret) F. Gould & C. A. Clark

P. brachyanthum E. von Steudel
P. bulbosum K. Kunth
Sy = P. bulbosum K. Kunth var. minus G. Vasey
P. capillare C. Linnaeus var. ***capillare***
Sy = P. capillare var. agreste A. Gattinger
P. capillare C. Linnaeus var. ***barbipulvinatum*** (G. Nash) R. McGregor
Sy = P. barbipulvinatum G. Nash
Sy = P. capillare subsp. barbipulvinatum (G. Nash) N. Tzvelev
P. capillare C. Linnaeus var. ***brevifolium*** G. Vasey *ex* P. Rydberg & C. Shear
Sy = P. capillare var. occidentale P. Rydberg
P. capillare C. Linnaeus var. ***sylvaticum*** J. Torrey
Sy = P. philadelphicum J. Bernhardi *ex* C. Nees von Esenbeck
Sy = P. philadelphicum var. tuckermanii (M. Fernald) J. Steyermark & H. M. Schmoll
Sy = P. tuckermanii M. Fernald
P. capillarioides G. Vasey
P. clandestinum C. Linnaeus
Sy = Dichanthelium clandestinum (C. Linnaeus) F. Gould
P. coloratum C. Linnaeus var. ***coloratum***
P. depauperatum G. H. Muhlenberg
Sy = Dichanthelium depauperatum (G. H. Muhlenberg) F. Gould
P. dichotomiflorum A. Michaux subsp. ***dichotomiflorum***
P. dichotomum C. Linnaeus var. ***dichotomum***
Sy = Dichanthelium dichotomum (C. Linnaeus) F. Gould
Sy = Panicum dichotomum subsp. roanokense (W. Ashe) R. Freckmann & M. Lelong
Sy = P. dichotomum subsp. yadkinense (W. Ashe) R. Freckmann & M. Lelong
Sy = Panicum dichotomum var. roanokense (W. Ashe) M. Lelong
Sy = P. roanokense W. Ashe
Sy = P. yadkinense W. Ashe
P. dichotomum C. Linnaeus var. ***lucidum*** (W. Ashe) M. Lelong
Sy = P. dichotomum subsp. lucidum (W. Ashe) R. Freckmann & M. Lelong
Sy = P. lucidum W. Ashe
Sy = P. sphagnicola G. Nash
P. dichotomum C. Linnaeus var. ***nitidum*** (J. de Lamarck) A. Wood
Sy = P. dichotomum subsp. nitidum (J. de Lamarck) R. Freckmann & M. Lelong
Sy = P. nitidum J. de Lamarck
P. dichotomum C. Linnaeus var. ***ramulosum*** (J. Torrey) M. Lelong
Sy = P. dichotomum subsp. microcarpon (G. H. Muhlenberg *ex* S. Elliott) R. Freckmann & M. Lelong
Sy = P. microcarpon G. H. Muhlenberg *ex* S. Elliott
Sy = P. nitidum J. de Lamarck var. barbatum J. Torrey
Sy = P. nitidum var. ramulosum J. Torrey
P. dichotomum C. Linnaeus var. ***unciphyllum*** (K. von Trinius) J. K. Wipff & S. D. Jones
Sy = Dichanthelium dichotomum (C. Linnaeus) F. Gould var. tenue (G. H. Muhlenberg) F. Gould & C. A. Clark
Sy = D. dichotomum var. unciphyllum (K. von Trinius) G. Davidse
Sy = D. ensifolium (W. Baldwin *ex* S. Elliott) F. Gould var. unciphyllum (K. von Trinius) B. F. Hansen & R. Wunderlin
Sy = Panicum dichotomum var. tenue (G. H. Muhlenberg) F. Zuloaga & O. Morrone
Sy = P. tenue G. H. Muhlenberg
Sy = P. unciphyllum K. von Trinius
Sy = P. unciphyllum var. implicatum (F. Lamson-Scribner) F. Lamson-Scribner & E. Merrill
P. divergens K. Kunth
Sy = Dichanthelium commutatum (J. A. Schultes) F. Gould
Sy = Panicum ashei G. Pearson *ex* W. Ashe
Sy = P. commutatum J. A. Schultes
Sy = P. commutatum var. ashei (G. Pearson *ex* W. Ashe) M. Fernald
Sy = P. commutatum var. joorii (G. Vasey) M. Fernald
Sy = P. divergens subsp. ashei (G. Pearson *ex* W. Ashe) R. Freckmann & M. Lelong
Sy = P. divergens subsp. commutatum (J. A. Schultes) R. Freckmann & M. Lelong
Sy = P. divergens subsp. equilaterale (F. Lamson-Scribner) R. Freckmann & M. Lelong
Sy = P. equilaterale F. Lamson-Scribner
Sy = P. joorii G. Vasey
P. ensifolium W. Baldwin *ex* S. Elliott var. ***ensifolium***
Sy = Dichanthelium dichotomum (C. Linnaeus) F. Gould var. ensifolium (S. Elliott) F. Gould
Sy = D. ensifolium (W. Baldwin *ex* S. Elliott) F. Gould & C. A. Clark
Sy = Panicum chamaelonche C. Trinius
Sy = P. curtifolium G. Nash
Sy = P. ensifolium subsp. curtifolium (G. Nash) R. Freckmann & M. Lelong
Sy = P. ensifolium var. curtifolium (G. Nash) M. Lelong

P. flexile (A. Gattinger) F. Lamson-Scribner
P. ghiesbreghtii E. Fournier
P. gymnocarpon S. Elliott
Sy = Phanopyrum gymnocarpon (S. Elliott) G. Nash
P. hallii G. Vasey var. ***filipes*** (F. Lamson-Scribner) F. Waller
Sy = P. diffusum O. Swartz
Sy = P. filipes F. Lamson-Scribner
Sy = P. hallii subsp. filipes (F. Lamson-Scribner) R. Freckmann & M. Lelong [**ined.**]
P. hallii G. Vasey var. ***hallii***
Sy = P. hallii subsp. hallii [**ined.**]
P. havardii G. Vasey
P. hemitomon J. A. Schultes
P. hians S. Elliott
Sy = Steinchisma hians (S. Elliott) G. Nash
P. hillmanii M. A. Chase
Sy = P. capillare C. Linnaeus subsp. hillmanii (M. A. Chase) R. Freckmann & M. Lelong [**ined.**]
P. hirsutum O. Swartz
P. hirticaule J. Presl var. ***hirticaule***
Sy = P. capillare C. Linnaeus var. hirticaule (J. Presl) F. Gould
Sy = P. hirticaule subsp. hirticaule [**ined.**]
Sy = P. pampinosum A. Hitchcock & M. A. Chase
P. laxiflorum J. de Lamarck
Sy = Dichanthelium laxiflorum J. de Lamarck
Sy = Panicum xalapense K. Kunth
P. linearifolium F. Lamson-Scribner *ex* G. Nash
Sy = Dichanthelium linearifolium (F. Lamson-Scribner *ex* G. Nash) F. Gould
Sy = Panicum perlongum G. Nash
Sy = P. werneri F. Lamson-Scribner
P. malacophyllum G. Nash
Sy = Dichanthelium malacophyllum (G. Nash) F. Gould
P. maximum N. von Jacquin
P. miliaceum C. Linnaeus var. ***miliaceum***
P. nodatum A. Hitchcock & M. A. Chase
Sy = Dichanthelium nodatum (A. Hitchcock & M. A. Chase) F. Gould
P. obtusum K. Kunth
P. oligosanthes J. A. Schultes var. ***oligosanthes***
Sy = Dichanthelium oligosanthes (J. A. Schultes) F. Gould
Sy = Panicum oligosanthes subsp. oligosanthes
P. oligosanthes J. A. Schultes var. ***scribnerianum*** (G. Nash) M. Fernald
Sy = Dichanthelium oligosanthes (J. A. Schultes) F. Gould var. scribnerianum (G. Nash) F. Gould
Sy = Panicum oligosanthes subsp. scribnerianum (G. Nash) R. Freckmann & M. Lelong
Sy = P. scribnerianum G. Nash
P. pedicellatum G. Vasey
Sy = Dichanthelium pedicellatum (G. Vasey) F. Gould
P. plenum A. Hitchcock
P. portoricense N. Desvaux *ex* W. Hamilton
Sy = Dichanthelium portoricense (N. Desvaux *ex* W. Hamilton) B. F. Hansen & R. Wunderlin
Sy = D. sabulorum (J. de Lamarck) F. Gould & C. A. Clark var. patulum (F. Lamson-Scribner & E. Merrill) F. Gould & C. A. Clark
Sy = Panicum acuminatum O. Swartz var. columbianum (F. Lamson-Scribner) M. Lelong
Sy = P. columbianum F. Lamson-Scribner
Sy = P. nashianum F. Lamson-Scribner
Sy = P. portoricense subsp. patulum (F. Lamson-Scribner & E. Merrill) R. Freckmann & M. Lelong
Sy = P. portoricense N. Desvaux *ex* W. Hamilton var. nashianum (F. Lamson-Scribner) M. Lelong
Sy = P. webberianum G. Nash
P. ravenelii F. Lamson-Scribner & E. Merrill
Sy = Dichanthelium ravenelii (F. Lamson-Scribner & E. Merrill) F. Gould
P. repens C. Linnaeus
Sy = P. gouini E. Fournier
P. rigidulum C. Nees von Esenbeck var. ***combsii*** (F. Lamson-Scribner & C. Ball) M. Lelong
Sy = P. combsii F. Lamson-Scribner & C. Ball
Sy = P. longifolium J. Torrey var. combsii (F. Lamson-Scribner & C. Ball) M. Fernald
Sy = P. rigidulum subsp. combsii (F. Lamson-Scribner & C. Ball) R. Freckmann & M. Lelong
P. rigidulum C. Nees von Esenbeck var. ***elongatum*** (F. Lamson-Scribner) M. Lelong
Sy = P. agrostoides K. Sprengel var. elongatum F. Lamson-Scribner
Sy = P. elongatum F. Pursh, non R. A. Salisbury
Sy = P. rigidulum subsp. elongatum (F. Lamson-Scribner) R. Freckmann & M. Lelong
Sy = P. stipitatum G. Nash
P. rigidulum C. Nees von Esenbeck var. ***pubescens*** (G. Vasey) M. Lelong
Sy = P. anceps A. Michaux var. pubescens G. Vasey
Sy = P. longifolium J. Torrey
Sy = P. longifolium var. pubescens (G. Vasey) M. Fernald

Sy = P. rigidulum subsp. pubescens (G. Vasey) R. Freckmann & M. Lelong

P. rigidulum C. Nees von Esenbeck var. ***rigidulum***

Sy = P. agrostoides K. Sprengel

Sy = P. agrostoides var. ramosisus (C. Mohr) M. Fernald

Sy = P. condensum G. Nash

Sy = P. rigidulum subsp. rigidulum

P. scabriusculum S. Elliott

Sy = Dichanthelium scabriusculum (S. Elliott) F. Gould & C. A. Clark

P. scoparium J. de Lamarck

Sy = Dichanthelium scoparium (J. de Lamarck) F. Gould

P. sphaerocarpon S. Elliott var. ***isophyllum*** (F. Lamson-Scribner) R. Angelo

Sy = Dichanthelium sphaerocarpon (S. Elliott) F. Gould var. isophyllum (F. Lamson-Scribner) F. Gould & C. A. Clark

Sy = D. sphaerocarpon (S. Elliott) F. Gould var. polyanthes (J. A. Schultes) F. Gould

Sy = Panicum microcarpon G. H. Muhlenberg var. isophyllum F. Lamson-Scribner

Sy = P. polyanthes J. A. Schultes

Sy = P. sphaerocarpon var. polyanthes (J. A. Schultes) A. S. Sherif

P. sphaerocarpon S. Elliott var. ***sphaerocarpon***

Sy = Dichanthelium sphaerocarpon (S. Elliott) F. Gould

P. strigosum G. H. Muhlenberg *ex* S. Elliott var. ***glabrescens*** (A. Grisebach) M. Lelong

Sy = Dichanthelium leucoblepharis (K. von Trinius) F. Gould & C. A. Clark var. glabrescens (A. Grisebach) F. Gould & C. A. Clark

Sy = D. strigosum (G. H. Muhlenberg *ex* S. Elliott) R. Freckmann var. glabrescens (A. Grisebach) R. Freckmann

Sy = Panicum dichotomum C. Linnaeus var. glabrescens A. Grisebach

Sy = P. strigosum subsp. glabrescens (A. Grisebach) R. Freckmann & M. Lelong

P. strigosum G. H. Muhlenberg *ex* S. Elliott var. ***leucoblepharis*** (K. von Trinius) M. Lelong

Sy = Dichanthelium leucoblepharis (K. von Trinius) F. Gould & C. A. Clark

Sy = D. strigosum (G. H. Muhlenberg *ex* S. Elliott) R. Freckmann var. leucoblepharis (K. von Trinius) R. Freckmann

Sy = Panicum ciliatifolium K. Kunth

Sy = P. ciliatum, non G. Maerklin

Sy = P. leucoblepharis K. von Trinius

Sy = P. strigosum subsp. leucoblepharis (K. von Trinius) R. Freckmann & M. Lelong

P. strigosum G. H. Muhlenberg *ex* S. Elliott var. ***strigosum***

Sy = Dichanthelium leucoblepharis (K. von Trinius) F. Gould & C. A. Clark var. pubescens (G. Vasey) F. Gould & C. A. Clark

Sy = D. strigosum (G. H. Muhlenberg *ex* S. Elliott) R. Freckmann

Sy = Panicum ciliatum S. Elliott var. pubescens (G. Vasey) R. Freckmann

Sy = P. laxiflorum J. de Lamarck var. pubescens G. Vasey

Sy = P. strigosum subsp. strigosum

P. tenerum H. Beyrich *ex* K. von Trinius

P. trichoides O. Swartz

P. verrucosum G. H. Muhlenberg

P. virgatum C. Linnaeus

Sy = P. virgatum C. Linnaeus var. cubense A. Grisebach

PAPPOPHORUM J. von Schreber

P. bicolor E. Fournier

P. vaginatum S. Buckley

PARAPHOLIS C. Hubbard

P. incurva (C. Linnaeus) C. Hubbard

Sy = Aegilops incurva C. Linnaeus

Sy = Pholiurus incurvus (C. Linnaeus) H. Schinz & A. Thellung

PASCOPYRUM Á. Löve

P. smithii (P. Rydberg) Á. Löve

Sy = Agropyron smithii P. Rydberg

Sy = Elymus smithii (P. Rydberg) F. Gould

Sy = Elytrigia smithii (P. Rydberg) S. Nevski

PASPALIDIUM O. Stapf

P. geminatum (P. Forsskål) O. Stapf var. ***geminatum***

Sy = Panicum geminatum P. Forsskål

Sy = Setaria geminata (P. Forsskål) J. Veldkamp

P. geminatum (P. Forsskål) O. Stapf var. ***paludivagum*** (A. Hitchcock & M. A. Chase) F. Gould

Sy = Panicum paludivagum A. Hitchcock & M. A. Chase

Sy = Paspalidium paludivagum (A. Hitchcock & M. A. Chase) L. Parodi

Sy = Setaria geminata (P. Forsskål) J. Veldkamp var. paludivaga (A. Hitchcock & M. A. Chase) R. D. Webster

PASPALUM C. Linnaeus
*P. **acuminatum*** G. Raddi
*P. **almum*** M. A. Chase
*P. **bifidum*** (A. Bertoloni) G. Nash
Sy = P. bifidum var. projectum M. Fernald
*P. **boscianum*** J. Flüggé
*P. **conjugatum*** P. J. Bergius
*P. **convexum*** F. von Humboldt & A. Bonpland *ex* J. Flüggé
*P. **dilatatum*** J. Poiret
*P. **dissectum*** (C. Linnaeus) C. Linnaeus
Sy = Panicum dissectum C. Linnaeus
*P. **distichum*** C. Linnaeus var. ***distichum***
Sy = P. paspalodes (A. Michaux) F. Lamson-Scribner
*P. **distichum*** C. Linnaeus var. ***indutum*** L. Shinners
*P. **floridanum*** A. Michaux var. ***floridanum***
*P. **floridanum*** A. Michaux var. ***glabratum*** G. Engelmann *ex* G. Vasey
*P. **hartwegianum*** E. Fournier
*P. **hydrophilum*** J. Henrard
Sy = P. texanum J. Swallen
*P. **laeve*** A. Michaux var. ***circulare*** (G. Nash) M. Fernald
Sy = P. circulare G. Nash
*P. **laeve*** A. Michaux var. ***laeve***
*P. **laeve*** A. Michaux var. ***pilosum*** F. Lamson-Scribner
Sy = P. longipilum G. Nash
*P. **langei*** (E. Fournier) G. Nash
Sy = Dimorphostachys langei E. Fournier
*P. **lividum*** K. von Trinius
*P. **malacophyllum*** K. von Trinius
*P. **minus*** E. Fournier
*P. **monostachyum*** G. Vasey
*P. **notatum*** J. Flüggé var. ***latiflorum*** J. Döll
*P. **notatum*** J. Flüggé var. ***saurae*** L. Parodi
*P. **plicatulum*** A. Michaux var. ***plicatulum***
*P. **praecox*** T. Walter
Sy = P. lentiferum J. de Lamarck
Sy = P. praecox var. curtissianum (E. von Steudel) G. Vasey
*P. **pubiflorum*** F. Ruprecht *ex* E. Fournier var. ***glabrum*** G. Vasey *ex* F. Lamson-Scribner
*P. **pubiflorum*** F. Ruprecht *ex* E. Fournier var. ***pubiflorum***
*P. **repens*** P. J. Bergius var. ***fluitans*** (S. Elliott) J. Wipff & S. D. Jones
Sy = Ceresia fluitans S. Elliott
Sy = Paspalum fluitans (S. Elliott) K. Kunth
*P. **scrobiculatum*** C. Linnaeus
*P. **separatum*** L. Shinners
*P. **setaceum*** A. Michaux var. ***ciliatifolium*** (A. Michaux) G. Vasey
*P. **setaceum*** A. Michaux var. ***muhlenbergii*** (G. Nash) D. Banks
*P. **setaceum*** A. Michaux var. ***setaceum***
Sy = P. debile A. Michaux
*P. **setaceum*** A. Michaux var. ***stramineum*** (G. Nash) D. Banks
Sy = P. stramineum G. Nash
*P. **unispicatum*** (F. Lamson-Scribner & E. Merrill) G. Nash
Sy = Panicum unispicatum F. Lamson-Scribner & E. Merrill
*P. **urvillei*** E. von Steudel
*P. **vaginatum*** O. Swartz
*P. **virgatum*** C. Linnaeus

PENNISETUM L. C. Richard
*P. **alopecuroides*** (C. Linnaeus) K. Sprengel **[cultivated]**
Sy = Panicum alopecuroides C. Linnaeus
Sy = Pennisetum alopecuroides var. purpurascens (C. Thunberg) J. Ohwi
Sy = P. alopecuroides var. viridescens (F. Miquel) J. Ohwi
*P. **ciliare*** (C. Linnaeus) J. Link var. ***ciliare***
Sy = Cenchrus ciliaris C. Linnaeus
*P. **ciliare*** (C. Linnaeus) J. Link var. ***setigerum*** (M. A. Vahl) G. Leeke **[cultivated]**
Sy = Cenchrus setigerus M. A. Vahl
*P. **flaccidum*** A. Grisebach **[cultivated]**
Ma = P. incomptum *auct.*, non C. Nees von Esenbeck *ex* E. von Steudel: Greenlee (1992) and Darke (1994)
*P. **glaucum*** (C. Linnaeus) R. Brown **[cultivated]**
Sy = Panicum americanum C. Linnaeus
Sy = P. glaucum C. Linnaeus
Sy = P. lutescens C. von Weigel
Sy = Pennisetum americanum (C. Linnaeus) G. Leeke
Sy = P. typhoides (J. Burmann) O. Stapf & C. Hubbard
Sy = Setaria glauca (C. Linnaeus) A. Palisot de Beauvois
Sy = S. lutescens (C. von Weigel) F. Hubbard
*P. **macrostachyum*** (A. T. de Brongniart) K. von Trinius **[cultivated]**
Sy = Gymnothrix macrostachys A. T. de Brongniart
Hn = Pennisetum cv. "Burgundy Giant"

P. nervosum (C. Nees von Esenbeck) K. von Trinius
Sy = Cenchrus nervosus (C. Nees von Esenbeck) K. E. O. Kuntze
P. orientale C. von Willdenow *ex* L. C. Richard **[cultivated]**
P. purpureum C. Schumacher
P. setaceum (P. Forsskål) E. Chiovenda **[cultivated]**
Sy = P. ruppelii E. von Steudel
Sy = Phalaris setacea P. Forsskål
P. sp. **[cultivated]**
Ma = P. macrostachyum *auct.*, non (A. T. de Brongniart) K. von Trinius: recent American authors
Hn = P. cv. "Eaton Canyon"
Hn = P. setaceum cv. "Rubrum"
Hn = P. setaceum cv. "Rubrum Dwarf"
P. villosum R. Brown *ex* J. Fresenius **[cultivated]**
Sy = Cenchrus longisetus M. A. Johnston

PHALARIS C. Linnaeus
P. angusta C. Nees von Esenbeck *ex* K. von Trinius
P. aquatica C. Linnaeus
Sy = P. stenoptera E. Hackel
Sy = P. tuberosa C. Linnaeus
Sy = P. tuberosa var. hirtiglumis G. Battarra & L. Trabut
Sy = P. tuberosa var. stenoptera (E. Hackel) A. Hitchcock
P. arundinacea C. Linnaeus **[cultivated]**
Sy = P. arundinacea var. picta C. Linnaeus
P. brachystachys J. Link
P. canariensis C. Linnaeus
P. caroliniana T. Walter
P. minor A. Retzius

PHLEUM C. Linnaeus
P. pratense C. Linnaeus

PHRAGMITES M. Adanson
P. australis (A. Cavanilles) K. von Trinius *ex* E. von Steudel subsp. australis
Sy = Arundo phragmites C. Linnaeus
Sy = Phragmites communis K. von Trinius
P. karka (A. Retzius) E. von Steudel
Sy = P. vallatoria (L. Plukenet *ex* C. Linnaeus) J. Veldkamp
Sy = Arundo karka A. Retzius
Sy = A. vallatoria L. Plukenet *ex* C. Linnaeus
Sy = Phragmites australis (A. Cavanilles) K. von Trinius *ex* E. von Steudel var. berlandieri (E. Fournier) C. F. Reed
Sy = P. berlandieri E. Fournier
Sy = P. communis K. von Trinius var. berlandieri (E. Fournier) M. Fernald

PHYLLOSTACHYS P. von Siebold & J. Zuccarini, *nomen conservandum*
P. angusta F. McClure **[cultivated]**
P. arcana F. McClure **[cultivated]**
P. aurea É. Carrière *ex* A. Rivière & C. Rivière
P. aureosulcata F. McClure **[cultivated]**
P. bambusoides P. von Siebold & J. Zuccarini **[cultivated]**
P. bissetii F. McClure **[cultivated]**
P. congesta A. Rendle **[cultivated]**
P. decora F. McClure **[cultivated]**
P. dulcis F. McClure **[cultivated]**
P. edulis (É. Carrière) J. Houzeau de Lehaie **[cultivated]**
Sy = Bambusa heterocycla É. Carrière
Sy = Phyllostachys heterocycla (É. Carrière) A. Freeman-Mitford
Sy = P. pubescens E. Mazel *ex* J. Houzeau de Lehaie
Ma = P. mitis *auct.*, non A. Rivière & C. Rivière: 19th-century Japanese authors
P. flexuosa A. Rivière & C. Rivière **[cultivated]**
P. glauca F. McClure **[cultivated]**
P. heteroclada D. Oliver **[cultivated]**
Sy = P. cerata F. McClure
Sy = P. congesta A. Rendle
Sy = P. purpurata F. McClure
P. makinoi B. Hayata **[cultivated]**
P. meyeri F. McClure **[cultivated]**
P. nidularia W. Munro **[cultivated]**
P. nigra (C. Loddiges) W. Munro var. ***henonis*** (A. Freeman-Mitford) O. Stapf *ex* A. Rendle **[cultivated]**
Sy = P. henonis A. Freeman-Mitford
P. nigra (C. Loddiges) W. Munro var. ***nigra*** **[cultivated]**
P. nuda F. McClure **[cultivated]**
P. purpurata F. McClure **[cultivated]**
P. rubromarginata F. McClure **[cultivated]**
P. sulphurea (É. Carrière) A. Rivière & C. Rivière var. ***sulphurea*** **[cultivated]**
Sy = Bambusa sulphurea É. Carrière
P. sulphurea (É. Carrière) A. Rivière & C. Rivière var. ***viridis*** R. A. Young **[cultivated]**
Sy = P. mitis A. Rivière & C. Rivière
P. viridi-glaucescens (É. Carrière) A. Rivière & C. Rivière **[cultivated]**
Sy = Bambusa viridi-glaucescens É. Carrière

P. viridis (R. A. Young) F. McClure [**cultivated**]
P. vivax F. McClure [**cultivated**]

PIPTATHERUM A. Palisot de Beauvois
P. micranthum (K. von Trinius & F. Ruprecht) M. Barkworth
Sy = Oryzopsis micrantha (K. von Trinius & F. Ruprecht) G. Thurber
Sy = Urachne micrantha K. von Trinius & F. Ruprecht

PIPTOCHAETIUM J. Presl, *nomen conservandum*
P. avenaceum (C. Linnaeus) L. Parodi
Sy = Stipa avenacea C. Linnaeus
P. fimbriatum (K. Kunth) A. Hitchcock
Sy = P. fimbriatum var. confine I. M. Johnston
Sy = Stipa fimbriata K. Kunth
P. pringlei (F. Lamson-Scribner) L. Parodi
Sy = Stipa pringlei F. Lamson-Scribner

PLEURAPHIS J. Torrey
P. jamesii J. Torrey
Sy = Hilaria jamesii (J. Torrey) G. Bentham
P. mutica S. Buckley
Sy = Hilaria mutica (S. Buckley) G. Bentham

POA C. Linnaeus
P. annua C. Linnaeus
Sy = P. annua var. aquatica P. Ascherson
Sy = P. annua var. reptans H. Haussknecht
P. arachnifera J. Torrey
P. arida G. Vasey
P. autumnalis G. H. Muhlenberg *ex* S. Elliott
P. bigelovii G. Vasey & F. Lamson-Scribner
P. bulbosa C. Linnaeus
P. chapmaniana F. Lamson-Scribner
P. compressa C. Linnaeus
P. fendleriana (E. von Steudel) G. Vasey subsp. *fendleriana*
P. interior P. Rydberg
Sy = P. nemoralis C. Linnaeus subsp. interior (P. Rydberg) W. A. Weber
Sy = P. nemoralis var. interior (P. Rydberg) F. Butters & E. Abbe
P. occidentalis G. Vasey
P. pratensis C. Linnaeus
P. reflexa G. Vasey & F. Lamson-Scribner.
NOTE: We have been unable to obtain and examine voucher specimens (Burgess and Northington 1981), housed in the herbarium at Texas Tech University, to verifiy specimens identified as *P. reflexa*. However, these specimens are likely *P. leptocoma* K. von Trinius, a similar species which occurs in the mountains of southern New Mexico, whereas the southernmost distribution of *P. reflexa* is in the mountains of northern New Mexico at 3050–3660 m (Soreng 1985).
P. strictiramea A. Hitchcock
Sy = P. involuta A. Hitchcock
P. sylvestris A. Gray
P. trivialis C. Linnaeus

POLYPOGON R. Desfontaines
P. elongatus K. Kunth
P. monspeliensis (C. Linnaeus) R. Desfontaines
Sy = Alopecurus monospeliensis C. Linnaeus
P. viridis (A. Gouan) M. Breistroffer
Sy = Agrostis semiverticillata (P. Forsskål) C. Christenson
Sy = A. verticillata D. Villars
Sy = Polypogon semiverticillatus (P. Forsskål) N. Hylander

PSATHYROSTACHYS S. Nevski
P. juncea (F. von Fischer) S. Nevski
Sy = Elymus junceus F. von Fischer

PSEUDOROEGNERIA (S. Nevski) Á. Löve
P. arizonica (F. Lamson-Scribner & J. G. Smith) Á. Löve
Sy = Agropyron arizonicum F. Lamson-Scribner & J. G. Smith
Sy = A. spicatum (F. Pursh) F. Lamson-Scribner & J. G. Smith var. arizonicum (F. Lamson-Scribner & J. G. Smith) M. E. Jones
Sy = Elymus arizonicus (F. Lamson-Scribner & J. G. Smith) F. Gould
Sy = Elytrigia arizonica (F. Lamson-Scribner & J. G. Smith) D. Dewey
P. spicata (F. Pursh) Á. Löve
Sy = Agropyron inerme F. Lamson-Scribner & J. G. Smith
Sy = A. spicatum F. Pursh
Sy = Pseudoroegneria spicata subsp. inermis (F. Lamson-Scribner & J. G. Smith) Á. Löve

PSEUDOSASA T. Makino *ex* T. Nakai
P. japonica (P. von Siebold & J. Zuccarini *ex* E. von Steudel) T. Makino *ex* T. Nakai var. *japonica* [**cultivated**]

Sy = Arundinaria japonica P. von Siebold & J. Zuccarini *ex* E. von Steudel
Sy = Sasa japonica (P. von Siebold & J. Zuccarini *ex* E. von Steudel) T. Makino
P. japonica (P. von Siebold & J. Zuccarini *ex* E. von Steudel) T. Makino *ex* T. Nakai var. ***tsutsumiana*** K. Yanagita [**cultivated**]

REDFIELDIA G. Vasey
R. flexuosa (G. Thurber) G. Vasey

ROSTRARIA K. von Trinius
R. cristata (C. Linnaeus) N. Tzvelev
Sy = Festuca cristata C. Linnaeus
Sy = Koeleria gerardii (D. Villars) L. Shinners
Sy = K. phleoides (D. Villars) C. Persoon
Sy = Lophochloa cristata (C. Linnaeus) N. Hylander

ROTTBOELLIA C. von Linné, *nomen conservandum*
R. cochinchinensis (J. de Loureiro) W. Clayton [**federal noxious weed**]
Sy = Aegilops exaltata C. Linnaeus
Sy = Manisuris exaltata (C. Linnaeus) K. E. O. Kuntze
Sy = Rottboellia exaltata (C. Linnaeus) C. von Linné

SACCHARUM A. Michaux
S. alopecuroideum (C. Linnaeus) T. Nuttall
Sy = Andropogon alopecuroides C. Linnaeus
Sy = A. divaricatus C. Linnaeus
Sy = Erianthus alopecuroides (C. Linnaeus) S. Elliott
Sy = E. alopecuroides var. hirsutis G. Nash
Sy = E. divaricatus (C. Linnaeus) A. Hitchcock
Sy = E. tracyi G. Nash
S. baldwinii K. Sprengel
Sy = Andropogon dura (K. von Trinius) E. von Steudel
Sy = Erianthus strictus W. Baldwin
Sy = Pollinia dura K. von Trinius
Sy = Saccharum strictum (W. Baldwin) T. Nuttall
S. brevibarbe (A. Michaux) C. Persoon var. ***brevibarbe***
Sy = Erianthus alopecuroides (C. Linnaeus) S. Elliott var. brevibarbis (A. Michaux) A. Chapman
Sy = E. brevibarbis A. Michaux
Sy = E. saccharoides A. Michaux subsp. brevibarbis (A. Michaux) E. Hackel
S. brevibarbe (A. Michaux) C. Persoon var. ***contortum*** (T. Nuttall) R. Webster
Sy = Calamagrostis rubra L. Bosc ex K. Kunth
Sy = Erianthus alopecuroides (C. Linnaeus) S. Elliott var. contortus (S. Elliott) A. Chapman
Sy = E. contortus W. Baldwin *ex* S. Elliott
Sy = E. saccharoides A. Michaux subsp. contortus (S. Elliott) E. Hackel
Sy = E. smallii G. Nash
Sy = Saccharum contortum (W. Baldwin *ex* S. Elliott) T. Nuttall
S. coarctatum (M. Fernald) R. Webster
Sy = Erianthus coarctatus M. Fernald
Sy = E. coarctatus var. elliotianus M. Fernald
S. giganteum (T. Walter) C. Persoon
Sy = Anthoxanthum giganteum T. Walter
Sy = Erianthus compactus G. Nash
Sy = E. giganteus (T. Walter) F. Hubbard
Sy = E. giganteus var. compactus (G. Nash) M. Fernald
Sy = E. laxus G. Nash
Sy = E. saccharoides A. Michaux var. compactus (G. Nash) M. Fernald
Sy = E. saccharoides var. michauxii E. Hackel
S. officinarum C. Linnaeus [**cultivated**]
S. ravennae (C. Linnaeus) J. Murray [**cultivated**]
Sy = Andropogon ravennae C. Linnaeus
Sy = Erianthus ravennae (C. Linnaeus) A. Palisot de Beauvois

SACCIOLEPIS G. Nash
S. indica (C. Linnaeus) M. A. Chase
Sy = Aira indica C. Linnaeus
S. striata (C. Linnaeus) G. Nash
Sy = Holcus striatus C. Linnaeus

SASA T. Makino & K. Shibata
S. masumuneana (T. Makino) C. S. Chao & S. Renvoize [**cultivated**]
Sy = Pleioblastus masumuneana T. Makino
Sy = Sasaella masumuneana (T. Makino) Hatsusima and H. Muroi
S. palmata (F. Burbidge) E. Camus [**cultivated**]
Sy = Bambusa palmata F. Burbidge
S. ramosa (T. Makino) T. Makino & K. Shibata [**cultivated**]
Sy = Arundinaria ramosa T. Makino
Sy = Sasaella ramosa (T. Makino) T. Makino
S. veitchii (É. Carrière) A. Rehder [**cultivated**]
Sy = Bambusa veitchii É. Carrière

SCHEDONNARDUS E. von Steudel
S. paniculatus (T. Nuttall) W. Trelease

SCHISMUS A. Palisot de Beauvois
S. barbatus (P. Löfling *ex* C. Linnaeus) A. Thellung
Sy = Festuca barbata P. Löfling *ex* C. Linnaeus

SCHIZACHYRIUM C. Nees von Esenbeck
S. cirratum (E. Hackel) E. Wooton & P. Standley
Sy = Andropogon cirratus E. Hackel
S. sanguineum (A. Retzius) A. Alston
Sy = Andropogon hirtiflorus (C. Nees von Esenbeck) K. Kunth
Sy = Schizachyrium hirtiflorum C. Nees von Esenbeck
Sy = S. sanguineum var. hirtiflorum (C. Nees von Esenbeck) S. Hatch
S. scoparium (A. Michaux) G. Nash var. ***divergens*** (E. Hackel) F. Gould
Sy = Andropogon divergens (E. Hackel) N. Andersson *ex* A. Hitchcock
Sy = A. scoparius A. Michaux var. divergens E. Hackel
Sy = A. scoparius var. virilis L. Shinners
Sy = Schizachyrium scoparium subsp. divergens (E. Hackel) K. Gandhi & F. Smeins
Sy = S. scoparium var. virile (L. Shinners) F. Gould
S. scoparium (A. Michaux) G. Nash var. ***littorale*** (G. Nash) F. Gould
Sy = Andropogon littoralis (G. Nash) A. Hitchcock
Sy = A. scoparius A. Michaux var. littoralis (G. Nash) A. Hitchcock
Sy = Schizachyrium littorale (G. Nash) E. Bicknell
Sy = S. scoparium subsp. littorale (G. Nash) K. Gandhi & F. Smeins
S. scoparium (A. Michaux) G. Nash var. ***neomexicanum*** (G. Nash) F. Gould
Sy = Andropogon neomexicanus (G. Nash) A. Hitchcock
Sy = A. scoparius A. Michaux var. neomexicanus (G. Nash) A. Hitchcock
Sy = Schizachyrium scoparium subsp. neomexicanum (G. Nash) K. Gandhi & F. Smeins
S. scoparium (A. Michaux) G. Nash var. ***scoparium***
Sy = Andropogon scoparius A. Michaux
Sy = A. scoparius var. frequens F. Hubbard
Sy = Schizachyrium scoparium var. frequens (F. Hubbard) F. Gould
Sy = S. scoparium var. polycladum (F. Lamson-Scribner & C. Ball) C. F. Reed
S. tenerum C. Nees von Esenbeck
Sy = Andropogon tener (C. Nees von Esenbeck) K. Kunth

SCLEROCHLOA A. Palisot de Beauvois
S. dura (C. Linnaeus) A. Palisot de Beauvois
Sy = Cynosurus durus C. Linnaeus

SCLEROPOGON R. Philippi
S. brevifolius R. Philippi
Sy = S. longisetus A. Beetle

SECALE C. Linnaeus
S. cereale C. Linnaeus [**cultivated**]
Sy = Triticum cereale (C. Linnaeus) R. A. Salisbury

SEMIARUNDINARIA T. Makino *ex* T. Nakai
S. fastuosa (M. Latour-Marliac *ex* A. Freeman-Mitford) T. Makino [**cultivated**]
Sy = Bambusa fastuosa M. Latour-Marliac *ex* A. Freeman-Mitford

SETARIA A. Palisot de Beauvois, *nomen conservandum*
S. corrugata (S. Elliott) J. A. Schultes
Sy = Chaetochloa corrugata (S. Elliott) F. Lamson-Scribner
Sy = Panicum corrugatum S. Elliott
S. firmula (A. Hitchcock & M. A. Chase) R. Pilger
Sy = Panicum firmulum A. Hitchcock & M. A. Chase
S. grisebachii E. Fournier
S. italica (C. Linnaeus) A. Palisot de Beauvois
Sy = Chaetochloa italica (C. Linnaeus) F. Lamson-Scribner
Sy = Panicum italicum C. Linnaeus
Sy = Setaria italica var. metzeri (F. Körnicke) S. Jávorke
Sy = S. italica var. stramineofructa (F. Hubbard) L. Bailey
S. leucopila (F. Lamson-Scribner & E. Merrill) K. Schumann
Sy = Chaetochloa leucopila F. Lamson-Scribner & E. Merrill
S. macrostachya K. Kunth
S. magna A. Grisebach
Sy = Chaetochloa magna (A. Grisebach) F. Lamson-Scribner
S. paniculiferia (E. von Steudel) E. Fournier *ex* W. Hemsley [**cultivated**]

S. parviflora (J. Poiret) M. Kerguélen
Sy = Cenchrus parviflorus J. Poiret
Sy = Chaetochloa geniculata (J. de Lamarck) C. Millspaugh & M. A. Chase
Sy = Panicum geniculatum J. de Lamarck
Sy = P. geniculatum C. von Willdenow, non J. de Lamarck
Sy = P. imberbe J. Poiret
Sy = Setaria geniculata A. Palisot de Beauvois, non J. de Lamarck
Sy = S. gracilis K. Kunth

S. poiretiana (J. A. Schultes) K. Kunth [**cultivated**]
Sy = Chaetochloa poiretiana (J. A. Schultes) A. Hitchcock
Sy = Panicum poiretianum J. A. Schultes

S. pumila (J. Poiret) J. J. Römer & J. A. Schultes
Sy = Panicum pumilum J. Poiret
Sy = Setaria pallide-fusca (C. Schumacher) O. Stapf & C. Hubbard [**federal noxious weed**]
Sy = S. pumila (J. Poiret) J. J. Römer & J. A. Schultes subsp. pallide-fusca (C. Schumacher) B. Simon
Ma = S. glauca *auct.*, non (C. Linnaeus) A. Palisot de Beauvois: Correll and Johnston (1970); Gould (1975b)
Ma = S. lutescens *auct.*, non (C. von Weigel) F. Hubbard: Silveus (1933); Chase (1951)

S. ramiseta (F. Lamson-Scribner) R. Pilger
Sy = Panicum ramisetum F. Lamson-Scribner

S. reverchonii (G. Vasey) R. Pilger
Sy = Panicum reverchoni G. Vasey (orthographic error)

S. scheelei (E. von Steudel) A. Hitchcock
Sy = Panicum scheelei E. von Steudel

S. texana W. Emery

S. verticillata (C. Linnaeus) A. Palisot de Beauvois var. ***respiciens*** (A. Richard) A. Braun
Sy = Panicum adhaerans P. Forsskål
Sy = Pennisetum respiciens A. Richard
Sy = Setaria adhaerans (P. Forsskål) E. Chiovenda
Sy = S. respiciens (A. Richard) W. Walpers

S. verticillata (C. Linnaeus) A. Palisot de Beauvois var. ***verticillata***
Sy = Chaetochloa verticillata (C. Linnaeus) F. Lamson-Scribner
Sy = Panicum verticillatum C. Linnaeus

S. villosissima (F. Lamson-Scribner & E. Merrill) K. Schumann
Sy = Chaetochloa villosissima F. Lamson-Scribner & E. Merrill

S. viridis (C. Linnaeus) A. Palisot de Beauvois var. ***major*** (J. Gaudin) E. Pospichal
Sy = Panicum viride C. Linnaeus var. major J. Gaudin

S. viridis (C. Linnaeus) A. Palisot de Beauvois var. ***viridis***
Sy = Chaetochloa viridis (C. Linnaeus) F. Lamson-Scribner
Sy = Panicum viride C. Linnaeus
Sy = Setaria viridis var. breviseta (J. Döll) A. Hitchcock
Sy = S. viridis var. weinmannii (J. J. Römer & J. A. Schultes) V. von Borbás

SHIBATAEA T. Nakai

S. kumasaca (H. Zollinger *ex* E. von Steudel) T. Makino *ex* T. Nakai [**cultivated**]
Sy = Bambusa kumasaca H. Zollinger *ex* E. von Steudel

SORGHASTRUM G. Nash

S. elliottii (C. Mohr) G. Nash
Sy = Chrysopogon elliottii C. Mohr

S. nutans (C. Linnaeus) G. Nash
Sy = Andropogon nutans C. Linnaeus
Sy = Sorghastrum avenaceum (A. Michaux) G. Nash

SORGHUM C. Moench, *nomen conservandum*

S. bicolor (C. Linnaeus) C. Moench [**cultivated**]
Sy = Holcus bicolor C. Linnaeus
Sy = Sorghum vulgare C. Persoon

S. halepense (C. Linnaeus) C. Persoon
Sy = Holcus halepensis C. Linnaeus

S. × ***almum*** L. Parodi [***bicolor*** × ***halepense***] [**cultivated**]

S. × ***drummondii*** (E. von Steudel) C. Millspaugh & M. A. Chase [***arundinaceum*** × ***bicolor***] [**cultivated**]
Sy = S. bicolor (C. Linnaeus) C. Moench subsp. drummondii (E. von Steudel) J. de Wet
Sy = S. sudanense (C. Piper) O. Stapf

SPARTINA J. von Schreber

S. alterniflora J. Loiseleur-Deslongchamps var. ***glabra*** (J. Bigelow) M. Fernald

S. bakeri E. Merrill

S. cynosuroides (C. Linnaeus) A. Roth
Sy = S. cynosuroides var. polystachya (A. Michaux) W. Beal *ex* M. Fernald

S. patens (W. Aiton) G. H. Muhlenberg
Sy = Dactylis patens W. Aiton

Sy = Spartina juncea (A. Michaux) C. von Willdenow
Sy = S. patens var. juncea (A. Michaux) A. Hitchcock
Sy = S. patens var. monogyna (M. A. Curtis) M. Fernald
S. pectinata J. Link
Sy = S. pectinata var. suttiei (O. Farwell) M. Fernald
S. spartinae (K. von Trinius) E. Merrill *ex* A. Hitchcock
Sy = Vilfa spartinae K. von Trinius

SPHENOPHOLIS F. Lamson-Scribner
S. filiformis (A. Chapman) F. Lamson-Scribner
S. interrupta (S. Buckley) F. Lamson-Scribner
Sy = Trisetum interruptum S. Buckley
S. nitida (J. Biehler) F. Lamson-Scribner
S. obtusata (A. Michaux) F. Lamson-Scribner var. ***major*** (J. Torrey) E. Erdman
Sy = S. intermedia (P. Rydberg) P. Rydberg
Sy = S. longiflora (G. Vasey) A. Hitchcock
S. obtusata (A. Michaux) F. Lamson-Scribner var. ***obtusata***
Sy = Aira obtusa A. Michaux

SPOROBOLUS R. Brown
S. airoides (J. Torrey) J. Torrey subsp. ***airoides***
Sy = Agrostis airoides J. Torrey
S. buckleyi G. Vasey
S. compositus (J. Poiret) E. Merrill var. ***clandestinus*** (J. Biehler) J. Wipff & S. D. Jones
Sy = S. asper (P. de Beauvois) K. Kunth var. canovirens (G. Nash) L. Shinners
Sy = S. asper var. clandestinus (J. Biehler) L. Shinners
Sy = S. canovirens G. Nash
Sy = S. clandestinus (J. Biehler) A. Hitchcock
S. compositus (J. Poiret) E. Merrill var. ***compositus***
Sy = S. asper (P. de Beauvois) K. Kunth var. asper
Sy = S. asper var. drummondii (K. von Trinius) G. Vasey
Sy = S. asper var. hookeri (K. von Trinius) G. Vasey
Sy = S. compositus var. drummondii (K. von Trinius) J. Kartesz & K. Gandhi
S. compositus (J. Poiret) E. Merrill var. ***macer*** (K. von Trinius) J. Kartesz & K. Gandhi
Sy = S. asper (P. de Beauvois) K. Kunth var. macer (K. von Trinius) L. Shinners
Sy = S. macer (K. von Trinius) A. Hitchcock
S. contractus A. Hitchcock
S. cryptandrus (J. Torrey) A. Gray
Sy = Agrostis cryptandra J. Torrey
Sy = Sporobolus cryptandrus subsp. fuscicola (W. Hooker) E. K. Jones & N. Fassett
Sy = S. cryptandrus var. fuscicola (W. Hooker) R. Pohl
Sy = S. cryptandrus var. occidentalis E. K. Jones & N. Fassett
S. flexuosus (G. Thurber *ex* G. Vasey) P. Rydberg
S. giganteus G. Nash
S. indicus (C. Linnaeus) R. Brown var. ***indicus***
Sy = Agrostis indica C. Linnaeus
Ma = Sporobolus poiretii *auct.*, non (J. J. Römer & J. A. Schultes) A. Hitchcock
S. junceus (A. Palisot de Beauvois) K. Kunth
Sy = S. poiretii (J. J. Römer & J. A. Schultes) A. Hitchcock
S. nealleyi G. Vasey
S. purpurascens (O. Swartz) W. Hamilton
Sy = Agrostis purpurascens O. Swartz
S. pyramidatus (J. de Lamarck) A. Hitchcock
Sy = Agrostis pyramidata J. de Lamarck
Sy = Sporobolus pulvinatus J. Swallen
S. silveanus J. Swallen
S. texanus G. Vasey
S. tharpii A. Hitchcock
S. vaginiflorus (J. Torrey *ex* A. Gray) A. Wood var. ***neglectus*** (G. Nash) F. Lamson-Scribner
Sy = S. neglectus G. Nash
Sy = S. ozarkanus M. Fernald
S. vaginiflorus (J. Torrey *ex* A. Gray) A. Wood var. ***vaginiflorus***
Sy = S. vaginiflorus var. inaequalis M. Fernald
S. virginicus (C. Linnaeus) K. Kunth
Sy = Agrostis virginica C. Linnaeus
S. wrightii W. Munro *ex* F. Lamson-Scribner
Sy = S. airoides var. wrightii (F. Lamson-Scribner) F. Gould

STENOTAPHRUM K. von Trinius
S. secundatum (T. Walter) K. E. O. Kuntze
Sy = Ischaemum secundatum T. Walter
Sy = Stenotaphrum secundatum var. variegatum A. Hitchcock [**cultivated**]

STIPA C. Linnaeus
S. arida M. E. Jones
Sy = Achnatherum aridum (M. E. Jones) M. Barkworth
S. clandestina E. Hackel
Sy = Achnatherum clandestinum (E. Hackel) M. Barkworth

S. curvifolia J. Swallen
Sy = Achnatherum curvifolium (J. Swallen) M. Barkworth
S. eminens A. Cavanilles
Sy = Achnatherum eminens (A. Cavanilles) M. Barkworth
S. hymenoides J. J. Römer & J. A. Schultes
Sy = Achnatherum hymenoides (J. J. Römer & J. A. Schultes) M. Barkworth
Sy = Oryzopsis hymenoides (J. J. Römer & J. A. Schultes) P. Ricker *ex* C. Piper
S. lobata J. Swallen
Sy = Achnatherum lobatum (J. Swallen) M. Barkworth
S. perplexa (P. Hoge & M. Barkworth) J. Wipff & S. D. Jones
Sy = Achnatherum perplexum P. Hoge & M. Barkworth
Ma = Stipa columbiana *auct.*, non J. Macoun: Texas authors
S. robusta (G. Vasey) F. Lamson-Scribner
Sy = Achnatherum robustum (G. Vasey) M. Barkworth
Sy = Stipa viridula K. von Trinius var. robusta G. Vasey

THEMEDA P. Forrskål
T. triandra P. Forrskål [**cultivated**]

THINOPYRUM Á. Löve
T. intermedium (N. Host) M. Barkworth & D. R. Dewey subsp. ***barbulatum*** (P. Schur) M. Barkworth & D. R. Dewey
Sy = Agropyron glaucum R. Desfontaines *ex* A. P. de Candolle subsp. barbulatum (P. Schur) K. Richter
Sy = A. intermedium (N. Host) A. P. de Beauvois subsp. trichophorum (J. Link) P. Ascherson & K. Graebner
Sy = A. trichophorum (J. Link) K. Richter
Sy = Elymus hispidus (P. Opiz) A. Melderis subsp. barbulatus (P. Schur) A. Melderis
Sy = Elytrigia intermedia (N. Host) S. Nevski subsp. barbulata (P. Schur) Á. Löve
Sy = E. intermedia subsp. trichophora (J. Link) Á. Löve & D. Löve
Sy = E. trichophora (J. Link) S. Nevski
T. intermedium (N. Host) M. Barkworth & D. R. Dewey subsp. ***intermedium***
Sy = Agropyron intermedium (N. Host) A. P. de Beauvois
Sy = Elymus hispidus (P. Opiz) A. Melderis
Sy = Elytrigia intermedia (N. Host) S. Nevski
Sy = Triticum intermedium N. Host
T. ponticum (J. Podpěra) M. Barkworth & D. R. Dewey
Sy = Elytrigia pontica (J. Podpěra) J. Holub
Sy = Triticum ponticum J. Podpěra
Ma = Agropyron elongatum *auct.*, non (N. Host) A. P. de Beauvois: American authors
Ma = Elymus elongatus *auct.*, non (N. Host) H. Runemark: American authors
Ma = Elytrigia elongata *auct.*, non (N. Host) S. Nevski: American authors
Ma = Lophopyrum elongatum *auct.*, non (N. Host) Á. Löve: American authors

TRACHYPOGON C. Nees von Esenbeck
T. spicatus (C. von Linné) K. E. O. Kuntze
Sy = Stipa spicata C. von Linné
Sy = Trachypogon montufari (K. Kunth) C. Nees von Esenbeck
Sy = T. plumosus (F. von Humboldt & A. Bonpland *ex* C. von Willdenow) C. Nees von Esenbeck
Sy = T. secundus (J. Presl) F. Lamson-Scribner

TRAGUS A. von Haller, *nomen conservandum*
T. berteronianus J. A. Schultes
Ma = T. racemosus *auct.*, non (C. Linnaeus) C. Allioni: Texas authors

TRICHLORIS E. Fournier *ex* G. Bentham
T. crinita (M. Lagasca y Segura) L. Parodi
Sy = Chloris crinita M. Lagasca y Segura
Sy = C. mendocina R. Philippi
Sy = Trichloris mendocina (R. Philippi) F. Kurtz
T. pluriflora
Sy = Chloris pluriflora (E. Fournier) W. Clayton

TRICHONEURA N. Andersson
T. elegans J. Swallen

TRIDENS J. A. Schultes & J. H. Schultes
T. albescens (G. Vasey) E. Wooton & P. Standley
Sy = Rhombolytrum albescens (G. Vasey) G. Nash
T. ambiguus (S. Elliott) J. A. Schultes
* ***T. buckleyanus*** (L. Dewey) G. Nash [**TOES: V**]
T. congestus (L. Dewey) G. Nash
T. eragrostoides (G. Vasey & F. Lamson-Scribner) G. Nash
Sy = Triodia eragrostoides G. Vasey & F. Lamson-Scribner

T. flavus (C. Linnaeus) A. Hitchcock var. ***chapmanii*** (J. K. Small) L. Shinners
Sy = T. chapmanii (J. K. Small) M. A. Chase
Sy = Triodia chapmanii (J. K. Small) B. Bush
T. flavus (C. Linnaeus) A. Hitchcock var. ***flavus***
Sy = Triodia flavus (C. Linnaeus) B. Smyth
T. muticus (J. Torrey) G. Nash var. ***elongatus*** (S. Buckley) L. Shinners
Sy = T. elongatus (S. Buckley) G. Nash
Sy = Triodia elongata (S. Buckley) F. Lamson-Scribner
T. muticus (J. Torrey) G. Nash var. ***muticus***
Sy = T. muticus (J. Torrey) G. Nash f. effusus M. C. Johnston
Sy = Triodia mutica (J. Torrey) F. Lamson-Scribner
T. strictus (T. Nuttall) G. Nash
Sy = Triodia stricta (T. Nuttall) G. Bentham *ex* G. Vasey
T. texanus (S. Watson) G. Nash

TRIPLASIS A. Palisot de Beauvois
T. purpurea (T. Walter) A. Chapman var. ***purpurea***

TRIPOGON J. A. Schultes & J. H. Schultes
T. spicatus (C. Nees von Esenbeck) E. Ekman
Sy = Bromus spicatus C. Nees von Esenbeck

TRIPSACUM C. Linnaeus
T. dactyloides (C. Linnaeus) C. Linnaeus var. ***dactyloides***

× **TRITICOSECALE** M. Wittmarck [***Triticum*** × ***Secale***]
Sy = × Triticale A. Müntzing
× ***T.*** [***Triticum aestivum*** × ***Secale cearale***] [**cultivated**]

TRITICUM C. Linnaeus
T. aestivum C. Linnaeus [**cultivated**]
T. spelta C. Linnaeus [**cultivated**]

UNIOLA C. Linnaeus
U. paniculata C. Linnaeus

UROCHLOA A. Palisot de Beauvois
U. arizonica (F. Lamson-Scribner & E. Merrill) O. Morrone & F. Zuloaga
Sy = Brachiaria arizonica (F. Lamson-Scribner & E. Merrill) S. T. Blake
Sy = Panicum arizonicum F. Lamson-Scribner & E. Merrill
U. brizantha (C. F. Hochstetter *ex* A. Richard) R. D. Webster
Sy = Brachiaria brizantha (C. F. Hochstetter *ex* A. Richard) O. Stapf
Sy = Panicum brizanthum C. F. Hochstetter *ex* A. Richard
U. ciliatissima (S. Buckley) R. D. Webster
Sy = Brachiaria ciliatissima (S. Buckley) M. A. Chase
Sy = Panicum ciliatissimum S. Buckley
U. fasciculata (O. Swartz) R. D. Webster
Sy = Brachiaria fasciculata (O. Swartz) L. Parodi
Sy = Panicum fasciculatum O. Swartz
Sy = P. fasciculatum var. reticulatum (J. Torrey) W. Beal
U. mosambicensis (E. Hackel) J. Dandy
Sy = U. pullulans O. Stapf
Sy = U. rhodesiensis S. Stent
U. mutica (P. Forsskål) T. Q. Nguyen
Sy = Brachiaria mutica (P. Forsskål) O. Stapf
Sy = B. purpurascens (G. Raddi) J. Henrard
Sy = Panicum barbinode K. von Trinius
Sy = P. muticum P. Forsskål
Sy = P. purpurascens G. Raddi
U. panicoides A. Palisot de Beauvois var. ***panicoides*** [**federal noxious weed**]
U. plantaginea (J. Link) R. D. Webster
Sy = Brachiaria plantaginea (J. Link) A. Hitchcock
Sy = Panicum plantagineum J. Link
U. platyphylla (W. Munro *ex* C. Wright) R. D. Webster
Sy = Brachiaria platyphylla (W. Munro *ex* C. Wright) G. Nash
Sy = B. extensa M. A. Chase
Sy = Panicum platyphyllum W. Munro *ex* C. Wright
Sy = Paspalum platyphyllum A. Grisebach, non J. A. Schultes
U. ramosa (C. Linnaeus) R. D. Webster
Sy = Brachiaria ramosa (C. Linnaeus) O. Stapf
Sy = Panicum ramosum C. Linnaeus
U. reptans (C. Linnaeus) O. Stapf
Sy = Brachiaria reptans (C. Linnaeus) C. A. Gardner & C. Hubbard
Sy = Panicum prostratum J. de Lamarck
Sy = P. reptans C. Linnaeus
U. texana (S. Buckley) R. D. Webster
Sy = Brachiaria texana (S. Buckley) S. T. Blake
Sy = Panicum texanum S. Buckley

VASEYOCHLOA A. Hitchcock
V. multinervosa (G. Vasey) A. Hitchcock
Sy = Melica multinervosa G. Vasey

VETIVERIA Bory de Saint-Vincent
V. zizanioides (C. Linnaeus) G. Nash [**cultivated**]
Sy = Anatherum zizanioides (C. Linnaeus) A. Hitchcock & M. A. Chase
Sy = Phalaris zizanioides C. Linnaeus

VULPIA K. Gmelin
V. bromoides (C. Linnaeus) S. Gray
Sy = Bromus dertonensis C. Allioni
Sy = Festuca bromoides C. Linnaeus
Sy = F. dertonensis (C. Allioni) P. Ascherson & K. Graebner
Sy = Vulpia dertonensis (C. Allioni) G. Gola
V. myuros (C. Linnaeus) K. Gmelin var. ***hirsuta*** E. Hackel
Sy = Festuca megalura T. Nuttall
V. myuros (C. Linnaeus) K. Gmelin var. ***myuros***
Sy = Festuca myuros C. Linnaeus
V. octoflora (T. Walter) P. Rydberg var. ***glauca*** (T. Nuttall) M. Fernald
Sy = Festuca octoflora T. Walter var. glauca (T. Nuttall) M. Fernald
Sy = F. octoflora var. tenella (C. von Willdenow) M. Fernald
Sy = F. tenella C. von Willdenow
Sy = Vulpia octoflora var. tenella (C. von Willdenow) M. Fernald
V. octoflora (T. Walter) P. Rydberg var. ***hirtella*** (C. Piper) J. Henrard
Sy = Festuca octoflora T. Walter subsp. hirtella C. Piper
Sy = F. octoflora var. hirtella (C. Piper) C. Piper *ex* A. Hitchcock
V. octoflora (T. Walter) P. Rydberg var. ***octoflora***
Sy = Festuca octoflora T. Walter
Sy = F. octoflora var. aristulata J. Torrey *ex* L. Dewey
V. sciurea (T. Nuttall) J. Henrard
Sy = Festuca sciurea T. Nuttall
Sy = Vulpia elliotea (C. Rafinesque-Schmaltz) M. Fernald

WILLKOMMIA E. Hackel
W. texana A. Hitchcock var. ***texana***

ZEA C. Linnaeus
Z. mays C. Linnaeus [**cultivated**]

ZIZANIA C. Linnaeus
* ***Z. texana*** A. Hitchcock [**TOES: I**]

ZIZANIOPSIS J. Döll
Z. miliacea (A. Michaux) J. Döll & P. Ascherson
Sy = Zizania miliacea A. Michaux

ZOYSIA C. von Willdenow, *nomen conservandum*
Z. japonica E. von Steudel [**cultivated**]
Z. matrella (C. Linnaeus) E. Merrill [**cultivated**]
Sy = Agrostis matrella C. Linnaeus
Sy = Zoysia matrella var. tenuifolia (C. von Willdenow *ex* C. E. Thiele) Y. Sasaki
Sy = Z. pungens C. von Willdenow
Sy = Z. tenuifolia C. von Willdenow *ex* C. E. Thiele
Z. pacifica (P. Goudswaard) M. Hotta & S. Kuroki [**cultivated**]
Sy = Z. matrella (C. Linnaeus) E. Merrill var. pacifica P. Goudswaard
Ma = Z. matrella var. tenuifolia *auct.*, non (C. von Willdenow *ex* C. E. Thiele) Y. Sasaki
Ma = Z. tenuifolia *auct.*, non C. von Willdenow *ex* C. E. Thiele

PONTEDERIACEAE, *nomen conservandum*

EICHHORNIA K. Kunth, *nomen conservandum*
E. crassipes (K. von Martius) C. Solms-Laubach
Sy = Piaropus crassipes (K. von Martius) C. Rafinesque-Schmaltz

HETERANTHERA H. Ruiz López & J. Pavón, *nomen conservandum*
H. dubia (N. von Jacquin) C. MacMillan
Sy = Commelina dubia N. von Jacquin
Sy = Heteranthera graminea M. H. Vahl
Sy = H. liebmannii (P. Magnus) L. Shinners
Sy = Zosterella dubia (N. von Jacquin) J. K. Small
H. limosa (O. Swartz) C. von Willdenow
Sy = Pontederia limosa O. Swartz
H. mexicana S. Watson
Sy = Eurystemon mexicanus (S. Watson) E. Alexander
H. reniformis H. Ruiz López & J. Pavón

PONTEDERIA C. Linnaeus
P. cordata C. Linnaeus
Sy = P. cordata var. angustifolia (F. Pursh) J. Torrey & S. Elliott
Sy = P. cordata var. lanceolata (T. Nuttall) A. Grisebach
Sy = P. cordata var. lancifolia (G. H. Muhlenberg *ex* S. Elliott) J. Torrey
Sy = P. lanceolata T. Nuttall

POTAMOGETONACEAE, *nomen conservandum*

POTAMOGETON C. Linnaeus
P. bicupulatus M. Fernald
Sy = P. diversifolius C. Rafinesque-Schmaltz var. trichophyllus T. Morong
* ***P. clystocarpus*** M. Fernald [**TOES: I**]
P. crispus C. Linnaeus
P. diversifolius C. Rafinesque-Schmaltz
Sy = P. diversifolius var. multidenticulatus (T. Morong) P. Ascherson & K. Graebner
P. foliosus C. Rafinesque-Schmaltz var. ***foliosus***
Sy = P. foliosus var. genuinus M. Fernald
Sy = P. foliosus var. macellus M. Fernald
P. illinoënsis T. Morong
P. latifolius (J. Robbins) T. Morong
Sy = P. filiformis C. Persoon var. latifolius (J. Robbins) J. Reveal
P. nodosus J. Poiret
Sy = P. americanus A. von Chamisso & D. von Schlechtendal
P. pectinatus C. Linnaeus
Sy = P. columbianus W. Suksdorf
Sy = P. interior P. Rydberg
P. pulcher E. Tuckerman
P. pusillus C. Linnaeus var. ***pusillus***
Sy = P. pusillus var. minor (A. de Bivona-Bernardi) M. Fernald & B. Schubert
P. pusillus C. Linnaeus var. ***tenuissimus*** F. Mertens & W. Koch
Sy = P. berchtoldii F. Fieber
Sy = P. berchtoldii var. tenuissimus (F. Mertens & W. Koch) M. Fernald
Sy = P. pusillus var. mucronatus (F. Fieber) K. Graebner
P. × ***faxonii*** T. Morong [***illinoënsis*** × ***nodosus***]

RUPPIACEAE, *nomen conservandum*

RUPPIA C. Linnaeus
R. maritima C. Linnaeus
Sy = R. curvicarpa A. Nelson
Sy = R. maritima var. brevirostris C. Agardh
Sy = R. maritima var. exigua M. Fernald & K. Wiegand
Sy = R. maritima var. intermedia (K. Thedenius) P. Ascherson & K. Graebner
Sy = R. maritima var. longipes J. Hagström
Sy = R. maritima var. obliqua (P. Schur) P. Ascherson & K. Graebner
Sy = R. maritima var. occidentalis (S. Watson) K. Graebner
Sy = R. maritima var. rostrata C. Agardh
Sy = R. maritima var. spiralis E. Morris
Sy = R. maritima var. subcapitata M. Fernald & K. Wiegand

SMILACACEAE, *nomen conservandum*

SMILAX C. Linnaeus
S. bona-nox C. Linnaeus var. ***bona-nox***
S. bona-nox C. Linnaeus var. ***hederifolia*** (H. Beyrich) M. Fernald
S. glauca T. Walter
Sy = S. glauca var. genuina S. F. Blake
Sy = S. glauca var. leurophylla S. F. Blake
S. lasioneura W. Hooker
Sy = Nemexia lasioneura (W. Hooker) P. Rydberg
Sy = Smilax herbacea C. Linnaeus var. lasioneura (W. Hooker) A. P. de Candolle
S. laurifolia C. Linnaeus
S. pumila T. Walter
S. renifolia J. K. Small
S. rotundifolia C. Linnaeus
Sy = S. rotundifolia var. crenulata J. K. Small & A. A. Heller
Sy = S. rotundifolia var. quadrangularis (G. H. Muhlenberg *ex* C. von Willdenow) A. Wood
S. smallii T. Morong
Sy = S. domingensis C. von Willdenow
Sy = S. lanceolata C. Linnaeus
S. tamnoides C. Linnaeus
Sy = S. hispida G. H. Muhlenberg *ex* J. Torrey
Sy = S. hispida var. australis J. K. Small
Sy = S. hispida var. montana W. C. Coker
Sy = S. tamnoides var. hispida (G. H. Muhlenberg *ex* J. Torrey) M. Fernald
S. walteri F. Pursh

SPARGANIACEAE, *nomen conservandum*

SPARGANIUM C. Linnaeus
S. americanum T. Nuttall
S. androcladum (G. Engelmann) T. Morong
Sy = S. lucidum M. Fernald & E. H. Eames

STRELITZIACEAE, *nomen conservandum*

STRELITZIA W. Aiton
S. nicolai E. von Regel & K. Koch [**cultivated**]
S. reginae J. Banks [**cultivated**]

TYPHACEAE, *nomen conservandum*

TYPHA C. Linnaeus
  T. domingensis C. Persoon
    Sy = T. angustata Bory de Saint-Vincent & L. Chaubard
    Sy = T. angustifolia C. Linnaeus var. domingensis (C. Persoon) A. Grisebach
    Sy = T. bracteata E. Greene
    Sy = T. truxillensis K. Kunth
    Ma = T. angustifolia *auct.*, non C. Linnaeus: Texas authors
  T. latifolia C. Linnaeus
    Sy = Massula latifolia J. Dulac

XYRIDACEAE, *nomen conservandum*

XYRIS C. Linnaeus
  X. ambigua H. Beyrich *ex* K. Kunth
  X. baldwiniana J. A. Schultes
    Sy = X. baldwiniana var. tenuifolia (A. Chapman) G. Malme
    Sy = X. juncea W. Baldwin
  X. caroliniana T. Walter
    Sy = X. arenicola J. K. Small
    Sy = X. flexuosa G. H. Muhlenberg *ex* S. Elliott
    Sy = X. pallescens (C. Mohr) J. K. Small
  X. difformis A. Chapman var. ***curtissii*** (G. Malme) R. Kral
    Sy = X. bayardii M. Fernald
    Sy = X. curtissii G. Malme
    Sy = X. neglecta J. K. Small
    Sy = X. papillosa N. Fassett
  X. difformis A. Chapman var. ***difformis***
    Sy = X. elata A. Chapman var. difformis A. Chapman
* ***X. drummondii*** G. Malme [**TOES: V**]
  X. elliottii A. Chapman
  X. fimbriata S. Elliott
  X. jupicai L. C. Richard
    Sy = X. arenicola F. Miquel
    Sy = X. communis K. Kunth
    Sy = X. elata A. Chapman
  X. laxifolia K. von Martius var. ***iridifolia*** (A. Chapman) R. Kral
    Sy = X. iridifolia A. Chapman
  X. louisianica E. Bridges & S. Orzell
  X. platylepis A. Chapman
* ***X. scabrifolia*** R. Harper [**TOES: V**]
  X. smalliana G. Nash
    Sy = X. caroliniana T. Walter var. olneyi A. Wood
    Sy = X. congdonii J. K. Small
    Sy = X. smalliana var. olneyi (A. Wood) H. Gleason
  X. stricta A. Chapman
  X. torta J. E. Smith
    Sy = X. bulbosa K. Kunth
    Sy = X. flexuosa A. Chapman
    Sy = X. indica C. Linnaeus
    Sy = X. torta var. macropoda M. Fernald
    Sy = X. torta var. occidentalis G. Malme

ZANNICHELLIACEAE, *nomen conservandum*

ZANNICHELLIA C. Linnaeus
  Z. palustris C. Linnaeus var. ***major*** (C. Hartman) C. von Bönninghausen *ex* W. Koch

ZINGIBERACEAE, *nomen conservandum*

ALPINIA W. Roxburgh, *nomen conservandum*
  A. zerumbet (C. Persoon) B. D. Burtt & R. M. Smith [**cultivated**]
    Sy = A. nutans G. Andréanszky
    Sy = A. speciosa (H. Wendland) K. Schumann
    Sy = Catimbium speciosum (H. Wendland) R. Holttum
    Sy = Languas speciosa (H. Wendland) E. Merrill
    Sy = Zerumbet speciosum H. Wendland

CURCUMA C. Linnaeus, *nomen conservandum*
  C. petiolata W. Roxburgh [**cultivated**]
  C. roscoeana N. Wallich [**cultivated**]

HEDYCHIUM K. König
  H. coronarium K. König [**cultivated**]
  H. gardnerianum J. Ker-Gawler [**cultivated**]

ZINGIBER G. Böhmer, *nomen conservandum*
  Z. officinale W. Roscoe [**cultivated**]
    Sy = Amomum zingiber C. Linnaeus
    Sy = Zingiber zingiber (C. Linnaeus) G. Karsten
  Z. zerumbet (C. Linnaeus) J. E. Smith [**cultivated**]
    Sy = Amomum zerumbet C. Linnaeus

Bibliography

Adams, P. 1973. Clusiaceae of the southeastern United States. Journal of the Elisha Mitchell Scientific Society 89:62–71.

Adams, P., and N. K. B. Rolson. 1961. A re-evaluation of the generic status of *Ascyrum* and *Crookea* (Guttiferae). Rhodora 63:10–16.

Adams, R. P. 1972. Chemosystematic and numerical studies of natural populations of *Juniperus pinchotii* Sudworth. Taxon 21:407–427.

Adams, R. P. 1973. Reevaluation of the biological status of *Juniperus deppeana* var. *sperryi* Correll. Brittonia 25:284–289.

Adams, R. P. 1975. Numerical-chemosystematic studies of infraspecific variation in *Juniperus pinchotii*. Biochemical Systematics and Ecology 3:71–74.

Adams, R. P. 1986. Geographical variation in *Juniperus silicola* and *J. virginiana* in the southeastern United States: Multivariate analysis of morphology and terepenoids. Taxon 35:61–75.

Adams, R. P., and J. R. Kistler. 1991. Hybridization between *Juniperus erythrocarpa* Cory and *Juniperus pinchotii* Sudworth in the Chisos Mountains, Texas. Southwestern Naturalist 36:295–301.

Adams, R. P., and T. A. Zanoni. 1979. The distribution, synonymy, and taxonomy of three junipers of southwestern United States and northern México. Southwestern Naturalist 24:323–329.

Aiken, S. G. 1981. A conspectus of *Myriophyllum* (Haloragaceae) in North America. Brittonia 33:57–69.

Aiken, S. G., and L. P. Lefkovitch. 1993. On the separation of two species within *Festuca* subg. *Obtusae* (Poaceae). Taxon 42(2):323–337.

Ajilvsgi, G. 1979. *Wild Flowers of the Big Thicket, East Texas, and Western Louisiana*. College Station: Texas A&M University Press.

Ajilvsgi, G. 1984. *Wildflowers of Texas*. Bryan, Texas: Shearer Publishing.

Alderson, J. S., and W. C. Sharp. 1994. *Grass Varieties in the United States*. Agriculture Handbook No. 170, United States Department of Agriculture—Soil Conservation Service.

Allred, K. W. 1984a. Morphological variation and classification of the North American *Aristida purpurea* complex (Gramineae). Brittonia 36(4):382–395.

Allred, K. W. 1984b. Studies in the *Aristida* (Gramineae) of the southeastern United States. I. Spikelet variation in *A. purpurescens, A. tenuispica*, and *A. virgata*. Rhodora 86:73–77.

Allred, K. W. 1985a. Studies in the genus *Aristida* (Gramineae) of the southeastern United States. II. Morphometric analysis of *A. intermedia* and *A. longespica*. Rhodora 87(850):137–145.

Allred, K. W. 1985b. Studies in the *Aristida* (Gramineae) of the southeastern United States. III. Nomenclature and a taxonomic comparison of *A. lanosa* and *A. palustris*. Rhodora 87(850):147–155.

Allred, K. W. 1986. Studies in the *Aristida* (Gramineae) of the southeastern United States. IV. Key and conspectus. Rhodora 88(855):367–387.

Allred, K. W. 1993a. *A Field Guide to the Grasses of New Mexico*. Las Cruces: New Mexico State University, Department of Agriculture Communications.

Allred, K. W. 1993b. *Bromus*, section *Pnigma*, in New Mexico, with a key to the bromegrass of the state. Phytologia 74(4):319–345.

Allred, K. W., and F. W. Gould. 1978. Geographic variation in the *Dichanthelium aciculare* complex (Poaceae). Brittonia 30:497–504.

Allred, K. W., and F. W. Gould. 1983. Systematics of the *Bothriochloa saccharoides* complex (Poaceae: *Andropogoneae*). Systematic Botany 8(2):168–184.

Al-Shehbaz, I. A. 1973. The biosystematics of the genus *Thelypodium*. Contributions of the Asa Gray Herbarium of Harvard University 204:3–148.

Al-Shehbaz, I. A. 1984. The tribes of Cruciferae (Brassicaceae) in the southeastern United States. Journal of the Arnold Arboretum 65:343–373.

Al-Shehbaz, I. A. 1985a. The genera of *Thelepodieae* (Cruciferae: Brassicaceae) in the southeastern United States. Journal of the Arnold Arboretum 66(1):95–111.

Al-Shehbaz, I. A. 1985b. The genera of Brassicaceae (Cruciferae: Brassicaceae) in the southeastern United States. Journal of the Arnold Arboretum 66(3):279–351.

Al-Shehbaz, I. A. 1986. The genera of *Lepidieae* (Cruciferae: Brassicaceae) in the southeastern United States. Journal of the Arnold Arboretum 68:185–240.

Al-Shehbaz, I. A. 1987. The genera of *Alysseae* in the southeastern United States. Journal of the Arnold Arboretum 68(2):185–240.

Al-Shehbaz, I. A. 1988a. *Cardamine dissecta*, a new combination replacing *Dentaria multifida* (Cruciferae). Journal of the Arnold Arboretum 69(2):81–84.

Al-Shehbaz, I. A. 1988b. The genera of *Arabideae* (Cruciferae: Brassicaceae) in the southeastern United States. Journal of the Arnold Arboretum 69(2):85–166.

Al-Shehbaz, I. A. 1988c. The genera of *Anchonieae (Hesperideae)* (Cruciferae: Brassicaceae) in the southeastern United States. Journal of the Arnold Arboretum 69(3):193–212.

Al-Shehbaz, I. A. 1988d. The genera of *Sisymbrieae* (Cruciferae: Brassicaceae) in the southeastern United States. Journal of the Arnold Arboretum 69:213–237.

Al-Shehbaz, I. A., and V. Bates. 1987. *Armoracia lacustris* (Brassicaceae), the correct name for the North American lake cress. Journal of the Arnold Arboretum 68(3):357–359.

Al-Shehbaz, I. A., and B. G. Schubert. 1989. The Dioscoreaceae in the southeastern United States. Journal of the Arnold Arboretum 70(1):57–95.

Alston, A. H. G. 1955. The heterophyllous *Selaginellae* of Continental North America. Bulletin of the British Museum (Natural History), Botany 1(8):219–274.

Amos, B. B., and F. R. Gehlbach. 1988. *Edwards Plateau Vegetation*. Waco, Texas: Baylor University Press.

Anderberg, A. A. 1991. Taxonomy and phylogeny of the tribe *Gnaphalieae* (Asteraceae). Opera Botanica 104:1–195.

Anderson, D. E. 1961. Taxonomy and distribution of the genus *Phalaris*. Iowa State College Journal of Science 36:1–96.

Anderson, D. E. 1971. New names and combinations in *Trixis* (Compositae). Brittonia 23:347–353.

Anderson, D. E. 1972. A monograph of the Mexican and Central American species of *Trixis* (Compositae). Memoirs of the New York Botanical Garden 22: 1–68.

Anderson, D. E. 1974. Taxonomy of the genus *Chloris* (Gramineae). Brigham Young University Science Bulletin, Biological series 19(2):1–133.

Anderson, D. E. 1975. *Chloris* Swartz. In *The Grasses of Texas*, ed. F. W. Gould. College Station: Texas A&M University Press.

Anderson, D. E. 1977. Studies on *Bigelovia* (Asteraceae): III. Cytotaxonomy and biogeography. Systematic Botany 2:209–218.

Anderson, D. E. 1980. Morphology and biogeography of *Chrysothamnus nauseosus* ssp. *texensis* (Asteraceae): A new Guadalupe Mountains endemic. Southwestern Naturalist 25:197–206.

Anderson, L. C. 1970. Studies on *Bigelowia* (*Astereae*, Compositae). I. Morphology and taxonomy. Sida 3: 451–463.

Anderson, W. R. 1971. The correct spelling of the generic name *Mitracarpus* (Rubiaceae). Taxon 20:643.

Andrews, J. 1984. *Peppers: The Domesticated Capsicums*. Austin: University of Texas Press.

Angelo, R. 1991. A new combination in *Panicum* (Poaceae) subgenus *Dichanthelium*. Phytologia 71(2): 85–86.

Angerstein, M. B., and D. E. Lemke. 1994. First reports of the aquatic weed *Hygrophila polysperma* (Acanthaceae) from Texas. Sida 16(2):365–371.

Anonymous. 1950. *Native flora of Texas*. Texas Highway Department, (Division of Maintenance Operations), Landscape. Operations Division, Austin.

Anton, A. M. 1981. The genus *Tragus* (Gramineae). Kew Bulletin 36(1):55–61.

Anton, A. M., and A. T. Hunziker. 1978. El género *Munroa* (Poaceae): Sinopsis morfológica y taxonómica. Boletín de la Academia Nacional de Ciencias 52:229–252.

Argus, G. W. 1986. The genus *Salix* (Salicaceae) in the southeastern United States. Systematic Botany Monograph 9:1–170.

Arnow, L. A. 1987. Gramineae A. L. Jussieu, Grass Family. In *A Utah flora*, ed. S. L. Welsh, N. D. Atwood, S. Goodrich, and L. C. Higgins. Memoirs No. 9, Great Basin Naturalist.

Arnow, L. A. 1994. *Koeleria macrantha* and *K. pyramidata* (Poaceae): Nomenclatural problems and biological distinctions. Systematic Botany 19(1):6–20.

Arridge, R. E., and P. J. Fonteyn. 1981. Naturalization of *Colocasia esculenta* (Araceae) in the San Marcos River, Texas. Southwestern Naturalist 26:210–211.

Atwood, D. N. 1975. A revision of the *Phacelia crenulata* group (Hydrophyllaceae) for North America. Great Basin Naturalist 35:127–190.

Austin, D. F. 1976. Varieties of *Ipomoea trichocarpa* (Convolvulaceae). Sida 6:216–220.

Austin, D. F. 1977. *Ipomoea carnea* Jacq. vs. *Ipomoea fistulosa* Mart. *ex* Choisy. Taxon 26:235–238.

Austin, D. F., and G. W. Staples. 1985. *Petrogenia* as a synonym of *Bonamia* (Convolvulaceae) with comments on allied species. Brittonia 37:310–316.

Averett, J. E. 1973. Biosystematic study of *Chamaesaracha* (Solanaceae). Rhodora 75:325–365.

Ayers, Brother B. 1946. The genus *Cyperus* in México. Catholic University of America, Biological series 1: xii, 103.

Babel, W. K. 1943. The variations of *Brachyelytrum erectum*. Rhodora 45:260–262.

Bacon, J. D. 1974. Chromosome numbers and taxonomic notes on the genus *Nama* (Hydrophyllaceae). Brittonia 26:101–105.

Bacon, J. D. 1978. Taxonomy of *Nerisyrenia* (Cruciferae). Rhodora 80:159–227.

Baden, C., and R. von Bother. 1994. A taxonomic revision of *Hordeum* section *Critseion*. Nordic Journal of Botany 14:117–136.

Bailey, D. K., and F. G. Hawksworth. 1979. Piñyons of the Chihuahuan Desert Region. Phytologia 44: 129–133.

Baker, J. G. 1883. A synopsis of the genus *Selaginella*. Part 1. Journal of Botany 21:1–5.

Baker, R. J., and B. L. Turner. Taxonomy of *Flyriella* (Asteraceae-*Eupatorieae*). Sida 11:300–317.

Ball, C. R. 1961. Salicaceae: *Salix*. Flora of Texas, Vol. 3, part 6. Texas Research Foundation, Renner, Texas.

Ballard, R. 1986. *Bidens pilosa* complex (Asteraceae) in North and Central America. American Journal of Botany 73:1452–1465.

Banks, D. J. 1966. Taxonomy of *Paspalum setaceum* (Gramineae). Sida 2:269–284.

Barber, S. C. 1982. Taxonomic studies in the *Verbena stricta* complex (Verbenaceae). Systematic Botany 7:433–456.

Barkley, F. A. 1943. Anacardiaceae. In C. L. Lundell, Flora of Texas, Vol. 3, part 2. Texas Research Foundation, Renner, Texas.

Barkley, F. A. 1944. *Shinus* L. Brittonia 5(2):160–198.

Barkley, T. M. 1978. *Senecio*. In North American Flora: II. (10):50–139.

Barkley, T. M. 1980. Taxonomic notes on *Senecio tomentosus* and its allies (Asteraceae). Brittonia 32: 291–308.

Barkley, T. M. 1985. Generic boundaries in the *Senecioneae*. Taxon 34:17–21.

Barkworth, M. E. 1982. Embryological characters and the taxonomy of the *Stipeae* (Gramineae). Taxon 31(2):233–243.

Barkworth, M. E. 1986. *Piptochaetium* Presl (Gramineae: *Stipeae*) in North America and Mesoamerica: Taxonomic and distributional observations. Brenesia 25–26:169–178.

Barkworth, M. E. 1990. *Nassella* (Gramineae: *Stipeae*): Revised interpretation and nomenclatural changes. Taxon 39:597–614.

Barkworth, M. E. 1993. North American *Stipeae* (Gramineae): Taxonomic changes and other comments. Phytologia 74(1):1–25.

Barkworth, M. E., and R. J. Atkins. 1984. *Leymus* Hochst. (Gramineae: *Triticeae*) in North America: Taxonomy and distribution. American Journal of Botany 71:609–625.

Barkworth, M. E., and J. Everett. 1987. Evolution in the *Stipeae*: Identification and relationships of its monophyletic taxa. In *Grass systematics and evolution*, ed. T. R. Soderstrom et al. Washington, D.C.: Smithsonian Press.

Barkworth, M. E., J. Valdés-R., and R. Q. Landers. 1989. *Stipa clandestina*: New weed threat on southwestern rangelands. Weed Technology 3:699–702.

Barneby, R. C. 1977. *Dalea imagines*. Memoirs of the New York Botanical Garden 27:1–891.

Barneby, R. C. 1986. Notes on the *Mimosae* (Leguminosae: *Mimosoideae*) of the Chihuahuan Desert akin to *M. zygophylla*. Brittonia 38:4–8.

Barneby, R. C. 1991. Sensitive *Censitae*. A description of the genus *Mimosa* Linnaeus (Mimosaceae) in the New World. Memoirs of the New York Botanical Garden 65:1–835.

Barneby, R. C., and D. Isely. 1986. Reevaluation of *Mimosa biuncifera* and *M. texana* (Leguminosae: *Mimosoideae*). Brittonia 38:119–122.

Bates, D. M. 1974. *Fryxellia*, a new genus of North American Malvaceae. Brittonia 26:95–100.

Bates, D. M. 1978a. *Bastardiastrum*, a segregate from *Wissadula* (Malvaceae). Gentes Herbarium 11:311–328.

Bates, D. M. 1978b. *Allowissadula*. Gentes Herbarium 11:329–354.

Baum, B. R. 1967. Introduced and naturalized Tamarisks in the United States and Canada (Tamaricaceae). Baileya 15:19–25.

Baum, B. R. 1978. The genus *Tamarix*. Israel Academy of Science and Humanities, 1–209.

Baum, B. R., and L. G. Bailey. 1990. Key and synopsis of North American *Hordeum* species. Canadian Journal of Botany 68(11):2433–2442.

Bayer, R. J., and G. L. Stebbins. 1982. A revised classification of *Antennaria* (Asteraceae: *Inuleae*) of the eastern United States. Systematic Botany 7:300–313.

Beal, E. O., J. W. Wooten, and C. R. Kaul. 1982. Review of the *Sagittaria engelmanniana* complex (Alismataceae) with environmental correlations. Systematic Botany 7:417–432.

Beatley, J. C. 1973. Russian-thistle (*Salsola*) species in western United States. Journal of Range Management 26:225–226.

Beaty, H. E. 1978. *A checklist of flora and fauna, Central and West Bell County, Texas*. Temple, Tex: Author.

Beauchamp, M. 1980. New names in American *Acacia*. Phytologia 46:5–9.

Becker, K. 1975. New combinations in *Wedelia* Jacq. (Asteraceae). Phytologia 31:25.

Beetle, A. A. 1942. Studies in the genus *Scirpus* L.: IV. The section *Bolboschoenus* Pall. American Journal of Botany 42:82–88.

Beetle, A. A. 1943. The North American variations of *Distichlis spicata*. Bulletin of the Torrey Botanical Club 70:638–650.

Beetle, A. A. 1955. The grass genus *Distichlis*. Revista Agricultura y Agronómica 22:86–94.

Benedict, R. C. 1909. The genus *Ceratopteris*: A preliminary revision. Bulletin of the Torrey Botanical Club 36:463–476.

Benham, D. M. 1982. Biology of *Cheilanthes leucopoda* and *Cheilanthes kaulfussii* (Filicales). Master's thesis, Angelo State University, San Angelo, Texas.

Benham, D. M. 1989. A biosystematic revision of the fern genus *Astrolepis* (Adiantaceae). Ph.D. dissertation, Northern Arizona University, Flagstaff.

Benham, D. M. 1992. Additional taxa in *Astrolepis*. American Fern Journal 82:59–62.

Benham, D. M., and M. D. Windham. 1992. Generic affinities of the star-scaled cloak ferns. American Fern Journal 82:47–58.

Benson, L. 1954. Supplement to a treatise on the North American *ranunculi*. American Midland Naturalist 52:328–369.

Benson, L. 1969. Cactaceae. Flora of Texas, Vol. 2, part 2. Texas Research Foundation, Renner, Texas.

Benson, L. 1982. *The Cacti of the United States and Canada*. Stanford, California: Stanford University Press.

Benson, L., and R. A. Darrow. 1954. *Trees and Shrubs of the Southwestern Deserts*. 2d ed. Tucson: University of Arizona Press.

Benson, L., and R. A. Darrow. 1981. *Trees and Shrubs of the Southwestern Deserts*. 3d ed. Tucson: University of Arizona Press.

Bentley, H. L. 1898. A report upon the grasses and forage plants of Central Texas. Bulletin No. 10. United States Department of Agriculture, Division of Agrostology.

Berlandier, J. L. 1980. *Journey to Mexico During the Years 1826 to 1834*, Vols. I and II. Translated by S. M. Ohlendorf, J. M. Bigelow, and M. M. Standifer. The Texas State Historical Association in cooperation with the Center of Studies in Texas History, University of Texas at Austin.

Bierner, M. W. 1972. Taxonomy of *Helenium* sect. *Tetrodus* and a conspectus of North American *Helenium* (Compositae). Brittonia 24:331–355.

Bierner, M. W. 1989. Taxonomy of *Helenium* sect. *Amarum* (Asteraceae). Sida 13(4):453–459.

Bierner, M. W., J. G. Diaz, B. Barba, and W. Herz. 1992. *Tetraneuris linearifolia* var. *arenicola* (Asteraceae: *Heliantheae*): A new variety from South Texas. Sida 15(2):231–239.

Bigony, M.-Love. 1995. State symbol quiz. Texas Parks and Wildlife 53(1):48–53.

Black, G. A. 1963. Grasses of the genus *Axonopus*. Advancing Frontiers of Plant Sciences 5:1–186.

Blackwell, W. H., M. D. Beachle, and G. Williamson. 1978. Synopsis of *Kochia* (Chenopodiaceae) in North America. Sida 7:248–254.

Blasdell, R. F. 1963. A monographic study of the fern genus *Cystopteris*. Memoirs of the Torrey Botanical Club 21:1–102.

Bobrov, A. E. 1967. The family Osmundaceae (R. Br.) Kaulf. Its taxonomy and geography. Botanicnyj zurnal (Moscow & Leningrad) 52:1600–1610.

Bogle, A. L. 1970. The genera of Molluginaceae and Aizoaceae in the southeastern United States. Journal of the Arnold Arboretum 51:257–309.

Boom, B. M. 1982. Synopsis of *Isoëtes* in the southeastern United States. Castanea 47:38–59.

Boonbundarl, S. 1985. A biosystematic study of the *Digitaria leucites* complexes in North America. Ph.D. dissertation, Texas A&M University, College Station.

Bothmer, R. von, N. Jacobsen, and O. Seberg. 1993. Variation and taxonomy in *Hordeum depressum* and in the *H. brachyantherum* complex (Poaceae). Nordic Journal of Botany 13(1):3–17.

Boyle, W. S. 1945. A cytotaxonomic study of the North American species of *Melica*. Madroño 8:1–32.

Bradley, T. R. 1975. Hybridization between *Triodanis perfoliata* and *Triodanis biflora* (Campanulaceae). Brittonia 27:110–114.

Brainerd, E. 1921. *Violets of North America*. Bulletin No. 224. Vermont State Agricultural Experiment Station. Burlington, Vermont: Free Press Printing Co.

Brandenburg, D. M., W. H. Blackwell, and J. W. Thieret. 1991. Revision of the genus *Cinna* (Poaceae). Sida 14(4):581–596.

Brandenburg, D. M., J. R. Estes, and S. L. Collins. 1991. A revision of *Diarrhena* (Poaceae) in the United States. Bulletin of the Torrey Botanical Club 118(2): 128–136.

Bray, W. L. 1905. Vegetation of the sotol country in Texas. Bulletin No. 60, Scientific series, No. 6. University of Texas, Austin.

Breckenridge, F. G., III, and J. M. Miller. 1982. Pollination biology, distribution and chemotaxonomy of the *Echinocereus enneacathus* complex (Cactaceae). Systematic Botany 7(4):365–378.

Breeden, J. O. 1994. *A long ride in Texas, the explorations of John Leonard Riddell.* Edited by J. O. Breeden. College Station: Texas A&M University Press.

Bretting, P. K. 1983. The taxonomic relationship between *Proboscidea louisianica* and *Proboscidea fragrans* (Martyniaceae). Southwestern Naturalist 28: 445–449.

Bridges, E. L., and S. L. Orzell. 1987. A new species of *Xyris* (sect. *Xyris*) from the Gulf Coastal Plains. Phytologia 64:56–61.

Bridges, E. L., and S. L. Orzell. 1989a. Additions and noteworthy vascular plant collections from Texas and Louisiana, with historical, ecological and geographical notes. Phytologia 66(1):12–69.

Bridges, E. L., and S. L. Orzell. 1989b. A new species of *Carex* (sect. *Oligocarpae*) from the Edwards Plateau of Texas. Phytologia 67:148–154.

Brinkler, R. R. Monograph of *Schoenocaulon*. Annals of the Missouri Botanical Garden 29(4):287–315.

Britton, N. L. 1884. A list of Cyperaceae. Bulletin of the Torrey Botanical Club 11:85–87.

Britton, N. L. 1887. A list of plants collected by Miss Mary B. Croft, 1884–85, at San Diego, Texas, near the headwaters of the Rio Dulce. Transactions of the New York Academy of Sciences 7:7–14.

Britton, N. L. 1890a. Contributions to Texas botany: Additions to the list of plants collected by Miss Mary B. Croft at San Diego, Texas. Transactions of the New York Academy of Sciences 9:181–183.

Britton, N. L. 1890b. Contributions to Texas botany: Note on some plants collected by Mr. Frank Tweedy in Tom Greene Co., Texas. Transactions of the New York Academy of Sciences 9:183–185.

Brooks, A. R. 1991. The woody vegetation of wet creek bottom communities in East Texas. Master's thesis, Stephen F. Austin State University, Nacogdoches, Texas.

Brooks, R. E. 1983. New combinations in *Delphinium* and *Rhus*. Phytologia 54:8.

Brown, C. A. 1972. *Wildflowers of Louisiana and Adjoining States*. Baton Rouge: Louisiana State University Press.

Brown, D. F. M. 1964. A monographic study of the fern genus *Woodsia*. Beichefte zur Nova Hedwigia 16:1–154.

Brown, L. E. 1969. A biosystematic study of the *Chloris cucullata-Chloris verticillata* complex. Ph.D. dissertation, Texas A&M University, College Station.

Brown, L. E. 1985. *Campanula rapunculoides* (Campanulaceae) new to Texas. Sida 11:102.

Brown, L. E. 1986a. *Thaspium trifoliatum* (Apiaceae) and *Ranunculus marginatus* (Ranunculaceae) new to Texas. Sida 11:488.

Brown, L. E. 1986b. A new species of *Rudbeckia* (Asteraceae-*Heliantheae*) from hillside bogs in East Texas. Phytologia 61:367–371.

Brown, L. E. 1991. Brazos Bend State Park plant list. Texas Parks and Wildlife Department, Austin. Mimeo.

Brown, L. E. 1992. *Cayratia japonica* (Vitaceae) and *Paederia foetida* (Rubiaceae) adventive in Texas. Phytologia 72(1):45–47.

Brown, L. E. 1993. The deletion of *Sporobolus heterolepis* (Poaceae) from the Texas and Louisiana floras, and the addition of *Sporobolus silveanus* to the Oklahoma flora. Phytologia 74(5):371–381.

Brown, L. E. 1995. A checklist of the vascular plants of the Houston area. Mimeographed.

Brown, L. E., and K. N. Gandhi. 1989. Notes on the flora of Texas with additions, range extensions, and one correction. Phytologia 67:394–399.

Brown, L. E., and C. D. Peterson. 1984. *Carex rosea* (Cyperaceae), *Trifolium lappaceum* (Fabaceae), and *Aira caryophyllea* (Poaceae) new to Texas. Sida 10:363–364.

Brown, L. E., and C. D. Peterson. 1995. A checklist of the vascular plants of the Davis Hill State Park. Mimeographed.

Brown, L. E., and J. Schultz. 1991. *Arthraxon hispidus* (Poaceae), new to Texas. Phytologia 71(5):379–381.

Brown, L. E., and J. R. Ward. 1986. *Scutellaria minor* (Lamiaceae) new to North America. Sida 11:489.

Brown, R. C. 1978. Biosystematics of *Psilostrophe* (Compositae: *Helenieae*): II. Artificial hybridization and systematic treatment. Madroño 25:187–201.

Bruce, J. G. 1975. Systematics and morphology of subgenus *Lepidotus* of the genus *Lycopodium* (Lycopodiaceae). Ph.D. dissertation, University of Michigan, Ann Arbor.

Bruce, J. G. 1976. Comparative studies of *Lycopodium carolinianum*. American Fern Journal 66:125–137.

Brunken, J. N. 1977. A systematic study of *Pennisetum* sect. *Pennisetum* (Gramineae). American Journal of Botany 64(2):161–176.

Bryson, C. T. 1980. A revision of the North American *Carex* section *laxiflorae* (Cyperaceae). Ph.D. dissertation, Mississippi State University, Starksville.

Buchenau. F. 1906. Juncaceae. In Engler, *Das Pflanzenreich IV* 36(Heft 25):1–284. Leipzig: Wilhelm Engelmann.

Buckley, S. B. 1860. Description of several new species of plants. Proceedings of the Academy of Natural Sciences of Philadelphia, 443–445.

Buckley, S. B. 1861. Description of new plants from Texas. Proceedings of the Academy of Natural Sciences of Philadelphia, 448–463.

Buckley, S. B. 1862a. Descriptions of new plants from Texas. No. 2. Proceedings of the Academy of Natural Sciences of Philadelphia, 5–10.

Buckley, S. B. 1862b. Descriptions of plants. No. 3. Gramineae. Proceedings of the Academy of Natural Sciences of Philadelphia, 88–100.

Buckley, S. B. 1883. Some new Texan plants. Bulletin of the Torrey Botanical Club 10:90–91.

Burgess, T. L. 1979. *Agave*-Complex of the Guadalupe Mountains National Park; putative hybridization between members of different subgenera. In *Biological Investigation in the Guadalupe Mountains National Park*, ed. H. H. Genoways and R. J. Baker. Proceedings series No. 4. National Park Service.

Burgess, T. L., and D. K. Northington. 1981. *Plants of the Guadalupe Mountains and Carlsbad Caverns National Parks. An Annotated Checklist*. Contribution No. 107. Chihuahuan Desert Research Institute, Alpine, Texas.

Burkart, A. 1969. *Gramineas*. In *Flora ilustrada de Entre Rios (Argentina)*, vol. 6, pt. II. Colección Científica del I.N.T.A., Buenos Aires, Argentina.

Burkart, A. 1976. A monograph of the genus *Prosopis* (Leguminosae subfamily *Mimosoideae*). Journal of the Arnold Arboretum 57:219–249.

Burkhalter, R. E. 1992. The genus *Nyssa* (Cornaceae) in North America: A revision. Sida 15(2):323–342.

Burton, G. W. 1947. Breeding bermuda grass for the southeastern United States. Agronomy Journal 39(7):551–569.

Burton, G. W. 1948. Coastal bermuda grass. Circular 10 (revised):1–12. Georgia Coastal Plain Experiment Station.

Burt-Utley, K., and J. F. Utley. 1987. Contribution toward a revision of *Hectia* (Bromeliaceae). Brittonia 39:37–43.

Bush, B. F. 1903. A list of the ferns of Texas. Bulletin of the Torrey Botanical Club 30(6):343–358.

Bush, B. F. 1905. The North American species of *Fuirena*. Annual report of the Missouri Botanical Garden 16:87–99.

Butterwick, M. 1979. A survey of the flora of Enchanted Rock and vicinity, Llano and Gillespie Counties, Texas. In *Enchanted Rock, a Natural Area Survey* no. 14. University of Texas, LBJ School of Public Affairs. Austin.

Butterwick, M. 1980. *Cucurbita digitata* (Cucurbitaceae) in Texas. Sida 8:315.

Butterwick, M., and J. M. Poole. 1980. *Hesperaloe funifera* (Agavaceae) in Texas. Sida 8:314–315.

Cabrera, A. L. 1977. Mutsieae-systematic review. In *The Biology and Chemistry of the Compositae*, ed. V. H. Heywood, J. B. Harborne, and B. L. Turner, vol. 2, 1039–1066.

Campbell, C. S. 1983. Systematics of the *Andropogon virginicus* complex (Gramineae). Journal of the Arnold Arboretum 64:171–254.

Canne, J. M. 1969. A revision of the genus *Galinsoga* (Compositae: *Helenieae*). University of Kansas Science Bulletin 48:225–267.

Cantino, P. D. 1982. A monograph of the genus *Physostegia* (Labiatae). Contribution of the Asa Gray Herbarium of Harvard University 211:1–105.

Carleton, M. A. 1892. Observations of the native plants of Oklahoma Territory and adjacent districts. Contributions from the United States National Herbarium. Smithsonian Institution 1(6):220–232.

Carlquist, S. 1976. Tribal interrelationships and phylogeny of the Asteraceae. Aliso 8:465–492.

Carlson, T. J., and W. H. Wagner, Jr. 1982. The North American distribution of the genus *Dryopteris*. Contributions from the University of Michigan Herbarium 15:141–162.

Carr, B. 1988. *Cyperus difformis* L. (Cyperaceae) new to Texas. Sida 13:25–26.

Carr, B., N. Damude, and M. Lindsay, eds. 1993. *Endangered, Threatened and Watch Lists of Texas Plants*. 3d rev. Publication No. 9. Texas Organization for Endangered Species. Austin.

Carr, B., and D. Hernández. 1993. *Polygonum caespitosum* var. *longisetum* (Polygonaceae) new to Texas. Sida 15(4):656.

Carr, B., and P. McNeal. 1992. *Manihot subspicata* (Euphorbiaceae) new to Texas. Sida 15(2):347.

Carr, W. R., and M. H. Mayfield. 1993. *Chamaesyce velleriflora* (Euphorbiaceae) new to Texas. Sida 15(3): 550–551.

Carter, J. R. 1984. A systematic study of the New World species of section *Umbellati* of *Cyperus*. Ph.D. dissertation, Vanderbilt University, Nashville, Tennessee.

Carter, R. 1990. *Cyperus entrerianus* (Cyperaceae), an overlooked species in temperate North America. Sida 14(1):69–77.

Carter, R., and R. Kral. 1990. *Cyperus echinatus* and *Cyperus croceus*, the correct names for North American *Cyperus ovularis* and *Cyperus globulosus* (Cyperaceae). Taxon 39(2):322–327.

Carulli, J. P., A. O. Tucker, and N. H. Dill. 1988. *Aeschynomene rudis* Benth. (Fabaceae) in the United States. Bartonia 54:18–20.

Catling, P. M., and K. C. McIntosh. 1979. Rediscovery of *Spiranthes parksii* Correll. Sida 8:188–193.

Celarier, R. P., and J. R. Harlan. 1958. The cytogeography of the *Bothriochloa ischaemum* complex. Gramineae: I. Taxonomy and Geographic Distribution. Journal of the Linnean Society. Botany 55(363): 755–760.

Center for Plant Conservation. 1995. *1995 Plant Conservation Directory*. St. Louis: Missouri Botanical Garden, Center for Plant Conservation.

Chambers, K. L. 1973. In The genera of *Lactuceae* (Compositae) in the southeastern United States, ed. B. S. Vuilleumier. Journal of the Arnold Arboretum 54:52.

Chambless, L. F. 1972. The woody vegetation of the Angelina River bottom in Nacogdoches County, Texas. Master's thesis, Stephen F. Austin State University, Nacogdoches, Texas.

Channel, R. B., and C. E. Wood, Jr. 1987. The Buxaceaea in the southeastern United States. Journal of the Arnold Arboretum 68(2):241–257.

Chapman, G. C., and S. B. Jones. 1978. Biosystematics of the Texanae Vernonias (*Vernonieae*: Compositae). Sida 7:264–281.

Chase, A. 1921. The North Amerian species of *Pennisetum*. Contributions from the United States National Herbarium. Smithsonian Institution 22(4):209–234.

Chase, A. 1929. The North American species of *Paspalum*. Contributions from the United States National Herbarium. Smithsonian Institution 29(1): 1–310.

Chase, A. 1951. *Manual of the grasses of the United States*. 2d ed. Misc. Pub. No. 200. United States Department of Agriculture. Washington, D.C.: United States Government Printing Office.

Chiang, C. F. 1981. A taxonomic study of the North American species of *Lycium* (Solanaceae). Ph.D. dissertation, University of Texas, Austin.

Christensen, C. 1930. The genus *Cyrtomium*. American Fern Journal 20:41–52.

Clare, M. M. 1907. A flora of Bexar County, Texas. Ph.D. dissertation, The Catholic University of America, Washington, D.C.

Clark, C. 1975. Ecogeographic races of *Lesquerella engelmannii* (Cruciferae); distribution, chromosome numbers, and taxonomy. Brittonia 27:263–278.

Clark, C. 1978. Systematic studies of *Eschscholzia* (Papaveraceae): I. The origin and affinities of *E. mexicana*. Systematic Botany 3:374–385.

Clark, R. B. 1942. A revision of the genus *Bumelia* in the United States. Annals of the Missouri Botanical Garden 29:155–182.

Clausen, R. T. 1938. A monograph of the Ophioglossaceae. Memoirs of the Torrey Botanical Club 19(2):1–177.

Clausen, R. T. 1946. *Selaginella* subgenus *Euselaginella*, in the southeastern United States. American Fern Journal 36:65–82.

Clausen, R. T. 1975. *Sedum* of North America, North of the Mexican Plateau. Ithaca, New York: Cornell University Press.

Clayton, W. D. 1965. The *Sporobolus indicus* complex. Kew Bulletin 19:287–293.

Clayton, W. D. 1969. A revision of the genus *Hyparrhenia*. Kew Bulletin 2:1–196.

Clayton, W. D. 1970. Studies in the Gramineae: XXI. *Coelorachis* and *Rhytachne*: A study in numerical taxonomy. Kew Bulletin 24(2):309–314.

Clayton, W. D. 1974. *Cynodon*. In *Flora of Tropical East Africa*, ed. W. D. Clayton, S. M. Philips, and S. A. Renoize, pt. 2: *Gramineae*, 316–321. London: Crown Agents for Overseas Governments and Administrations.

Clayton, W. D. 1980. Proposal to conserve or reject: (533) Proposal to conserve *Rottboellia* L. f. 1781 against *Rottboellia* L. f. 1779. Taxon 29:691–692.

Clayton, W. D. 1981. Notes on the tribe Andropogoneae (Gramineae). Kew Bulletin 35(4):813–818.

Clayton, W. D., and J. R. Harlan. 1970. The genus *Cynodon* L. C. Rich. in Tropical Africa. Kew Bulletin 24: 185–189.

Clayton, W. D., and S. A. Renvoize. 1986. *Genera Graminium: Grasses of the World*. Kew Bulletin, additional series XIII. London: Her Majesty's Stationary Office.

Clewell, A. F. 1966. Native North American species of *Lespedeza* (Leguminosae). Rhodora 68:359–405.

Clewell, A. F. 1985. *Guide to the Vascular Plants of the Florida Panhandle*. Tallahassee: University Press of Florida.

Clewell, A. F., and J. W. Wooten. 1971. A revision of *Ageratina* (Compositae). Brittonia 23:123–143.

Clover, E. 1937. Vegetational survey of the lower Rio Grande Valley, Texas. Madroño 4:41–72.

Coffee, J. V., and S. B. Jones, Jr. 1980. Biosystematics of *Lysimachia* section *Seleucia* (Primulaceae). Brittonia 32:309–322.

Coffey, C. R. 1986. A floristic study of the La Copita Research Area in Jim Wells County, Texas. Master's thesis, Texas A&M University, College Station.

Coffey, C. R., and J. Valdés-R. 1986. *Monerma cylindrica* (Poaceae: *Monermeae*) new to Texas. Sida 11: 352–353.

Coffey, V. J., and S. B. Jones, Jr. 1980. Biosystematics of *Lysimachia* section *Seleucia* (Primulaceae). Brittonia 32:309–322.

Collins, S. L., and W. H. Blackwell. 1979. *Bassia* (Chenopodiaceae) in North America. Sida 8:57–64.

Collins, T. 1973. Revision of *Orobanche* (Orobanchaceae). Ph.D. dissertation, University of Wisconsin, Milwaukee.

Comeaux, B. L. 1986. Overview of the native grapes of Texas. Proceedings of the Texas Grape Growers Association 10:15.

Conert, H. J., and A. M. Turpe. 1974. Revision der Gattung *Schismus* (Poaceae: *Arundinoideae: Danthonieae*). Abhandlungen her von der Senckenbergischen Naturforschenden Gesellschaft 532:1–81.

Conrad, J. H. 1978. *Texas Wildflower Bibliography*. 2d ed. East Texas State University, James G. Gee Library. Commerce.

Constance, C. 1949. A revision of *Phacelia* subgenus *Cosmanthus*. Contributions of the Asa Gray Herbarium of Harvard University. CLXVIII:1–48.

Cook, C. D. K., and R. Lüönd. 1982. A revision of the genus *Hydrilla* (Hydrocharitaceae). Aquatic Botany 13(4):485–504.

Cook, C. D. K., and K. Urmi-König. 1984. A revision of the genus *Egeria* (Hydrocharitaceae). Aquatic Botany 19(1/2):73–96.

Core, E. L. 1936. The American species of *Scleria*. Brittonia 2:1–105.

Correll, D. S. 1944. Orchidaceae. Flora of Texas, Vol. 3, part 3. Texas Research Foundation, Renner, Texas.

Correll, D. S. 1949. A preliminary survey of the distribution of Texas Pteriodphyta. Wrightia 1(5): 247–278.

Correll, D. S. 1950. *Native Orchids of North America North of México*. Waltham, Massachusetts: Chronica Botanica Co.

Correll, D. S. 1956. *Ferns and Fern Allies of Texas*. Renner, Texas: Texas Research Foundation.

Correll, D. S. 1961. Salicaceae: *Populus*. Flora of Texas, Vol. 3, part 6. Texas Research Foundation, Renner, Texas.

Correll, D. S. 1965. Some additions and corrections to the flora of Texas. Wrightia 3(7):126–140.

Correll, D. S. 1966a. Two new plants in Texas. Wrightia 3(8):188–191.

Correll, D. S. 1966b. Some additions and corrections to the flora of Texas: II. Brittonia 18(4):306–310.

Correll, D. S. 1966c. Some additions and corrections to the flora of Texas: III. Rhodora 68(776):420–428.

Correll, D. S. 1968. Some additions and corrections to the flora of Texas. Wrightia 6:74–78.

Correll, D. S. 1972. Manual of the vascular plants of Texas: I. Additions and corrections. American Midland Naturalist 88:490–496.

Correll, D. S., and H. B. Correll. 1975. *Aquatic and Wetland Plants of Southwestern United States*. Stanford, California: Stanford University Press.

Correll, D. S., and M. C. Johnston. 1970. *Manual of the Vascular Plants of Texas*. Texas Research Foundation, Renner, Texas.

Cory, V. L. 1931. A new *Selenia* from the Edwards Plateau of Texas. Rhodora 33:142–144.

Cory, V. L. 1937. Some new plants from Texas. Rhodora 39(466):417–423.

Cory, V. L. 1938. Notes on *Ephedera* in Texas. Rhodora 40(473):215–218.

Cory, V. L. 1939. Notes on Texas plants. Rhodora 41(492):561–563.

Cory, V. L. 1943. A new *Streptanthus* from the Big Bend of Texas. Rhodora 45:258–260.

Cory, V. L. 1944a. *Forestiera* in southern and southwestern Texas. Madroño 7(8):252–255.

Cory, V. L. 1944b. *Paronychia* in Central and western Texas. Rhodora 46:278–281.

Cory, V. L. 1945. A new *Argythamnia* from Texas. Madroño 8(3):91–92.

Cory, V. L. 1946. The genus *Palafoxia* in Texas. Rhodora 48:84–86.

Cory, V. L. 1947. A new *Dyssodia* from Texas. Rhodora 49:161–163.

Cory, V. L., and H. B. Parks. 1937. *Catalogue of the Flora of the State of Texas*. Bulletin No. 550. Texas Agricultural Experiment Station, College Station.

Coulter, J. M. 1891–1894. *Botany of Western Texas: A Manual of the Phaenograms and Pteriophytes of Western Texas*. Contributions from the United States National Herbarium. Smithsonian Institution, vol. 2. Washington, D.C.

Cox, P. B., and L. E. Urbatsch. 1989. *Rudbeckia texana,* a taxon worthy of specific status (Asteraceae: *Heliantheae*). Phytologia 67:366–367.

Cox, P. B., and L. E. Urbatsch. 1994. A taxonomic revision of *Rudbeckia* subg. macroline (Asteraceae: *Heliantheae; Rudbeckiinae*). Castanea 59:300–318.

Crawford, D. J. 1975. Systematic relationships in the narrow-leaved species of *Chenopodium* of the western United States. Brittonia 27:279–288.

Crawford, D. J. 1977. A study of morphological variability in *Chenopodium incanum* (Chenopodiaceae) and the recognition of two new varieties. Brittonia 29:291–296.

Crins, W. J. 1989. The Tamaricaceae in the southeastern United States. Journal of the Arnold Arboretum 70(3):403–425.

Cronquist, A. 1945. Studies in the Sapotaceae, III, *Dipholis* and *Bumelia*. Journal of the Arnold Arboretum 26:447–470.

Cronquist, A. 1964. *Salix*. In *Vascular Plants of the Pacific Northwest*, ed. C. L. Hitchcock and A. Cronquist, vol. 2, 37–70.

Cronquist, A. 1968. *The Evolution and Classification of Flowering Plants*. Boston: Houghton Mifflin.

Cronquist, A. 1977. The Compositae revisited. Brittonia 29:137–153.

Cronquist, A. 1980. *Vascular Flora of the Southeastern United States*. Vol. 1, *Asteraceae*. Chapel Hill: University of North Carolina Press.

Cronquist, A. 1981. *An Integrated System of Classification of Flowering Plants*. New York: Columbia University Press.

Cronquist, A., T. M. Barkley, G. Morton, R. I. Ediger, R. W. Pippen, and J. L. Strother. 1978. Compositae tribe *Senecioneae*. North American Flora II 10: 14–179.

Crow, G. E. 1978. A taxonomic revision of *Sagina* (Caryophyllaceae) in North America. Rhodora 80:1–91.

Curry, M. G. 1976. *Elephantopus* (Compositae) of Louisiana: Taxonomy, distribution, and field key. Proceedings of the Louisiana Academy of Sciences 39: 50–53.

Cutler, H. C. 1939. Monograph of the North American species of the genus *Ephedra*. Annals of the Missouri Botanical Garden 26:373–429.

Cutler, H. C., and E. Anderson. 1941. A preliminary survey of the genus *Tripsacum*. Annals of the Missouri Botanical Garden 28:249–269.

Dahling, G. V. 1978. Systematics of *Garrya*. Contributions of the Gray Herbarium of Harvard University 209:1–104.

Daniel, T. F. 1980a. The genus *Justicia* in the Chihuahuan Desert. Contributions from the University of Michigan Herbarium 14:61–67.

Daniel, T. F. 1980b. Range extensions of *Carlowrightia* (Acanthaceae) and a key to the species of the United States. Southwestern Naturalist 25:425–426.

Daniel, T. F. 1983. *Carlowrightia* (Acanthaceae). Flora Neotropica. Monograph 34:1–116.

Daniel, T. F. 1984a. The Acanthaceae of the southwestern United States. Desert Plant Life 5:162–179.

Daniel, T. F. 1984b. A revision of *Stenandrium* (Acanthaceae) in México and adjacent regions. Annals of the Missouri Botanical Garden 71:1028–1043.

Daniel, T. F. 1986. Systematics of *Tetramerium* (Acanthaceae). Systematic Botany Monograph No. 11.

Daoud, H. S., and R. L. Wilbur. 1965. A revision of the North American species of *Helianthemum* (Cistaceae). Rhodora 67:63–312.

Darbyshire, S. J. 1993. Realignment of *Festuca* subgenus *Schedonorus* with the genus *Lolium* (Poaceae). Novon 3:239–243.

Darke, R. 1994. *Manual of Grasses. A series from The New Royal Horticultural Dictionary of Gardening*. M. Griffiths, series ed. Portland, Oregon: Timber Press.

Darlington, J. 1934. A monograph of the genus *Mentzelia*. Annals of the Missouri Botanical Garden 21: 103–220.

Davenport, L. J. 1988. A monograph of *Hydrolea* (Hydrophyllaceae). Rhodora 90:169–208.

Davey, J. C., and W. D. Clayton. 1978. Some multiple discriminant function studies on *Oplismenus* (Gramineae). Kew Bulletin 33(1):147–157.

Davidse, G., M. Sousa S., A. O. Chater, eds. 1994. *Flora Mesoamericana*, vol. 6. Universidad Nacional Autónoma de México, Instituto de Biología. Ciudad Universitaria, México, D. F.

Davies, R. S. 1980. Introgression between *Elymus canadensis* L., and *E. virginicus* L. (*Triticeae*, Poaceae) in South Central United States. Ph.D. dissertation, Texas A&M University, College Station.

Davila, P. D. 1988. Systematic revision of the genus *Sorghastrum* (Poaceae: *Andropogoneae*). Ph.D. dissertation, Iowa State University, Ames.

DeJong, D. C. D. 1965. A systematic study of the genus *Astranthium* (Compositae, *Asterae*). Michigan State University Museum Biological Series 2:431–528.

Delgado-Salinas, A. 1981. *Macroptilium gibbosifolium* (Ortega) A. Delgado, the correct name for *M. heterophyllum* (Willd.) Maréchal & Baudet (Fabaceae: *Phaseolinae*). Systematic Botany 6:294–296.

Delgado-Salinas, A. 1985. Systematics of the genus *Phaseolus* in North and Central America. Ph.D. dissertation, University of Texas, Austin.

DeLisle, D. G. 1963. Taxonomy and distribution of the genus *Cenchrus*. Iowa State College Journal of Science 37(3):259–351.

Dempster, L. T. 1973. The polygamous species of the genus *Galium* (Rubiaceae) section *Laphogalium* of México and southwestern United States. University of California Publications in Botany 64:1–36.

Dempster, L. T. 1976. *Galium mexicanum* (Rubiaceae) of Central America and western North America. Madroño 23:378–386.

Dennis, W. M. 1979. *Clematis pitcheri* T. & G. var. *dicyota* (Greene) Dennis, comb. *nov*. (Ranunculaceae). Sida 8:194–195.

Dennis, W. M., and D. H. Webb. 1981. The distribution of *Pilularia americana* R. Br. (Marsileaceae) in North America and México. Sida 9:19–24.

Denton, M. F. 1973. A monograph of *Oxalis*, section *Ionoxalis* (Oxalidaceae). Biological series 4(10): 460–615. Publications of the Museum. Michigan State University, East Lansing.

Denton, M. F. 1978. A taxonomic treatment of the *Luzulae* group of *Cyperus*. Contributions from the University of Michigan Herbarium 11(4):197–271.

Detling, L. E. 1939. A revision of the North American species of *Descurainia*. American Midland Naturalist 22:481–520.

de Wet, J. M. J., and J. R. Harlan. 1968. Taxonomy of *Dichanthium* section *Dichanthium* (Gramineae). Boletín de la Sociedad Argentina de Botánica 12: 206–277.

de Wet, J. M. J., and J. R. Harlan. 1970. Biosystematics of *Cynodon* L. C. Richard (Gramineae). Taxon 19: 565–569.

Diamond, D. D., D. H. Riskind, and S. L. Orzell. 1988. A framework for plant community classification and conservation in Texas. Texas Journal of Science 39(3):203–221.

Dietrich, W., and W. L. Wagner. 1987. A new combination and new subspecies in *Oenothera elata* Kunth (Onagraceae). Annals of the Missouri Botanical Garden 74:151–152.

Dietrich, W., and W. L. Wagner. 1988. Systematics of *Oenothera* section *Oenothera* subsection *Raimannia* and subsection *Nutantig* (Onagraceae). Systematic Botany Monograph 24:1–91.

Dillon, M. O. 1984. A systematic study of *Flourensia* (Asteraceae, *Heliantheae*). Fieldiana: Botany 16: 1–66.

Donoghue, M. 1982. Systematic studies in the genus *Viburnum*. Ph.D. dissertation, Harvard University.

Dorn, R. D. 1988. *Chenopodium simplex*, an older name for *C. gigantospermum* (Chenopodiaceae). Madroño 35:162.

Dorn, R. P. 1976. Synopsis of American *Salix*. Canadian Journal of Botany 54:2769–2789.

Dorr, L. J. 1990. A revision of the North American genus *Callirhoë* (Malvaceae). Memoirs of the New York Botanical Garden 56:1–74.

Dorr, L. J., and C. Barnett. 1986. The identity of *Nephropetalum* (Sterculiaceae). Taxon 35:163–164.

Dorr, L. J., and K. C. Nixon. 1985. Typification of the oak (*Quercus*) taxa described by S. B. Buckley (1809–1884). Taxon 34:211–228.

Dorr, L. J., and J. Olsen. 1983. Lectotypification of *Verbesina texana* Buckl. (Asteraceae). Sida 10:82–85.

Dudley, T. R., and F. S. Santamour, Jr. 1992. *Cornus florida* subsp. *urbiniana* (Rose) Rickett from México: The correct name for "*C. florida* var. *pringlei*." Phytologia 73(3):169–179.

Duncan, T. A. 1980. A taxonomic study of the *Ranunculus hispidus* Michx. complex in the Western Hemisphere. University of California Publications in Botany 77:1–125.

Duncan, T. A., and C. S. Keener. 1991. A classification of the Ranunculaceae with special reference to the Western Hemisphere. Phytologia 70(1):24–27.

Duncan, W. H. 1964. New *Elatine* (Elatinaceae) populations in the southeastern United States. Rhodora 66:47–53.

Duncan, W. H. 1967. Woody vines of the southeastern states. Sida 3:1–76.

Duncan, W. H. 1979. Changes in *Galactia* (Fabaceae) of the southeastern United States. Sida 8(2):170–180.

Duncan, W. H., and M. B. Duncan. 1988. *Trees of the Southeastern United States*. Athens: University of Georgia Press.

Dunn, D. B. 1965. *Lupinus* notes: III. A reevaluation and redefinition of the *Lupinus perennis* L. complex. Leaflets of Western Botany 10:151–154.

Dyksterhuis, E. J. 1946. The vegetation of the Fort Worth Plains Prairie. Ecological Monographs 16: 1–29.

Dyksterhuis, E. J. 1948. The vegetation of Western Cross Timbers. Ecological Monographs 18(3): 325–376.

Eckenwalder, J. E. 1977. North American cottonwoods (*Populus*, Salicaceae) of sections *Abaso* and *Aigeiros*. Journal of the Arnold Arboretum 58:193–208.

Ediger, R. I. 1970. Revision of section *Suffruticosi* of the genus *Senecio* (Compositae). Sida 3:504–524.

Egger, M. 1995. New natural hybrid combinations and comments on interpretation of hybrid populations in *Castilleja* (Scrophulariaceae). Phytologia 77(5): 381–389. [Note: This volume is marked 1994 but was not published until March 1995.]

Eggers, D. M. 1969. A revision of *Valerianella* in North America. Ph.D. dissertation, Vanderbilt University, Nashville.

Eifert, I. J. 1972. New combinations in *Hoffmannseggia* Cav., and *Caesalpinia* L. Sida 5:43–44.

Elias, T. S. 1971. The genera of Fagaceae in the southeastern United States. Journal of the Arnold Arboretum 52:159–195.

Elisens, W. J. 1985. Monograph of the *Maurandyinae* (Scrophulariaceae-*Antirrhineae*). Systematic Botany Monograph 5:1–97.

Elisens, W. J., R. D. Boyd, and A. D. Wolfe. 1992. Genetic and morphological divergence among varieties of *Aphanostephus skirrhobasis* (Asteraceae-*Astereae*) and related species of different chromosome numbers. Systematic Botany 17(3):380–394.

Emery, W. H. P. 1957. A cyto-taxonomic study of *Setaria macrostachya* (Gramineae) and its relatives in the southwestern United States and México. Bulletin of the Torrey Botanical Club 84:95–105.

Emery, W. H. 1977. Current status of Texas wildrice. Southwestern Naturalist 22:393–394.

Endrizzi, J. E. 1957. Cytological studies of some species and hybrids in the Eu-sorghums. The Botanical Gazette 119(1):1–10.

Enquist, M. 1987. *Wildflowers of the Texas Hill Country*. Austin, Texas: Lone Star Botanical.

Epling, C. 1942. The American species of *Scutellaria*. University of California Publications in Botany 20: 1–141.

Erdman, K. S. 1965. Taxonomy of the genus *Sphenopholis* (Gramineae). Iowa State Journal of Science 39(3):289–336.

Erickson, R. O. 1943. Taxonomy of *Clematis* section *Viorna*. Annals of the Missouri Botanical Garden 30:1–60.

Ernst, W. R. 1963. The genera of Hamamelidaceae and Platanaceae in the southeastern United States. Journal of the Arnold Arboretum 44:193–210.

Essig, F. B. 1990. The *Clematis virginiana* (Ranunculaceae) complex in the southeastern United States. Sida 14(1):49–68.

Essig, F. B., and C. E. Jarvis. 1989. Lectotypification of *Clematis virginiana* L. Taxon 38:271–277.

Estes, J. R., and R. J. Tyrl. 1982. The generic concept and generic circumscription in the *Triticeae*: and end paper. In *Grasses and Grasslands: Systematics and Ecology*, ed. J. R. Estes, R. J. Tyrl, and J. N. Brunken. Norman: University of Oklahoma Press.

Everitt, J. H., and D. L. Drawe. 1993. *Trees, Shrubs & Cacti of South Texas*. Lubbock: Texas Tech University Press.

Faden, R. B. 1993. The misconstrued and rare species of *Commelina* (Commelinaceae) in the eastern United States. Annals of the Missouri Botanical Garden 80:208–218.

Faden, R. B., and D. R. Hunt. 1991. The classification of the Commelinaceae. Taxon 40(1):19–31.

Fairey, J. E. 1967. The genus *Scleria* in the southeastern United States. Castanea 32:37–71.

Fairey, J. E. 1969. *Scleria pauciflora* Muhl., and its varieties in North America. Castanea 34:87–90.

Fairey, J. E. 1975. Nomenclatural changes in the North American species of *Scleria* (Cyperaceae), sections *Hypoporum* and *Scleria*. Association of Southeastern Biologists Bulletin 22(2):52.

Farmer, J., and C. R. Bell. 1985. A new combination in *Asclepias*. Phytologia 57:380.

Fassett, N. C. 1951. *Callitriche* in the New World. Rhodora 53:137–222.

Faust, W. Z. 1972. A biosystematic study of the interior species group of the genus *Vernonia* (Compositae). Brittonia 24:363–378.

Feddema, C. 1972. *Sclerocarpus uniserialis* (Compositae) in Texas and México. Phytologia 23:201–209.

Federal Register. 1979. Fish and Wildlife Service, endangered and threatened wildlife and plants 44(209):1–916. Washington, D.C.: United States Government Printing Office.

Ferguson, D. J. 1986. *Opuntia chisoensis* (Anthony) comb. *nov*. Cactus and Succulent Journal 58: 165–177.

Ferguson, D. J. 1989. Revision of the U.S. members of the *Echinocereus triglochidiatus* group. Cactus and Succulent Journal 61:217–224.

Ferguson, D. J. 1991. In defense of the genus *Glandulicactus* Backeb. Cactus and Succulent Journal 63: 87–91.

Ferguson, D. J. 1992. The genus *Echinocactus* Link & Otto, subgenus *Homalcephala* (Britton & Rose) *stat. nov*. Cactus and Succulent Journal 64:169–172.

Fernald, M. L. 1943. The identity of *Scleria setacea* of Poiret. Rhodora 45:296–297.

Fernald, M. L. 1950. *Gray's Manual of Botany*. 8th ed. Portland, Oregon: Dioscorides Press.

Fischer, P. C. 1980. The varieties of *Coryphantha vivipara*. Cactus and Succulent Journal 52:186–191.

Fleetwood, R. J. 1973a. Plants of Laguna Atascosa National Wildlife Refuge, Cameron County, Texas. Report RF-2352000-11. United States Department of Interior, Fish and Wildlife Service.

Fleetwood, R. J. 1973b. Plants of Santa Ana National Wildlife Refuge, Hidalgo County, Texas. Report RF-2354600-11. United States Department of Interior, Fish and Wildlife Service.

Fleetwood, R. J. 1973c. Plants of Brazoria/San Bernard National Wildlife Refuges, Brazoria County, Texas. United States Department of Interior, Fish and Wildlife Service. Mimeographed.

Flook, J. M. 1975. Additions and corrections to the flora of Texas. Sida 6(2):114.

Flora of North America Editorial Committee. 1993. *Flora of North America North of México*, vol. 2, *Pteridophytes and Gymnosperms*. New York: Oxford University Press.

Flyr, D. L. 1970. A systematic study of thc tribe *Leucophylleae* (Scrophulariaceae). Ph.D. dissertation, University of Texas, Austin.

Forbes, Jr., I. 1952. Chromosome numbers and hybrids in *Zoysia*. Agronomy Journal 44:194–199.

Fosberg, F. R. 1976. *Ipomoea indica* taxonomy: A tangle of morning glories. Botaniska notiser 129:35–38.

Fosberg, F. R. 1977. *Paspalum distichum* again. Taxon 26:201–202.

Foster, R. C. 1945. A revision of the North American species of *Nemastylis* Nutt. Contributions from the Asa Gray Herbarium of Harvard University 155: 26–44.

Fowler, B. A., and B. L. Turner. 1977. Taxonomy of *Selinocarpus* and *Ammocodon* (Nyctaginaceae). Phytologia 37:177–208.

Fraser-Jenkins, C. R. 1989. A classification of the genus *Dryopteris* (Pteriophyta: Dryopteridaceae). Bulletin of the British Museum (Natural History), Botany 18: 323–477.

Freckman, R. W. 1981. Realignments in the *Dichanthelium acuminatum* complex (Poaceae). Phytologia 48: 99–110.

Freeman, J. D. 1975. Revision of *Trillium* subgenus *Phyllantherum* (Liliaceae). Brittonia 27:1–62.

Friedland, S. 1942. The American species of *Hemicarpha*. American Journal of Botany 28(10):855–861.

Friend, W. H. 1942. *Plants of Ornamental Value for the Rio Grande Valley of Texas*. Bulletin No. 609. Texas Agricultural Experiment Station, College Station.

Fryxell, J. E. 1983. A revision of *Abutilon* sect. *Oligocarpae* (Malvaceae), including a new species from México. Madroño 30:84–92.

Fryxell, P. A. 1969. The genus *Cienfuegosia* Cav. (Malvaceae). Annals of the Missouri Botanical Garden 56: 179–250.

Fryxell, P. A. 1974. The North American Malvellas (Malvaceae). Southwestern Naturalist 19:97–103.

Fryxell, P. A. 1975. *Batesimalva* y *Meximalva*, dos géneros nuevos de Malvaceas mexicanas. Boletín de la Sociedad Botánica de México 35:23–36.

Fryxell, P. A. 1978. Neotropical segregates from *Sida* L. (Malvaceae). Brittonia 30:447–462.

Fryxell, P. A. 1979a. *The Natural History of the Cotton Tribe*. College Station: Texas A&M University Press.

Fryxell, P. A. 1979b. *Sidus sidarum*-III. *Sida rzedowskii* sp. *nov*. including a preliminary discussion of the *Sida elliottii* species group. Sida 8:123–127.

Fryxell, P. A. 1980. A revision of the American species of *Hibiscus* section *Bombicella* (Malvaceae). Technical Bulletin No. 1624. United States Department of Agriculture.

Fryxell, P. A. 1982. *Billieturnera* (Malvaceae), a new genus from Texas and México. Sida 9:195–200.

Fryxell, P. A. 1985. *Sidus sidarum*-V. The North and Central American species of *Sida*. Sida 11:62–91.

Fryxell, P. A. 1987. Revision of the genus *Anoda* (Malvaceae). Aliso 11:485–522.

Fryxell, P. A. 1988. Malvaceae of México. Systematic Botany Monograph, vol. 25.

Furlow, J. J. 1987. The *Carpinus caroliniana* complex in North America: II. Systematics. Systematic Botany 12:416–434.

Furlow, J. J. 1990. The genera of Betulaceae in the southeastern United States. Journal of the Arnold Arboretum 71(1):1–67.

Gabel, M. L. 1982. A biosystematic study of the genus *Imperata* (Gramineae: *Andropogoneae*). Ph.D. dissertation, Iowa State University, Ames.

Gale, S. 1944a. *Rhynchospora*, section *Eurhynchospora*, in Canada, the United States and the West Indies. Rhodora 46(544):89–134.

Gale, S. 1944b. *Rhynchospora*, section *Eurhynchospora*, in Canada, the United States and the West Indies (continued). Rhodora 46(545):159–197, 821–826.

Gale, S. 1944c. *Rhynchospora*, section *Eurhynchospora*, in Canada, the United States and the West Indies (continued). Rhodora 46(546):207–249, 827–831.

Gale, S. 1944d. *Rhynchospora*, section *Eurhynchospora*, in Canada, the United States and the West Indies (continued). Rhodora 46(547):255–278, 832–835.

Galloway, L. A. 1972. *Abronia macrocarpa* (Nyctaginaceae): A new species from Texas. Brittonia 24: 148–149.

Galloway, L. A. 1975. Systematics of the North American desert species of *Abronia* and *Tripterocalyx* (Nyctaginaceae). Brittonia 27:328–347.

Gandhi, K. N., and L. E. Brown. 1989. A nomenclatural note on *Vitis cinerea* and *V. berlandieri* (Vitaceae). Sida 13(4):506.

Gandhi, K. N., and B. E. Dutton. 1993. Palisot de Beauvois, the correct combining author of *Erianthus giganteus* (Poaceae). Taxon 42:855–856.

Gandhi, K. N., and P. A. Fryxell. 1990. A nomenclatural note on *Eupatorium fistulosum* (Asteraceae). Sida 14(1):129–131.

Gandhi, K. N., and S. L. Hatch. 1988. Nomenclatural changes in *Acalypha* (Euphorbiaceae) and *Chamacrista* (Fabaceae). Sida 13:122–123.

Gandhi, K. N., and R. D. Thomas. 1989. *Asteraceae of Louisiana*. Sida Miscellany No. 4:1–202.

Gandhi, K. N., R. D. Thomas, and S. L. Hatch. 1987. Cuscutaceae of Louisiana. Sida 12:361–379.

Garay, L. A. 1980. A generic revision of the *Spiranthinae*. Harvard University. Botanical Museum Leaflets 28(4):277–426.

Gardner, C. S. 1983. A systematic study of *Tillandsia* subgenus *Tillandsia*. Ph.D. dissertation, Texas A&M University, College Station.

Garrett, J. H. 1994. *Plants of the Metroplex III*. Austin: University of Texas Press.

Gastony, G. J. 1988. The *Pellaea glabella* complex: Electrophoretic evidence for the derivations of the agamosporous taxa and a revised taxonomy. American Fern Journal 78:44–67.

Gehlbach, F. R., and R. C. Gardner. 1983. Relationships of sugarmaples (*Acer saccharum* and *A. grandidentatum*) in Texas and Oklahoma with special reference to relict populations. Texas Journal of Science 35: 231–237.

Geiser, S. W. 1930. Naturalists of the Frontier: VII. Thomas Drummond. Southwest Review 15:478–512.

Geiser, S. W. 1935. Charles Wright's 1849 botanical collecting-trip from San Antonio to El Paso; with type-localities for new species. Field and Laboratory 4:23–32.

Geiser, S. W. 1942. John Allen Veatch. In Texas Collection (W. P. Webb). The Southwestern Historical Quarterly 46:167–174.

Geiser, S. W. 1947. Greenleaf Cilley Nealley (1846–1896), Texas Botanist. Field and Laboratory 15: 41–46.

Geiser, S. W. 1948. *Naturalists of the Frontier*. 2d ed. Dallas: University Press in Dallas.

Geiser, S. W. 1958. Men of Science in Texas, 1820–1880. Field and Laboratory 26:86–139.

Geiser, S. W. 1959a. Men of Science in Texas, 1820–1880: II. Field and Laboratory 27:20–48.

Geiser, S. W. 1959b. Men of Science in Texas, 1820–1880: III. Field and Laboratory 27:81–96.

Geiser, S. W. 1959c. Men of Science in Texas, 1820–1880: IV. Field and Laboratory 27:110–160.

Geiser, S. W. 1959d. Men of Science in Texas, 1820–1880. Field and Laboratory 27:163–256.

Gentry, A. H. 1992. Bignoniaceae—Part II (Tribe *Tecomeae*). Flora Neotropica. Monograph 25(2):1–370.

Gentry, H. S. 1982. *Agaves of Continental North America*. Tucson: University of Arizona Press.

Gentry, J. L. 1972. The *Agave* family in Sonora. Agricultural Handbook No. 399. United States Department of Agriculture. Washington, D.C.: United States Government Printing Office.

Gentry, J. L. 1974. Studies in the genus *Hackelia* (Boraginaceae) in the western United States and México. Southwestern Naturalist 19:139–146.

Gentry, J. L., and R. L. Carr. 1976. A revision of the genus *Hackelia* (Boraginaceae) in North America, north of México. Memoirs of the New York Botanical Garden 26:121–227.

Gillespie, T. S. 1976. The flowering plants of Mustang Island, Texas—an annotated checklist. Texas Journal of Science 27(1):131–148.

Gillis, W. T. 1971. The systematics and ecology of poison-ivy and the poison-oaks (*Toxicodendron*, Anacardiaceae). Rhodora 73:72–159, 161–237, 371–443, 465–540.

Gillis, W. T. 1977. *Pluchea* revisited. Taxon 26: 587–591.

Ginzbarg, S. 1992. A new disjunct variety of *Croton alabamensis* (Euphorbiaceae) from Texas. Sida 15(1): 41–52.

Gleason, H. A., and A. Cronquist. 1991. *Manual of Vascular Plants of Northeastern United States and Adjacent Canada*. 2d ed. Bronx, New York: The New York Botanical Garden.

Godfrey, R. K., and J. W. Wooten. 1978. *Aquatic and Wetland Plants of Southeastern United States. Monocotyledons*. Athens: University of Georgia Press.

Godfrey, R. K., and J. W. Wooten. 1981. *Aquatic and Wetland Plants of Southeastern United States. Dicotyledons.* Athens: University of Georgia Press.

Goldberg, A. 1967. The genus *Melochia* L. (Sterculiaceae). Contributions from the United States National Herbarium. Smithsonian Institution 34: 191–363.

Goldblatt, P. 1975. Revision of the bulbous Iridaceae of North America. Brittonia 27:373–385.

Goldblatt, P. 1977. *Herbertia* (Iridaceae) reinstated as a valid generic name. Annals of the Missouri Botanical Garden 64:378–379.

Goldman, D. H. 1995. A new species of *Calopogon* from the midwestern United States. Lindleyana 10: 37–42.

Gonsoulin, G. J. 1974. A revision of *Styrax* (Styracaceae) in North America, Central America, and the Carribbean. Sida 5:191–258.

Gonzalez, G. M. 1974. The lectotype of *Potamogeton illinoensis* (Potamogetonaceae). Taxon 36(1): 112–113.

Goospeed, T. H. 1954. The genus *Nicotiana*. Chronica Botanica 16:536.

Goudswaard, P. C. 1980. The genus *Zoysia* (Gramineae) in Malesia. Blumea 26:169–175.

Gould, F. W. 1950. *Eriochloa* in Arizona. Leaflets of Western Botany 6:50–51.

Gould, F. W. 1955. *Parapholis incurva* and *Chloris polydactyla* in Texas. Field and Laboratory 23:83–84.

Gould, F. W. 1962. *Texas Plants—A Checklist and Ecological Summary*. MP-585. Texas Agricultural Experiment Station, College Station.

Gould, F. W. 1963. *Woody Plants of the Rob & Bessie Welder Wildlife Refuge*. Contribution No. 85. Welder Wildlife Foundation, Sinton, Texas.

Gould, F. W. 1967. The grass genus *Andropogon* in the United States. Brittonia 19:70–76.

Gould, F. W. 1969. *Texas Plants—A Checklist and Ecological Summary*. MP-585 Revised. Texas Agricultural Experiment Station, College Station.

Gould, F. W. 1975a. *Texas Plants—A Checklist and Ecological Summary*. MP-585 Revised. Texas Agricultural Experiment Station, College Station.

Gould, F. W. 1975b. *The Grasses of Texas*. College Station: Texas A&M University Press.

Gould, F. W. 1979. The genus *Bouteloua* (Poaceae). Annals of the Missouri Botanical Garden 66(3): 348–416.

Gould, F. W., M. A. Ali, and D. E. Fairbrothers. 1972. A revision of *Echinochloa* in the United States. American Midland Naturalist 87(1):36–59.

Gould, F. W., and T. W. Box. 1965. *Grasses of the Texas Coastal Bend*. College Station: Texas A&M University Press.

Gould, F. W., and C. A. Clark. 1978. *Dichanthelium* (Poaceae) in the United States and Canada. Annals of the Missouri Botanical Garden 65(4):1088–1132.

Govindarajalu, E. 1970. Studies in Cyperaceae III. Novelties in *Scleria* Berg. Proceedings of the Indian Academy of Science 71(6):221–225.

Graham, S. A. 1975. Taxonomy of the Lythraceae in the southeastern United States. Sida 6:80–103.

Graham, S. A. 1977. The American species of *Nesaea* (Lythraceae) and their relationship to *Heimia* and *Decodon*. Systematic Botany 2:61–71.

Graham, S. A. 1979a. The origin of *Ammannia x coccinea* Rottboell. Taxon 28:164–178.

Graham, S. A. 1979b. *Cuphea carthagenensis* (Jacquin) Macbride: The correct orthography. Sida 8: 114–115.

Graham, S. A. 1985. A revision of *Ammania* (Lythraceae) in the Western Hemisphere. Journal of the Arnold Arboretum 66(4):395–420.

Grant, V. 1956. A synopsis of *Ipomopsis*. Aliso 3: 351–362.

Grant, V., and K. A. Grant. 1979. Systematics of the *Opuntia phaeacantha* group in Texas. Botanical Gazette 140:199–207.

Grashoff, J. L. 1972. A systematic study of the North and Central American species of *Stevia*. Ph.D. dissertation, University of Texas, Austin.

Grashoff, J. L. 1974. Novelties in *Stevia*. Brittonia 26: 347–384.

Grear, J. W. 1978. A revision of the New World species of *Rhynchosia* (Leguminosae-*Faboideae*). Memoirs of the New York Botanical Garden 31:1–168.

Great Plains Flora Association. 1986. *Flora of the Great Plains*. Lawrence: University Press of Kansas.

Greenlee, J. 1992. *The Encyclopedia of Ornamental Grasses: How to Grow and Use Over 250 Beautiful and Versatile Plants*. Emmaus, Pennsylvania: Rodale Press.

Greuter, W. 1995. (1147) Proposal to conserve the name *Silene gallica* L. (Caryophyllaceae) against several synonyms of equal priority. Taxon 44(1): 102–104.

Greuter, W., F. R. Barrie, H. M. Burdet, W. G. Chaloner, V. Demoulin, D. L. Hawksworth, P. M. Jørgensen, D. H. Nicolson, P. C. Silva, P. Trehane, and J. McNeill, eds. 1994. *International Code of Botanical Nomenclature (Tokyo Code)*. Königstein, Germany: Koeltz Scientific Books.

Grimes, J. W. 1988. Systematics of New World *Psoraleeae* (Leguminosae: *Papillionoideae*). Ph.D. dissertation, University of Texas, Austin.

Grimes, J. W. 1990. A revision of the New World species of *Psoraleeae* (Leguminosae: *Papilionoideae*). Memoirs of the New York Botanical Garden 61: 1–113.

Grimmer, G. G. *The ABC's of Texas Wildflowers*. Austin, Texas: Eakin Press.

Hall, D. W. 1978. The grasses of Florida. Ph.D. dissertation, University of Florida, Tampa.

Hall, H. M. 1928. The genus *Haplopappus*. Publication No. 389. Carnegie Institution of Washington.

Ham, H. 1984. *South Texas Wildflowers*. Conner Museum, Texas A&I University, Kingsville.

Hanesworth, V. 1993. *Crepis setosa* (Asteraceae) a newly established introductant in Central Texas. Sida 15(4):658.

Hansen, B. F., and R. P. Wunderlin. 1988. Synopsis of *Dichanthelium* (Poaceae) in Florida. Annals of the Missouri Botanical Garden 75:1637–1657.

Harcombe, P. A., and P. L. Marks. 1979. Forest vegetation of the Big Thicket National Preserve: Report. Report to Office of Natural Sciences, Southwest Region, National Park Service, Santa Fe, New Mexico.

Hardin, J. W. 1957. A revision of the American *Hippocastanaceae*: II. Brittonia 9(4):173–195.

Hardin, J. W. 1971. Studies of the southeastern United States flora: II. The gymnosperms. Journal of the Elisha Mitchell Scientific Society 87:43–50.

Hardin, J. W. 1972. Studies of the southeastern United States flora: III. Magnoliaceae and Illiceaceae. Journal of the Elisha Mitchell Scientific Society 88: 30–32.

Hardin, J. W. 1990. Variation patterns and recognition of varieties of *Tilia americana*. Systematic Botany 15(1):33–48.

Harlan, J. R., and J. M. J. de Wet. 1970. A guide to the species of *Cynodon* (Gramineae). Oklahoma Agricultural Experiment Station Bulletin B-673. Oklahoma State University.

Harlan, J. R., J. M. J. de Wet, and W. L. Richardson. 1969. Hybridization studies with species of *Cynodon* from East Africa and Malagasy. American Journal of Botany 56(8):944–950.

Harms, L. J. 1970. *Juncus kansanus*: A synonym of *J. brachyphyllus* (Juncaceae). Sida 3:525–528.

Harms, L. J. 1972. Cytotaxonomy of the *Eleocharis tenuis* complex. American Journal of Botany 59: 483–487.

Hart, J. A. , and R. A. Price. 1990. The genera of Cupressaceae (including Taxodiaceae) in the southeastern United States. Journal of the Arnold Arboretum 71(3):275–322.

Hart, T. W., and W. H. Eshbaugh. 1976. The biosystematics of *Cardamine bulbosa* (Muhlenb.) B.S.P., and *C. douglassii* Britt. Rhodora 78:329–419.

Hartman, R. L. 1979. *Cerastium clawsonii* (Caryophyllaceae): A synonym of *Linum hudsonioides* (Linaceae). Rhodora 81:283.

Hartman, R. L., J. D. Bacon, and C. F. Bohnstedt. 1975. Biosystematics of *Draba cuneifolia* and *D. platycarpa* (Cruciferae) with emphasis on volatile and flavonoid constituents. Britonnia 27:317–327.

Hartman, R. L., and M. A. Lane. 1987. A new species of *Machaeranthera* section *Psilactis* (Asteraceae-*Astereae*) from Central Texas. Brittonia 39:253–257.

Hartog, C. D. 1959. A key to the species of *Halophila* (Hydrocharitaceae), with descriptions of the American species. Acta Botanica Neerlandica 8(4): 484–489.

Hartog, C. D. 1964. An approach to the taxonomy of the sea-grass genus *Halodule* Endl. (Potamogetonaceae). Blumea 12(2):289–312.

Hartog, C. D. 1970. The sea-grasses of the world. Mémoires de l'academie royale des sciences coloniales. Classe des sciences naturelles et médicales. Verhandelingen. Koninklike academie voor koloniale wetenschappen. Klasse der natuur—en geneeskundige wetenschappen. 59(1):1–275.

Hartog, C. D., and F. V. D. Plas. 1970. A synopsis of the Lemnaceae. Blumea 18:355–368.

Harvey, L. H. 1948. *Eragrostis* in North and Middle America. Ph.D. dissertation, University of Michigan, Ann Arbor.

Hatch, S. L. 1975. A biosystematic study of the *Schizachyrium cirratum-S. sanguineum* complex. Ph.D. dissertation, Texas A&M University, College Station.

Hatch, S. L. 1978. Nomenclatural changes in *Schiazchyrium* (Poaceae). Brittonia 30(4):496.

Hatch, S. L. 1984. A new combination in Poaceae. Sida 10(4):321.

Hatch, S. L. 1995. *Salvinia minima* (Salvinaceae), new to Texas. Sida 16(3):595.

Hatch, S. L., and D. A. Bearden. 1983. *Stipa curvifolia* (Poaceae)—studies on a rare taxon. Sida 10(2): 184–187.

Hatch, S. L., and C. A. Clark. 1977. New plant distribution and extension records for Texas and the U.S. Southwestern Naturalist 22:139–140.

Hatch, S. L., K. N. Gandhi, and L. E. Brown. 1990. *Checklist of the Vascular Plants of Texas*. MP-1655. Texas Agricultural Experiment Station, College Station.

Haufler, C. H. 1979. A biosystematic revision of *Bommeria*. Journal of the Arnold Arboretum 60: 445–476.

Hauke, R. L. 1963. A taxonomic monograph of *Equisetum* subgenus *Hippochaete*. Beichefte zur Nova Hedwigia 8:1–123.

Hauke, R. L. 1966. A systematic study of *Equisetum arvense*. Beichefte zur Nova Hedwigia 13:81–109.

Hauke, R. L. 1978. A taxonomic monograph of *Equisetum* subgenus *Equisetum*. Beichefte zur Nova Hedwigia 30:385–455.

Hayes, T., comp. 1992. Endangered, threatened, and watch list of natural communities of Texas. Publication No. 8. Texas Organization for Endangered Species, Austin.

Haynes, R. R. 1969. *Conopholis alpina* Liebmann var. *mexicana* (Gray *ex* Watson) Haynes, comb. *nov*. (Orobanchaceae). Sida 3:347.

Haynes, R. R. 1971. A monograph of the genus *Conopholis* (Orobanchaceae). Sida 4:246–264.

Haynes, R. R. 1977. The Najadaceae in the southeastern United States. Journal of the Arnold Arboretum 58(2):161–170.

Haynes, R. R. 1978. The Potamogetonaceae in the southeastern United States. Journal of the Arnold Arboretum 59(2):170–191.

Haynes, R. R. 1979. Revision of North and Central American *Najas* (Najadaceae). Sida 8(1):34–56.

Haynes, R. R. 1985. A revision of the clasping-leaved *Potamogeton* (Potamogetonaceae). Sida 11(2): 173–188.

Haynes, R. R. 1986. Typification of Linnaean species of *Potamogeton* (Potamogetonaceae). Taxon 35(3): 325–332.

Haynes, R. R. 1987. The Zannichelliaceae in the southeastern United States. Journal of the Arnold Arboretum 68(2):259–268.

Haynes, R. R., and L. B. Holm-Nielsen. 1994. The Alismataceae. Flora Neotropica. Monograph 64:1–112.

Heard, S. B., and J. C. Semple. 1988. The *Solidago rigida* complex (Compositae: *Astereae*): A multivariate morphometric analysis and chromosome numbers. Canadian Journal of Botany 66:1800–1807.

Heath, P. V. 1992. The type of *Ancistrocactus* (Cactaceae). Taxon 41(2):329–331.

Heckard, L. R. 1973. Studies in Orobanchaceae. Madroño 22:41–70.

Heep, M. R., and R. I. Lonard. 1986. *Esenbeckia berlandieri* (Rutaceae) rediscovered in extreme southern Texas. Southwestern Naturalist 31:259–260.

Heil, K., B. Armstrong, and D. Schleser. 1981. A review of the genus *Pediocactus*. Cactus and Succulent Journal 53:17–39.

Heil, K. D., and S. Brack. 1986. The cacti of Guadalupe Mountains National Park. Cactus and Succulent Journal 58:165–177.

Heil, K. D., and S. Brack. 1988. The cacti of Big Bend National Park. Cactus and Succulent Journal 60: 17–34.

Heiser, C. B., D. M. Smith, S. B. Clevenger, and W. C. Martin, Jr. 1969. The North American sunflowers (*Helianthus*). Memoirs of the Torrey Botanical Club 22:1–218.

Hennen, J. E. 1950. The true clovers (*Trifolium*) of Texas. Field and Laboratory 18:159–164.

Henrard, J. T. 1950. Monograph of the genus *Digitaria*. Leiden: Universitare Pers Leiden.

Henrickson, J. 1972. A taxonomic revision of the Fouquieriaceae. Aliso 7:439–537.

Henrickson, J. 1983. A revision of *Samolus ebracteatus* (*sensu lato*) (Primulaceae). Southwestern Naturalist 28:303–314.

Henrickson, J. 1985. A taxonomic revision of *Chilopsis* (Bignoniaceae). Aliso 11:179–197.

Henrickson, J. 1986a. *Anisacanthus quadrifidus sensu lato* (Acanthaceae). Sida 11:286–299.

Henrickson, J. 1986b. Notes on Rosaceae. Phytologia 60:468.

Henrickson, J. 1987a. Notes on *Cynanchum* (Asclepiadaceae). Sida 12:91–99.

Henrickson, J. 1987b. A taxonomic reevaluation of *Gossypianthus* and *Guilleminea* (Amaranthaceae). Sida 12:307–337.

Henrickson, J. 1988. A revision of the *Atriplex acanthocarpa* complex (Chenopodiaceae). Southwestern Naturalist 33:451–463.

Henrickson, J., and T. F. Daniel. 1979. Three new species of *Carlowrightia* (Acanthaceae) from the Chihuahuan Desert Region. Madroño 26:26–36.

Henrickson, J., and R. A. Hilsenbeck. 1979. New taxa and combinations in *Siphonoglossa* (Acanthaceae). Brittonia 31:373–378.

Henrickson, J., and E. J. Lott. 1982. New combinations in Chihuahuan Desert *Anisacanthus* (Acanthaceae). Brittonia 34:170–176.

Henrickson, J., and S. Sunberg. 1986. On the submersion of *Dicraurus* into *Iresine* (Amaranthaceae). Aliso 11:355–364.

Herman, W. C. 1915. The botany of Texas: An account of botanical investigations in Texas and adjoining territory. Bulletin of the University of Texas, No. 18, Austin.

Hermann, F. J. 1970. *Manual of the Carices of the Rocky Mountains and Colorado Basin*. Agricultural Handbook 374. United States Department of Agriculture, Washington, D.C.

Hermann, F. J. 1975. *Manual of the Rushes (Juncus spp.) of the Rocky Mountains and Colorado Basin*. Forest Service General Technical Report RM-18, United States Department of Agriculture, Fort Collins, Colorado.

Herndon, A. 1992. Nomenclatural notes on North American *Hypoxis* (Hypoxidaceae). Rhodora 94(877):43–47.

Hess, W. J., and J. Henrickson. 1987. A taxonomic revision of *Vauquelinia* (Rosaceae). Sida 12:101–163.

Hevly, R. H. 1965. Studies of the sinuous cloak fern (*Notholaena sinuata*) complex. Journal of the Arizona Academy of Science 3:205–208.

Hewitson, W. 1962. Comparative morphology of the Osmundaceae. Annals of the Missouri Botanical Garden 49:57–93.

Heywood, V. H., J. B. Harborne, and B. L. Turner, eds. 1977. *The Biology and Chemistry of the Compositae*, vols. I –II. London: Academic Press.

Heywood, V. H., and C. J. Humphries. 1977. In V. H. Heywood, J. B. Harborne, and B. L. Turner (eds.). *Anthemideae*-systematics review. *The Biology and Chemistry of the Compositae* 2:851–898.

Hicks, R. R., and G. K. Stephenson. 1978. *Woody Plants of the Western Gulf Region*. Dubuque, Iowa: Kendall/Hunt Publishing Company.

Higgins, L. C. 1971. A revision of *Cryptantha* subgenus *Oreocarya*. Brigham Young University, Science Bulletin, Biological series 12:1–63.

Higgins, L. C. 1974. *Rumex venosus* (Polygonaceae) new to Texas. Southwestern Naturalist 19:329.

Highnight, K. W., J. K. Wipff, and S. L. Hatch. 1988. *Grasses (Poaceae) of the Texas Cross Timbers and Prairies*. MP-1657. Texas Agricultural Experiment Station, College Station.

Hill, S. R. 1979. A revision of the genus *Malvastrum* A. Gray (Malvaceae: *Malveae*). Ph.D. dissertation, Texas A&M University, College Station.

Hill, S. R. 1980. A new county record for *Pilularia americana* in Texas. American Fern Journal 70:28.

Hill, S. R. 1981. Supplement to flora of the Texas Coastal Bend by F. B. Jones. Sida 9:43–54.

Hill, S. R. 1982. Distributional and nomenclatural notes on the flora of the Texas Coastal Bend. Sida 9: 309–326.

Hitchcock, A. S. 1916. The scope and relations of taxonomic botany. Science 43:331–342.

Hitchcock, A. S. 1935. *Manual of the Grasses of the United States*. Misc. Pub. No. 200. United States Department of Agriculture. Washington, D.C.: United States Government Printing Office.

Hitchcock, C. L. 1936. The genus *Lepidium* in the United States. Madroño 3:265–300.

Hoff, V. J., and E. S. Nixon. 1977. Extension of *Gaura demareei* into East Texas. Southwestern Naturalist 22:135.

Holm, R. W. 1950. American species of *Sarcostemma*. Annals of the Missouri Botanical Garden 37: 480–590.

Holm, S. F. 1972. A field guide to the trees, shrubs, and wood vines of Tamberwood, Newton County, Texas. Master's thesis, Texas A&M University, College Station.

Holmes, W. C. 1983. The distribution of *Habenaria integra* (Nutt.) Spreng. (Orchidaceae) in Mississippi, Louisiana, and Texas. Southwestern Naturalist 24: 451–456.

Holmes, W. C., and B. L. Lipscomb. 1992. First documentation of *Mikania cordifolia* (Compositae) in Texas. Sida 15(1):163.

Holmes, W. C., and C. J. Wells. 1980. The distribution of *Habranthus tubispathus* (L'Her.) Taub in North America, Texas and Louisiana. Sida 8:328–333.

Holmgren, P. K. 1971. A biosystematic study of North American *Thlaspi montanum* and its allies. Memoirs of the New York Botanical Garden 21:1–106.

Holzinger, J. M. 1892. List of plants collected by C. S. Sheldon and M. A. Carleton in the Indian Territory in 1891: II. M. A. Cartleton's collection. Contributions from the United States National Herbarium. Smithsonian Institution 1(6):202–219.

Hopkins, C. O., and W. H. Blackwell, Jr. 1977. Synopsis of *Suaeda* (Chenopodiaceae) in North America. Sida 7:147–173.

Hornberger, K. L. 1991. The blue-eyed grasses (*Sisyrinchium*: Iridaceae) of Arkansas. Sida 14:597–604.

Horton, J. H. 1972. Studies in the southeastern United States Flora: IV. Polygonaceae. Journal of the Elisha Mitchell Scientific Society 88:92–102.

Hotta, M., and S. Kuroki. 1994. Taxonomical notes on plants of southern Japan: I. Proposal of *Zoyzia pacifica* stat. *nov*. (Poaceae). Acta Phytotaxonomica et geobotanica 45:67–74.

Hunt, D. R. 1975. The reunion of *Setcreasea* and *Spathotheca* with *Tradescantia*. American Commelinaceae: I. Kew Bulletin 30:443–458.

Hunt, D. R. 1980. Sections and series in *Tradescantia*. American Commelinaceae: IX. Kew Bulletin 41(1): 437–442.

Hunziker, A. T., and A. M. Anton. 1979. A synoptical revision of *Blepharidachne*. Brittonia 31:446–453.

Iltis, H. H. 1958. Studies in the Capparidaceae IV. *Polanesia* Raf. Brittonia 10(2):33–58.

Ingram, J. 1980. A revision of *Argythamnia* subgenus *Chiropetalum* (Euphorbiaceae). Gentes Herbarium 11:437–468.

Irving, R. S. 1970. Novelties in *Hedeoma* (Labiatae). Brittonia 22:338–345.

Irving, R. S. 1980a. Sections and series 1972. A revision of the genus *Poliomintha* (Labiatae). Sida 5(1):8–22.

Irving, R. S. 1980b. The systematics of *Hedeoma* (Labiatae). Sida 8:218–295.

Irwin, H. S. 1961. *Roadside Flowers of Texas*. Austin: University of Texas Press.

Irwin, H. S., and R. C. Barneby. 1975. Notes preliminary to an account of *Cassia* in the Chihuahuan Desert. Sida 6:7–18.

Irwin, H. S., and R. C. Barneby. 1976. Nomenclatural notes on *Cassia* Linnaeus (Leguminosae: *Caesalpinioideae*). Brittonia 28:435–442.

Irwin, H. S., and R. C. Barneby. 1979. New names in *Senna* P. Mill., and *Chamaecrista* Moench (Leguminosae: *Caesalpinioideae*) precursory to the Chihuahuan Desert flora. Phytologia 44:499–501.

Irwin, H. S., and R. C. Barneby. 1982. The American *Cassiinae*: A synoptic revision of Leguminosae tribe *Cassieae* subtribe *Cassiinae* in the New World. Memoirs of the New York Botanical Garden 35(1):1–918.

Isely, D. 1969. Legumes of the United States: I. Native *Acacia*. Sida 3:365–386.

Isely, D. 1970. Legumes of the United States: II. *Desmanthus* and *Neptunia*. Iowa State College Journal of Science 44:495–511.

Isely, D. 1971. Legumes of the United States: III. *Schrankia*. Sida 4:232–245.

Isely, D. 1972. Legumes of the U.S.: VI. *Calliandra, Pithecellobium, Prosopis*. Madroño 21:273–298.

Isely, D. 1973. Leguminosae of the Unites States: I. Subfamily *Mimosoidae*. Memoirs of the New York Botanical Garden 25:1–152.

Isely, D. 1975. Leguminosae of the United States: II. Subfamily *Caesalpinioideae*. Memoirs of the New York Botanical Garden 25:1–228.

Isely, D. 1978. New varieties and combinations in *Lotus, Baptisia, Thermopsis*, and *Sophora* (Leguminosae). Brittonia 30:466–472.

Isely, D. 1981. Leguminosae of the United States: III. Subfamilies *Papilionoideae*: Tribes *Sophoreae, Podalyrieae, Loteae*. Memoirs of the New York Botanical Garden 25:1–264.

Isely, D. 1986a. Notes on Leguminosae: *Papilionoideae* of the southeastern United States. Brittonia 38:352–359.

Isely, D. 1986b. Notes about *Psoralea sensu auct., Amorpha, Baptisia, Sesbania*, and *Chamaecrista* (Leguminosae) in the southeastern United States. Sida 11: 429–440.

Isely, D. 1988. Two legume emendations. Sida 13(1): 121–122.

Isely, D. 1990. *Vascular Flora of the Southeastern United States*, 3(2), *Leguminosae (Fabaceae)*. Chapel Hill: University of North Carolina Press.

Ivey, R. D. 1986. *Flowering Plants of New Mexico*. 2d ed. Albuquerque, New Mexico: Author.

Jansen, R. K. 1985. The Systematics of *Acmella* (Asteraceae-*Heliantheae*). Systematic Botany Monograph 8: 1–115.

Jeffrey, C. 1978. Further notes on the Cucurbitaceae. IV. Some New World taxa. Kew Bulletin 33: 347–380.

Johnson, B. L. 1945. Cyto-taxonomic studies in *Oryzopsis*. Botanical Gazette 107(1):1–32.

Johnson, B. L. 1962. Amphiploidy and introgression in *Stipa*. American Journal of Botany 49:253–262.

Johnson, B. L. 1972. Polyploidy as a factor in the evolution and distribtuion of grasses. In *The Biology and Utilization of Grasses*, ed. V. B. Younger and C. M. McKell. New York: Academic Press.

Johnson, D. M. 1986. Systematics of the New World Species of *Marsilea* (Marsileaceae). Systematic Botany Monograph 11:1–87.

Johnson, E. H. 1931. *The Natural Regions of Texas*. Research Monograph No. 8. Bulletin No. 3113. University of Texas, Bureau of Business Research. Austin.

Johnson, G. P. 1988. Revision of *Castanea* Sect. *Balanocastanon* (Fagaceae). Journal of the Arnold Arboretum 69(1):25–49.

Johnson, R. R. 1969. Monograph of the plant genus *Porophyllum* (Compositae: *Helenieae*). University of Kansas Science Bulletin 48:225–267.

Johnston, L. A. 1975. Revision of the *Rhamnus serrata* complex. Sida 6:67–69.

Johnston, M. C. 1957a. Synopsis of the United States species of *Forestiera* (Oleaceae). Southwestern Naturalist 1(4):140–151.

Johnston, M. C. 1957b. *Phoradendron serotinum* for *P. flavescens* (Loranthaceae): Nomenclatural corrections. Southwestern Naturalist 2:45–47.

Johnston, M. C. 1958. The Texas species of *Croton*. Southwestern Naturalist 3:175–203.

Johnston, M. C. 1967. *Ericameria austrotexana* M. C. Johnston (Compositae), *nomen novum*. Southwestern Naturalist 12:106–109.

Johnston, M. C. 1971. Revision of *Colubrina* (Rhamnaceae). Brittonia 23:2–53.

Johnston, M. C. 1974. *Acacia emoryana* in Texas and México and its relationship to *A. berlandieri* and *A. greggii*. Southwestern Naturalist 19:331–333.

Johnston, M. C. 1975. Studies of the *Euphorbia* species of the Chihuahuan Desert Region and adjacent areas. Wrightia 5:120–143.

Johnston, M. C. 1981. *Andropogon spadiceus* (Poaceae), a Coahuilan species now known from Texas. Southwestern Naturalist 25:557–558.

Johnston, M. C. 1982. *Bouteloua rigideseta* var. *chihuahuana* (Poaceae), new variety from the chihuahuan desert region. Southwestern Naturalist 27(1):29–31.

Johnston, M. C. 1984. *Rhynchosia senna* var. *texana*, new combination made necessary by the Demoulin Rule. Phytologia 54:474.

Johnston, M. C. 1988. *The Vascular Plants of Texas: A List, Updating the Manual of the Vascular Plants of Texas*. Austin, Texas: Author.

Johnston, M. C. 1990. *The Vascular Plants of Texas: A List, Updating the Manual of the Vascular Plants of Texas*. 2d ed. Austin, Texas: Author.

Johnston, M. C., and L. A. Johnston. 1969. Rhamnaceae. Flora of Texas, Vol. 2, part 2. Texas Research Foundation, Renner, Texas.

Jones, A. G. 1978. The taxonomy of *Aster* section *Multiflori* (Asteraceae): I. Nomenclatural review and formal presentation of taxa. Rhodora 80:319–357.

Jones, A. G. 1980. A classification of the New World species of *Aster* (Asteraceae). Brittonia 32:230–239.

Jones, A. G. 1981. *Aster laevis* (Asteraceae) new for Texas—a significant range extension and a new variety. Sida 9:171–175.

Jones, A. G. 1984. Nomenclatural notes on *Aster* (Asteraceae): II. New combinations and some transfers. Phytologia 55:373–387.

Jones, A. G. 1985. Nomenclatural change in *Aster* (Asteraceae). Bulletin of the Torrey Botanical Club 110: 39–42.

Jones, A. G. 1987. New combinations and status changes in *Aster* (Asteraceae). Phytologia 63: 131–133.

Jones, A. G. 1992. *Aster & Brachyactis (Asteraceae) in Oklahoma*. Sida, Botanical Miscellany No. 8.

Jones, F. B. 1961. *Flowering Plants and Ferns of the Texas Coastal Bend Counties, Sinton, Texas*. Welder Wildlife Foundation, Sinton, Texas.

Jones, F. B. 1975. *Flora of the Texas Coastal Bend*. Welder Wildlife Foundation, Sinton, Texas.

Jones, F. B. 1977. *Flora of the Texas Coastal Bend*. 2d ed. Welder Wildlife Foundation, Sinton, Texas.

Jones, F. B. 1982. *Flora of the Texas Coastal Bend*. 3d ed. Welder Wildlife Foundation, Sinton, Texas.

Jones, F. B., C. M. Rowell, Jr., and M. C. Johnston. 1961. *Flowering Plants and Ferns of the Texas Coastal Bend Counties*. Welder Wildlife Foundation, Sinton, Texas.

Jones, G. D., and S. D. Jones. 1991. *Sarcostemma clausum*, series *clausa* (Asclepiadaceae), new to Texas. Phytologia 71(2):160–162.

Jones, M. E. 1933. *Fuirena*. Contributions to Western Botany 18:25.

Jones, R. L. 1983. A systematic study of *Aster* section *patentes* (Asteraceae). Sida 10:41–81.

Jones, S. B., Jr. 1982. The genera of *Vernonieae* (Compositae) in the southeastern United States. Journal of the Arnold Arboretum 63(4):489–507.

Jones, S. B., Jr., and W. Z. Faust. 1978. Compositae tribe *Vernonieae*. North American Flora: II. 10:180–202.

Jones, S. D. 1994a. A taxonomic study of the *Carex muhlenbergii* and *C. cephalophora* complexes (Cyperaceae: *Phaestoglochin*). Ph.D. diss., Texas A&M University, College Station.

Jones, S. D. 1994b. A new species of *Carex* (Cyperaceae: *Phaestoglochin*) from Oklahoma and Texas; typification of section *Phaestoglochin*, and notes on sections *Bracteosae* and *Phaestoglochin*. Sida 16(2):341–353.

Jones, S. D., and S. L. Hatch. 1990. Synopsis of *Carex* section *Lupulinae* (Cyperaceae) in Texas. Sida 14: 87–99.

Jones, S. D., and G. D. Jones. 1990. *Rhynchospora capillacea* (Cyperaceae), new to Texas. Sida 14(1): 134–135.

Jones, S. D., and G. D. Jones. 1992. *Cynodon nlemfuënsis* (Poaceae: *Chlorideae*) previously unreported in Texas. Phytologia 72(2):93–95.

Jones, S. D., and G. D. Jones. 1993. A new species of *Carex* (Cyperaceae: *Triquetrae*) from the Chisos Mountains, Texas, and a key to species of section *Triquetrae*. Sida 15:509–518.

Jones, S. D., G. D. Jones, and S. L. Hatch. 1991. The deletion of *Carex stipata* (Cyperaceae) from the Texas flora. Phytologia 71(1):1–4.

Jones, S. D., G. D. Jones, and J. K. Wipff. 1990a. *Carex fissa*, section *Multiflorae* (Cyperaceae), new to Texas. Phytologia 68(6):47–50.

Jones, S. D., G. D. Jones, and J. K. Wipff. 1990b. The rediscovery of *Carex lupuliformis*, section *Lupulinae* (Cyperaceae) in Texas. Phytologia 69(5):346–347.

Jones, S. D., G. D. Jones, and J. K. Wipff. 1991. *Kosteletzkya depressa*, section *Kosteletzkya* (Malvaceae), new to Texas. Phytologia 71(5):387–389.

Jones, S. D., and A. A. Reznicek. 1991. *Carex bicknellii* "Bicknell's sedge," new in Texas and a key to species of section *Ovales*. Phytologia 70(2):115–118.

Jones, S. D., and J. K. Wipff. 1992. *Bulbostylis barbata* (Cyperaceae) previously unreported for Texas. Phytologia 73(5):381–383.

Jones, S. D., J. K. Wipff, and G. D. Jones. 1991. The rediscovery of *Carex gigantea* (Cyperaceae) in Texas. Sida 14(3):511–512.

Jordan, R. 1991. Putative natural hybrid of *Eupatorium capillifolium* × *E. glaucescens* from Hardin County, Texas. Phytologia 71(5):360–361.

Judd, W. S. 1981a. A monograph of *Lyonia* (Ericaceae). Journal of the Arnold Arboretum 62(1):63–128.

Judd, W. S. 1981b. A monograph of *Lyonia* (Ericaceae). Journal of the Arnold Arboretum 62(2):129–209.

Judd, W. S. 1981c. A monograph of *Lyonia* (Ericaceae). Journal of the Arnold Arboretum 62(3):315–436.

Judziewicz, E. J. 1990a. *Flora of the Guianas*, ed. Gorts-Van Rijn. Series A: Phaerograms, Fascicle 8, No. 187, Gramineae (Poaceae). Koenigstein, Germany: Koeltz Scientific Books.

Judziewicz, E. J. 1990b. A new South American species of *Sacciolepis* (Poaceae: *Panicoideae: Paniceae*), with a summary of the genus in the New World. Systematic Botany 15(3):415–420.

Kam, Y. K. 1974. Developmental studies of the floret in *Oryzopsis virescens* and *O. hymenoides* (Gramineae). Canadian Journal of Botany 52:125–149.

Kam, Y. K., and J. Maze. 1974. Studies on the relationships and evolution of supraspecific taxa utilizing developmental data: II. Relationships and evolution of *Oryzopsis hymenoides, O. virescens, O. kingii, O. micrantha*, and *O. asperifolia*. Botanical Gazette 135(3): 227–247.

Kartesz, J. T. 1994a. *A Synonymized Checklist of the Vascular Flora of the United States, Canada, and Greenland*. 2d ed. Vol. 1—Checklist. Portland, Oregon: Timber Press.

Kartesz, J. T. 1994b. *A Synonymized Checklist of the Vascular Flora of the United States, Canada, and Greenland*. 2d ed. Vol. 2—Thesaurus. Portland, Oregon: Timber Press.

Kartesz, J. T., and K. N. Gandhi. 1989. Nomenclatural notes for the North American flora: I. Phytologia 67(6):461–467.

Kartesz, J. T., and K. N. Gandhi. 1990a. Nomenclatural notes for the North American flora: II. Phytologia 68(6):421–427.

Kartesz, J. T., and K. N. Gandhi. 1990b. Nomenclatural notes for the North American flora: III. Phytologia 69(3):129–137.

Kartesz, J. T., and K. N. Gandhi. 1990c. Nomenclatural notes for the North American flora: IV. Phytologia 69(4):301–312.

Kartesz, J. T., and K. N. Gandhi. 1991a. Nomenclatural notes for the North American flora: V. Phytologia 70(3):194–208.

Kartesz, J. T., and K. N. Gandhi. 1991b. Nomenclatural notes for the North American flora: VI. Phytologia 71(1):58–65.

Kartesz, J. T., and K. N. Gandhi. 1991c. Nomenclatural notes for the North American flora: VII. Phytologia 71(2):87–100.

Kartesz, J. T., and K. N. Gandhi. 1991d. Nomenclatural notes for the North American Flora: VIII. Phytologia 71(4):269–280.

Kartesz, J. T., and K. N. Gandhi. 1992a. *Chloris barbata* Sw., and *C. elata* Desvaux (Poaceae), the earlier names for *C. inflata* Link and *C. dandyana* Adams. Rhodora 94:135–140.

Kartesz, J. T., and K. N. Gandhi. 1992b. Nomenclatural notes for the North American Flora: IX. Phytologia 72(1):17–30.

Kartesz, J. T., and K. N. Gandhi. 1992c. Nomenclatural notes for the North American Flora: X. Phytologia 72(2):80–92.

Kartesz, J. T., and K. N. Gandhi. 1992d. Nomenclatural notes for the North American Flora: XI. Phytologia 73(2):124–136.

Kartesz, J. T., and K. N. Gandhi. 1993. Nomenclatural notes for the North American Flora: XII. Phytologia 74(1):43–55.

Kartesz, J. T., and K. N. Gandhi. 1994. Nomenclatural notes for the North American Flora: XIII. Phytologia 76(6):441–457.

Kartesz, J. T., and K. N. Gandhi. 1995. Nomenclatural notes for the North American Flora: XIV. Phytologia 78(1):1–17. [Note: This volume is marked January 1995, but was not published until July 1995.]

Kartesz, J. T., and R. Kartesz. 1980. *A Synonymized Checklist of the Vascular Flora of the United States, Canada, and Greenland*. Chapel Hill: University of North Carolina Press.

Kearney, T. H. 1951. The American genera of *Malvaceae*. American Midland Naturalist 46(1):93–131.

Kearney, T. H. 1954. A tentative key to the North American species of *Sida* L. Leaflets of Western Botany 7(6):138–150.

Kearney, T. H. 1955a. *Malvastrum*, A. Gray—A redefinition of the genus. Leaflets of Western Botany 7(10): 238–241.

Kearney, T. H. 1955b. A tentative key to the North American species of *Abution* Miller. Leaflets of Western Botany 7(10):241–254.

Kearney, T. H., R. H. Peebles, and collaborators. 1960. *Arizona Flora*. 2d ed., with supplement. Berkeley: University of California Press.

Kearns, D. M. 1994. The genus *Ibervillea* (Cucurbitaceae): An enumeration of the species and two new combinations. Madroño 41(1):13–22.

Keener, C. S. 1975a. Studies in the Ranunculaceae of the southeastern United States: I. *Anemone* L. Castanea 40:36–44.

Keener, C. S. 1975b. Studies in the Ranunculaceae of the southeastern United States: III. *Clematis* L. Sida 6(1):33–47.

Keener, C. S. 1976. Studies in the Ranunculaceae of the southeastern United States: IV. Genera with zygomorphic flowers. Castanea 41:12–20.

Keener, C. S. 1977. Studies in Ranunculaceae of the southwestern United States: VI. Miscellaneous genera. Sida 7:1–12.

Keener, C. S. 1978. Nomenclatural correction in *Clematis*. Sida 7:397.

Keener, C. S. 1979. New state records for Ranunculaceae in the southeastern United States. Sida 8:114.

Keener, C. S., and W. M. Dennis. 1982. The subgeneric classification of *Clematis* (Ranunculaceae) in temperate North America north of México. Taxon 31: 37–44.

Keener, C. S., and B. E. Dutton. 1994. A new species of *Anemone* (Ranunculaceae) from Central Texas. Sida (16)1:191–202.

Keener, C. S., and S. B. Hoot. 1987. *Ranunculus* section *Echinella* (Ranunculaceae) in the southeastern United States. Sida 12:57–68.

Keeney, T., and M. Enquist. 1990. *Crataegus desertorum* (Rosaceae) rediscovered. Phytologia 69(6):471–476.

Keeney, T., and B. L. Lipscomb. 1985. Notes on two Texas plants. Sida 11:102–103.

Keil, D. J. 1975. *Pectis cylindrica* (Compositae) established as a member of the Texas flora and confirmed as a distinct species. Southwestern Naturalist 20: 286–287.

Keil, D. J. 1977. A revision of *Pectis* section *Pectothrix* (Compositae: *Tageteae*). Rhodora 79:32–78.

Keller, S. 1979. A revision of the genus *Wislizenia* (Capparidaceae) based on population studies. Brittonia 31:333–351.

Kessler, J. W. 1983. Cyperaceae new to Texas and Louisiana. Sida 10:190–191.

Kessler, J. W. 1987. A treatment of *Scleria* (Cyperaceae) for North America north of México. Sida 12:391–407.

Kessler, J. W., and L. E. Brown. 1984. *Scirpus cernuus* Vahl var. *californicus* (Torr.) Beetle (Cyperaceae) new to Texas. Sida 10:322.

Kessler, J. W., and T. Starbuck. 1983. Cyperaceae new to Texas and Louisiana. Sida 10(2):190–191.

Kim, K.-J., and B. L. Turner. 1992. Systematic overview of *Krigia* (Asteraceae-*Lactuceae*). Brittonia 44(2): 173–198.

Kim, K.-J., B. L. Turner, and R. K. Jansen. 1992. Phylogenetic and evolutionary implications of inter specific chloroplast DNA variation in *Krigia* (Asteraceae-*Lactuceae*). Systematic Botany 17:449–469.

King, B. L., and S. B. Jones. 1975. The *Vernonia lindheimeri* complex (Compositae). Brittonia 27:74–86.

King, R. M., and H. Robinson. 1970a. Studies in the *Eupatorieae* (Compositae): XVII. New combinations in *Fleischmannia*. Phytologia 19:201–207.

King, R. M., and H. Robinson. 1970b. Studies in the *Eupatorieae* (Compositae): XIX. New combinations in *Ageratina*. Phytologia 19:208–229.

King, R. M., and H. Robinson. 1970c. Studies in the *Eupatorieae* (Compositae): XII. A new genus *Shinnersia*. Phytologia 19:297–298.

King, R. M., and H. Robinson. 1970d. Studies in the *Eupatorieae* (Compositae): XIII. The genus *Conoclinum*. Phytologia 19:299–300.

King, R. M., and H. Robinson. 1970e. Studies in the *Eupatorieae* (Compositae): XXV. A new genus *Eupatoriadelphus*. Phytologia 19:431–432.

King, R. M., and H. Robinson. 1970f. Studies in the *Eupatorieae* (Compositae): XXIX. The genus *Chromolaena*. Phytologia 20:196–209.

King, R. M., and H. Robinson. 1971a. Studies in the *Eupatorieae* (Asteraceae): LXIV. The genus *Koanophyllon*. Phytologia 22:147–152.

King, R. M., and H. Robinson. 1971b. Studies in the *Eupatorieae* (Asteraceae): LVIII. A new genus *Tamaulipa*. Phytologia 22:153–155.

King, R. M., and H. Robinson. 1972a. Studies in the *Eupatorieae* (Asteraceae): LXXVIII. A new genus *Brickelliastrum*. Phytologia 24:63–64.

King, R. M., and H. Robinson. 1972b. Studies in the *Eupatorieae* (Asteraceae): LXXX. A new genus, *Flyriella*. Phytologia 24:67–69.

King, R. M., and H. Robinson. 1987. *The genera of the Eupatorieae*. Missouri Botanical Garden Monograph 22:1–581.

Kirkbride, J. H., Jr. 1993. *Biosystematic Monograph of the Genus Cucumis (Cucurbitaceae)*. Boone, North Carolina: Parkway Publishers.

Kirkpatrick, Z. M. 1992. *Wildflowers of the Western Plains: A Field Guide*. Austin: University of Texas Press.

Koch, S. D. 1974. The *Eragrosts pectinacea-pilosa* complex in North and Central America (Gramineae: *Eragrostoideae*). Illinois Biological Monograph 48: 1–74. University of Illinois Press.

Koch, S. D. 1978. Notes on the genus *Eragrostis* (Gramineae) in the southeastern United States. Rhodora 80(823):390–403.

Koch, S. D., and I. Sanchez-V. 1985. *Eragrostis mexicana, E. neomexicana, E. orcuuttiana*, and *E. virescens*: The resolution of a taxonomic problem. Phytologia 58(6):377–381.

Kopp, L. E. 1966. A taxonomic revision of the genus *Persea* in the Western Hemisphere. (*Persea*-

Lauraceae). Memoirs of the New York Botanical Garden 14:1–117.

Koyama, T. 1961. Classification of the family Cyperaceae: I. Journal of the Faculty of Science, University of Tokyo 8:37–148.

Koyama, T. 1962. The genus *Scirpus* Linn. Some North American aphylloid species. Canadian Journal of Botany 40:913–937.

Koyama, T. 1963. The genus *Scirpus* Linn. Critical species in the section *Pterolepis*. Canadian Journal of Botany 41:1107–1131.

Koyama, T., and S. Kawano. 1964. Critical taxa of grasses with North American and eastern Asiatic distribution. Canadian Journal of Botany 42:859–884.

Kral, R. 1955. Populations of *Linaria* (Scrophulariaceae) in northeastern Texas. Field and Labratory XXIII: 74–77.

Kral, R. 1960. A revision of *Asimina* and *Deeringothamnus* (Annonaceae). Brittonia 12:233–278.

Kral, R. 1966a. *Xyris* (Xyridaceae) of the Continental United States and Canada. Sida 2(3):177–260.

Kral, R. 1966b. Eriocaulaceae of Continental North America north of México. Sida 2(4):285–332.

Kral, R. 1971. A treatment of *Abildgaardia, Bulbostylis* and *Fimbristylis* (Cyperaceae) for North America. Sida 4(2):57–227.

Kral, R. 1978. A synopsis of *Fuirena* (Cyperaceae) for the Americas north of South America. Sida 7: 309–354.

Kral, R. 1983. The Xyridaceae in the southeastern United States. Journal of the Arnold Arboretum 64(3):421–429.

Kral, R. 1989. The genera of Eriocaulaceae in the southeastern United States. Journal of the Arnold Arboretum 70(1):131–142.

Kral, R., and P. E. Bostick. 1969. The genus *Rhexia* (Melastomataceae). Sida 3:387–440.

Kramer, K. U. 1990. *Pteris*. In 1990+. The families and genera of vascular plants, ed. K. Kubitzki et al. 1+Vol. Berlin 1:250–252.

Kron, K. A. 1987. A taxonomic revision of *Rhododendron* L. section *Pentanthera* G. Don. Ph.D. dissertation, University of Florida.

Kron, K. A. 1989. *Azalea rosea* Loiseleur is a superfluous name. Sida 13(3):331–333.

Kruckeberg, A. R., J. E. Rodman, and R. D. Worthington. 1982. Natural hybridization between *Streptanthus arizonicus* and *S. carinatus* (Cruciferae). Systematic Botany 7:291–299.

Kuijt, J. 1982. The Viscaceae of the southeastern United States. Journal of the Arnold Arboretum 63(4):401–410.

Kükenthal, G. 1909. Cyperaceae: *Caricoideae*. Das Pflanzenreich: IV. 20 (Heft 38): 296–353. Leipzig: W. Engelmann.

Kükenthal, G. 1935–1936. Cyperaceae-*Scirpoideae-Cypereae*. Subgen. I. *Eucyperus* (Griseb.) C. B. Clarke; Subgen. II. *Juncellus* (Griseb.) C. B. Clarke. In Engler, *Das Pflanzenreich*: IV. 20(Heft 101):42–315. Leipzig: Wilhelm-Engelmann.

Kükenthal, G. 1936. Cyperaceae-*Scirpoideae-Cypereae*. In Engler, *Das Pflanzenreich*: IV. 20(Heft 101, pt. 2). Stuttgart: Engelmann-Cramer.

Kükenthal, G. 1949. Vorarbeiten zu einer Monographie der *Rhynchosporideae*. *Rhynchospora*. Botanische Jahribücher für Systematik 74:375–509.

Kükenthal, G. 1950a. Vorarbeiten zu einer Monographie der *Rhynchosporideae*. Botanische Jahribücher für Systematik 75(1):90–126.

Kükenthal, G. 1950b. Vorarbeiten zu einer Monographie der *Rhynchosporideae*. Botanische Jahribücher für Systematik 75(2):127–195.

Kükenthal, G. 1951. Vorarbeiten zu einer Monographie der *Rhynchosporideae*. Botanische Jahribücher für Systematik 75(3):273–314.

Kunth, C. S. 1816. In *Nova genera et species plantarum*, vol. 1, ed. F. W. H. A. von Humboldt, A. J. Bonpland, and C. S. Kunth. Paris: Sumptibus librariae graecolantini-germanicae.

La Duke, J. C. 1985. A new species of *Sphaeralcea* (Malvaceae). Southwestern Naturalist 30:433–436.

La Duke, J. C., and D. K. Northington. 1978. The systematics of *Sphaeralcea coccinea* (Nutt.) Rydb. (Malvaceae). Southwestern Naturalist 23:651–660.

La Frankie, J. V., Jr. 1986a. Morphology and taxonomy of the New World species of *Maianthemum* (Liliaceae). Journal of the Arnold Arboretum 67(4): 371–439.

La Frankie, J. V., Jr. 1986b. Transfer of the species of *Smilacina* to *Maianthemum* (Liliaceae). Taxon 35: 584–589.

Lamotte, C. 1940. *Pilularia* in Texas. American Fern Journal 30:99–101.

Landolt, E. 1986. The family of Lemnaceae—A monographic study. 2 vols. Veröffentlichungen des Geobotanischen Institutes Rübel in Zürich 71:1–566.

Landrum, L. R. 1986. *Campomanesia, Pimenta, Blepharocalyx, Legrandia, Acca, Myrrhinium,* and *Luma* (Myrtaceae). Flora Neotropica. Monograph 45:1–178.

Landry, P. A revised synopsis of the pines 5: The subgenera of *Pinus*, and their morphology and behavior. Phytologia 76(1):73–79.

Lane, M. A. 1979. Taxonomy of the genus *Amphiachyris* (Asteraceae: *Astereae*). Systematic Botany 4: 178–189.

Lane, M. A. 1982. Generic limits of *Xanthocephalum, Gutierrezia, Amphiachyris, Gymnosperma, Greenella,* and *Thurovia* (Compositae: *Astereae*). Systematic Botany 7:405–416.

Lane, M. A. 1983. Taxonomy of *Xanthocephalum* (Compositae: *Astereae*). Systematic Botany 8:305–316.

Lane, M. A. 1985. Taxonomy of *Gutierrezia* (Compositae: *Astereae*) in North America. Systematic Botany 10:2–28.

Lasseter, J. S. 1984. Taxonomy of the *Vicia ludoviciana* complex (Leguminosae). Rhodora 86:475–505.

LaVala, V., and S. Sabato. 1983. Nomenclature and typification of *Ipomoea imperati* (Convolvulaceae). Taxon 32:110–114.

Lawrence, G. H. M., A. F. G. Buchheim, G. S. Daniels, and A. Dolezal. (eds.). 1968. *B-P-H: Botanico-Periodicum-Huntianum*. Pittsburg, Pennsylvania: Hunt Botanical Library.

Lazarine, P. 1980 (1981). *Common Wetland Plants of Southeast Texas*. United States Army Corps of Engineers, Galveston District, Galveston, Texas.

Le Duc, F. A. 1993. Systematic study of *Mirabilis* section *Mirabilis* (Nyctaginaceae). Ph.D. dissertation, University of Texas, Austin.

Lellinger, D. B. 1985. *A Field Manual of the Ferns & Fern-allies of the United States & Canada*. Washington, D.C.: Smithsonian Institution Press.

Lelong, M. G. 1984. New combinations for *Panicum* subgenus *Panicum* and subgenus *Dichanthelium* (Poaceae) of the southeastern United States. Brittonia 36(3):262–273.

Lelong, M. G. 1986. A taxonomic treatment of the genus *Panicum* (Poaceae) in Mississippi. Phytologia 61(4):251–269.

Lemke, D. E. 1987. First record of *Lechea pulchella* (Cistaceae) for Texas. Southwestern Naturalist 32: 278–279.

Lemke, D. E. 1991. The genus *Solanum* (Solanaceae) in Texas. Phytologia 71(5):362–378.

Lemke, D. E. 1992. *Schinus terebinthifolius* (Anacardiaceae) in Texas. Phytologia 72(1):42–44.

Lemke, D. E., and E. L. Schneider. 1988. *Xanthosoma sagittifolium* (Araceae) new to Texas. Southwestern Naturalist 33:498–499.

Lemke, D. E., and V. Wesby. 1989. *Anchusa azurea* (Boraginaceae), new to Texas. Sida 13(4):516.

Lemke, D. E., and R. D. Worthington. 1991. *Brassica* and *Rapistrum* (Brassicaceae) in Texas. Southwestern Naturalist 36(2):194–197.

Leonard, E. C. 1927. The North American species *Scutellaria*. Contributions from the United States National Herbarium. Smithsonian Institution 22: 703–748.

Les, D. H. 1985. The phytogeography of *Ceratophyllum demersum* and *C. echinatum* (Ceratophyllaceae) in glaciated North America. Canadian Journal of Botany 64:498–509.

Leuenberger, B. E. 1991. Interpretation and typification of *Cactus ficus-indica* L., and *Opuntia ficus-indica* (L.) Miller (Cactaceae). Taxon 40(4):621–627.

Leuenberger, B. E. 1993. Interpretation and typification of *Cactus opuntia* L., *Opuntia vulgaris* Mill., and *O. humifusa* (Rafin.) Rafin. (Cactaceae). Taxon 42(2): 419–429.

Lewis, I. M. 1915. *The Trees of Texas*. Bulletin of the University of Texas, No. 22. Austin.

Lewis, W. A. 1972. *Hedyotis correllii* (Rubiaceae): A new Texas species. Brittonia 24:395–397.

Lichvar, R. W. 1983. Evaluation of varieties in *Stanleya pinnata* (Cruciferae). Great Basin Naturalist 43: 684–686.

Liede, S., and F. Albers. 1994. Tribal disposition of genera in the Asclepiadaceae. Taxon 43(2):201–231.

Liew, F. S. 1972. Numerical taxonomic studies on North American lady ferns and their allies. Taiwania 17:190–221.

Lipscomb, B. L. 1978. Additions to the Texas flora. Sida 7:393–394.

Lipscomb, B. L. 1984. New additions or otherwise noteworthy plants of Texas. Sida 10:326–327.

Lipscomb, B. L., and G. Ajilvsgi. 1982. *Bellardia trixago* (L.) All. (Scrophulariaceae) adventive in Texas. Sida 9:370–374.

Lipscomb, B. L., and E. B. Smith. 1977. Morphological integradation of varieties of *Bidens aristosa* (Compositae) in northern Arkansas. Rhodora 79: 203–213.

Little, E. L., Jr. 1969. Two varietal transfers in *Carya* (hickory). Phytologia 19:186–190.

Littlejohn, R. O. 1979. Woody vegetation associated with six oxbow lakes in East Texas. Master's thesis, Stephen F. Austin State University, Nacogdoches, Texas.

Lloyd, R. M. 1974. Systematics of the genus *Ceratopteris* Brongn. (Parkeriaceae): II. Taxonomy. Brittonia 26(2):139–160.

Lonard, R. I. 1974. An artificial key to cultivated perennial plants on the West Campus of Pan American University, Edinburg, Texas. Mimeo.

Lonard, R. I. 1993. *Guide to the Grasses of the Lower Rio Grande Valley, Texas*. Edinburg, Texas: University of Texas—Pan American Press.

Lonard, R. I., J. H. Everitt, and F. W. Judd. 1991. *Woody Plants of the Lower Rio Grande Valley, Texas*. Miscellaneous Publication No. 7. University of Texas, Texas Memorial Museum. Austin.

Lonard, R. I., and F. W. Gould. 1974. The North American species of *Vulpia* (Gramineae). Madroño 22: 217–230.

Lonard, R. I., and F. W. Judd. 1981. *The Terrestrial Flora of South Padre Island*. Miscellaneous Papers No. 6. University of Texas, Texas Memorial Museum. Austin.

Lonard, R. I., F. W. Judd, and S. L. Sides. Annotated checklist of the flowering plants of South Padre Island, Texas. Southwestern Naturalist 23(3): 497–510.

Lonard, R. I., and F. R. Waller, Jr. 1971. *Helianthus simulans* (Compositae) in Robertson County new to Texas. Southwestern Naturalist 16:121.

Long, R. W. 1970. The genera of Acanthaceae in the southeastern United States. Journal of the Arnold Arboretum 51:257–309.

Lott, E. J., B. M. Boom, and F. Chiang. 1982. *Isoëtes butleri* (Isoetaceae) in Texas. Sida 9:264–266.

Lott, E. J., B. M. Boom, and F. Chiang. 1985. New combinations in Chihuahuan Desert *Aquilegia* (Ranunculaceae). Phytologia 58:488.

Lott, E. J., and M. L. Butterwick. 1980. Notes on the flora of the Chinati Mountains, Presidio County, Texas. Sida 8:348–351.

Loughmiller, C., and L. Loughmiller. 1984. *A Field Guide: Texas Wildflowers*. Austin: University of Texas Press.

Lourteig, A. 1952. Mayacaceae. Notulae Systematicae (Paris) 14(4):234–248.

Lowden, R. M. 1973. Revision of the genus *Pontederia* L. Rhodora 75(803):426–487.

Lowden, R. M. 1978. Studies on the submerged genus *Ceratophyllum* L. in the Neotropics. Aquatic Botany 4:127–142.

Lowden, R. M. 1982. An approach to the taxonomy of *Vallisneria* (Hydrocharitaceae). Aquatic Botany 13(3):269–298.

Lowry, P. P., and A. G. Jones. 1984. Systematics of *Osmorhiza* Raf. (Apiaceae: *Apioideae*). Annals of the Missouri Botanical Garden 71:1128–1171.

Luckow, L. 1993. Monograph of *Desmanthus* (Leguminosae-*Mimosoideae*). Systematic Botany Monograph 38:1–166.

Luer, C. A. 1972. *The Native Orchids of Florida*. Bronx, New York: The New York Botanical Garden.

Luer, C. A. 1975. *The Native Orchids of the United States and Canada Excluding Florida*. Bronx, New York: The New York Botanical Garden.

Lundell, C. L. 1961. *Flora of Texas*, vol. 3. Texas Research Foundation, Renner, Texas.

Lundell, C. L. 1966. *Flora of Texas*, vol. 1. Texas Research Foundation, Renner, Texas.

Lundell, C. L. 1969. *Flora of Texas*, vol. 2. Texas Research Foundation, Renner, Texas.

Lundell, C. L. 1977. Studies of American plants: XIV. Wrightia 5(9):331–351.

Luteyn, J. L. 1976. Revision of *Limonium* (Plumbaginaceae) in eastern North America. Brittonia 28: 303–317.

Luteyn, J. L. 1990. The Plumbaginaceae in the flora of the southeastern United States. Sida 14(2):169–178.

Lynch, D., Brother. 1981. *Native and Naturalized Woody Plants of Austin and the Hill Country*. Saint Edward's University, Austin, TX.

Mackenzie, K. K. 1931–1935. North American Flora. 18. Cyperaceae, Tribe 2, *Caricae*. Bronx, New York: The New York Botanical Garden.

MacRoberts, D. T. 1977. Notes on *Tradescantia*—*T. diffusa* Bush and *T. pedicellata* Celarier. Phytologia 38: 227–228.

MacRoberts, D. T. 1980. Notes on *Tradescantia*. IV. (Commelinaceae): The distinction between *T. virginiana* and *T. hirsutiflora*. Phytologia 46:409–416.

Maddox, E. 1986. *Homalocephala texensis* "Texas Horse Crippler." Cactus and Succulent Journal 58: 218–221.

Maddox, E., and C. Glass. 1991. Unique cacti unique to Texas. Cactus and Succulent Journal 63:22–26.

Mahler, W. F. 1966. *Keys to the Embryophyta of Taylor County, Texas*. Dallas, Texas: Southern Methodist University Bookstore.

Mahler, W. F. 1971. *Keys to the Vascular Plants of the Black Gap Wildlife Management Area Brewster County, Texas*. Dallas, Texas: Author.

Mahler, W. F. 1973. Botanical survey of the Lake Monticella Area. Dallas, Texas: Southern Methodist University, Department of Anthropology.

Mahler, W. F. 1974. *Gnaphalium helleri* Britton (Compositae-*Inuleae*) in the Texas flora. Southwestern Naturalist 19:329.

Mahler, W. F. 1975. *Pyrrhopappus rothrockii* Gray (Compositae-*Inuleae*) in Texas. Southwestern Naturalist 20:139.

Mahler, W. F. 1979. *Rubus trivialis* Michx. var. *duplaris* (Shinners) Mahler, comb. *nov*. (Rosaceae). Sida 8: 211–212.

Mahler, W. F. 1981a. Notes on rare Texas and Oklahoma plants. Sida 9(1):76–86.

Mahler, W. F. 1981b. Field studies of Texas endemics. Sida 9(2):176–181.

Mahler, W. F. 1983. Rediscovery of *Hymenoxys texana* and notes on two other Texas endemics. Sida 10: 87–92.

Mahler, W. F. 1984. *Shinners' Manual of the North Central Texas Flora*. Southern Methodist University Herbarium, Dallas.

Mahler, W. F. 1987a. *Leavenworthia texana* (Brassicaceae), a new species from Texas. Sida 12:239–242.

Mahler, W. F. 1987b. New combinations and notes on the North Central Texas flora. Sida 12:250–251.

Mahler, W. F. 1988a. *Shinners' Manual of the North Central Texas Flora*. Botanical Research Institute of Texas, Incorporated, Fort Worth.

Mahler, W. F. 1988b. *Amorpha roemeriana* Scheele (Fabaceae), an upland species. Sida 13(1):121.

Mahler, W. F. 1989. *Agrimonia incia* (Rosaceae) new to Texas. Sida 13(3):383.

Mahler, W. F., and B. Lipscomb. 1978. Additions and corrections to the flora of Texas. Sida 7(4):392–394.

Mahler, W. F., and U. T. Waterfall. 1964. *Baccharis* (Compositae) in Oklahoma, Texas, and New Mexico. Southwestern Naturalist 9:189–202.

Maihle, N. J., and W. H. Blackwell, Jr. 1978. A synopsis of North American *Corispermum* (Chenopodiaceae). Sida 7:382–391.

Malusa, J. 1992. Phylogeny and biogeography of the pinyon pines (*Pinus* subsect. *Cembroides*). Systematic Botany 17(1):42–66.

Marcks, B. G. 1972. Population studies in North American *Cyperus* Section *Laxiglumi* (Cyperaceae). Ph.D. dissertation, University of Wisconsin, Madison.

Marcks, B. G. 1974. Preliminary reports on the flora of Wisconsin. No. 66. Cyperaceae. II—Sedge family. II. The genus *Cyperus*—the umbrella sedges. Wisconsin Academy of Sciences, Arts and Letters 62:261–284.

Marietta, K. L. 1979. Vegetation of three upland communities in East Texas. Master's thesis, Stephen F. Austin State University, Nacogdoches, Texas.

Marks, P. L., and P. A. Harcombe. 1981. Forest vegetation of the Big Thicket, Southeast Texas. Ecological Monographs 3:247–297.

Martin, R. F. 1940. A review of the cruciferous genus *Selenia*. American Midland Naturalist 23:455–462.

Martin, W. C., and C. R. Hutchins. 1980. *A Flora of New Mexico*. 2 vols. Hirschberg: Strauss & Cramer GmbH Publishers.

Massey, J. R. 1975. *Fatoua villosa* (Moraceae), additional notes on distribution in the southeastern United States. Sida 6:116.

Mathias, M. E., and L. Constance. 1951. Umbelliferae. Flora of Texas, Vol. 3, part 5. Texas Research Foundation, Renner, Texas.

Matoon, W. R., and C. B. Webster. 1953. *Forest Trees of Texas: How to Know Them*. 6th ed., ed. H. E. Weaver. Texas Forest Service, College Station.

Matos, J. A., and D. C. Rudolf. 1985. The vegetation of the Roy E. Larsen Sandylands Sanctuary in the Big Thicket of Texas. Castanea 50(4):228–249.

Matthews, J. F., D. W. Ketron, and S. F. Zane. 1992. The reevaluation of *Portulaca pilosa* and *P. mundula* (Portulacaceae). Sida (15)1:71–89.

Matthews, J. F., D. W. Ketron, and S. F. Zane. 1993. The biology and taxonomy of the *Portulaca oleracea* L. complex in North America. Rhodora 95(882): 166–183.

Matthews, J. F., J. F. Levins, and P. A. Levins. 1985. *Portulaca pilosa* L., *P. mundula* I. M. Johnst., and *P. parvula* Gray in the Southwest. Sida 11:45–61.

Maxon, W. R. 1912. Notes on the North American species of *Phaenrophlebia*. Bulletin of the Torrey Botanical Club 39:23–28.

Mayfield, M. 1991. *Euphorbia johnstonii* (Euphorbiaceae), a new species from Tamaulipas, México, with notes on *Euphorbia* subsections *Acutae*. Sida 14(4):573–579.

Mayfield, M. 1993. New combinations in *Chamaesyce* A. Gray (Euphorbiaceae) from Texas and the Chihuahuan Desert. Phytologia 75(2):178–183.

Maze, J. 1972. Notes on the awn anatomy of *Stipa* and *Oryzopis* (Gramineae). Syesis 5:169–171.

McAlister, W. H. 1988. An annotated list of the plants of the Aransas National Wildlife Refuge. Mimeo.

McClure, F. A. 1946. The genus *Bambusa* and some of its first-known species. Blumea [Supplement III: Henrard Jubilee], 90–117.

McClure, F. A. 1956. New species in the bamboo genus *Phyllostachys* and some nomenclatural notes. Journal of the Arnold Arboretum 37:180–196.

McClure, F. A. 1957. Bamboos of the genus *Phyllostachys* under cultivation in the United States. Agricultural Handbook 114. United States Department of Agriculture.

McDonald, A. 1984. *Ipomoea dumetorum* (Convovulaceae): An amphitropical disjunct morning glory in the Southwest U.S. Sida 10:252–254.

McDonald, C. B. 1980. A biosystematic study of the *Polygonum hydropiperoides* (Polygonaceae) complex. American Journal of Botany 67:664–670.

McDougall, W. B., and O. E. Sperry. 1951. *Plants of Big Bend National Park*. Washington, D. C.: United States Government Printing Office.

McGivney, M. V. D. P. 1938. A revision of the subgenus *Eucyperus* found in the United States. Catholic University of America, Biological series 26.

McGregor, R. L. 1968. The taxonomy of the genus *Echinacea* (Compositae). University of Kansas Science Bulletin 48:113–142.

McGregor, R. L. 1984. *Camelina rumelica*, another weedy mustard established in North America. Phytologia 55:227–228.

McKelvey, S. D. 1938. *Yuccas of the southwestern United States*, pt. 1. The Arnold Arboretum of Harvard University, Jamaica Plain, Massachusetts.

McKelvey, S. D. 1947. *Yuccas of the Southwestern United States*, pt. 2. The Arnold Arboretum of Harvard University, Jamaica Plain, Massachusetts.

McKinney, L. E. 1992. *A Taxonomic Revision of the Acaulescent Blue Violets (Viola) of North America*. Sida, Botanical Miscellany No. 7.

McLaughlin, S. P. 1982. A revision of the southwestern species of *Amsonia* (Apocynaceae). Annals of the Missouri Botanical Garden 69:336–350.

McLeod, C. A. 1975. Southwestern limit of *Fagus grandifolia* Ehrh. Texas Journal of Science 26:179–184.

McNeill, J. 1979. *Diplachne* and *Leptochloa* (Poaceae) in North America. Brittonia 31: 399–404.

McNeill, J. 1980. The delimitation of *Arenaria* (Caryophyllaceae) and related genera in North America, with 11 new combinations in *Minuartia*. Rhodora 82:495–502.

McNeill, J. 1981. Nomenclatural problems in *Polygonum*. Taxon 30:630–641.

McNeill, J., I. J. Bassett, and C. W. Crompton. 1977. *Suaeda calceoliformis*, the correct name for *Suaeda depressa auct.* Rhodora 79:133–137.

McVaugh, R. 1946. The southwestern travels and plant-collections of G. C. Nealley, 1887–1892. Field and Laboratory 14:70–88.

McVaugh, R. 1947. The Travels and Botanical Collections of Dr. Melines Conkling Leavenworth. Field and Laboratory 15(2): 57–70.

McVaugh, R. 1951. Campanulaceae. Flora of Texas, Vol. 3, part 5. Texas Research Foundation, Renner, Texas.

McVaugh, R. 1952. Remarks on the genus *Cercocarpus* in Texas. Field and Laboratory 20(1):35–40.

Mears, J. A. 1975. The taxonomy of *Parthenium* section *Partheniastrum* DC. (Asteraceae-*Ambrosiinae*). Phytologia 31:463–482.

Mears, J. A. 1980. The Linaean species of *Gomphrena* L. (Amaranthaceae). Taxon 29:85–95.

Meeuse, A. D., J. Smit, and A. Smit. 1971. A new combination in *Krascheninnikovia* (Chenopodiaceae). Taxon 20:644.

Mellen, G. 1991. The *Echiocereus fendleri* controversy. Cactus and Succulent Journal 63:208–212.

Melville, R. 1958. Notes on *Alternanthera*. Kew Bulletin 1758:171–175.

Menapace, F. J., D. E. Wujeck, and A. A. Reznicek. 1986. A systematic revision of the genus *Carex* (Cyperaceae) with respect to the section *Lupulinae*. Canadian Journal of Botany 64:2785–2788.

Mennema, J. 1989. A taxonomic revision of *Lamium* (Lamiaceae). Leiden Botanical series 11:1–196.

Mickel, J. T. 1962. Monographic study of the fern genus *Anemia* subgenus *Coptophyllum*. Iowa State College Journal of Science 36:349–482.

Mickel, J. T. 1979. The fern genus *Cheilanthes* in Continental United States. Phytologia 41:431–437.

Mickel, J. T. 1981. Revision of *Anemia* subgenus *Anemiorrhiza* (Schizaeaceae). Brittonia 33:413–429.

Miller, G. N. 1955. The genus *Fraxinus*, the ashes, in North America, North of México. Cornell Experiment Station Memoir No. 335. Cornell University, Ithaca, New York.

Miller, N. G. 1971. The genera of the Urticaceae in the southeastern United States. Journal of the Arnold Arboretum 52:40–68.

Miller, N. G. 1982. The Caricaceae in the southeastern United States. Journal of the Arnold Arboretum 63(4):401–427.

Miller, N. G. 1990. The genera of the Meliaceae in the southeastern United States. Journal of the Arnold Arboretum 71(4):453–486.

Millspaugh, C. F., and E. E. Sherff. 1919a. New species of *Xanthium* and *Solidago*. Publications of the Field Museum of Natural History. Botanical Series 4(1): 1–7.

Millspaugh, C. F., and E. E. Sherff. 1919b. Revision of the North American species of *Xanthium*. Publications of the Field Museum of Natural History. Botanical Series 4(1):9–54.

Mitchell, R. J. 1964. A quantitative investigation of the perennial vegetation of Bastrop State Park. Master's thesis, University of Texas, Austin.

Mobberly, D. G. 1956. Taxonomy and distribution of the genus *Spartina*. Iowa State College Journal of Science 30:471–574.

Moldenke, H. N. 1942. Eriocaulaceae. Flora of Texas, Vol. 3, part 1. Texas Research Foundation, Renner, Texas.

Moldenke, H. N. 1980. A sixth summary of the Verbenaceae, Avicenniaceae, Stilbaceae, Chloanthaceae, Symphoremaceae, Nycanthaceae, and Eriocaulaceae of the world as to valid taxa, geographic distribution and synonomy. Phytologia Memoirs 2:1–629.

Montgomery, F. H. 1955. Preliminary studies in the genus *Dentaria* in eastern North America. Rhodora 57: 161–173.

Montgomery, J. D. 1982. *Dryopteris* in North America. Fiddlehead Forum 8:25–31.

Moore, M. O. 1991. Classification and systematics of eastern North American *Vitis* L. (Vitaceae) North of México. Sida 14(3):339–367.

Morden, C. W. 1995. A new combination in *Muhlenbergia* (Poaceae). Phytologia 79(1):28–30. Note: This volume is marked July 1995 but was not published until February 1996.

Morden, C. W., and S. L. Hatch. 1981. *Polypogon elongatus* H.B.K. (Poaceae) new in Texas. Sida 9:187–188.

Morden, C. W., and S. L. Hatch. 1987. Anatomical study of the *Muhlenbergia repens* complex (Poaceae: *Chloridoidea: Eragrostideae*). Sida 12(2):347–359.

Morden, C. W., and S. L. Hatch. 1989. An analysis of morphological variation in *Muhlenbergia capillaris* (Poaceae) and its allies in the southeastern United States. Sida 13(3):303–314.

Morgan, D. R. 1993. A molecular systematic study and taxonomic revision of *Psilactis* (Asteraceae: *Astereae*). Systematic Botany 18(2):290–308.

Mosquin, T. 1971. Biosystematic studies in the North American species of *Linum*, section *Adenolinum* (Linaceae). Canadian Journal of Botany 49:1379–1388.

Moyer, J. A., and B. L. Turner. 1994. Systematics study of Texas populations of *Phacelia patuliflora* (Hydrophyllaceae). Sida 16(2):245–252.

Müller, C. H. 1951. *The Oaks of Texas*. Contributions from the Texas Research Foundation 1:21–312.

Müller, C. H. 1979. A new combination in *Pithecellobium*. Phytologia 41:384–386.

Müller, C. H. 1980. *Journey to Mexico During the Years 1826 to 1834*, Vol. I, pp. xi–xxxvi. By J. L. Berlandier, translated by S. M. Ohlendorf, J. M. Bigelow, and M. M. Standifer. The Texas State Historical Association in cooperation with the Center of Studies in Texas History, University of Texas at Austin.

Mulligan, G. A., and Frankton. 1962. Taxonomy of the genus *Cardaria* with particular reference to the species introduced into North America. Canadian Journal of Botany 40:1411–1425.

Mulligan, G. A., and D. B. Munro. 1989. Taxonomy of species of North American *Stachys* (Labiatae) found north of México. Revue d'ecologie et Systamatic 1126:35–51.

Munz, P. A. 1944. Onagraceae. Flora of Texas, Vol. 3, part 4. Texas Research Foundation, Renner, Texas.

Naczi, R. F. C. 1990. The taxonomy of *Carex bromoides* (Cyperaceae). Contributions from the University of Michigan Herbarium 17:215–222.

Naczi, R. F. C. 1992. Systematics of *Carex* section *Griseae* (Cyperaceae) (phylogenetic systematics, cytology). Ph.D. dissertation, University of Michigan, Ann Arbor.

Naczi, R. F. C., and C. T. Bryson. 1990. Noteworthy records of *Carex* (Cyperaceae) from the southeastern United States. Bartonia 56:49–58.

Nealley, G. C. 1888. Report of an investigation of the forage plant of western Texas. In Report of an Investigation of the grasses of the arid districts of Texas, New Mexico, Arizona, Nevada, and Utah in 1887, ed. G. Vasey. Department of Agriculture Botany Division Bulletin 6:30–47.

Nelson, J. B. 1980. *Mitreola* vs. *Cynoctonum*, and a new combination. Phytologia 46:338–340.

Nesom, G. L. 1979. *Erigeron geiseri* (Compositae) in Oklahoma. Southwestern Naturalist 24:386–387.

Nesom, G. L. 1985. New combinations in *Erigeron* (Asteraceae). Sida 11:249.

Nesom, G. L. 1988. Synopsis of *Chaetopappa* (Compositae: *Astereae*) with a new species and the inclusion of *Leucelene*. Phytologia 64:448–456.

Nesom, G. L. 1989a. Infrageneric taxonomy of New World *Erigeron* (Compositae: *Astereae*). Phytologia 67:67–93.

Nesom, G. L. 1989b. Further definition of *Conyza* (Asteraceae: *Astereae*). Phytologia 68:229–233.

Nesom, G. L. 1989c. The *Solidago canadensis* (Asteraceae: *Astereae*) complex in Texas with a new species from Texas and México. Phytologia 67:441–450.

Nesom, G. L. 1990a. Synopsis of the species of *Omphalodes* (Boraginaceae) native to the New World. Sida 13(1):25–30.

Nesom, G. L. 1990b. Taxonomy of *Erigeron bellidiastrum* (Asteraceae: *Astereae*), with a new variety. Phytologia 69(3):163–168.

Nesom, G. L. 1990c. Taxonomy of *Heterotheca* sect. *Heterotheca* (Asteraceae: *Astereae*) in México, with comments on the taxa of the United States. Phytologia 69(4):282–294.

Nesom, G. L. 1990d. Studies in the systematics of Mexican and Texan *Grindelia* (Asteraceae: *Astereae*). Phytologia 68:303–332.

Nesom, G. L. 1990e. Taxonomy of *Solidago petiolaris* (*Astereae*: Asteraceae) and related Mexican species. Phytologia 69(6):445–456.

Nesom, G. L. 1991a. Taxonomy of *Isocoma* (Compositae: *Astereae*). Phytologia 70(2):69–114.

Nesom, G. L. 1991b. Union of *Bradburia* with *Chrysopsis* (Asteraceae: *Astereae*), with a phylogenetic hypothesis for *Chrysopsis*. Phytologia 71(2):109–121.

Nesom, G. L. 1992a. A new species of *Castilleja* (Scrophulariaceae) from Southcentral Texas with comments on other Texas taxa. Phytologia 72(3):209–230.

Nesom, G. L. 1992b. New species and taxonomic evaluations of Mexican *Castilleja* (Scrophulariaceae). Phytologia 72(3):231–252.

Nesom, G. L. 1992c. Species rank for the varieties of *Grindelia microcephala* (Asteraceae: *Astereae*). Phytologia 73(4):326–329.

Nesom, G. L. 1993. Taxonomic infrastructure of *Solidago* and *Oligoneuron* (Asteraceae: *Astereae*) and observations on their phylogenetic position. Phytologia 75(1):1–44.

Nesom, G. L. 1995a. Review of the taxonomy of *Aster sensu lato* (Asteraceae: *Astereae*), emphasizing the New World species. Phytologia 77(3):141–297. [Note: This volume is marked 1994, but was not published until February 1995.]

Nesom, G. L. 1995b. Revision of *Chaptalia* (Asteraceae: *Mutisieae*) from North America and continental Central America. Phytologia 78(3):153–188. [Note: This volume is marked March 1995, but was not published until August 1995.]

Nesom, G. L., and G. I. Baird. 1993. Completion of *Ericameria* (Asteraceae: *Astereae*), diminution of *Chrysothamnus*. Phytologia 75(1):74–93.

Nesom, G. L., Y. Suh, D. Morgan, and B. B. Simpson. 1990. *Xylothamia* (Asteraceae: *Astereae*), a new genus related to *Euthamia*. Sida 14(1):101–116.

Nesom, G. L., Y. Suh, D. R. Morgan, S. D. Sundberg, and B. B. Simpson. 1991. *Chloracantha*, a new genus of North American *Astereae* (Asteraceae). Phytologia 70(5):371–381.

Nesom, G. L., Y. Suh, and B. B. Simpson. 1993. *Prionopsis* (Asteraceae: *Astereae*) united with *Grindelia*. Phytologia 75(5):341–346.

Nesom, G. L., and S. Sunberg. 1985. New combinations in *Erigeron* (Asteraceae). Sida 11:249–250.

Newsom, V. M. 1929. A revision of the genus *Collinsia* (Scrophulariaceae). Botanical Gazette 87:260–301.

Niehaus, T. F. 1984. *A Field Guide to Southwestern and Texas Wildflowers*. Boston: Houghton Mifflin.

Niles, W. E. 1970. Taxonomic investigations in the genera *Perityle* and *Laphamia* (Compositae). Memoirs of the New York Botanical Garden 21:1–82.

Nipper, V. M. 1940. *Plants of East Texas*. Technical Bulletin 1:1–9. Stephen F. Austin State Teacher's College.

Nixon, E. S. 1985. *Trees, Shrubs, and Woody Vines of East Texas*. Nacogdoches, Texas: Bruce Lyndon Cunningham Productions.

Nixon, E. S., S. C. Damuth, and M. McCrary. 1987. Five additions to the Texas flora. Sida 12:421–422.

Nixon, E. S., K. L. Marietta, and M. McCrary. 1980. *Brachyletrum erectum* and *Talinum rugospermum* new species to Texas and notes on *Schoenolirion wrightii*. Sida 8:355–356.

Nixon, E. S., J. R. Sullivan, J. T. Brown, J. Lacey, and J. D. Freeman. 1970. Notes on the distribution of *Trillium gracile* and *Trillium recurvatum* (Liliaceae) in Texas. Sida 3:528–530.

Nixon, E. S., and J. R. Ward. 1981. Distribution of *Schoenolirion wrightii* (Liliaceae) and *Bartonia texana* (Gentianaceae). Sida 9:64–69.

Nixon, E. S., and J. R. Ward. 1982. *Rhynchospora miliacea* and *Scirpus divaricatus* new to Texas. Sida 9: 367.

Nixon, E. S., J. R. Ward, and B. L. Lipscomb. 1983. Rediscovery of *Lesquerella pallida* (Cruciferae). Sida 10: 167–175.

Nixon, K. C. 1984. A biosystematic study of *Quercus* series *virentes* (the liveoaks) with phylogenetic analysis of *Fagales*, Fagaceae, and *Quercus*. Ph.D. dissertation, University of Texas, Austin.

Nixon, K. C., and C. H. Muller. 1992. The taxonomic resurrection of *Quercus laceyi* Small (Fagaceae). Sida 15(1):57–69

Nordenstam, B. 1977. *Senecioneae* and *Liabeae*-systematic review. In *The Biology and Chemistry of the Compositae*, ed. V. H. Heywood, J. B. Harborne, and B. L. Turner. 2:799–830.

Northington, D. K. 1973. A new combination in *Pyrrhopappus* (Compositae: *Cichorieae*). Southwestern Naturalist 18:343.

Northington, D. K. 1974. Systematics studies of the genus *Pyrrhopappus* (Compositae: *Cichorieae*). Special Publication 6. Texas Tech University, The Museum. Lubbock.

Norton, J. B. S. 1898. Joseph F. Joor. Botanical Gazette.

Nowack, R. 1995. *Eustachys caribaea* and *E. paspaloides* (Gramineae). Bulletin du museum d'histoire naturelle. Paris, 4ᵉ sér., 17. Section B, Adansonia, 1–2: 53–57.

Ockendon, D. J. 1965. A taxonomic study of *Psoralea* subgenus *Pediomelum* (Leguminosae). Southwestern Naturalist 10:81–124.

Oka, H. I. 1988. *Origin of Cultivated Rice*. Developments in Crop Science 14. Tokyo: Japanese Science Society Press.

O'Kennon, B. 1991a. *Paliurus spina-christi* (Rhamnaceae) new for North America in Texas. Sida 14(4): 606–609.

O'Kennon, B. 1991b. *Euphorbia lathyris* (Euphorbiaceae) new for Texas. Sida 14(4):609–610.

O'Kennon, B., and G. Nesom. 1988. First report of *Cirsium vulgare* (Asteraceae) from Texas. Sida 13(1): 115–116.

Orgaard, M. 1991. The genus *Cabomba* (Cabombaceae)—A taxonomic study. Nordic Journal of Botany 11(2):179–204.

Ownbey, G. B. 1958. Monograph of the genus *Argemone* for North America and the West Indies. Memoirs of the Torrey Botanical Club 21(1):1–159.

Ownbey, G. B. 1959. Monograph of the North American species of *Corydalis*. Annals of the Missouri Botanical Garden 34(3):187–259.

Ownbey, M. 1950. The genus *Allium* in Texas. Research studies of the State College of Washington 18(4): 181–222.

Page, C. N. 1976. The taxonomy of phytogeography of bracken—a review. Journal of the Linnean Society. Botany 73:1–34.

Palmer, P. G. 1975. A biosystematic study of the *Panicum amarum–P. amarulum* complex (Gramineae). Brittonia 27:142–150.

Palmer, R. 1988. A field checklist: National Audubon Society's Sabal Palm Grove Sanctuary. Mimeographed.

Parker, K. F. 1981. New combinations in *Tetraneura* Greene (*Heliantheae*, Asteraceae). Phytologia 45: 467.

Parks, H. B. 1937. *Valuable Plants Native to Texas*. Bulletin No. 551. Texas Agricultural Experiment Station, College Station.

Parks, H. B. 1949. A list of plants reported to occur in Brazos County, Texas. (Compiled between 1945 and 1949). Mimeographed.

Parks, H. B., and V. L. Cory. 1938. *The Fauna and Flora of the Big Thicket Area*. 2d ed. Huntsville, Texas: Sam Houston State Teaching College.

Parks, H. B., and V. L. Cory. 1958. *Biological Survey of the East Texas Big Thicket Area*. Huntsville, Texas: Sam Houston State Teaching College.

Parodi, L. R. 1947. Las especies de gramíneas del género *Nassella* de la Argentina y Chile. De Darwiniana 7:369–395.

Paton, A. 1990. A global taxonomic investigation of *Scutellaria* (Labiatae). Kew Bulletin 45(3):399–450.

Payson, E. B. 1921. A monograph of the genus *Lesquerella*. Annals of the Missouri Botanical Garden 8: 103–236.

Pedersen, T. M. 1972. *Cyperus rigens* Presl subsp. *cephalanthus* (Torrey & Hooker) T. M. Pedersen. De Darwiniana 17:539.

Pennell, F. W. 1935. *The Scrophulariaceae of Eastern Temperate North America*. Lancaster, Pennsylvania: Wickersham Printing Co.

Pennell, F. W. 1940. Scrophulariaceae of Trans-Pecos Texas. Proceedings of the Academy of Natural Sciences of Philadelphia 92:289–308.

Pennington, T. D. 1990. Sapotaceae. Flora Neotropica. Monograph 52:1–770.

Perry, G., and J. McNeill. 1986. The nomenclature of *Eragrostis cilianensis* (Poaceae) and the contribution of Bellardi to Allioni's *Flora Pedemontana*. Taxon 35: 696–701.

Peterson, C. D., and L. E. Brown. 1983. *Vascular Flora of the Little Thicket Nature Sanctuary San Jacinto County, Texas*. Houston: Outdoor Nature Club.

Peterson, K. M., and W. W. Payne. 1973. The genus *Hymenoclea* (Compositae: *Ambrosieae*). Brittonia 25: 243–256.

Peterson, P. M., and C. R. Annable. 1990. A revision of *Blepharoneuron* (Poaceae: *Eragrostideae*). Systematic Botany 15(4):515–525.

Peterson, P. M., and C. R. Annable. 1991. Systematics of the annual species of *Muhlenbergia* (Poaceae-*Eragrostideae*). Systematic Botany Monograph, vol. 31.

Pfeifer, H. W. 1966. Revision of the North and Central American hexandrous species of *Aristolochia* (Aristolochiaceae). Annals of the Missouri Botanical Garden 53:115–196.

Pfeifer, H. W. 1970. Revision of pentandrous *Aristolochia*. Annals of the Missouri Botanical Garden 53: 115–196.

Pfeifer, N. E. 1922. Monograph of the Isoëtaceae. Annals of the Missouri Botanical Garden 9:79–232.

Phillips, R. C., C. McMillan, H. R. Bittaker, and R. Heiser 1974. *Halodule wrightii* Ascherson in the Gulf of México. Marine Science 18:257–261.

Phipps, J. B. 1988. *Crataegus* (*Maloideae*, Rosaceae) of the southeastern United States, I. Introduction and Series *aestivales*. Journal of the Arnold Arboretum 69(4):401–431.

Phipps, J. B. 1990. *Crataegus secreta* (Rosaceae), a new species of hawthorn from the Edwards Plateau, Texas. Sida 14(1):13–19.

Pilbeam, J. 1981. *Mammillaria, A Collector's Guide*. New York: Universe Books.

Pilger, R. 1931. Benerkungen zu *Panicum* und verwandten Gattungen. Notizblatt des Königlichen botanischen Gartens und Museums zu Berlin 11:246.

Pilz, G. E. 1978. Systematics of *Mirabilis* subgenus *Quamoclidion* (Nyctaginaceae). Madroño 25: 113–132.

Ping-Sheng, H., S. Kurita, Yuzhi-Zhou, and L. Jin-Zhen. 1994. Synopsis of the genus *Lycoris* (Amaryllidaceae). Sida 16(2):301–331.

Pinkava, D. J., and B. D. Parfitt. 1988. Nomenclatural changes in Chihuahuan Desert *Opuntia* (Cactaceae). Sida 13(2):125–130.

Pinkava, D. J., B. D. Parfitt, M. A. Baker, and R. D. Worthington. Chromosome numbers in some cacti of western North America—VI, with nomenclatural changes. Madroño 39(2):98–113.

Pinson, J. N., Jr., and W. T. Batson. 1971. The status of *Muhlenbergia filipes* Curtis (Poaceae). Journal of the Elisha Mitchell Scientific Society 87:188–191.

Pippen, R. W. 1978. *Cacalia*. North American Flora: II. 10:151–159.

Plumb, G. A. Vegetation classifications of Big Bend National Park, Texas. Texas Academy of Science 44(4): 375–387.

Pohl, R. W. 1972. New taxa of *Hierochloë, Pariana*, and *Triplasis* from Costa Rica. Iowa State Journal of Research 47(1):71–78.

Pohl, R. W. 1980. Gramineae. In *Flora Costaricensis*, ed. Burger. Family # 15 W. Fieldiana: Botany, new series, No. 4. Field Museum of Natural History, Chicago.

Porter, D. M. 1969. The genus *Kallstroemia*. Contributions from the Gray Herbarium of Harvard University 198:1–153.

Porter, D. M. 1974. Disjunct distributions in the New World Zygophyllaceae. Taxon 23:339–346.

Powell, A. M. 1963. An emended description of the monotypic genus *Bartlettia* A. Gray (*Senecioneae*) with distributional notes. Southwestern Naturalist 8:117–120.

Powell, A. M. 1967. Novelties in *Perityle* (Compositae). Sida 3:177–180.

Powell, A. M. 1969. Taxonomy of *Perityle* section *Pappothrix* (Compositae-*Peritylinae*). Rhodora 71: 58–93.

Powell, A. M. 1973a. Taxonomy of *Perityle* section *Laphamia* (Compositae-*Helenieae-Peritylinae*). Sida 5: 61–128.

Powell, A. M. 1973b. Taxonomy of *Pericome* (Compositae-*Peritylinae*). Southwestern Naturalist 18(3): 335–339.

Powell, A. M. 1974. Taxonomy of *Perityle* section *Perityle* (Compositae-*Peritylinae*). Rhodora 76(806): 229–306.

Powell, A. M. 1978. Systematics of *Flaveria* (*Flaveriinae*-Asteraceae). Annals of the Missouri Botanical Garden 65:590–636.

Powell, A. M. 1988. *Trees and Shrubs of Trans-Pecos Texas*. Big Bend Natural History Association, Big Bend National Park, Texas.

Powell, A. M. 1994. *Grasses of the Trans-Pecos and Adjacent Areas*. Austin: University of Texas Press.

Powell, A. M., S. Powell, and A. S. Tomb. 1977. Cytotypes in *Cevallia sinuata* (Loasaceae). Southwestern Naturalist 21:433–441.

Powell, A. M., and B. L. Turner. 1976. New gypsophilic species of *Pseudoclappia* and *Sartwellia* (Asteraceae) from West Texas and eastern Chihuahua. Sida 6: 317–320.

Powell, A. M., and B. Wauer. 1990. A new species of *Viola* (Violaceae) from the Guadalupe Mountains, Trans-Pecos Texas. Sida 14:1–6.

Powell, A. M., A. D. Zimmerman, and R. A. Hilsenbeck. 1991. Experimental documentation of natural hybridization in Cactaceae: Origin of Lloyd's hedgehog cactus, *Echinocereus* ×. *lloydii*. Plant Systematics and Evolution 178:107–122.

Pringle, J. S. 1971. Taxonomy and distribution of *Clematis* section *Atragene* (Ranunculaceae) in North America. Brittonia 23:361–393.

Puff, C. 1976. The *Galium trifidum* group (*Galium* sect. *Aparinoides* Rubiaceae). Canadian Journal of Botany 54:1911–1925.

Puff, C. 1977. The *Galium obtusum* group, *Galium* sect. *Aparinoides* (Rubiaceae). Bulletin of the Torrey Botanical Club 104:202–208.

Puff, C. 1991. Revision of the genus *Paederia* L. (Rubiaceae-*Paederieae*) in America. Opera Botanica Belgique 3:325–333.

Pyrah, G. L. 1969. Taxonomic and distributional studies in *Leersia* (Gramineae). Iowa State Journal of Science 44(2):215–270.

Rabeler, R. K. 1985. *Petrorhagia* (Caryophyllaceae) of North America. Sida 11:6–44.

Rabeler, R. K. 1992. A new combination in *Minuartia* (Caryophyllaceae). Sida 15(1):95–96.

Rabeler, R. K., and J. W. Thieret. 1988. Comments on the Caryophyllaceae of the southeastern United States. Sida 13:149–156.

Radford, A. E., H. E. Ahles, and C. R. Bell. 1968. *Manual of the Vascular Flora of the Carolinas*. Chapel Hill: University of North Carolina Press.

Rajhathy, T., and H. Thomas. 1974. *Cytogenetics of Oats (Avena* L.). Misc. Pub. no. 2 of the Genetics Society of Canada. Ottawa, Ontario.

Ramamoorthy, T. P., and B. L. Turner. 1992. *Nomaphila stricta* (Acanthaceae), a newly discerned aquatic weed in Texas, and the first report for North America. Sida 15(1):115–117.

Raven, P. H., and D. P. Gregory. 1972. A revision of the genus *Gaura* (Onagraceae). Memoirs of the Torrey Botanical Club 23:1–96.

Raven, P. H., and D. R. Parnell. 1970. Two new species and some nomenclatural changes in *Oenothera* subgenus *Hartmannia* (Onagraceae). Madroño 20: 146–149.

Raynal, J. 1976a. Notes cyperologiques: 26. Le Genre Schoenoplectus. II. L'amphicarpie et la sect. Supini. Adansonia, series 2, 16:119–155.

Raynal, J. 1976b. Notes cyperologiques: 27. Identification de deux *Scleria* de Poiret. Adansonia, series 2, 16:211–217.

Reed, C. F. 1965. *Isoëtes* in southeastern United States. Phytologia 12:369–400.

Reed, C. F. 1969a. Chenopodiaceae. Flora of Texas, Vol. 2, part 1. Texas Research Foundation, Renner, Texas.

Reed, C. F. 1969b. Amaranthaceae. Flora of Texas, Vol. 2, part 1. Texas Research Foundation, Renner, Texas.

Reed, C. F. 1969c. Nyctaginaceae. Flora of Texas, Vol. 2, part 1. Texas Research Foundation, Renner, Texas.

Reed, C. F. 1981. *Cypripedium kentuckiense* Reed, a new species of orchid in Kentucky. Phytologia 48: 426–428.

Reed, C. F. 1989. New combinations required for the flora of Central eastern United States: III. Phytologia 67(6):451–453.

Reed, P. B., Jr. 1988. National list of plant species that occur in wetlands: South Plains (Region 6). Biological Report 88(26.6). United States Fish and Wildlife Service, St. Petersburg, Florida.

Reeder, C. G. 1985. The genus *Lycurus* (Gramineae) in North America. Phytologia 57(4):283–291.

Reeder, J. R. 1986. Another look at *Eragrostis tephrosanthes* (Gramineae). Phytologia 60(2):153–154.

Reeder, J. R., and C. G. Reeder. 1980. Systematics of *Bouteloua breviseta* and *B. ramosa* (Gramineae). Systematic Botany 5(3):312–321.

Reeder, J. R., and C. G. Reeder. 1988. *Hilaria annua* (Gramineae), a new species from México. Madroño 35(1):6–9.

Reeder, J. R., and L. J. Toulin. 1987. *Scleropogon* (Gramineae), a monotypic genus with disjunct distribution. Phytologia 62:267–275.

Reeder, J. R., and L. J. Toulin. 1989. Notes on *Pappophorum* (Gramineae: *Pappophoreae*). Systematic Botany 14:349–358.

Reeves, R. G. 1972. *Flora of Central Texas*. Dallas: Grant Davis.

Reeves, R. G. 1977. *Flora of Central Texas* (rev. ed.). Dallas: Grant Davis.

Reeves, R. G., and D. C. Bain. 1947. *Flora of South Central Texas*. College Station: Texas A&M University, The Exchange Store.

Reeves, T. 1979. A monograph of the fern genus *Cheilanthes* subgenus *Physapteris* (Adiantaceae). Ph.D. dissertation, Arizona State University, Tempe.

Renvoize, S. A. 1978. Studies in *Elionurus* (Gramineae). Kew Bulletin 32(3):665–672.

Rettig, J. H. 1988. A biosystematic study of the *Carex pensylvanica* group (section *Acrosystis*) in North America. Ph.D. dissertation, University of Georgia, Athens.

Rettig, J. H. 1989. Nomenclatural changes in the *Carex pensylvanica* group (section *Acrocystis*, Cyperaceae) of North America. Sida 13(4):449–452.

Rettig, J. H. 1990. Correct names for the varieties of *Carex albicans—C. emmonsii*. Sida 14(1):132–133.

Reveal, J. L. 1968. Notes on the Texas Eriogonums (Polygonaceae). Sida 3:195–205.

Reveal, J. L. 1990. The neotypification of *Lemna minuta* Humb., Bonpl. & Kunth, an earlier name for *Lemna minuscula* Herter (Lemnaceae). Taxon 39:328–330.

Reveal, J. L. 1993. On the valid publication of *Collinsia violacea* Nuttall (Scrophulariaceae). Phytologia 74(3): 190–192.

Reveal, J. L., and N. H. Holmgren. 1972. *Ceratoides*, an older generic name for *Eurotia*. Taxon 21:209.

Reveal, J. L., and M. C. Johnston. 1989. A new combination in *Phoradendron* (Viscaceae). Taxon 38(1): 107–108.

Reveal, J. L., and C. S. Keener. 1981. *Virgula* Raf., an earlier name for *Lasallea* Greene (Asteraceae). Taxon 30:648–651.

Reveal, J. L., and R. M. King. 1973. Re-establishment of *Acourtia* D. Don (Asteraceae). Phytologia 27: 228–232.

Reverchon, J. 1903. Fern flora of Texas. Fern Bulletin 2: 33–38.

Reznicek, A. A. 1990. Evolution in sedges (*Carex*: Cyperaceae). Canadian Journal of Botany 68: 1409–1432.

Reznicek, A. A., and P. W. Ball. 1980. The taxonomy of *Carex* section *Stellulatae* in North America north of México. Contributions from the University of Michigan Herbarium 14:153–203.

Reznicek, A. A., and R. F. C. Naczi. 1993. Taxonomic status, ecology, and distribution of *Carex hyalina* (Cyperaceae). Contributions from the University of Michigan Herbarium 19:141–147.

Richards, E. L. 1968. A monograph of the genus *Ratibida*. Rhodora 70:348–393.

Richardson, A. 1976. Reinstatement of the genus *Tiquilia* (Boraginaceae: *Ehretioideae*) and descriptions of four new species. Sida 6:235–240.

Richardson, A. 1977. Monograph of the genus *Tiquilia* (*Coldenia, sensu lato*), Boraginaceae: *Ehretioideae*. Rhodora 79:467–572.

Richardson, A. 1990. *Plants of Southernmost Texas*. Brownsville, Texas: Gorgas Science Foundation.

Richardson, A. 1995. *Plants of the Rio Grande Delta*. Austin: University of Texas Press.

Rickett, H. W. 1969. *Wildflowers of the United States*, vol. 3, pts. 1 and 2: *Texas*. New York: Macgraw-Hill.

Riggins, R. 1977. A biosystematic study of the *Sporobolus asper* complex (Gramineae). Iowa State Journal of Research 51:287–321.

Riskind, D. H. 1978. Noteworthy vascular plant records from Texas. Sida 7(4):394–396.

Roach, A. W., and B. B. Harris. 1952. Sand hill ferns of Henderson County, Texas. American Fern Journal 42:13–15.

Robertson, K. R. 1972. The Malpighiaceae in the southeastern United States. Journal of the Arnold Arboretum 53:101–112.

Robertson, K. R. 1974a. The genera of Rosaceae in the southeastern United States. Journal of the Arnold Arboretum 55:303–332, 344–401, 611–662.

Robertson, K. R. 1974b. The genera of Amaranthaceae in the southeastern United States. Journal of the Arnold Arboretum 62(3):267–314.

Robinson, E. A. 1964. Notes on *Scleria*: III. *Scleria hirtella* Sw., and some allied species: a transatlantic group. Kirkia 4:175–184.

Robinson, H. 1974. Studies in *Senecioneae* (Asteraceae): VI. The genus *Arnoglossum*. Phytologia 28(3): 294–295.

Robinson, H., and R. D. Brettell. 1973. Studies in the *Senecioneae* (Asteraceae): IV. The genera *Mesadenia, Syneilesis, Miracalia, Koyamacalia*, and *Sinacalia*. Phytologia 27:265–276.

Robinson, H., and J. Cuatrecasas. 1973. The generic limits of *Pluchea* and *Tessaria* (*Inuleae*, Asteraceae). Phytologia 27:277–285.

Robinson, H., and J. Cuatrecasas. 1992. *Thelechitonia* Cuatrecasas, an older name for *Complaya* Strother (*Eliptinae-Heliantheae*-Asteraceae). Phytologia 72(2): 141–143.

Robson, N. K. B. 1980. The Linnean species of *Ascyrum* (Guttiferae). Taxon 29:267–274.

Rock, H. F. 1957. A revision of the vernal species of *Helenium*. Rhodora 59:73–116, 128–158, 168–178, 203–216.

Rodman, J. E. 1974. Systematics and evolution of the genus *Cakile* (Cruciferae). Contributions from the Asa Gray Herbarium of Harvard University 205: 3–146.

Rodriguez, S. A., and A. Gomez-Pompa. 1976. Variability in *Ambrosia cumanensis* (Compositae). Systematic Botany 1:363–372.

Rogers, C. M. 1966. Yellow-flowered species of *Linum* in Central America and western North America. Brittonia 20: 107–135.

Rogers, C. M. 1979. A new species of *Linum* from southern Texas and adjacent México. Sida 8: 181–187.

Rogers, C. M. 1981. Linaceae. North American Flora: II. 12: 1–58.

Rogers, D. J., and S. G. Appan. 1973. Flora Neotropica. *Manihot* and *Manihotoides* (Euphorbiaceae). Monograph No. 13. New York: Hafner Press.

Rogers, G. K. 1983. The genera of Alismataceae in the southeastern United States. Journal of the Arnold Arboretum 64(3):491–510.

Rogers, G. K. 1984. The Zingiberales (Cannaceae, Marantaceae, and Zingiberaceae) in the southeastern United States. Journal of the Arnold Arboretum 65(1):5–55.

Rogers, G. K. 1985. The genera of Phytolaccaceae in the southeastern United States. Journal of the Arnold Arboretum 66(1):1–37.

Rogers, G. K. 1986. The genera of Loganiaceae in the southeastern United States. Journal of the Arnold Arboretum 67(2):143–185.

Rogers, G. K. 1987. The genera of *Cinchonoideae* (Rubiaceae) in the southeastern United States. Journal of the Arnold Arboretum 68(2):137–183.

Rollins, R. C. 1942. A systematic study of *Iodanthus*. Contributions from the Dudley Herbarium of Stanford University 3: 209–215.

Rollins, R. C. 1955. The auriculate-leaved species of *Lesquerella*. Rhodora 5: 241–264.

Rollins, R. C. 1957. Miscellaneous Cruciferae of México and western Texas. Rhodora 59: 61–71.

Rollins, R. C. 1959. The genus *Synthlipsis*. Rhodora 61: 253–264.

Rollins, R. C. 1974. Systematic and evolutionary study of the genus *Cakile*. Contributions from the Asa Gray Herbarium of Harvard University 205: 1–146.

Rollins, R. C. 1979. *Dithyrea and a related genus (Cruciferae)*, pp. 3–32. The Bussey Institute of Harvard University.

Rollins, R. C. 1980. The genus *Pennelia* (Cruciferae) in North America. Contributions from the Asa Gray Herbarium of Harvard University 210: 5–21.

Rollins, R. C. 1981. Weeds of the Cruciferae (Brassicaceae) in North America. Journal of the Arnold Arboretum 62: 517–540.

Rollins, R. C. 1982. *Thelepodiopsis* and *Schoenocrambe* (Cruciferae). Contributions from the Asa Gray Herbarium of Harvard University 212: 71–102.

Rollins, R. C. 1993. The Cruciferae of Continental North America. Stanford, California: Stanford University Press.

Rollins, R. C., and E. A. Shaw. 1973. The genus *Lesquerella* (Cruciferae) in North America. Cambridge, Massachusetts: Harvard University Press.

Rominger, J. M. 1962. Taxonomy of *Setaria* in North America. Illinois Biological Monograph (29):1–132.

Rosatti, T. J. 1984. The Plantaginaceae in the southeastern United States. Journal of the Arnold Arboretum 65(4):533–562.

Rosatti, T. J. 1986. The genera of Sphenocleaceae and Campanulaceae) in the southeastern United States. Journal of the Arnold Arboretum 67(1):1–64.

Rosatti, T. J. 1987. The genera of the Pontederiaceae in the southeastern United States. Journal of the Arnold Arboretum 68(1):35–71.

Rosatti, T. J. 1989a. The genera of suborder Apocynineae (Apocynaceae and Asclepiadaceae) in the southeastern United States. Journal of the Arnold Arboretum 70(3):307–401.

Rosatti, T. J. 1989b. The genera of suborder Apocynineae (Apocynaceae and Asclepiadaceae) in the southeastern United States. Journal of the Arnold Arboretum 70(4):443–514.

Rose, F. L., and R. W. Strandtmann. 1986. *Wildflowers of the Llano Estacado*. Dallas: Taylor Publishing Company.

Rothrock, P. E. 1991. The identity of *Carex albolutescens, C. festucacea*, and *C. longii* (Cyperaceae). Rhodora 93(873):51–66.

Rowell, C. M., Jr. 1949. A preliminary report on the floral composition of a sphagnum bog in Robertson County, Texas. Texas Journal of Science 1: 50–53.

Rowell, C. M., Jr. 1958. Provisional checklist of the flora of the Texas Panhandle. Mimeo.

Ruffin, J. 1977. A new combination in *Grindelia* (Compositae-*Astereae*). Rhodora 79:583–585.

Runemark, H. 1962. A revision of *Parapholis* and *Monerma* in the Mediterranean. Botaniska Notiser 115(1):1–17.

Russell, N. H. 1965. Violets (*Viola*) of central and eastern United States: An introductory survey. Sida 2:1–113.

Sanders, R. W. 1981. New taxa and combinations in *Agastache* (Lamiaceae). Brittonia 33:194–197.

Sanders, R. W. 1987. Taxonomy of *Agastache* section *Brittonastrum* (Lamiaceae-*Nepeteae*). Systematic Botany Monograph 15:1–92.

Sargent, C. S. 1922. *Manual of the Trees of North America*. Boston: Houghton, Mifflin, and Co.

Sargent, C. S. 1965. *Manual of the Trees of North America*, vol. 1–2, 2d ed. New York: Dover Publications.

Sauer, J. D. 1972. Revision of *Stenotaphrum* (Gramineae: *Paniceae*) with attention to its historical geography. Brittonia 24(2):202–222.

Saur, J. 1955. Revision of the dioecious *Amaranthus*. Madroño 13:5–46.

Schilling, E. E. 1981. Systematics of *Solanum* sect. *Solanum* (Solanaceae) in North America. Systematic Botany 6:171–185.

Schippers, P., S. J. Ter Borg, & J. J. Bos. 1995. A revision of the Infraspecific taxonomy of *Cyperus esculentus* (yellow nutsedge) with an experimentally evaluated character set. Systematic Botany 20(4):461–481.

Schlessman, M. A. 1984. Systematics of tuberous *Lomatium* (Umbelliferae). Systematic Botany Monograph No. 4.

Scholz, U. 1981. Monographie der gattung *Oplismenus* (Gramineae). Phaerogamarum Monographieae, Tomas XIII. Germany: J. Cramer.

Schulz, E. D. 1922. *500 Wild Flowers of San Antonio and Vicinity*. San Antonio, Texas: Author.

Schulz, E. D. 1928. *Texas Wild Flowers*. Chicago: Laidlaw Brothers Publishers.

Schulz, E. D., and R. Ruyon. 1930. *Texas Cacti: A Popular and Scientific Account of the Cacti Native of Texas*. San Antonio, Texas: Texas Academy of Science Publishers.

Schuyler, A. E. 1966. The taxonomic delineation of *Scirpus lineatus* and *Scirpus pendulus*. Notulae Naturae of the Academy of Natural Sciences of Philadelphia 3:1–3.

Schuyler, A. E. 1974. Typification and application of the names *Scirpus americana* Pers., *S. olneyi* Gray, and *S. pungens* Vahl. Rhodora 76(805):51–52.

Scora, R. W. 1967. Interspecific relationships in the genus *Monarda* (Labiatae). University of California Publications in Botany 41:1–59.

Scott, R. W. 1990. The genera of *Cardueae* (Compositae; Asteraceae) in the southeastern United States. Journal of the Arnold Arboretum 71(4):391–451.

Seigler, D. S., and T. E. Lockwood. 1975. *Blechnum occidentale*, new to Texas. American Fern Journal 65:96.

Seiler, G. J. 1981. New and interesting distribution records for *Helianthus paradoxus* Heiser (Asteraceae). Southwestern Naturalist 26:431–432.

Semple, J. C. 1978. A revision of the genus *Borrichia* (Compositae). Annals of the Missouri Botanical Garden 65:681–683.

Semple, J. C. 1981. A revision of the goldenaster genus *Chrysopsis* (Nutt.) Ell. *nom. cons.* (Compositae-*Astereae*). Rhodora 83:325–384.

Semple, J. C. 1985. New names and combinations in Compositae, tribe *Astereae*. Phytologia 58:429–431.

Semple, J. C. 1987. New names, combinations, and lectotypes in *Heterotheca* (Compositae: *Astereae*). Brittonia 39:379–386.

Semple, J. C. 1994. New combinations in the *Heterotheca villosa* (Pursh) Shinners complex (Compositae: *Astereae*). Novon 4:53–54.

Semple, J. C., V. C. Blok, and P. Heiman. 1980. Morphological, anatomical, habit, and habitat differences among the goldenaster genera *Chrysopsis, Heterotheca*, and *Pityopsis* (Compositae-*Astereae*). Canadian Journal of Botany 58:147–163.

Semple, J. C., and F. D. Bowers. 1985. A revision of the goldenaster genus *Pityopsis* Nutt. (Compositae-

Astereae). University of Waterloo, Biological series no. 28:1–34.

Semple, J. C., and J. Brouillet. 1980. A synopsis of North American asters: The subgenera, sections, and subsections of *Aster* and *Lasallea*. American Journal of Botany 67:1010–1026.

Semple, J. C., and C. C. Chinnappa. 1984. Observations on the cytology, morphology, and ecology of *Bradburia hirtella* (Compositae-*Astereae*). Systematic Botany 9:95–101.

Semple, J. C., and J. G. Chmielewski. 1987. Revision of *Aster lanceolatus* complex, including *A. simplex* and *A. hesperius* (Compositae-*Astereae*): A multivariate morphometric study. Canadian Journal of Botany 65:1047–1062.

Settle, W. J., and T. R. Fisher. 1970. The varieties of *Silphium integrifolium*. Rhodora 72:536–543.

Shaw, R. B., J. Dodd, and J. Durrance. 1976. *Vegetative Key to the Compositae of the Rio Grande Plain of Texas*. MP-1274. Texas Agricultural Experiment Station, College Station.

Shaw, R. B., and R. D. Webster. 1987. The genus *Eriochloa* (Poaceae: *Paniceae*) in North and Central America. Sida 12(1):165–207.

Shechter, Y. 1965. Morphologic and electrophoretic evidence of introgression in *Oryzopsis hymenoides* (Gramineae). Ph.D. dissertation, University of California, Los Angeles.

Shechter, Y., and B. L. Johnston. 1968. The probable origin of *Oryzopsis contracta*. American Journal of Botany 5:611–618.

Sherf, A. S. 1983. *Panicum sphaerocarpon* Ell. var. *polanthes* (Schultes) A. S. Sherif (Poaceae) comb. *nov.* Sida 10:191.

Sherman, H. L. 1964. A systematic study of the genus *Schoenolirion* (Liliaceae). Ph.D. dissertation, Vanderbilt University, Nashville, Tennessee.

Sherman, H. L. 1979. Evidence of misapplication of the name *Schoenolirion texanum* (Scheele) Gray (Liliaceae). Southwestern Naturalist 24:123–126.

Sherman, H. L., and R. W. Becking. 1991. The generic distinctness of *Schoenolirion* and *Hastingsia*. Madroño 38(2):130–138.

Shing, K. H. 1965. A taxonomical study of the genus *Cyrtomium* Presl. Acta Phytotaxonomica Sinica, suppl. 1:1–48.

Shinners, L. H. 1943. A revision of the *Liatris scariosa* complex. The American Midland Naturalist 29(1): 27–41.

Shinners, L. H. 1946a. Revision of the genus *Chaetopappa* DC. Wrightia 1(2):63–87.

Shinners, L. H. 1946b. Revision of the genus *Leucelene* Greene. Wrightia 1(2):82–89.

Shinners, L. H. 1946c. Revision of the genus *Aphanostephus* DC. Wrightia 1(2):95–121.

Shinners, L. H. 1946d. Revision of the genus *Kuhnia* L. Wrightia 1(2):122–144.

Shinners, L. H. 1947a. Two anomalous new species of *Erigeron* L. from Texas. Wrightia 1(3):183–186.

Shinners, L. H. 1947b. Revision of the genus *Krigia* Schreber. Wrightia 1(3):187–206.

Shinners, L. H. 1948. The vetches and pea vines (*Vicia* and *Lathyrus*) of Texas. Field and Laboratory 26(1): 18–29.

Shinners, L. H. 1949a. Notes of Texas Compositae: I. Field and Laboratory 17(1):23–30.

Shinners, L. H. 1949b. Early plant collections return to Texas. Field and Laboratory 17(2):66–68.

Shinners, L. H. 1949c. New names of Texas *Chamaesyces*. Field and Laboratory 17(2):69–70.

Shinners, L. H. 1949d. Transfer of Texas species of *Petalostemum* to *Dalea* (Leguminosae). Field and Laboratory 17(3):81–85.

Shinners, L. H. 1949e. The genus *Dalea* (including *Petalostemum*) in North-Central Texas. Field and Laboratory 17(3):85–89.

Shinners, L. H. 1949f. The Texas species of *Conyza* (Compositae). Field and Laboratory 17(4):142–144.

Shinners, L. H. 1949g. Notes on Texas Compositae: III. Field and Laboratory 17(4):170–176.

Shinners, L. H. 1950a. Notes on Texas Compositae: IV. Field and Laboratory 18:25–32.

Shinners, L. H. 1950b. The species of *Matelea* (including *Gonolobus*) in North Central Texas (Asclepiadaceae). Field and Laboratory 18(2):73–78.

Shinners, L. H. 1950c. The North Texas species of *Plantago*. Field and Laboratory 18(3):113–119.

Shinners, L. H. 1951a. Two new varieties of *Solidago* from North Texas. Field and Laboratory 19(1): 34–35.

Shinners, L. H. 1951b. The North Texas species of *Heterotheca*, including *Chrysopsis* (Compositae). Field and Laboratory 19(2):66–71.

Shinners, L. H. 1951c. Notes on Texas Compositae: VII. Field and Laboratory 19(2):74–82.

Shinners, L. H. 1951d. The Texas species of *Evax* (Compositae). Field and Laboratory 19(3):125–126.

Shinners, L. H. 1952. Addenda on Texas *Chamaesyce* (Euphorbiaceae). Field and Laboratory 20(1):24–26.

Shinners, L. H. 1953a. Synopsis of the United States species of *Lythrum*. Field and Laboratory 21(2): 80–89.

Shinners, L. H. 1953b. The bluebonnets (*Lupinus*) of Texas. Field and Laboratory 21:149–153.

Shinners, L. H. 1954. Notes on north Texas grasses. Rhodora 56:25–38.

Shinners, L. H. 1956. The Texas species of *Limonium*. Field and Laboratory 24(3):105–106.

Shinners, L. H. 1957. Synopsis of the genus *Eustoma* (Gentianaceae). Southwestern Naturalist 2(1): 38–43.

Shinners, L. H. 1958. *Spring Flora of the Dallas–Fort Worth Area, Texas*. Dallas: Author.

Shinners, L. H. 1962a. Annual *Sysyrinchium* (Iridaceae) in the United States. Sida 1:32–42.

Shinners, L. H. 1962b. *Drosera* (Droseraceae) in the southeastern United States: An interim report. Sida 1:53–59.

Shinners, L. H. 1962c. Texas Asclepiadaceae other than *Asclepias*. Sida 1:358–367.

Shinners, L. H. 1962d. *Rhododendron nudiflorum* and *R. roseum* (Ericaceae): Illegitimate names. Castanea 27:94–95.

Shinners, L. H. 1966. *Verbena pulchella* Sweet var. *gracilior* (Tronc.) Shinners, comb. *nov*. (Verbenaceae). Sida 2:266.

Shinners, L. H. 1971. *Kuhnia* L. transferred to *Brickellia* Ell. (Compositae). Sida 4:274.

Shinners, L. H. 1972. *Shinners' Spring Flora of the Dallas–Fort Worth Area, Texas*, 2d ed., ed. W. F. Mahler. Fort Worth: Prestige Press.

Sieren, D. J. 1981. The taxonomy of the genus *Euthamia*. Rhodora 83:551–579.

Silveus, W. A. 1933. *Texas Grasses: Classification and Description of Grasses: Descriptive Systematic Agrostology*. San Antonio, Texas: Author.

Simon, B. K. 1993. *A Key to Australian Grasses*, 2d ed. Brisbane: Queensland Department of Primary Industries.

Simpson, B. B. 1989. Krameriaceae. Flora Neotropica. Monograph 49:1–108.

Simpson, B. B., and L. C. Anderson. 1978. Compositae tribe *Mustisieae*. North American Flora: II. 10:1–13.

Simpson, B. J. 1991. *Symphoricarpos occidentalis* (Caprifoliaceae), new to Texas. Sida 14(3):512–513.

Simpson, B. J., J. P. Karges, and J. M. Carpenter. 1992. *Quercus polymorpha* (Fagaceae) new to Texas and the United States. Sida 15(1):153.

Sinnott, Q. P. 1985. A revision of *Ribes* L. subg. *Grossularia* (Mill.) Pers. sect. *Grossularia* (Mill.) Nutt. (Grossulariaceae) in North America. Rhodora 87:189–296.

Small, E., and A. Cronquist. 1976. A practical and natural taxonomy for *Cannabis*. Taxon 25:405–435.

Smeins, F. E., R. B. Shaw, and R. Blaine. 1978. *Natural Vegetation of Texas and Adjacent Areas, 1675–1975: A Bibliography*. College Station: Texas Agricultural Experiment Station.

Smith, A. I. 1979. *A Guide to Wildflowers of the Mid-South West Tennessee into Central Arkansas and South through Alabama and into East Texas*. Memphis: Memphis State University Press.

Smith, A. R. 1971. Systematics of the neotropical species of *Thelypteris* section *Cyclosorus*. University of California Publication in Botany. Berkeley, California.

Smith, E. B. 1974. *Coreopsis nuecensis* (Compositae) and a related new species from southern Texas. Brittonia 26:161–171.

Smith, E. B. 1976. A biosystematic survey of *Coreopsis* in eastern United States and Canada. Sida 6(3): 123–215.

Smith, E. B. 1978. *An Atlas and Annotated List of the Vascular Plants of Arkansas*. Fayetteville: University of Arkansas Press.

Smith, E. B. 1981. New combinations in *Croptilon* (Compositae-*Astereae*). Sida 9:59–63.

Smith, E. B. 1994. *Keys to the Flora of Arkansas*. Fayetteville: University of Arkansas Press.

Smith, E. B., and H. M. Parker. 1971. A biosystematic study of *Coreopsis tinctoria* and *C. cardaminefolia* (Compositae). Brittonia 23:161–170.

Smith, J. M. 1976. A taxonomic study of *Acleisanthes* (Nyctaginaceae). Wrightia 5:261–276.

Smith, J. P., Jr. 1971. Taxonomic revision of the genus *Gymnopogon* (Gramineae). Iowa State Journal of Science 45(3):319–385.

Smith, L. B. 1944. Bromeliaceae. Flora of Texas, Vol. 3, part 4. Texas Research Foundation, Renner, Texas.

Smith, R. R., and D. B. Ward. 1976. Taxonomy of the genus *Polygala* series *Decurrentes* (Polygalaceae). Sida 6:284–310.

Smith, S. G. 1995. New combinations in North American *Schoenoplectus, Bolboschoenus, Isolepis*, and *Trichophorum* (Cyperaceae). Novon 5:97–102.

Snow, N., and G. Davidse. 1993. *Leptochloa mucronata* (Michx.) Kunth is the correct name for *Leptochloa filiformis* (Poaceae). Taxon 42:413–417.

Soderstrom, T. R. 1967. Taxonomic study of subgenus *Podosemum* and section *Epicampes* of *Muhlenbergia* (Gramineae). Contributions from the United States National Herbarium. Smithsonian Institution 34(4): 75–203.

Soderstrom, T. R., and H. F. Decker. 1965. *Allolepis*: A new segregate genus of *Distichlis* (Gramineae). Madroño 18 (2):33–64.

Sohns, E. R. 1956. The genus *Hilaria*. Journal of the Washington Academy of Science 46(10):311–321.

Soják, J. 1972. Doplnky k nomenklature nekterych rodu (phanerogamae). Casopis Národního Muzea. Oddíl Prírodvedný 141(1/2):61–63.

Soreng, R. J. 1991. Systematics of the "Epiles" group of *Poa* (Poaceae). Systematic Botany 16(3):507–528.

Soreng, R. J., and S. L. Hatch. 1983. A comparison of *Poa tracyi* and *Poa occidentalis* (Poaceae: *Poaeae*). Sida 10(2):123–141.

Sperry, N. 1991. *Neil Sperry's Complete Guide to Texas Gardening*, 2d ed. Dallas: Taylor Publishing Company.

Spongberg, S. A. 1974. A review of deciduous-leaved species of *Stewartia* (Theaceae). Journal of the Arnold Arboretum 53:182–214.

Stafleu, F. A., and R. S. Cowan. 1976. *Taxonomic Literature*, vol. 1: A–G. Boston: Bohn, Scheltema & Holkema.

Stafleu, F. A., and R. S. Cowan. 1979. *Taxonomic Literature*, vol. 2: H–Le. Boston: Bohn, Scheltema & Holkema.

Stafleu, F. A., and R. S. Cowan. 1981. *Taxonomic Literature*, vol. 3: Lh–O. Boston: Bohn, Scheltema & Holkema.

Stafleu, F. A., and R. S. Cowan. 1983. *Taxonomic Literature*, vol. 4: P–Sak. Boston: Bohn, Scheltema & Holkema.

Stafleu, F. A., and R. S. Cowan. 1985. *Taxonomic Literature*, vol. 5: Sal–Ste. Boston: Bohn, Scheltema & Holkema.

Stafleu, F. A., and R. S. Cowan. 1986. *Taxonomic Literature*, vol. 6: Sti–Vuy. Boston: Bohn, Scheltema & Holkema.

Stafleu, F. A., and R. S. Cowan. 1988. *Taxonomic Literature*, vol. 7: W–Z. Boston: Bohn, Scheltema & Holkema.

Standley, P. C. 1920–1926. *Trees and Shrubs of Mexico*. Two vols. (parts 1–5). Contributions from the

United States National Herbarium. Smithsonian Institution. Vol. 23, Washington, D.C.

Stanford, J. W. 1976. *Keys to the Vascular Plants of the Texas Edwards Plateau and Adjacent Areas*. Howard Payne University, Brownwood, Texas: Author.

Stanford, N., and B. L. Turner. 1988. The natural distribution and biological status of *Helenium amarum* and *H. badium* (Asteraceae, *Heliantheae*). Phytologia 65:141–146.

Starbuck, T. J. 1984. The vascular flora of Robertson County, Texas. Master's thesis, Texas A&M University, College Station.

Staten, R. D., and E. C. Holt. 1965. Summer-annual forage grasses for Texas: Sudangrass, sudan-sorghum hybrids and millet. Progress Report No. 2379. Texas Agricultural Experiment Station, College Station.

Stephenson, S. N. 1971. The biosystematics and ecology of the genus *Brachyelytrum* (Gramineae) in Michigan. The Michigan Botanist 10: 19–33.

Steyermark, J. A. 1934. Studies in *Grindelia*: II. A monograph of the North American species of the genus *Grindelia*. Annals of the Missouri Botanical Garden 21: 433–608.

Steyermark, J. A. 1962. *Flora of Missouri*. Ames: Iowa State University Press.

St. John, H. 1916. A revision of the North American species of *Potamogeton* of the section *Coleophylli*. Rhodora 18(210):121–138.

Straley, G. B. 1977. Systematics of *Oenothera* sect. *Kneiffea* (Onagraceae). Annals of the Missouri Botanical Garden 64: 381–424.

Strong, M. T. 1994. New combinations in *Schoenoplectus* (Cyperaceae). Novon 3: 202–203.

Strother, J. L. 1978. Taxonomy and geography of *Nicolletia* (Compositae: *Tageteae*). Sida 7: 369–374.

Strother, J. L. 1986. Renovation of *Dyssodia* (Compositae: *Tageteae*). Sida 11: 371–378.

Strother, J. L., and L. E. Brown. 1987. Dysploidy in *Hymenoxys texana* (Compositae). American Journal of Botany 75(7):1097–1098.

Strother, J. L., and G. Ritz. 1975. Taxonomy of *Psathyrotes*. Madroño 23: 24–40.

Stuckey, R. L. 1972. Taxonomy and distribution of the genus *Rorippa* (Cruciferae) in North America. Sida 4: 279–430.

Stuessy, T. F. 1971. Systematic relationships in the white-rayed species of *Melampodium*. Brittonia 23: 177–190.

Stuessy, T. F. 1972. Revision of *Melampodium* (Compositae: *Heliantheae*). Rhodora 74: 1–70; 161–219.

Stuessy, T. F. 1990. *Plant Taxonomy: The Systematic Evaluation of Comparative Data*. New York: Columbia University Press.

Sullivan, J. R. 1985. Systematics of the *Physalis viscosa* complex (Solanaceae). Systematic Botany 10: 426–444.

Sullivan, V. I. 1976. Putative hybridization in the genus *Eupatorium* (Compositae). Rhodora 80: 513–527.

Sundberg, S. D. 1986. The systematics of *Aster* subg. *Oxytrifolium* (Compositae) and historically allied species. Ph.D. dissertation, University of Texas, Austin.

Sundberg, S. D. 1991. Infraspecific taxonomy of *Chloracantha spinosa* (Asteraceae: *Astereae*). Phytologia 70: 382–391.

Sundell, E. 1981. Systematics of *Cynanchum* (Asclepiadaceae). Evolution Monograph 5: 1–63.

Svenson, H. K. 1944. The New World species of *Azolla*. American Fern Journal 34: 69–84.

Svenson, H. K. 1957. Poales; Cyperaceae; Scirpeae (Continuatio): *Fuirena* Rottb. North American Flora 18(9):505–507.

Swallen, J. R. 1937. The grass genus *Cathestecum*. Journal of the Washington Academy of Science 27(12): 495–501.

Swallen, J. R. 1965. The grass genus *Luziola*. Annals of the Missouri Botanical Garden 52(3):472–475.

Tamura, M. 1968. Morphology, ecology and phylogeny of the Ranunculaceae VII. Science reports of South College, North College of Oska University, Japan 16: 21–43.

Tateoka, T. 1961. A biosystematic study of *Tridens* (Gramineae). American Journal of Botany 48: 565–573.

Taylor, C. E. S. 1975. *Euthamia gymnospermoides* (Compositae). Ph.D. dissertation. University of Oklahoma, Norman.

Taylor, C. S., and R. J. Taylor. 1981. Plants new to Arkansas, Oklahoma, and Texas. Sida 10:223–251.

Taylor, C. S., and R. J. Taylor. 1983. New species, new combinations, and notes on the goldenrods (*Euthamia* and *Solidago*-Asteraceae). Sida 10:176–183.

Taylor, C. S., and R. J. Taylor. 1984. *Solidago* (Asteraceae) in Oklahoma and Texas. Sida 9:223–251.

Taylor, N. P. 1979. Notes on *Ferocactus* B. & R. Cactus and Succulent Journal of Great Britain 41:88–94.

Taylor, N. P. 1985. The genus *Echinocereus*. The Royal Botanical Gardens, Kew. Portland, Oregon: Timber Press.

Taylor, P. 1989. The genus *Utricularia*: A taxonomic monograph. Kew Bulletin, additional series XIV: 1–724.

Taylor, W. C. R., H. Mohlenbrock, and J. A. Murray. 1975. The spores and taxonomy of *Isoëtes butleri* and *I. melanopoda*. American Fern Journal 65: 33–38.

Terrell, A. P., ed. 1989. *A garden book for Houston and Texas Gulf Coast*, 4th ed. Houston: Gulf Publishing Company.

Terrell, E. E. 1968. A taxonomic revision of the genus *Lolium*. Technical Bulletin No. 1392, United States Department of Agriculture.

Terrell, E. E. 1975. New combinations in *Houstonia* (Rubiaceae). Phytologia 31:425–426.

Terrell, E. E. 1979. New species and combinations in *Houstonia* (Rubiaceae). Brittonia 31:164–169.

Terrell, E. E. 1988. Nomenclatural notes on *Houstonia* (Rubiaceae). Phytologia 65:119–121.

Terrell, E. E. 1990. Synopsis of *Oldenlandia* (Rubiaceae) in the United States. Phytologia 68:125–133.

Terrell, E. E. 1991. Overview and annotated list of North American species of *Hedyotis, Houstonia, Oldenlandia* (Rubiaceae), and related genera. Phytologia 71(3):212–243.

Terrell, E. E., and H. Robinson. 1974. *Luziolinae*, a new subtribe of oryzoid grasses. Bulletin of the Torrey Botanical Club 101:235–245.

Tharp, B. C. 1926. *Structure of Texas Vegetation East of the 98th Meridian*. Bulletin No. 2606. Austin: University of Texas.

Tharp, B. C. 1939. *The Vegetation of Texas*. Nontechnical Publication Series. Austin: Texas Academy of Science.

Tharp, B. C. 1952. *Texas Range Grasses*. Plant Research Institute. Austin: University of Texas Press.

Tharp, B. C., and F. A. Barkley. 1949. The genus *Ruellia* in Texas. American Midland Naturalist 42:1–86.

Tharp, B. C., and M. C. Johnston. 1961. Recharacterization of *Dichondra* (Convolvulaceae) and a revision of the North American species. Brittonia 13(4): 346–360.

Thieret, J. W. 1966. Synopsis of the genus *Calamovilfa* (Gramineae). Castanea 31:145–152.

Thieret, J. W. 1970. *Nemophila microcalyx*, an incorrect name. Rhodora 72:399–400.

Thomas, R. D. 1979. First record of *Botrychium lunarioides* and *Ophioglossum nudicaule* var. *tenerum* (Ophioglossaceae) from Texas. Southwestern Naturalist 24:271–396.

Thomas, R. D., L. R. Briley, and N. Carroll. 1981. Additional collections of *Botrychium lunarioides* from Texas and Oklahoma and comments on its dormancy. Phytologia 48:276–278.

Thomas, W. W. 1984. The systematics of *Rhynchospora* section *Dichromena*. Memoirs of the New York Botanical Garden 37:1–116.

Thomas, W. W. 1992. A synopsis of *Rhynchospora* (Cyperaceae) in Mesoamerica. Brittonia 44:14–44.

Thompson, H. J., and A. M. Powell. 1981. Loasaceae of the Chihuahuan Desert region. Phytologia 49:16–32.

Thorne, R. F., and R. Scogin. 1978. *Forsellesia* Greene (*Glossopetalon* Gray), a third genus in the Crossomataceae, *Rosineae, Rosale*. Aliso 9:171–178.

Thurber, G. 1855. I. Plantae Novae Thurberianae. The characters of some new genera and species of plants in a collection made by George Thurber, Esq., of the late Mexican Boundary Commission, chiefly in New Mexico and Sonora, ed. A. Gray. Memoirs of the American Academy of Arts and Sciences, New Series 5(2):297–328.

Todsen, T. T. 1995. *Malaxis wendtii* (Orchidaceae) in the United States. Sida 16(3):591.

Tomb, A. S. 1970. Novelties in *Lygodesmia* and *Stephanomeria* (Compositae-*Cichorieae*). Brittonia 24:226.

Tomb, A. S. 1972. Re-establishment of the genus *Prenanthella* Rydb. (Compositae: *Cichorieae*). Brittonia 24:223–228.

Tomb, A. S. 1973. *Shinnersoseris* gen. *nov.* (Compositae: *Cichorieae*). Sida 5:183–189.

Tomb, A. S. 1974. *Hypochoeris* in Texas. Sida 5: 287–289.

Tomb, A. S. 1980. Taxonomy of *Lygodesmia* (Asteraceae). Systematic Botany Monograph 1:1–51.

Torrey, J. 1858 ("1859"). Botany of the boundary. In *Report on the United States and Mexican Boundary Survey*, vol. 2, ed. W. H. Emory. Washington, D.C.: A.O.P. Nicholson.

Torrey, J., and W. Hooker. 1836. In Monograph of North American Cyperaceae. Annals of the Lyceum of Natural History of New York, vol. 3, ed. J. Torrey.

Towner, H. F. 1977. The biosystematics of *Calylophus* (Onagraceae). Annals of the Missouri Botanical Garden 64:48–120.

Towner, H. F., and P. H. Raven. 1970. A new species and some new combinations in *Calylophus* (Onagraceae). Madroño 20:241–245.

Townsend, C. C. 1968. *Parietaria officinalis* and *P. judaica*. Watsonia 6:365–370.

Trelease, W. 1916. *The Genus Phoradendron: A Monographic Revision*. Urbana: University of Illinois Press.

Trent, J. S. 1985. A study of morphological variability in divaricate *Aristida* of the southwestern United States. Master's thesis, New Mexico State University, Las Cruces.

Tryon, A. F., R. M. Tryon, and F. Badré. 1980. Classification, spores, and nomenclature of the marsh fern. Rhodora 82:461–474.

Tryon, R. M. 1941. A revision of the genus *Pteridium*. Rhodora 43:1–31; 37–67.

Tryon, R. M. 1955. *Selaginella rupestris* and its allies. Annals of the Missouri Botanical Garden 42:1–99.

Tryon, R. M. 1956. A revision of the American species of *Notholaena*. Contributions from the Asa Gray Herbarium of Harvard University 179:1–106.

Tryon, R. M. 1957. A revision of fern genus *Pellaea* section *Pellaea*. Annals of the Missouri Botanical Garden 44:125–193.

Tryon, R. M. 1960. A review of the genus *Dennstaedtia* in America. Contributions from the Asa Gray Herbarium at Harvard University 187:23–52.

Tryon, R. M., and K. W. Allred. 1990. A taxonomic comparison of *Aristida ternipes* and *Aristida hamulosa* (Gramineae). Sida 14(2):251–261.

Tucker, G. C. 1983. The taxonomy of *Cyperus* (Cyperaceae) in Costa Rica and Panama. Systematic Botany Monograph 2:1–85.

Tucker, G. C. 1984. Taxonomic notes on two common neotropical species of *Cyperus* (Cyperaceae). Sida 10:298–307.

Tucker, G. C. 1985a. The correct name for *Cyperus cayennensis* (*C. flavus*) Cyperaceae. Southwestern Naturalist 30:607–608.

Tucker, G. C. 1985b. *Cyperus flavicomus*, the correct name for *Cyperus albomarginatus*. Rhodora 87: 539–541.

Tucker, G. C. 1986. The genera of the Elatinaceae in the southeastern United States. Journal of the Arnold Arboretum 67(4):471–483.

Tucker, G. C. 1987. The genera of Cyperaceae in the southeastern United States. Journal of the Arnold Arboretum 68:361–445.

Tucker, G. C. 1988. The genera of *Bambusoideae* (Gramineae) in the southeastern United States. Journal of the Arnold Arboretum 69(3):239–273.

Tucker, G. C. 1989. The genera of Commelinaceae) in the southeastern United States. Journal of the Arnold Arboretum 70(1):97–130.

Tucker, G. C. 1990. The genera of *Arundinoideae* (Gramineae) in the southeastern United States. Journal of the Arnold Arboretum 71(2):145–177.

Tucker, G. C. 1994. Revision of the Mexican species of *Cyperus* (Cyperaceae). Systematic Botany Monograph 43:1–213.

Tull, D. 1991. *A Field Guide to Wildflowers, Trees & Shrubs of Texas*. Houston: Gulf Publishing Company.

Turner, B. L. 1950. Texas species of *Desmanthus* (Leguminosae). Field and Laboratory 18:54–65.

Turner, B. L. 1959. *The Legumes of Texas*. Austin: University of Texas Press.

Turner, B. L. 1964. A taxonomic study of the genus *Amphicarpaea* (Leguminosae). Southwestern Naturalist 9:207–218.

Turner, B. L. 1975. Taxonomy of *Haploesthes* (Asteraceae-*Senecioneae*). Wrightia 5:108–115.

Turner, B. L. 1977a. *Lepidospartum burgessii* (Asteraceae, *Senecioneae*) a remarkable new gypsophilic species from Trans-Pecos Texas. Wrightia 5: 354–355.

Turner, B. L. 1977b. A new species of *Gaillardia* (Asteraceae-*Heliantheae*) from Northcentral México and adjacent Texas. Southwestern Naturalist 21:539–541.

Turner, B. L. 1978. Taxonomic study of the scapiform species of *Acourtia* (Asteraceae-*Mustisiieae*). Phytologia 38:456–468.

Turner, B. L. 1979. *Gaillardia aestivalis* var. *winkleri* (Asteraceae), a white flowering tetraploid taxon endemic to southeastern Texas. Southwestern Naturalist 24:621–624.

Turner, B. L. 1983. The Texas species of *Paronychia* (Caryophyllaceae). Phytologia 54:9–23.

Turner, B. L. 1984. Taxonomy of the genus *Aphanostephus* (Asteraceae-*Astereae*). Phytologia 56:81–101.

Turner, B. L. 1987a. Taxonomic study of *Machaeranthera* sections *Machaeranthera* and *Hesperastrum* (Asteraceae). Phytologia 62:207–266.

Turner, B. L. 1987b. Taxonomy of *Carpochaete* (Asteraceae-*Eupatorieae*). Phytologia 64:145–162.

Turner, B. L. 1988a. A new variety of *Sclerocarpus uniserialis* (Asteraceae, *Heliantheae*) from southernmost Texas. Phytologia 64:341–343.

Turner, B. L. 1988b. A new variety of *Berlandiera lyrata* from northwestern México. Phytologia 64: 205–208.

Turner, B. L. 1988c. Comments upon, and new combinations in, *Heliopsis* (Asteraceae, *Heliantheae*). Phytologia 64:337–340.

Turner, B. L. 1989. An overview of the *Brickellia (Kuhnia) eupatorioides* (Asteraceae, *Eupatorieae*) complex. Phytologia 67:121–131.

Turner, B. L. 1990. Taxonomy of *Varilla* (Asteraceae, *Heliantheae*). Phytologia 68(6):4–13.

Turner, B. L. 1991a. Texas species of *Ruellia* (Acanthaceae). Phytologia 71(4):281–299.

Turner, B. L. 1991b. An overview of the North American species of *Menodora* (Oleaceae). Phytologia 71(5):340–356.

Turner, B. L. 1992. Taxonomic overview of the genus *Cologania* (Fabaceae, *Phaeoleae*). Phytologia 73(4): 281–301.

Turner, B. L. 1993a. Lectotypification of *Senecio neomexicanus* A. Gray. Phytologia 75(3):221–223.

Turner, B. L. 1993b. New species and combinations in *Selinocarpus* (Nyctaginaceae). Phytologia 75(3): 239–242.

Turner, B. L. 1993c. The Texas species of *Centaurium* (Gentianaceae). Phytologia 75(3):259–275.

Turner, B. L. 1993d. Texas species of *Mirabilis* (Nyctaginaceae). Phytologia 75(6):432–451.

Turner, B. L. 1994a. Taxonomic overview of *Gilia*, sect. *Giliastrum* (Polemoniaceae) in Texas and México. Phytologia 76(1):52–68.

Turner, B. L. 1994b. Species of *Lupinus* (Fabaceae) occurring in northeastern México (Nuevo Leon and closely adjacent States). Phytologia 76(4):290–302.

Turner, B. L. 1994c. Native species of *Bauhinia* (Caesalpiniaceae) occurring in northeastern México. Phytologia 76(4):333–343.

Turner, B. L. 1994d. A taxonomic overview of *Scutellaria*, section *Resinosa* (Lamiacaceae). Phytologia 76(5):345–382.

Turner, B. L. 1994e. Taxonomic status of *Brickelliastrum villarreallii* R. M. King & H. Robins. (Asteraceae, *Eupatorieae*). Phytologia 76(5):389–390.

Turner, B. L. 1994f. Taxonomic study of the *Stachys coccinea* (Lamiaceae) complex. Phytologia 76(5): 391–401.

Turner, B. L. 1994g. Texas species of *Schrankia* (Mimosaceae) transferred to the genus *Mimosa*. Phytologia 76(5):412–420.

Turner, B. L. 1994h. Regional variation in the North American elements of *Oxalis corniculata* (Oxalidaceae). Phytologia 77(1):1–7.

Turner, B. L. 1994i. Revisionary study of the genus, *Allionia* (Nyctaginaceae). Phytologia 77(1):45–55.

Turner, B. L. 1994j. Taxonomic treatment of *Monarda* (Lamiaceae) for Texas and México. Phytologia 77(1): 56–79.

Turner, B. L. 1994k. *Mimosa rupertiana* B. L. Turner, a new name for *M. occidentalis* (Wooton & Standley) B. L. Turner, not *M. occidentalis* Britton & Rose. Phytologia 77(2):81–82.

Turner, B. L. 1995a. Synopsis of the genus *Onosmodium* (Boraginaceae). Phytologia 78(1):39–69. [Note: This volume is marked January 1995, but was not published until July 1995.]

Turner, B. L. 1995b. Synoptical study of *Rhododon* (Lamiaceae). Phytologia. 78(6):448–451. Note: This volume is marked June 1995 but was not published until December 1995.

Turner, B. L. 1995c. Taxonomic overview of *Hedyotis nigricans* (Rubiaceae) and closely allied taxa. Phytologia 79(1):12–21. Note: This volume is marked July 1995 but was not published until February 1996.

Turner, B. L., and J. Andrews. 1986. Lectotypification of *Lupinus subcarnosis* and *L. texensis* (Fabaceae). Sida 11:255–257.

Turner, B. L., and A. Birdsong. 1980. New combinations in the genus *Aphanostephus* (*Astereae*-Asteraceae). Phytologia 45:501.

Turner, B. L., and C. C. Cowan. 1993. Taxonomic overview of *Stemodia* (Scrophulariaceae) for North America and the West Indies. Phytologia 74(2):61–103.

Turner, B. L., and D. Dawson. 1980. Taxonomy of *Tetragonotheca* (Asteraceae-*Heliantheae*). Sida 8: 296–303.

Turner, B. L., and R. Hartman. 1976. Infraspecific categories of *Machaeranthera pinnatifida* (Compositae). Wrightia 5:308–315.

Turner, B. L., and Ki-Joong Kim. 1990. An overview of the genus *Pyrrhopappus* (Asteraceae: *Luctuceae*) with emphasis on chloroplast DNA restriction site data. American Journal of Botany 77:847–850.

Turner, B. L., and M. L. Morris. 1976. Systematics of *Palafoxia* (Asteraceae: *Helenieae*). Rhodora 78: 567–628.

Turner, B. L., and A. M. Powell. 1972. A new gypsophile *Sophora* (Leguminosae) from north central México and adjacent Texas. Phytologia 22:419–423.

Turner, B. L., M. W. Turner, and J. C. Crutchfield. 1988. Populational analysis and new combinations in *Psilostrophe tagetina* and *P. gnaphalodes* (Asteraceae, *Heliantheae*). Phytologia 65:231–236.

Turner, B. L., and M. Whalen. 1976. Taxonomic study of *Gaillardia pulchella* (Asteraceae: *Helenieae*). Wrightia 5:189–192.

Turner, M. W. 1993. Systematic study of the genus *Baileya* (Asteraceae: *Helenieae*). Sida 15(3):491–508.

Tutin, T. G. 1980a. × *Agropogon*. In Tutin et al. (eds.). *Flora Europaea*, vol. 5: Alismataceae to Orchidaceae (Monocotyledones). Cambridge: Cambridge University Press.

Tutin, T. G. 1980b. *Aira*. In Tutin et al. (eds.). *Flora Europaea*, vol. 5: Alismataceae to Orchidaceae (Monocotyledones). Cambridge: Cambridge University Press.

Tveton, J. L. 1993. *Wildflowers of Houston*. Houston: Rice University Press.

Umber, R. E. 1979. The genus *Glandularia* (Verbenaceae) in North America. Systematic Botany 41: 72–102.

Urbatsch, L. E. 1972. Systematic study of the *Altissimae* and *Giganteae* species groups of the genus *Vernonia* (compositae). Brittonia 24:229–238.

Urbatsch, L. E. 1978. The Chihuahuan Desert species of *Ericameria* (Compositae: *Astereae*). Sida 7: 298–303.

Uttal, L. J. 1987. The genus *Vaccinium* L. (Ericaceae) in Virginia. Castanea 52:231–255.

Uttal, L. J. 1988. Lectotypification of *Azalea rosea* Loisel. (Ericaceae) and a new combination in *Rhododendron periclymenoides* (Michx.) Shinners. Sida 13: 167–169.

Valdés-R., J. 1985. A biosystematic study of the genus *Erioneuron* Nash (Poaceae: *Eragrostideae*). Ph.D. dissertation, Texas A&M University, College Station.

Van den Borre, A., and L. Watson. The infrageneric classification of *Eragrostis* (Poaceae). Taxon 43(3): 383–422.

Vander Kloet, S. P. 1980. The taxonomy of the highbush blueberry. *Vaccinium corymbosum*. Canadian Journal of Botany 58:1187–1201.

Vandiver, V. V., Jr., D. W. Hall, and R. G. Westbrooks. 1992. Discovery of *Oryza rufipogon* (Poaceae: *Oryzeae*), new to the United States, with its implications. Sida 15(1):105–109.

Van Eseltine, G. P. 1918. The allies of *Selaginella rupestris* in the southeastern United States. Contributions from the United States National Herbarium. Smithsonian Institution 20(5):159–172.

Veldkamp, J. F. 1990. The true identity of *Sporobolus poiretii* (Gramineae). Taxon 39:327–328.

Veldkamp, J. F., R. De Koning, and M. S. M. Sosef. 1986. Generic delimitation of *Rottboellia* and related genera (Gramineae). Blumea 31:281–307.

Verhoek, S. 1978. Two new species and a new combination in *Manfreda* (Agavaceae). Brittonia 30: 165–171.

Veteto, G. H., C. E. Davis, R. V. Hart, and L. N. Lodwick (comps.). 1976. A tentative checklist of the vegetation of the Gus A. Engeling Wildlife Management Area and Anderson County, Texas. Wildlife 1/76. Texas Parks and Wildlife Department, Austin. Mimeographed.

Vines, R. A. 1953. *Native East Texas Trees*. Houston: Adco Press.

Vines, R. A. 1960. *Trees, Shrubs, and Woody Vines of the Southwest*. Austin: University of Texas Press.

Vines, R. A. 1977. *Trees of East Texas*. Austin: University of Texas Press.

Vines, R. A. 1982. *Trees of North Texas*. Austin: University of Texas Press.

Vines, R. A. 1984. *Trees of Central Texas*. Austin: University of Texas Press.

von Schweinitz, L. D. 1925. A monograph of the North American species of the genus *Carex*. Annals of the Lyceum of Natural History of New York 1(2): 283–373.

Vuilleumier, B. S. 1969a. The genera of *Senecioneae* in the southeastern United States. Journal of the Arnold Arboretum 50:104–123.

Vuilleumier, B. S. 1969b. The tribe *Mustisieae* (Compositae) in the southeastern United States. Journal of the Arnold Arboretum 50:620–625.

Vuilleumier, B. S. 1973. The genera of *Lactuceae* (Compositae) in the southeastern United States. Journal of the Arnold Arboretum 54:42–93.

Wagner, D. H. 1979. Systematics of *Polystichum* in western North America north of México. Pteridologia 1:1–64.

Wagner, W. L. 1983. New species and combinations in the genus *Oenothera* (Onagraceae). Annals of the Missouri Botanical Garden 70:194–196.

Wagner, W. L. 1984. Reconsideration of *Oenothera* subgenus *Gauropsis* (Onagraceae). Annals of the Missouri Botanical Garden 71:1114–1127.

Wagner, W. L. 1986. New taxa in *Oenothera* (Onagraceae). Annals of the Missouri Botanical Garden 73: 475–480.

Wagner, W. L., R. E. Stockhouse, and W. M. Klein. 1985. The systematics and evolution of the *Oenothera caespitosa* species complex (Onagraceae). Missouri Botanical Garden Monograph 12:2–103.

Walters, T. W., and R. Wyatt. 1982. The vascular flora of granite outcrops in the Central Mineral Region of Texas. Bulletin of the Torrey Botanical Club 109(3): 344–364.

Walther, E. 1972. *Echeveria*. San Francisco: California Academy of Sciences.

Ward, D. B. 1976. *Mitracarpus* (Rubiaceae), a genus new to Florida and eastern North America. Rhodora 78: 674–681.

Ward, D. B. 1977. *Nelumbo lutea*, the correct name for the American lotus. Taxon 26:227–234.

Warnock, B. H. 1970. *Wildflowers of the Big Bend Country Texas*. Alpine, Texas: Sul Ross State University Press.

Warnock, B. H. 1974. *Wildflowers of the Guadalupe Mountains and the Sand Dune Country, Texas*. Alpine, Texas: Sul Ross State University Press.

Warnock, B. H. 1977. *Wildflowers of the Davis Mountains and Marathon Basin, Texas*. Alpine, Texas: Sul Ross State University Press.

Warnock, B. H. 1982. A new three-awn grass from Trans-Pecos, Texas. Sida 9(4):358–359.

Warnock, B. H., and J. MacCarpenter. 1994. *Penstemon thurberi* (Scrophulariaceae) new to Texas. Sida 16(1): 207.

Warnock, M. H. 1981. Biosystematics of the *Delphinium carolinianum* complex (Ranunculaceae). Systematic Botany 6:38–54.

Warnock, M. H. 1987. An index to epithets treated by King and Robinson: *Eupatorieae* (Astereaceae). Phytologia 62:345–431.

Warren, S. D., and S. L. Hatch. 1984. *Bromus sterilis* L. (Poaceae) new to Texas. Sida 10(3):257–258.

Waterfall, U. T. 1951. The *Callirhoë* (Malvaceae) in Texas. Field and Laboratory 19(3):107–119.

Waterfall, U. T. 1958. A taxonomic study of the genus *Physalis* in North America, north of México. Rhodora 60:107–114, 128–142, 152–173.

Waterfall, U. T. 1979. *Keys to the Flora of Oklahoma*, 6th ed. Oklahoma State University, Stillwater, Oklahoma: Author.

Watson, T. J., Jr. 1977. The taxonomy of *Xylorhiza* (Asteraceae-*Astereae*). Brittonia 29:199–216.

Wauer, R. H. 1980. *Naturalist's Big Bend: An Introduction to the Trees and Shrubs, Wildflowers, Cacti, Mammals, Birds, Reptiles and Amphibians, Fish, and Insects*. College Station: Texas A&M University Press.

Webber, J. M., and P. W. Ball. 1984. The taxonomy of the *Carex rosea* group (section *Phaestoglochin*) in Canada. Canadian Journal of Botany 62:2058–2073.

Weber, W. A. 1985. The genus *Teloxys* (Chenopodiaceae). Phytologia 58:477–488.

Weber, W. A. 1991. New names and combinations, principally in the Rocky Mountain flora: VIII. Phytologia 70(4):231–233.

Webster, G. L. 1950. Observations on the vegetation and summer flora of the Stockton Plateau in northeastern Terrell County, Texas. Texas Journal of Science 5:158–177.

Webster, G. L. 1967. The genera of Euphorbiaceae in the southeastern United States. Journal of the Arnold Arboretum 48:303–430.

Webster, G. L. 1992. Realignments in American *Croton* (Euphorbiaceae). Novon 2:269–273.

Webster, G. L. 1993. A provisional synopsis of the sections of the genus *Croton* (Euphorbiaceae). Taxon 42(4):793–823.

Webster, G. L. 1994. Synopsis of the genera and suprageneric taxa of Euphorbiaceae. Annals of the Missouri Botanical Garden 81:33–144.

Webster, R. D. 1988. Genera of the North American *Paniceae* (Poaceae: *Panicoideae*). Systematic Botany 13(4):576–609.

Webster, R. D. 1995. Nomenclatural changes in *Setaria* and *Paspalidium* (Poaceae: *Paniceae*). Sida 16(3): 439–446.

Webster, R. D., and S. L. Hatch. 1990. Taxonomy of *Digitaria* section *Aequiglumae* (Poaceae: *Paniceae*). Sida 14(2):145–167.

Webster, R. D., J. H. Kirkbride, and J. V. Reyna. 1989. New world genera of the *Paniceae* (Poaceae: *Panicoideae*). Sida 13(4):393–417.

Webster, R. D., and R. B. Shaw. 1982. Relationship between *Digitaria milanjiana* (Poaceae: *Paniceae*) and the annual species of *Digitaria* section *Digitaria* in North America. Sida 9:333–343.

Webster, R. D., and R. B. Shaw. 1995. Taxonomy of the native North American species of *Saccharum* (Poaceae: *Andropogoneae*). Sida 16(3):551–580.

Welch, W. C. 1989. *Perennial Garden Color for Texas and the South*. Dallas: Taylor Publishing Company.

Wells, E. F. 1984. A revision of the genus *Heuchera* (Saxifragaceae) in eastern North America. Systematic Botany Monograph 3:45–121.

Welsh, S. L., N. D. Atwood, S. Goodrich, and L. C. Higgins. 1987. *A Utah Flora*. Great Basin Naturalist, Memoirs. No. 9.

Welzen, P. C. 1981. A taxonomic revision of the genus *Arthraxon* Beauv. (Graminieae). Blumea 27: 255–300.

Wemple, D. K. 1970. Revision of the genus *Petalostemon*. Iowa State College Journal of Science 45: 1–102.

Wendt, T. 1978. A systematic study of Polygalas sect. *Rhinotropis* (Polygalaceae). Ph.D. dissertation, University of Texas, Austin.

Wendt, T. 1979. Notes on the genus *Polygala* in the United States and México. Journal of the Arnold Arboretum 60(4):504–514.

Wendt, T. 1980. Notes on some *Pleopeltis* and *Polypodium* species of the Chihuahuan Desert Region. American Fern Journal 70:5–11.

Wendt, T. 1993. A new variety of *Ephedra torreyana* (Ephedraceae) from West Texas and Chihuahua, with notes on hybridization in the *E. torreyana* complex. Phytologia 74(2):141–150.

Wendt, T., and T. K. Todsen. 1982. A new variety of *Polygala rimulicola* (Polygalaceae) from Doña Ana County, New Mexico. Madroño 29(1):19–21.

Weniger, D. 1984. *Cacti of Texas and Neighboring States, A Field Guide*. Austin: University of Texas Press, Austin.

Werner, D., and W. L. Wagner. 1988. Systematics of *Oenothera* section *Oenothera* subsection *Raimannia* and subsection *Nutantigemma* (Onagraceae). Systematic Botany Monograph 24:1–91.

Whalen, M. 1977. Taxonomy of *Bebbia* (Compositae: *Heliantheae*). Madroño 24:112–113.

Whalen, M. 1980. A systematic revision of the New World species of *Frankenia* (Frankeniaceae). Ph.D. dissertation, University of Texas, Austin.

Whalen, M. A. 1976. New taxa of *Solanum*, section *Androceras* from México and adjacent United States. Wrightia 5:228–239.

Whalen, M. A. 1979. Taxonomy of *Solanum* section *Androceras*. Gentes Herbarium 11:359–426.

Whalen, M. A. 1987. Systematics of *Frankenia* (Frankeniaceae) in North and South America. Systematic Botany Monograph 16:1–93.

Whistler, R. L., and T. Hymowitz. 1979. *Guar: Agronomy, Production, Industrial Use, and Nutrition*. West Lafayette, Indiana: Purdue University Press.

Whitehouse, E. 1933. Plant succession on Central Texas granite. Ecology XIV (4):391–405.

Wiens, D. 1964. Revision of the Acataphyllous species of *Phoradendron*. Brittonia 16:11–54.

Wilbur, R. L. 1955. A revision of the North American genus *Sabatia* (Gentianaceae). Rhodora 57:1–36, 43–72, 78–104.

Wilbur, R. L. 1975. A revision of the North American genus *Amorpha* (Leguminosae-*Psoraleae*). Rhodora 77:337–409.

Wilbur, R. L. 1976. Illegitimate names: *Rhododendron nudiflorum* (L.) Torr., and *R. roseum* (Loisel.) Rehder. Taxon 25:178–179.

Wilbur, R. L. 1994. The Myricaceae of the United States and Canada: Genera, subgenera, and series. Sida 16(1):93–107.

Williams, J. G., and A. E. Williams. 1983. *Field Guide to Orchids of North America*. New York: Universe Books.

Williams, J. K. 1994. *Emilia fosbergii* (Asteraceae: *Senceioneae*), a new introduction to Texas. Sida 16(2):378.

Wilson, J. S. 1975. Variation of three taxonomic complexes of the genus *Cornus* in eastern United States. Transactions of the Kansas Academy of Science 67: 747–817.

Wilson, K. L. 1990. Typification and application of *Scirpus geniculatus* L. Cyperaceae Newsletter 7:6–8.

Windham, M. D. 1993. New taxa and nomenclatural changes in the North American fern flora. Contributions from the University of Michigan Herbarium 19:31–61.

Winkler, C. H. 1915. *The Botany of Texas, An Account of Botanical Investigation in Texas and Adjoining Territory*. Bulletin 1915, No. 18. University of Texas, Austin.

Wipff, J. K. In Press. *Gastridium*. In *The Manual of North American Grasses*, ed. M. E. Barkworth.

Wipff, J. K. In Press. *Tragus*. In *The Manual of North American Grasses*, ed. M. E. Barkworth.

Wipff, J. K. In Press. *Willkommia*. In *The Manual of North American Grasses*, ed. M. E. Barkworth.

Wipff, J. K., and S. L. Hatch. 1992. *Eustachys caribaea* in Texas. Sida 15(1): 160–162.

Wipff, J. K., and S. L. Hatch. 1994. A systematic study of *Digitaria* sect. *Pennatae* (Poaceae: *Paniceae*) in the New World. Systematic Botany 19(4):613–627.

Wipff, J. K., and S. D. Jones. 1995. Nomenclatural combinations in Poaceae and Cyperaceae. Phytologia 77(6):456–464. [Note: This volume is marked 1994, but was not published until April 1995.]

Wipff, J. K., and S. D. Jones. 1995. Nomenclatural combination in Poaceae. Phytologia 78(4):244–245. [Note: This volume is marked April 1995, but was not published until September 1995.]

Wipff, J. K., R. I. Lonard, S. D. Jones, and S. L. Hatch. 1993. The genus *Urochloa* (Poaceae: *Paniceae*) in Texas, including one previously unreported species for the state. Sida 15(3):405–413.

Wipff, J. K., and B. S. Rector. 1993. *Rottboellia cochinchinensis* (Poaceae: *Andropogoneae*) new to Texas. Sida 15(3):419–424.

Witherspoon, J. T. 1977. New taxa and combinations in *Eragrostis* (Poaceae). Annals of the Missouri Botanical Garden 64:324–329.

Wood, C. E., Jr. 1983. The genera of Burmanniaceae in the southeastern United States. Journal of the Arnold Arboretum 64(2):293–307.

Wood, C. E., Jr., and P. Adams. 1976. The genera of Guttiferae (Clusiaceae) in the southeastern United States. Journal of the Arnold Arboretum 57:74–90.

Wood, C. E., Jr., and R. E. Weaver. 1982. The genera of Gentianaceae in the southeastern United States. Journal of the Arnold Arboretum 63(4):441–487.

Woodland, D. W. 1982. Biosystematics of the perennial North American species of *Urtica* II. Taxonomy. Systematic Botany 7:282–290.

Woodland, D. W., I. J. Bassett, and C. W. Crampton. 1976. The annual species of stinging nettle (*Hesperocnide* and *Urtica*) in North America. Canadian Journal of Botany 54:374–383.

Woodson, R. E. 1941. The North American Asclepiadaceae. Annals of the Missouri Botanical Garden 28: 193–244.

Woodson, R. E. 1954. The North American species of *Asclepias* L. Annals of the Missouri Botanical Garden 41(1):1–211.

Wooten, J. W., and A. F. Clewell. 1971. *Fleischmannia* and *Conoclinium* in eastern North America. Rhodora 73:566–574.

Wooton, E. O., and P. C. Standley. 1915. *Flora of New Mexico*. Contributions from the United States National Herbarium. Smithsonian Institution, vol. 19. Washington, D.C.

Worthington, R. D. 1978. A revised floral inventory of the Franklin Mountains, El Paso County, Texas. Mimeo.

Worthington, R. D. 1982a. *Chloris submutica* (Poaceae) in Texas. Sida 9(4):368.

Worthington, R. D. 1982b. Noteworthy collection: New Mexico–Texas. Madroño 29(3):217.

Worthington, R. D. 1985. *Solidago spathulata* DC. var. *neomexicana* (Gray) Cronq. (Asteraceae) new to Texas. Sida 11:246.

Worthington, R. D. 1989. An annotated checklist of the native and naturalized flora of El Paso County, Texas. El Paso Southwestern Botany Miscellany No. 1.

Worthington, R. D. 1990. Additions to the flora of Texas from El Paso County. Sida 14(1):135–137.

Worthington, R. D. 1994. Checklist of the flora of the Big Bend Ranch State Natural Area with an account of UTEP Herbarium holdings. Mimeo.

Wujek, D. E., and F. J. Menapace. 1986. Taxonomy of *Carex* section *Folliculatae* using achene micromorphology. Rhodora 88:399–403.

Wunderlin, R. P. 1972. New combinations in Compositae. Annals of the Missouri Botanical Garden 59: 471–473.

Wunderlin, R. P. 1979. Notes on *Spermacoce* and *Mitracarpus* (Rubiaceae) in southeastern United States. Phytologia 41:313–316.

Wurdack, J. J., and R. Kral. 1982. The genera of Melastomataceae in southeastern United States. Journal of the Arnold Arboretum 63(4):429–439.

Wyatt, R., and L. N. Lodwick. 1981. Variation and taxonomy of *Aesculus pavia* L. (Hippocastanaceae) in Texas. Brittonia 33:39–51.

Yates, H. O. 1966a. Morphology and cytology of *Uniola* (Gramineae). Southwestern Naturalist 11:145–189.

Yates, H. O. 1966b. Revision of grasses traditionally referred to *Uniola*, I. *Uniola* and *Leptochloopsis*. Southwestern Naturalist 11:372–394.

Young, M. S. 1920. *The Seed Plants, Ferns, and Fern Allies of the Austin Region*. Austin: University of Texas.

Young, R. A. 1945a. Bamboos for American Horticulture (I). The National Horticultural Magazine, July, 171–196.

Young, R. A. 1945b. Bamboos for American Horticulture (II). The National Horticultural Magazine, Oct., 274–291.

Young, R. A. 1946a. Bamboos for American Horticulture (III). The National Horticultural Magazine, Jan., 40–64.

Young, R. A. 1946b. Bamboos for American Horticulture (IV). The National Horticultural Magazine, July, 257–283.

Young, R. A. 1946c. Bamboos for American Horticulture (V). The National Horticultural Magazine, Oct., 352–365.

Yuncker, T. G. 1943. Convolvulaceae: *Cuscuta*. Flora of Texas, Vol. 3, part 3. Texas Research Foundation, Renner, Texas.

Zardini, E., and P. H. Raven. 1992. A new section of *Ludwigia* (Onagraceae) with a key to the sections of the genus. Systematic Botany 17(3):481–485.

Zika, P. F. 1991. *Juncus marginatus* Rostk. var. *setosus* Coville (Juncaceae). Madroño 38(3):204–205.

Zimmerman, A. D. 1985. Systematics of the genus *Coryphantha* (Cactaceae). Ph.D. dissertation, University of Texas, Austin.

Zimmerman, A. D., and D. Zimmerman. 1977. A revision of the United States taxa of the *Mammillaria wrightii* complex with remarks upon the northern Mexican populations, pt. 2. Cactus & Succulent Journal 49:51–62.

Ziska, G. 1988. Revision der *Melinideae* Hitchcock (Poaceae, *Panicoideae*). Bibliotheca Botanica 138:1–149.

Zohary, M., and D. Heller. 1984. *The Genus Trifolium*. The Israel Academy of Sciences and Humanities, Jerusalem, Israel.

Zuloaga, F. O. 1986. Systematics of New World species of *Panicum* (Poaceae: *Paniceae*). In *Grasses: Systematics and Evolution*, ed. T. R. Soderstrom et al., 287–306. Washington D.C.: Smithsonian Press.

Zuloaga, F. O., R. P. Ellis, and O. Morrone. 1993. A revision of *Panicum* subg. *Dichanthelium* sect. *Dichanthelium* (Poaceae: *Panicoideae: Paniceae*) in Mesoamerica, the West Indies, and South America. Annals of the Missouri Botanical Garden 80:119–190.

Index